A
History
of the
Life
Sciences

THIRD EDITION, REVISED AND EXPANDED

A History of the Life Sciences

THIRD EDITION, REVISED AND EXPANDED

Lois N. Magner

Professor Emerita
Purdue University
West Lafayette, Indiana

CRC Press
Taylor & Francis Group
Boca Raton London New York

CRC Press is an imprint of the
Taylor & Francis Group, an informa business

CRC Press
Taylor & Francis Group
6000 Broken Sound Parkway NW, Suite 300
Boca Raton, FL 33487-2742

© 2002 by Taylor & Francis Group, LLC
CRC Press is an imprint of Taylor & Francis Group, an Informa business

International Standard Book Number-13: 978-0-8247-0824-5 (Hardcover)

Library of Congress Cataloging-in-Publication Data

Catalog record is available from the Library of Congress

Visit the Taylor & Francis Web site at
http://www.taylorandfrancis.com

and the CRC Press Web site at
http://www.crcpress.com

To Ki-Han and Oliver,
as always

Preface

An analysis of almost any scientific problem leads automatically to a study of its history.

—*Ernst Mayr*

My primary purpose in writing and revising this book has been to provide an updated introduction to the history of the life sciences. Although the text began as a "teaching assistant" for my own one-semester survey course, I hope that this new edition will also be of interest to the general reader, and to teachers who are trying to help their students think about the interplay between science and society.

As in previous editions of this book, I have tried to call attention to major themes in biology, the evolution of theories and methodologies, interactions among various disciplines, and the diverse attitudes and assumptions with which scientists have approached the phenomena of life. Questions that appear to reveal and reflect themes at the core of the life sciences are emphasized at the expense of areas—such as taxonomy, or the discovery of specific organs, hormones, enzymes, and metabolic pathways—that have consumed the energy and attention of generations of scientists. A survey must take a panoramic view of the forest, but it is also important to stop and examine some of the individual trees, those that might be considered typical and those that appear to be truly exceptional.

Since the second edition of this book was published, the first phase of the Human Genome Project was completed, and scientists have begun to plan a

Human Proteome Project. The chemistry of the gene was obscure in the 1950s, but by the end of the 20th century biologists had developed the tools that made it possible to sequence, clone, and manipulate genetic material. The social, ethical, and philosophical implications of the new genetic technologies are certainly profound, but all too often the discussion of their impact reveals a troubling lack of scientific literacy. Exploring the history of science could help us understand the scientific and social aspects of such challenges as gene therapy, new reproductive technologies, changing patterns of morbidity and mortality, overpopulation, the emergence of virulent new diseases and antibiotic-resistant pathogens, environmental degradation, genetically modified foods, conflicts between religious beliefs and scientific theories, and the threat of bioterrorism. Perhaps a better understanding of previous approaches to science will make it possible to recognize the sources of contemporary problems and the inherent limitations and liabilities of current paradigms.

The history of science can be enlightening for other reasons. Once beyond the elementary-school level of history, we realize that history is not so much the record of what happened, but rather the story of what survived. This is as true for the history of science as for any other branch of history. When removed from its social and cultural context, scientific knowledge can appear oversimplified and preordained, as if facts and theories had been handed down by some omniscient celestial experimentalist. More properly, we should see science as a human invention and a dynamic concept, encompassing a body of knowledge, a means of obtaining knowledge, and a way of envisioning the universe.

During the last half of the 20th century, the history of science underwent profound changes. From an almost exclusive focus on the evolution of modern theories and the ideas of major scientists, it turned its gaze toward new questions about the social, cultural, economic, and political context in which science and scientists are embedded. Profoundly influenced by concepts and techniques borrowed from sociology, psychology, anthropology, and demography, the new social historians of science emphasized factors such as race, class, gender, and institutional and professional affiliations. Although an abundance of contextual complexity sometimes seems to come at the expense of specific scientific content, recent studies of the history of science offer many insights that may well be valuable in coming to terms with the life sciences of the 21st century.

The tendency toward ultraspecialization in scholarship and teaching creates a problem in reaching the audiences that would most benefit from a general introductory synthesis. Historians and scientists are unlikely to reach a general audience unless they are willing to transcend narrow professional boundaries. I am therefore very pleased that Marcel Dekker, Inc., has been so willing to encourage the publication of general surveys of the history of science and medicine.

Once again I express my deep appreciation to John Parascandola, Vernard Foley, and the late Aaron J. Ihde for their advice, criticism, and encouragement during the preparation of previous editions of this book. You were my mentors in my metamorphosis from biochemist to historian of science. Many thanks also to the students who took my courses, read my books, and let me know what was clear and what was obscure. Of course, all remaining errors of omission and commission are my own. I would also like to acknowledge the History of Medicine Division of the National Library of Medicine for providing the illustrations used in this book.

Lois N. Magner

Contents

A
History
of the
Life
Sciences

1

THE ORIGINS OF
THE LIFE SCIENCES

Modern biology encompasses both the oldest and newest scientific disciplines, but the term biology was first introduced at the beginning of the nineteenth century to signify a departure from the ancient conventions of natural philosophy. Most simply defined as the science of living things, biology includes anatomy and physiology, embryology, cytology, genetics, molecular biology, evolution, and ecology. Like all the sciences, biology has roots reaching back into prehistory. Indeed, the most important lessons of all to human survival are aspects of biology—agriculture, animal husbandry, and the art of healing. Innovations in these areas among prehistoric peoples would involve the most ephemeral products of human endeavor. Stone tools, weapons, pottery, glass, and metals leave a more permanent record than a new understanding of the cycles of nature, or the relationship between the breath and the beat of the heart to life itself. Yet it is ultimately on such biological wisdom that human survival and cultural evolution depended.

Driven by necessity, the earliest human beings must have been keen observers of nature. Anthropologists and ethnobotanists have discovered that so-called primitive peoples establish fairly sophisticated classification schemes as a means of understanding and adapting to their environment. Objects may be divided into categories such as the raw and the cooked, the wet and the dry, but the subdivisions within such schemes may reflect keen insights into the medicinal properties of plants and animal products. Even such a seemingly simple division as edible and inedible requires considerable experimentation. Primitive people also learned from experience that a series of operations could transform

1

things from one class to another. For example, in the preparation of manioc both a food and a poison are produced.

Biology as the study of living things probably began with the emergence of *Homo sapiens sapiens* some 50,000 years ago. Ancient human beings living as hunter-gatherers during the Paleolithic Era, or Old Stone Age, manufactured crude tools made of chipped stones. Presumably these ancient people also produced useful inventions that were fully biodegradable and left no trace in the fossil record, such as digging sticks, ropes, bags, and baskets for carrying and storing food. But the inventions of the utmost importance in the separation of human beings from the ways of their ape-like ancestors were fire, speech, abstract thought, religion, and magic. At this stage, conscious of themselves as special entities, human beings could begin to confront the fundamental problems of existence—birth and death, health and disease, pain and hunger—in new ways.

Since the transition from "prepeople" to fully modern human status, cultural evolution, a phenomenon unique to humankind, has taken precedence over biological evolution, a process shared with the rest of the organic world. Another great transition took place about 10,000 years ago when agriculture was invented. The transition from the hunter-gatherer mode of food production to farming and animal husbandry is known as the Neolithic Revolution. In the process of domesticating plants and animals, human beings, too, became domesticated and enmeshed in ways of living, thinking, and planning that revolved around the natural forces and cycles controlling their new enterprises. Scientists and historians were once most interested in the *when* and *where* of the emergence of agriculture, but given the fact that hunter-gatherers may enjoy a better diet and more leisure than agriculturists, the greater puzzle is actually the *how* and *why*. In other words, when *progress* became a concept to be analyzed rather than an inevitable step in human history, the Neolithic transformation could no longer be seen as an early and inevitable step along the road to modernity.

Anatomy, both human and animal, was one of the earliest components of biological knowledge. Among ancient peoples, fear and respect for the dead led to elaborate funerary rites involving the manipulation and perhaps mutilation of corpses. Bodies might be cremated, buried, or preserved after removal of certain organs. Burial customs may reflect compassion and respect, as well as attempts to placate the spirits of the dead and make it impossible for them to return, as they sometimes seemed to do in dreams. Once death could be anticipated and rationalized, attempts could be made to ward it off by both magical and rational means. Primitive attempts at wound management and surgery would provide knowledge of human anatomy and the location of vulnerable sites on and in the body. Observations made while caring for the sick and the dying would have provided important physiological information, such as the importance of the breath, heartbeat, pulse, blood, and body heat.

Knowledge of animal anatomy and behavior is important to the hunter, herdsman, butcher, cook, healer, and shaman. Animals have served as totems of tribes or clans, and animal organs and behavior have been used as omens. Because divination and prophecy depended on the appearance and behavior of various animals, any peculiarity of shape, color, or behavior could decide the fortunes of the tribe and the fortune-teller. Following the natural migrations of animals forces hunters and herders to study their natural behavior, but the domestication of animals would encourage more detailed observations and attempts to select and breed the best of the herd.

Human cultural evolution was profoundly influenced by the domestication animals, especially the horse. Dogs, pigs, cattle, sheep, and goats were probably domesticated long before humans successfully captured, domesticated, and bred horses. The earliest evidence of the domestication of horses appears about 6,000 years ago in central Asia. Horses were widely dispersed throughout the Eurasian steppes some 35,000 years ago. They began to decline about 10,000 years ago. Archaeological sites in Ukraine and Kazakhstan from about 6,000 years ago contain the remains of horses, including some with tooth wear that was probably caused by the use of bits. Although other animals were primarily valuable as sources of food, horses provided the power to transform work, transportation, trade, and warfare. The increased danger posed by men with horses might have stimulated the construction of cities as a way of providing protection from horse-based armies.

In search of the ancestors of domesticated horses, scientists have looked for a common ancestor, an equine Eve, among the diverse family lines of modern horses. Some scientists believe that the large diversity found in modern horses means that they did not derive from a single population. Studies of DNA from modern horses and DNA recovered from the bones of horses preserved in the Alaskan permafrost suggest highly diverse and ancient lines of descent. Interactions between humans and horses might, therefore, have taken place well before the first archaeological evidence of domestication.

BIOLOGY AND ANCIENT CIVILIZATIONS

The invention of science is generally credited to the Greek natural philosophers who lived during the sixth century B.C. This assumption may be largely a product of Eurocentric bias and ignorance about other ancient civilizations. It is not clear how much Greek civilization was influenced by the great civilizations that developed along the valleys of the Nile, Tigris-Euphrates, Yellow, and Indus rivers as much as 5000 years ago. Although these cultures left their mark in terms of archeological monuments, artifacts, and texts, only a small portion of what they achieved was recorded and preserved. Because of the paucity of sources, we undoubtedly have a greatly distorted view of their real accomplishments.

The emphasis on the traditional Western heritage from Greek culture has resulted in the neglect of earlier civilizations and other traditions. Traditionally, India and China were virtually excluded from the general history of science and medicine, except for accounts of exotic medical practices and the invention of printing and gunpowder. Over the course of thousands of years, these complex civilizations developed unique approaches to fundamental questions about the nature of human beings and their relationship to nature. Scholarly explorations of science and philosophy in the context of Chinese and Indian history are relatively new.

Pioneering studies of science and civilization in China led some scholars to suggest that no system of knowledge equivalent to Western "biology" appeared in classical Chinese science or medicine, but biology did not exist as a distinct field in Western science until the nineteenth century. Certainly observations and theories concerning human health and disease, plants and animals, the natural world in general can be found in ancient Chinese writings. They were, however, dispersed throughout a scholarly tradition very different from that of the West. Ideas about vital phenomena and chemical processes can be found in the writings of ancient Chinese alchemists, physicians, and philosophers. Chinese alchemy, especially the ideas of the Taoists, could be considered part of the science of life. Alchemists are often thought of as mystics engaged in the impossible quest for the philosopher's stone that would allow them to transform lead into gold. Taoist alchemists, however, were more concerned with the search for the great elixirs of life, drugs that would provide health, longevity, and even immortality. While the writings of the alchemists were often obscure, the knowledge accumulated by illiterate craftsmen, herbalists, folk healers, and fortune-tellers generally remained outside the scholarly tradition and has been lost to history.

Within Chinese classical science and philosophy, the body of ideas and observations belonging to medicine would come closest to providing knowledge corresponding to the life sciences. Classical scholars explained the unity of nature, life, health, disease, and medicine in terms of the theory of *yin* and *yang* and the *five phases*. The five phases are sometimes called the "five elements," but this is generally considered a mistranslation based on a false analogy with the four elements of ancient Greek science. The Chinese term actually implies transition, movement, or passage rather than the stable, homogeneous chemical entity implied by the term element. Accepting the impossibility of translating the terms yin and yang, scholars simply retain the terms as representations of all the pairs of opposites that express the dualism of the cosmos. Thus, yang corresponds to that which is masculine, light, warm, firm, heaven, full, and so forth, while yin is characterized as female, dark, cold, soft, earth, empty. Rather

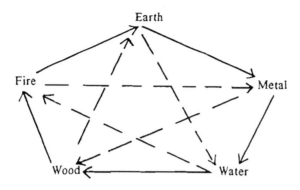

The five phases. As individual names or labels for the finer ramifications of yin and yang, the five phases represent aspects in the cycle of changes. The five phases are linked by relationships of generation and destruction. Patterns of destruction may be summarized as follows: water puts out fire; fire melts metal; a metal ax will cut wood; a wooden plow will turn the earth; an earthen dam will stop the flow of water. The cycle of generation proceeds as water produces the wood of trees; wood produces fire; fire creates ash, or earth; earth is the source of metal; when metals are heated they flow like water.

than simple pairs of opposites, yin and yang represent relational concepts; that is, members of a pair would not be hot or cold per se, but only in comparison to other entities or states. Applying these concepts to the human body, the inside is relatively yin, whereas the outside in relatively yang, and specific internal organs are associated with yang or yin.

According to scholarly tradition, the principle of yin-yang is the basis of everything in creation, the cause of all transformations, and the origin of life and death. The five phases—water, fire, metal, wood, and earth—may be thought of as finer textured aspects of yin and yang. The five phases represent aspects in the cycle of changes. Classical Chinese medical practice was based on the theory of systematic correspondences, which linked the philosophy of yin and yang and the five phases with a complex system that purportedly explained the functions of the constituents of the human body. Thus, the goal of classical Chinese anatomy was to explain the body's functional systems and the Western distinction between anatomy and physiology was essentially irrelevant. Within this philosophical system, Chinese scholars were able to accept the relationship between the heart and pulse and the ceaseless circulation of the river of blood within the human body. It was not until the seventeenth century that these concepts were incorporated into Western science through the experimental physiology of William Harvey (1578–1657).

Many aspects of the history of the Indian subcontinent are obscure before the fourth century B.C., but archaeologists have discovered evidence of the complex Indus River civilization that flourished from about 2700 to 1500 B.C. The sages of ancient India had little interest in chronology or the separation of mythology and history, but traces of their ideas about the nature of the universe and the human condition can be found in the Vedas, sacred books of divinely inspired knowledge. Brahma, the First Teacher of the Universe, one of the major gods of Hindu religion, was said to be the author of a great epic entitled *Ayurveda, The Science of Life*. After oral transmission through a long line of divine sages, fragments of the sacred *Ayurveda* were incorporated into the written texts that now provide glimpses into ancient Indian concepts of life, disease, anatomy, physiology, psychology, embryology, alchemy, and medical lore. A striking difference between Indian religions and those of the West is the Hindu concept of a universe of immense size and antiquity undergoing a continuous process of development and decay. Indian scholars excelled in astronomy and mathematics, especially the ability to manipulate the large numbers associated with mythic traditions concerning the duration of the cosmic cycle.

Health, illness, and physiological functions were explained in terms of three primary humors, which are usually translated as *wind*, *bile*, and *phlegm*, in combination with *blood*. The concept of health as a harmonious balance of the primary humors is similar to fundamental Greek concepts, but the Ayurvedic system provided further complications. The body was composed of a combination of the five elements—earth, water, fire, wind, and empty space—and the seven basic tissues. Five separate *winds*, plus the *vital soul*, and the *inmost soul* also mediated vital functions. Ayurvedic writers also expressed considerable interest in the fundamental question of conception and development. According to the Ayurvedic classics, living creatures fall into four categories as determined by their method of birth: from the womb, from eggs, from heat and moisture, and from seeds. Within the womb, conception occurred through the union of material provided by the male and female parents and the *spirit*, or external self. The development of the fetus from a shapeless jelly-like mass to the formed infant was described in considerable detail in association with the signs and symptoms characteristic of the stages of pregnancy.

The Ayurvedic surgical tradition presents an interesting challenge to the Western assumption that progress in surgery follows the development of systematic human dissection and animal vivisection. Religious precepts prohibited the use of the knife on human cadavers, but Ayurvedic surgeons performed major operations such as couching the cataract, amputation, trepanation, lithotomy, cesarean section, tonsillectomy, and plastic surgery. The study of human anatomy, or the science of being, was justified as a form of knowledge that helped explicate the relationship between human beings and the gods. Despite barriers to systematic dissection of human cadavers, Indian scholars developed a remarkably detailed map of the body including a complex system of *vital points*, or *marmas*. Injuries at such points could result in hemorrhage, permanent weakness, paralysis, chronic pain, deformity, loss of speech, or death. Obviously, knowledge of such points was critical to ancient physicians and surgeons, who had to take the system into account when performing operations, treating wounds, or prescribing bloodletting and cauterization. Knowledge of the system is also of interest to certain contemporary healers, such as those who use the points as a guide to therapeutic massage. Indeed, Ayurvedic medicine is a living tradition that brings comfort to millions of people, while the ancient classics provide valuable medical insights and inspiration for modern physicians and scientists.

MESOPOTAMIA AND EGYPT

Along the great river valleys of Mesopotamia and Egypt, human life and welfare depended on the cyclic flooding and retreat of the great rivers. Complex agricultural and engineering activities had to be planned, directed, and supervised by knowledgeable authorities. The priests and nobles who directed this system dominated the way their subject populations thought about nature. The ancient myths of the separation of chaos into dry land and water and the development of living things from the mud are in essence the life story of these first civilizations. These myths served powerful central governments in dealing with the mundane problems of agriculture—land surveying, irrigation and flood control, storing food for bad years—and in explaining the capriciousness of nature. For the privileged members of the priesthood, knowledge of biological and astronomical phenomena was a valuable professional monopoly, not the subject matter for secular scientific inquiry.

Unpredictable and dangerous waters played a prominent role in Mesopotamian cosmology, as might be expected in an area subject to unpredictable flooding. Early Mesopotamian myths described the earth and heavens as flat disks supported by water. Somewhat later the heavens were described as a hemispherical vault which rested on the waters that surrounded the earth. The heavenly bodies were gods who dwelt beyond the waters above the heavens and came out of their dwelling places for their daily journey through the heavens. Since the gods controlled events on earth, the motions of the heavenly bodies were closely studied to reveal the intentions and designs of the gods.

According to Egyptian cosmology, the world was rather like a rectangular box. The earth was at the bottom, making that part of the box somewhat concave. Four mountain peaks at the corners of the earth supported the sky, while the Nile was a branch of a universal river that flowed around the earth. This river carried the boat of the sun god on his journey across the sky. Seasonal changes and the annual Nile flood were explained by deviations in the path taken by the boat.

Creation myths explained how the world came into existence from a primeval chaos of waters. Matings between male and female gods in the time of chaos brought forth the heavens, earth, air, and the natural forces that were personified by various gods. The gods organized the universe out of the original chaos, separating the land from the waters much as the first settlers had reclaimed the earth from the waters. Egyptian and Mesopotamian civilizations, in their political organization, scientific interests, and religions, reflect the differences in the patterns of the great rivers. The Nile floods were predictable, beneficent, and eagerly anticipated. Egyptian dynasties were generally long-lived, rigid, and complacent. Great efforts were expended in creating appropriate tombs for the mummified remains of the Pharaohs. The floods of the Tigris and Euphrates, in contrast, were unpredictable and frightening events. Governments in Mesopotamia were more transient and the future always uncertain. Astrology

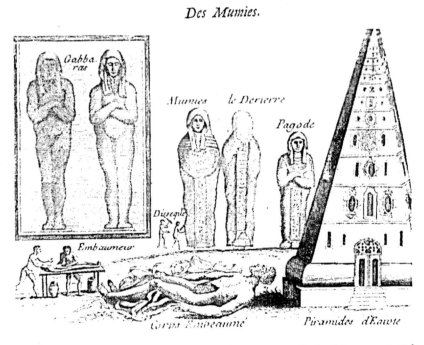

Egyptian mummies, pyramids, and the embalming process as depicted in a seventeenth-century French engraving

and divination were the tools and weapons with which Mesopotamian priests wrestled with the gods and the chaotic forces of nature.

Despite great accomplishments during the high Bronze Age, these civilizations remained essentially static while various technological innovations transformed cultures that had been at the fringes of Bronze Age civilization. The smelting of iron and the use of a simple alphabetic script created a new cultural revolution. Sometime in the second millennium B.C., a tribe in the Armenian mountains developed an efficient method of smelting iron. Originally quite secret, the method became generally known after about 1000 B.C. Iron was used in weapons and for agricultural implements, which led to more efficient exploitation of the lands to the north. With iron weapons, barbarians like the Greek tribes were able to conquer rich Bronze Age cultures.

Long after the civilizations of Mesopotamia, Egypt, and India had reached their peak and entered into periods of decline or chaos, Greek civilization seems to have emerged with remarkable suddenness, like Athena from the head of Zeus. This impression is certainly false, but it is difficult to correct because of the paucity of materials from the earliest stages of Greek history. What might be called the prehistory of Greek civilization can be divided into the Mycenaean period, from about 1500 to the collapse of Mycenaean civilization about 1100 B.C., and the so-called Dark Ages from about 1100 to 800 B.C. Fragments of the history of this era have survived in the great epic poems known as the *Iliad* and the *Odyssey*, traditionally attributed to the ninth-century poet known as Homer. Ancient concepts of life and death, the relationship between the gods and human beings, and the structure and vital functions of the parts of the body are encoded in these myths and legends. Later Greek history is conventionally divided into three periods: Archaic (before 480 B.C.), Classical (480–323 B.C.), and Hellenistic (after 323 B.C.). The Greek city-state was quite different from the centralized governments of Egypt and Mesopotamia. Moreover, unlike previous high cultures, that of the Greeks was not completely based on agriculture. Trade and shipping were important, and since Greece was relatively overpopulated in relation to cultivatable land, colonization and industry were encouraged.

THE PRESOCRATIC PHILOSOPHERS

As early as the sixth century B.C., Greek natural philosophers were involved in establishing a secular tradition of inquiry into the natural world. That is, they attempted to formulate new explanations of how and why the world and human beings came to be formed and organized as they were. Such questions are very old, but the Greek philosophers took them out of the realm of religion and mythology into the domain now occupied by science and philosophy. In place of gods and magic, these philosophers attempted to explain the workings of the universe in terms of everyday experience and by analogies with craft processes.

Natural philosophy became a significant part of the quest for a reliable means of understanding and controlling nature. Many of the earliest philosophers are known only through a few fragments of their work, but what has survived suggests that their ingenious questions and answers served as the seed crystals that were to stimulate the subsequent development of Western biology, medicine, astronomy, and physics.

Natural science developed first not in the Athens of Socrates, Plato, and Aristotle, but on the Aegean fringes of the mainland of Asia Minor known as Ionia. Miletus, home of the first of the natural philosophers, was the mother city of numerous colonies on the Black Sea and had commercial interests throughout the Mediterranean world as well as contacts with the ancient civilizations of Mesopotamia and Egypt. The first philosophers seem to have had a keen interest in the opportunities for involvement in trade, travel, and politics that characterized these dynamic commercial centers. Scholars in the ancient world were generally highly respected, as shown by the fact that various city-states would ask them to serve as lawgivers, rulers, or tutors to the sons of rulers. On the other hand, in times of political turmoil scholars might find themselves persecuted by hostile factions. Thus, while Greek philosophers had more potential freedom of inquiry than the priests and officials who formed the intellectual elite of other cultures, they lacked the protection that the intermediaries of the gods always claimed.

The Ionian philosophers were preoccupied with finding the natural regularities they believed existed beneath the apparent changes in the world; that is, the common primary cause that initiated the entire chain of cause and effects and the common element from which everything originated. According to the Egyptians and Babylonians, the primary material constituents of the world were water, earth, and air. Chinese philosophers believed the world was made up of yin-yang and the five phases. Greek philosophers constructed a world of earth, air, water, and fire.

While the universe of the early Greek philosophers was certainly still full of gods, their gods were very different from those who so frequently intervened in human affairs in the great epics of Homer. In essence, Greek philosophers from Thales to Aristotle posed the questions that still confound and intrigue philosophers and scientists. What is the grand design of the universe? Are there transcendent laws that explain why the universe is the way it is? Do such laws explain how and why matter and energy became formed, shaped, and organized, and ultimately generate conscious beings? Is the universe the way it is of necessity, that is, does it contain the reason for its existence within itself, or is it what philosophers call contingent, that is, dependent on something beyond itself? The answers presented in the surviving fragments might best be thought of in terms of a theme and its numerous variations. For example, Thales of Miletus (ca. 625 B.C.-ca. 547 B.C.) chose *water* as his primary element.

Thales, called the founder of Ionian natural philosophy by no less an authority than Aristotle, flourished at about the same time as Buddha, Confucius, Zoroaster, and Hippocrates. Although Thales apparently distinguished himself as a statesman, merchant, engineer, mathematician, astronomer, and a lover of knowledge, there was considerable uncertainty as to his written work even in antiquity. Some authorities said that he left no writings at all, but others claim that he was the author of "The Nautical Star Guide." Many legends grew up about his accomplishments and his travels to the older civilizations that represented timeless wisdom to the Greeks of his era. According to one such story, Thales invented a way of calculating the distance of ships at sea by means of the doctrine of similar triangles after he became familiar with Egyptian geometry during a business trip. Then, having learned Phoenician astronomy while in Mesopotamia, Thales is said to have predicted an eclipse of the sun in 585 B.C., transforming what had been a terrifying and mysterious phenomenon into a natural, predictable event.

Probably the most popular story associated with Thales demonstrates how astronomy and meteorology can be put to practical use. Through his study of the heavenly bodies, the philosopher predicted an unusually good olive crop. Before the harvest season, he rented all the olive presses in Miletus very cheaply and was able to make a great profit when the demand for the presses was greatest. Thus, he proved to those who said that philosophy was useless that those who understood nature could easily become rich, though it was not in the nature of philosophers to seek wealth. Aristotle suggested that this story was not literally true, but that it had became attached to Thales because of his reputation for wisdom. It is, therefore, rather like the story of young George Washington and the cherry tree, which serves as an emblem of his honesty. On the other hand, Thales was also depicted as the prototype of the absent-minded professor. According to Plato, Thales was mocked for falling into a well while he was observing the stars. Intently gazing upward, the philosopher ignored the hazards at his feet.

Thales attempted to create a complete cosmological theory in which water served as the primary material basis of all things. He described the earth as a disk that floated on water. Above the earth was a rarefied form of water that constituted the sky. Experience tells us that this is not unreasonable; landmasses are bordered by oceans and rivers, wells bring us water from below the earth, and rainwater falls from the sky. Whereas Homer said that earthquakes were caused by the god Poseidon, Thales argued that just as a ship at sea is rocked by the waves, perturbations in the waters beneath the earth rock the surface. Even though Thales did not dispute the doctrine that "all things are full of gods," he believed that philosophers would be able to explain natural phenomena without invoking supernatural agents. Perhaps Thales was influenced by the creation myths of older civilizations and reshaped these ideas into a more or less self-consistent and secular view of the universe, but Aristotle suggested that Thales's reasoning was primarily physiological. That is, because nutriment and semen were always moist and the warmth of life was a moist warmth, all things must come from and be composed of water.

In addition to a primary material substance, Thales's cosmology required two processes or forces: consolidation and expansion. Water could be expanded until it turned into air and consolidated until it turned into earth. Anyone who has observed the appearance of a kettle after hard water has boiled away will know that everyday experience appears to validate the idea that water can be transformed into steam (air) and mineral deposits (earth).

Other Greek philosophers agreed with Thales that the world could be explained in purely naturalistic terms, but disagreed with him about the nature of the primary element and the forces that shaped the cosmos. Anaximander (ca. 611–547 B.C.), who may have been a pupil and perhaps a nephew of Thales, argued that the primary material basis of the universe could not have any of the individual characteristics of the things that it becomes. To avoid the trap of the primary element being the same as any one of the four elements and a part of the other three, Anaximander's cosmology is based on the existence of *aperion*, a rather ill-defined universal "stuff." When acted on by heat and cold, this primary material was transformed into the four elements that made up the world: earth, air, fire, and water.

According to Anaximander, the formation of the universe out of the original chaos was brought about by the operation of some enigmatic primary force that caused the formation of a vortex. The motion of the vortex led to the separation of our world from the infinite and to the separation and stratification of the elements according to their density. Because earth is the heaviest element, it tends to remain at the center of the world. Water covers the earth and is in turn enveloped in a layer of mist. Fire, which is the lightest element, escaped to its natural place at the outside of the universe, where it formed the heavenly bodies. Eventually the whirling motion disrupted the orderly arrangement of the elements and created wheels of fire enclosed by tubes of mist. The sun and moon were ring-shaped bodies made of fire surrounded by air. Tubes in this mass of air allow light caused by the fires to reach the earth. Because the tubes are of various sizes, the sun appears to be larger than the moon, and the moon seems to be larger than the stars.

Eventually the action of the vortex caused the earth and its creatures to become formed and organized. The earth took on the shape of a rather flat cylinder. While the separation of the elements was occurring, fire acted on mud to produce dry land and mist. As the sun warmed the mud, which was a mixture of earth and water, it bubbled and heaved and brought forth various animals. The first living creatures were fish that formed when water predominated. As separation continued and dry land appeared, some of these fishy creatures came onto the land and changed as conditions on earth changed. Man developed from a fish-like creature that had adapted to life on land and discarded its fishy skin. Because man is originally helpless, he could not emerge as an infant, but had to be nurtured inside the fish skin until the land was ready to receive him. Anaximander did not believe that such changes were directed along a unidirectional path. On the contrary, change was a cyclical process in which the forces that led to the separation of the elements and the development of life could be reversed so that chaos would once again predominate. In the great cosmic cycle, worlds come into being, perish, and are ultimately reabsorbed into the eternal infinite.

Anaximenes (fl. 546/545 B.C.), the last of the Milesian philosophers, accepted Anaximander's idea that all that exists has come from the infinite, but Anaximenes chose *air*, or *mist*, as his primary element. Other substances were formed from air by *rarefaction* and *condensation*; that is, the amount of air packed into a particular place determined its form without altering its original nature. The process of felting, in which wool fibers are packed down to form a fabric, may have served as a simple model for this idea.

Air is always in motion because there can be no change without movement. When air is in its most uniform or evenly distributed state, it is invisible to our senses. When air is rarefied and made progressively finer, it becomes fire. Air can be revealed to us as it becomes condensed by means of movement, cold, warmth, and moisture in the form of wind, mist, and clouds. If air is progressively condensed it becomes water and earth. When it is condensed as much as possible, it becomes stone.

Anaximenes essentially formalized the general concept of the earth as a broad, flat, shallow disk floating on air and situated at the center of the universe. The heavenly bodies, which were created by the rarefaction of mist into fire, were also supported by air. An infinity of different worlds came into existence and eventually passed away to be resorbed into the infinite air that is all encompassing and in perpetual motion.

The speculations of the earliest natural philosophers were subjected to further analysis by Xenophanes and Heraclitus, who were also Ionians, and by Anaxagoras, Leucippus, and Democritus.

Xenophanes of Colophon (576–490 B.C.) apparently taught that everything, including man, originated from water and earth. Earth and sea, he suggested, had changed places in the past and would do so again. Fossil evidence supported this idea because shells were found inland, in the mountains, and in the quarries of Syracuse and other places. Fossils proved that the earth had once been covered with mud, so that impressions of fish and other marine objects remained when the mud dried out. Previous philosophers had also suggested that living creatures developed in the mud, but Xenophanes appears to be the first to call attention to the physical evidence provided by fossils.

Primarily interested in religion, Xenophanes recognized the tendency of man to make gods in his own image. In a famous passage, he called attention to the cultural relativity of human judgments on supposedly universal problems. "The Ethiopians say that their gods are snub-nosed and black, the Thracians that theirs have light blue eyes and red hair," he wrote, "but if cattle and horses or lions had hands, horses would draw the forms of the gods like horses, cattle like cattle." Although Xenophanes emphasized the limitations of human knowledge, he believed that by actively seeking the truth one could reach a higher state of wisdom. Not all of his contemporaries believed that he had reached such an exalted state. Heraclitus of Ephesus (556–469 B.C.) dismissed Xenophanes with the statement that "a great deal of learning does not bring wisdom."

Bright *fire* was the primary element of Heraclitus, but his contemporaries called him the "dark one" because they found his work difficult and obscure. The doctrines for which he is best known are those attributed to him by Plato and Aristotle, who considered his work important to the development of cosmology. Heraclitus helped establish the distinction between philosophers who thought of the universe in terms of change and those who thought about it in terms of stability. His detractors called him an arrogant misanthrope who adopted the style of the Delphic oracle in his collection of aphorisms. While Heraclitus may have impressed his contemporaries as a pretentious, supercilious aristocrat, he modestly referred to himself as the vehicle, rather than the author of revelations concerning the relationships that governed the natural world just as civil laws regulated relationships among citizens.

If Heraclitus's system seems more unified than those of his predecessors, perhaps this is merely because it is better preserved. His major premise was that there is a regularity and balance that governs all change in nature. Balance in the universe was based on pairs of opposites in a state of tension with each other, much as harmonious music is created by the tension and opposition of the bow on the strings of the lyre. This balance existed in the mundane world of business transactions as it did in cosmic transactions. "All things are an equal exchange for fire and fire for all things," wrote Heraclitus, "as goods are for gold and gold for goods." The *opposites* of the Heraclitean system seem to be different from the *contraries*, such as heat and cold, or wet and dry, that are found in most Presocratic philosophies. The Heraclitean opposites are pairs generally related to the properties of living things, such as life and death or health and disease, or verbal expressions, such as present and absent. Like a Zen master, Heraclitus confounded his audience by telling them that binary opposites such as youth and age, day and night, "the way up and the way down" are the same. Similarly, the ever-living fire that is the common element of all things both kindles and quenches.

It is not clear whether Heraclitus selected fire to represent the substance from which the whole world formed through the same kind of reasoning that led Thales to choose water, or whether he saw fire as a paradigm for explaining all natural processes. Fire is both an element that changes and the process that causes change, whether in cooking, pottery making, or metallurgy. When things are consumed by fire they are transformed into fire, smoke, and vapors, which can form liquids and solids when condensed. Images of fire and rivers appear frequently in the Heraclitean fragments, reflecting the principle that all things are in flux and nothing is stable. The world we think we are seeing is actually like a flowing river, and according to Heraclitus we can never step twice into the same river. The endless flux could be seen in the changing seasons and the human life cycle.

Heraclitean philosophy could explain everything from the changes in the external world to the complexities of human thought. Indeed, understanding nature was not separate from understanding man, because both were subject to the same laws. Life and health were associated with fire, while death and decline were associated with water. If men did not struggle towards an understanding of the true constitution of things, their souls, which were made of fire, would become excessively moistened. A dry soul was the wisest and best, and unfortunately for a civilization that treasured the fruit of the vine, wine in particular moistened the soul.

Weary of attempts at human discourse, Heraclitus withdrew from society to live in the mountains, where he ate grasses and plants. This caused him to develop dropsy, which is also called edema, a disease characterized by an excess of water. Seeking aid from physicians, he tested them with a riddle. Because they could not answer him, Heraclitus rejected their medicines and buried himself in a cow stall. Theoretically, the heat of the decaying manure should have evaporated the pathological excess of water in his body, but it did not. And so the philosopher died, buried up to his neck in cow manure. This bore out his statement that "it is death for souls to become water" as well as his comment that corpses are more worthless than dung.

All the speculative thinkers of the sixth and fifth can be referred to as the Presocratic philosophers, but there are significant differences between the group known as the Milesians, those referred to as the Pythagoreans, and those known as the Eleatic philosophers, who accepted the teachings of Parmenides of Elea (fl. 500 B.C.). Very little is known for certain about Pythagoras of Samos (ca. 582–500 B.C.), whose name has been immortalized through the Pythagorean theorem. He is said to have escaped from Samos during the reign of the tyrant Polycrates and to have settled in southern Italy at Croton, where he founded a brotherhood concerned with religious beliefs and practices, as well as mathematical inquiries. The Pythagorean community was remarkable in that men and women were admitted on equal terms. Eventually, the Pythagoreans seem to have split up into scientific and religious fractions. By the time of Plato, the Pythagorean legacy was already enveloped in mystery. Pythagoreans apparently believed in the immortality of souls and thought that human souls could be reincarnated in the form of other animals. This has been taken to mean that the Pythagoreans believed in the kinship of all living things.

Apparently inspired by mathematical principles, the Pythagoreans believed that all things in the universe could be thought of in terms of numbers. For the Pythagoreans, numbers seemed more fundamental to explaining the nature of the universe than fire, earth, and water. Pythagoras seems to have believed that religion and science were inseparable aspects of the complete life. But there is no direct reliable evidence about the scientific teaching of Pythagoras himself and no satisfactory way to separate his ideas from those of his followers.

Few members of the Pythagorean community are even known by name, but Alcmaeon of Croton (fl. ca. 500 B.C.), a physician interested in biological questions, was probably a pupil of Pythagoras. In addition to medical theory, physiology, and anatomy, Alcmaeon may have written on meteorology, astronomy, philosophy, and the nature of the soul. As would be expected of a Pythagorean, Alcmaeon believed that the soul was immortal and always in motion.

Alcmaeon may have been the first natural philosopher to carry out dissections and vivisections of animals for the sake of learning about their nature rather than for purposes of divination. Moreover, he may have introduced the practice of examining the developing chick egg as a way of studying embryology. In order to understand the nature of sense perception, Alcmaeon conducted dissections that led to the discovery of the optic nerves and the eustachian tubes (which were rediscovered by Eustachius in the sixteenth century). Despite attacks by his critics, Alcmaeon was generally regarded as one of the first to clearly define the difference between man and other animals. Man is the only creature capable of understanding the world; other animals see the world, but cannot understand. In keeping with Pythagorean ideas, Alcmaeon explained health in terms of a harmonious balance of contrary qualities, such as moist and dry, hot and cold. An imbalance of qualities caused disease.

An intriguing answer to the problems posed by competing speculations about the nature of the primary element, the reliability of sense perceptions, and the nature of change was proposed by Empedocles of Agrigentum in Sicily (ca. 492 to ca. 432 B.C.), who was probably influenced by both the Pythagoreans and by Parmenides (ca. 515–445 B.C.). Although only fragments of his poem "On Nature" have survived, many philosophers have recognized the questions raised by Parmenides as a great challenge. Ultimately, Parmenides argued, contrary to Heraclitus, change does not occur at all. Parmenides limited philosophical discourse to two general ideas: *Being* (that which exists) and *Not-Being*. According to Parmenides, True Being, or reality, can only be discovered through reason. That is, the phenomena that human beings assumed they knew through the senses did not really exist. According to Parmenides, Being cannot arise

Empedocles estoit un Philosophe, qui a l'imitation de Pythagore ne
vouloit point manger de tout ce qui vivoit et se mouvoit

Empedocles of Agrigentum

from Not-Being, and Being is not created or destroyed. Because all change must involve the addition of Not-Being to Being, change is logically impossible.

Empedocles was said to be the founder of the four-element theory of matter, the inventor of rhetoric, a great orator, prolific author, fervent democrat, pioneer of city planning, and the founder of public health medicine. Romantic poets later cast him in the role of heroic figure and champion of democracy. Like Pythagoras and Heraclitus, Empedocles inspired many legends. He allegedly boasted about his supernatural gifts, claiming to have the power to heal the sick, cure the infirmities of old age, raise the dead, change the direction of winds and rivers, and bring the rain and the sun. References to him in later medical writings suggest that he was a very successful physician, who reduced the burden of disease in his native city by draining the swamps, improving water supplies, and creating an opening in the mountains to admit the cool north wind. Another legend says that he was offered the kingship of Agrigentum but refused because of his commitment to democratic principles.

Only fragments of two poems by Empedocles have survived. "On Nature," in part a response to the metaphysics of Parmenides, was a complex physical explanation of the universe. "Purifications" dealt with the question of personal salvation in a manner influenced by the Pythagorean belief in transmigration of souls. Parmenides had taught that the real world must be "unborn and imperishable, one and indivisible, immobile, and a complete actuality." In contrast, that which is found in the world of sense perception was merely a man-made illusion.

In accordance with the doctrine that Being cannot come from Not-Being and that a primordial unity cannot produce a subsequent plurality, Empedocles taught that there were four eternal, unchanging, and distinct elements or roots: earth, air, fire, and water. Mixing these four eternal principles created new things, just as artists mix pigments to produce new colors. Creations and destructions are merely changes caused by mixing or separating the "roots of all things." But what produces the motions of the four elements and changes their local combinations? According to Empedocles these changes are due to two forces: *Love* and *Strife*. Love causes attraction, aggregation, and the creation of new worlds, but when Strife predominates, worlds are torn apart. Thus, in the Cosmic Cycle of Empedocles, the universe oscillates between stages of creation and destruction.

Unlike Parmenides, Empedocles believed that the senses could serve as a valid guide to philosophical truths if each sense was carefully and critically employed for its appropriate purpose. Perception was ascribed to the recognition of similar elements and forces; we see earth with earth, water with water, and Love with Love. This mechanism of perception was possible because all things, including plants and animals, earth, sea, and stones, give off subtle emanations of varying sizes. Effluences of particular sizes could fit into the pores or passages of the various sense organs so that the meeting of elements, which constituted perception, could occur within the body. Consciousness and thought were also products of this sorting of the elements. Thought processes take place mainly in the sea of blood surging back and forth around the heart. Because the four elements were most intimately blended in this blood, it was able to serve "as the chief seat of perception." A kinship among all living things is suggested by the assertion that "all things possess thought," however, Empedocles stipulated that not all things or all men possess the power of thought to the same degree.

Man and all other creatures, made up of the same four elements, were the products of the complex and peculiar evolution that characterized the cosmic cycle. There is considerable ambiguity about Empedoclean doctrines, but basically the cosmic cycle consisted of two polar and two transitional stages and two distinct stages in the evolution of living things during the transitions. In stage one, the four elements are completely mixed in a homogeneous sphere. Stage two is a transition in which the elements become increasingly separated under the influence of Hate. During stage three, the elements are completely separated. Stage four involves another transition during which the elements coalesce under the influence of Love.

At an earlier stage of the great cycle, the Earth had the power to create multitudes of living things. The first entities were, however, so incomplete and crudely made that arms wandered about without shoulders, faces without necks, and eyes were not attached to heads. In the second stage, parts began to unite in random combinations. Sometimes monstrosities were formed: man-faced oxen, ox-headed men, and sterile creatures that were part male and part female. A great variety of living forms appeared, including some "whole-natured forms." Still, there were many poorly constructed creatures that were unable to compete with those that had a better assortment of parts, and few creatures contained all the parts necessary to reproduce their own kind. As separation continued, the sexes eventually become distinct and separate, leading to the next stage. In the fourth stage, new organisms no longer arise directly from the Earth, but only from beings that had the power of generation. All the species that survived into the fourth stage were those with special courage or skills or speed that allowed them to protect, preserve, and reproduce themselves.

According to Empedocles, all living things are composed of a particular mixture of elements. Each kind of being seeks its natural place according to the mixture of elements it contains. Trees are still attached to the earth because they have more earth in their nature; fish obviously have more water, while birds have more air and fire in their composition. Different mixtures of the four elements produced specific kinds of compound substances. Bone was composed of fire, water, and earth in the ratio of 4:2:2. Unfortunately, the recipes for other substances did not survive.

In plants, as in the case of the whole-natured forms of the previous stage, the two sexes are combined in one individual, but human beings and the higher animals reproduce through sexual generation. In contrast to the prevalent doctrine that the child was the product of the father, Empedocles suggested that some parts of the embryo came from the father and others from the mother. Male offspring, however, were conceived in the warmer part of the womb and contained a greater proportion of warmth than females.

Imaginative speculation on a grand, even grandiose scale is part of the legacy of Empedocles. Another aspect is the use of careful observations of simple, ordinary events to provide insight into the workings of nature. One of Empedocles' most significant insights, in terms of the history of science, was his attempt to explain the process of respiration in terms of an experimental demonstration concerning the physical nature of air. Empedocles seemed to think that the blood oscillated in the body like the tides of the ocean, so that the motion of the blood drove air into and out of the body. Empedocles compared the movement of the air into and out of the body with the movement of liquid in a *clepsydra*, or "water-stealer," a device rather like a pipette. Essentially a hollow cylinder with a strainer at the bottom, the clepsydra was used to transfer small quantities of liquid from one vessel to another. In "On Nature" Empedocles described a young girl playing with a gleaming brass clepsydra. When she covered the mouth of the cylinder and dipped the device into a fluid, no liquid entered the vessel because air inside prevented liquid from entering. Uncovering the device allowed air to escape as liquid entered. These observations, Empedocles argued, proved that air is not simply empty space. Air, therefore, has the nature of a substantial although invisible body. Once aware that nature works by unseen bodies, we can overcome the limitations of the senses by combining reason and observation in order to learn about things that are not directly perceptible.

Exactly what this insight meant in terms of the history of science is controversial. Some historians of science see Empedocles as the originator of the experimental method, while others contend that he merely applied a common observation to a preconceived theory. If Empedocles had really thought that the experimental method was the key to understanding nature, he could have invented many other experimental tests of his theories. Perhaps, given the literary device employed in making his case, Empedocles actually deserves credit for observing that children of both sexes are natural experimentalists. Certainly many of Empedocles' theories can be thought of as fruitful, in terms of inspiring large new ideas about natural phenomena, which is often more important than being correct about small matters.

In any case, Empedocles was not the only philosopher to offer a defense of the use of the senses, in conjunction with reason and logic, as a guide to understanding the world of natural phenomena. Anaxagoras of Clazomenae (ca.

500 to ca. 428 B.C.) represents the new Athenian phase in the development of Greek philosophy and the dangers that would haunt those who pursued unconventional ideas. Despite his friendship with the great Athenian statesman Pericles, Anaxagoras was persecuted for impiety, allegedly for saying that the sun was only a red-hot rock rather than a god. Perhaps Anaxagoras had expressed other dangerous and unorthodox ideas. In a brief fragment dealing with the difference between humans and animals, Anaxagoras said that it was "possession of hands that makes man the wisest of living things." The philosopher was forced to leave Athens for Lampsacus. When asked what privilege he wished to be granted by his adopted city, the philosopher is said to have required a holiday for children.

In his defense of the value of evidence obtained by the senses through observation and experience, Anaxagoras resembles Empedocles. Like the early Ionian natural philosophers, Anaxagoras combined an interest in practical matters with a passion for understanding how the universe worked. Wrestling with the logical and abstract arguments of Parmenides concerning the impossibility of change, Anaxagoras proposed the perpetual existence of different substances that totally filled all of space. These eternal substances could aggregate and separate under the influence of Mind and Motion.

For Anaxagoras, the first principles were homogeneous entities, or *seeds*. The seeds that were the first principles in his system were infinite in both number and variety, and yet every one contained a little of all the qualities to which our senses respond. While all things contain a portion of everything, according to Anaxagoras, Mind alone was pure and unmixed, as well as infinite and self-ruled. Mind contained all knowledge and controlled all living things, and Mind alone possessed the power of initiating motion. In Anaxagoras's scheme Mind plays the role Empedocles ascribed to Love and Strife. In the beginning all things existed in a homogeneous, motionless mixture. The universe as we know it began when Mind initiated a localized rotational motion or vortex that continuously increased so that all things were mixed, separated, and divided. The dense was separated from the fine, hot from cold, light from dark, and moist from dry. Dense, cold, wet, dark materials were driven to the center by the vortex, while the hot, dry, light, fiery materials were thrown to the periphery. A disk-like Earth formed from the dense matter in the lower region, but the vortex separated out the sun, moon, and stars and sent them into the fiery upper regions.

One fragment associated with Anaxagoras poses the question: "How could hair come to be from what is not hair and flesh from what is not flesh?" His answer was that everything that is assimilated by human beings must preexist in their food because there are no simpler elements to form common natural substances by aggregation. The foods we consume must contain the seeds or entities that are capable of producing blood, muscle, bones, and so forth. These seeds must exist in foods in a form that is comprehended by means of reason, but invariably hidden from the senses. During the process of digestion, the seeds or elements are sorted out. Growth occurs by assimilating the seeds into their proper places by the natural attraction of like to like. Any given item must contain within itself a mixture of substances; the predominant constituent would determine its special characteristics. Empedocles, in contrast, held that flesh is a mixture of the four elements; if we divide a specimen until an absolutely minimal bit of flesh remains, any further division will yield particles of the four separate elements.

Like Empedocles, Anaxagoras called attention to observation as a means of understanding the nature of air. Moreover, he may have carried out experiments or demonstrations, such as blowing up bladders and demonstrating their resistance to compression. In opposition to Empedocles, Anaxagoras believed that perception occurs through opposites, because we do not notice things unless there are contrasts. For example, something that is warm or cold to the touch would command our attention, while something that was as warm or cold as we are would not warm or cool us. Nevertheless, Anaxagoras apparently proposed a "thought experiment" to show that the evidence presented by our senses might not reveal the truth. Imagine two vessels: one contains a white liquid and the other a black one. If either liquid is added drop by drop to the other, the eye cannot perceive any change in color until a significant quantity has been added. Although our senses tell us no change has occurred, reason knows that our senses have been deceived.

ATOMS AND THE VOID

The speculations of the early natural philosophers might be thought of as ingenious approaches to understanding the nature of the universe, the earth, and its living beings, but their theories appear to have no direct counterparts in modern science. Atomic theory, on the other hand, immediately appeals to modern scientists as a brilliant, if somewhat crude anticipation of one of the most fundamental concepts of modern chemistry and physics. When more closely examined, the atomic theory of the ancients emerges most clearly as an elegant solution to problems that had puzzled philosophers since the time of Thales, rather than a primitive precursor of John Dalton's (1766–1844) atomic theory.

Much is uncertain about the lives of the fifth-century atomists Leucippus (fl. ca. 440 B.C.) and Democritus (ca. 470–370 B.C.). Leucippus may have been the first philosopher to develop a cosmology in which atoms served as the first principles. Democritus, who refined the theory and explored its implications, might have been his pupil. Like Empedocles, Democritus wrestled with the puzzles posed by Parmenides in constructing his own cosmology. Both Leucippus and Democritus postulated a world composed of innumerable atoms in perpetual motion through the infinite void of empty space. The atoms are uncreated and eternal and do not undergo actual change or destruction, but things appear to come into being through their combinations, whereas their separation is interpreted as destruction. Atoms and the void constituted the material causes of all things. The differences in shape, arrangement, and position of the atoms accounted for all apparent modifications in the universe. In essence, varying the arrangement of atoms produced different materials, just as different combinations of letters produced different words. The idea that variety and change depend on the combination and separation of some primary material that was itself unchanging is found in Empedocles and Anaxagoras; these philosophers, however, had not postulated the existence of the void.

Atoms were compact, indivisible, and impenetrable because they had no void within them, but they were too small to impinge directly on the senses. Whether or not the first atomic theorists thought that weight was a primary characteristic of atoms is unclear. Compound bodies, on the other hand, were divisible because of the void between their constituent atoms. The number of shapes atoms could assume was probably infinite because there was no reason that any atom should be of one shape rather than another.

From the void and the atoms, Democritus believed that innumerable worlds came into existence through mechanical means. Some of these worlds had no sun and moon; in others there were suns and moons more numerous or larger than in our world. Sometimes worlds were destroyed by colliding with each other. Some worlds had living creatures, whereas others were devoid of plants, animals, and moisture. Exactly how things happened is unclear, but the atomists argued that nothing occurred at random, but only "for a reason and by necessity." Events that seemed to occur at random were actually the result of a chain of collisions between atoms.

Later critics ridiculed Democritus's explanation for sensation because it attempted to reduce all aspects of perception and sensation to touch, contact, or interaction with the emanations allegedly given off by all objects. According to Democritus, all perceptible things are arrangements of atoms that differ only in size and shape. We assign certain qualities to these arrangements and call them color, taste, odor, texture, and so forth. But, he said, these qualities are not in the bodies themselves, they are the effects of the bodies on our sense organs. That is, qualities such as sweet and bitter, hot and cold, and even colors exist only by convention. Sight, for example, was the product of emanations given off by various objects that caused the air between the eye and the object to become contracted and marked. The visual image, therefore, did not originate in the pupil, but in the air admitted to the eye. The other senses were explained in terms of the effects of different sizes and shapes of atoms on the appropriate sense organs. Thus, a bitter taste might be the result of contact between the tongue and smooth, rounded atoms, while atoms with jagged edges might cause a salty taste.

Democritus taught that the human being was a world in miniature, a *microcosm*, reflecting the whole universe, the *macrocosm*. As a reflection of the macrocosm, the microcosm contained atoms of every kind. Thus, even thought, consciousness, sleep, sickness, and death could be explained in terms of the properties of atoms. The nature of respiration was of particular interest to Democritus because *pneuma*, or "vital air," composed of light, highly mobile soul-atoms, served as the vehicle of life. Soul-atoms tended to escape the body, especially during sleep, but they were replaced by other atoms until respiration ceased and the soul flowed out of the body. Although the very mobile spherical atoms that comprised the soul were diffused throughout the body, the soul-atoms were concentrated in the mind. Thought was the product of the motion that resulted from collisions between appropriate atoms. Tradition has it that Democritus tested his own sense impressions by withdrawing into solitary places such as caves and tombs.

In order to understand the nature of living things and sense perception, Democritus carried out systematic dissections of various animals. He seems to have established the practice later adopted by Aristotle of dividing animals into two major categories: *sanguiferous* (red blooded, essentially the vertebrates) and *bloodless* (invertebrates). Democritus pursued a great range of physiological studies and may have written works on human anatomy, regimen, prognosis, reproduction, embryology, fever, and respiratory diseases and may have developed a theory about the origin of epidemic disease. Unfortunately, most of his biological writings were lost and his views are known primarily through Aristotle's polemics against them. Although this surely must distort his reasoning, glimpses of his original insights remain. In contrast to Aristotle, Democritus called the brain the organ of thought; Aristotle believed that the role of the brain was to cool the blood. Democritus apparently described the heart as the organ of courage and ascribed sensuality to the liver. He also called attention to the mule and its peculiar sterility, a topic of considerable interest in the history of genetics. We are perhaps fortunate to have even a few fragments of the work of Democritus, for one tradition holds that Plato tried to burn all of his writings.

Democritus was a contemporary of Hippocrates (ca. 460 to 361 B.C.), and, according to ancient tradition, the two met in a professional capacity. Convinced that excessive interest in anatomical dissections and experiments on sense deprivation were indicative of madness rather than signs of scholarly eccentricity, neighbors of Democritus are said to have sent for Hippocrates. But after speaking to Democritus, Hippocrates told the people of Abdera that they were more likely to be insane than Democritus was. Not only was the philosopher in complete control of his senses, according to Hippocrates, he was the wisest man living. Some historians argue that Hippocrates was neither the author of the Hippocratic texts nor even a real person. Nevertheless, it is pleasant to imagine such a meeting, for atomic theory along with the rational medical theories of the Hippocratic school were among the most enduring aspects of early Greek philosophy.

Hippocratic medicine emphasized the patient rather than the disease, observation rather than theory, respect for facts and experience rather than philosophical systems. One of the most important of the Hippocratic aphorisms was "at least do no harm." Above all, Hippocratic medicine was a craft in which success in helping nature cure the patient was the ultimate measure of the physician. Philosophy was less important to medical practice than technique, but the Hippocratic school did develop a theory of health and disease based on the four elements and the four humors. Like the macrocosm, the microcosm that is the human being was composed of the four elements: earth, air, fire, and water. The four elements were related to the four bodily humors—black bile, blood, yellow bile, phlegm—and the four associated qualities—hot, cold, moist, dry. Individual variations in composition determined temperament—melancholy, sanguine, choleric, phlegmatic—as well as vulnerability to disease. When the four humors

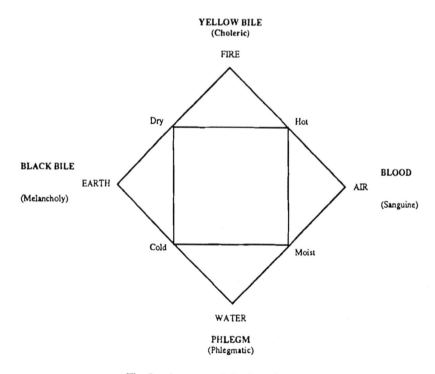

The four humors and the four elements

Hippocrates

were well balanced, a state of health existed. Deviations from perfect balance produced disease. For hundreds of years, humoral theory rationalized a therapeutic regimen designed to restore humoral equilibrium by bleeding, purging, and dietary management designed to remove putrid humors and prevent the formation of addition bad humors.

A modern physician might think it naive to reduce the complex phenomena of physiology and pathology to the balance of four humors, but humoral theory was enormously successful, eminently subtle, infinitely flexible, and capable of incorporating and resolving an otherwise bewildering variety of problems. For those who were skeptical about its theoretical rationale, physicians could provide justifications based on common observations. Phlegm could be related to mucus, yellow bile to the bitter fluid stored in the gallbladder, and black bile to the dark material sometimes found in vomit, urine, and feces.

Because the only clinical laboratory available to the physician until very recent times was the informed judgment of the five senses, the odor, and perhaps taste, of all the patient's excretions, secretions, and effluvia had to be tested by the earliest analytic chemists: the human nose and tongue. The linkage between theory and observation is also revealed by thinking about the appearance of clotted blood: the darkest part of the clot at the bottom of the collection vessel corresponds to black bile, the serum above the clot is apparently yellow bile, and the light material at the top is phlegm.

The Hippocratic conviction that humoral pathology could provide a natural explanation for mental as well as physical illnesses is expressed most forcefully in *On the Sacred Disease*, an account of the illness known as epilepsy, or the falling sickness. Rejecting all supernatural and magical accounts of the cause and treatment of the disease, Hippocrates suggested that epilepsy might be due to an abnormal accumulation of phlegm during gestation that injured the developing brain. Many philosophers thought that the heart was the seat of consciousness, but Hippocrates assigned that role to the brain. Afflictions of the brain, therefore, presented the most formidable challenge to the physician.

THE AGE OF SOCRATES, PLATO, AND ARISTOTLE

Greek philosophy entered into a new phase with the end of the Presocratic period. The changing climate of Greek thought is best appreciated from the vantage point of Athens, the turbulent scene of the next phase in the history of philosophy. While the work of Socrates (470–399 B.C.), Plato (429–347 B.C.), and Aristotle (384–322 B.C.) created the very core of Western philosophy, their effect on the history of science was almost equally profound.

According to Plato, philosophy treated only *physics* before Socrates added a second subject: *ethics*. Socrates was said to have rescued Greek thought from the abstract and materialistic concerns of his Ionian predecessors by bringing philosophy down from heaven to earth. His teacher, Archelaus of Athens, had been a pupil of Anaxagoras, and as a youth Socrates had studied the competing claims of the various natural philosophers, including Parmenides, who is said to have visited Athens in his old age.

But the mature Socrates is said to have warned his disciples that the study of science could drive men mad, as in the case of Anaxagoras. Presumably Socrates realized how dangerous the study of nature could be when he saw Anaxagoras banished from Athens for impiety. Unorthodox inquiries into ethics and politics were to prove no safer than speculations about nature. Accused of corrupting the youth of Athens, neglecting the official gods of the city, and introducing new and different gods, the 70-year-old philosopher was condemned to death for impiety. In responding to his prosecutors, Socrates argued that he was actually a public benefactor, who should be honored and supported by the state. Angered by the philosopher's apparent arrogance, the court of some 500 Athenian citizens reconfirmed the death sentence by an even larger majority. Although his friends urged him to escape, Socrates accepted the judgment of the court and drank the hemlock, the traditional potion of the condemned.

Primarily, the work of Socrates pertains to moral philosophy rather than science, but his ideas became critical components of both areas; Socrates taught that there was only one good, which is knowledge, and one evil, which is ignorance. He claimed that he had been given a divine mission to question the assumptions and beliefs of his fellow men. In carrying out his mission he professed to act the part of a midwife rather than a teacher, because by making others think, he helped them give birth to their thoughts. This image remains strikingly imaginative, even though it might reflect the fact that the philosopher's mother, Phaenarete, was a midwife. Socrates left no writings, but his doctrines survived in the work of Plato, who used Socrates as the protagonist in his own dialogues. A few additional insights into his ideas and personality are preserved in the stolid memoirs of Xenophon and the ribald comedies of Aristophanes.

Like Socrates, Plato's main concern was moral philosophy, encompassing ethics, politics, and theology. Nevertheless, Plato exerted a profound influence on the development of Western science and the characteristics and goals of scientific investigation. After the death of Socrates, Plato is said to have traveled widely to learn from other philosophers, mathematicians, and even the priests of Egypt. Returning to Athens, he founded the Academy in an olive grove on the outskirts of Athens. Here, surrounded by his pupils, he lectured, developed his own philosophical system, and composed accounts of the doctrines he attributed to Socrates. To physics and ethics as the components of philosophy, Plato added dialectics, and he taught his disciples that philosophy was a divine hunger for wisdom. Apparently, not all of his students were equally famished, for when Plato performed a reading of his dialogue *On the Soul*, only the faithful Aristotle is said to have stayed to the end.

The Academy provided the means by which Plato could begin his primary mission, the training of political philosophers who would further his vision of the ideal society. The task of the philosopher, according to Plato, was to elucidate the eternal values. Knowledge of these was reached not through direct observation of nature; the mind could only discover truth by the exercise of pure reason. Fundamentally, Plato instilled into philosophy a profound distrust of sensory experience. He taught that reason alone could reach beyond the imperfect shadow world of sensory perceptions into the world of transcendent eternal reality. Despite the considerable triumphs of empiricism, Plato's ideas continue to shape, direct, and challenge modern-day Platonists, that is, scientists and philosophers in search of the "grand design" and transcendent laws of the universe.

As revealed in the *Timaeus*, Plato's chief cosmological treatise, Plato's inquiries into nature were undertaken in support of his ethical precepts. In contrast to the cosmologies of the Ionian philosophers, Plato's account of the creation of the world is highly religious and teleological; it is an extended argument based on purposeful design in nature. Plato's world was created out of an original chaos by an eternal and perfect Creator who imposes order on disorderly matter in accordance with a rational, intelligent design.

In Plato's version of the creation, man was not derived from simpler creatures, but was the first being to appear. In this case, as in so many others, "man" definitely means man, not male and female human beings. Borrowing freely from the Pythagoreans, Alcmaeon, Empedocles, Hippocrates, and other predecessors, Plato presents a wealth of biological ideas and observations, but focuses almost exclusively on human beings. Accounts of the causes of disease, respiration, and human anatomy are used to explicate his all-pervasive teleological principle. Thus, the first part of man to be formed was the head because it was most nearly spherical and served as the organ of the immortal soul. Because of problems caused by the passive resistance of disorderly matter, the Creator apparently achieves the best possible results, rather than ideal designs. The covering of the head is a rather thin layer of bone, even though a thicker covering would offer more safety, because that would reduce sensitivity and intelligence.

According to Plato, the soul was tripartite, with the rational part located in the head, the heart the locus of the passionate part, and the liver the site of the appetitive part. To protect these vital organs, the lungs were designed to serve as a buffer against the pulsation of the heart and the spleen was designed to absorb impurities from the liver. In a remarkable reversal of modern evolutionary theory, Plato suggests that other animal species were created by degeneration from man. Even woman was a secondary production, being created from the souls of habitually cowardly and criminal men. With more wit than compassion for the predecessors from whom he had borrowed so much, Plato wrote that "harmless but light-witted men, who studied the heavens but in their simplicity supposed that the surest evidence in these matters is that of the eye" were turned into birds. Men who had no use for philosophy were transformed into four-legged beasts. Creatures that lived in the water were derived from the most foolish and stupid of men, that is, from philosophers like Anaximander who speculated about men evolving from fish-like ancestors. These souls were so corrupt that the gods found them unfit to breathe pure air. This was Plato's judgment for philosophers who tried to develop a naturalistic explanation of the universe, based, at least in part, on perceptions obtained through the senses. In the world of Plato, true reality is found only in the sphere of abstract thought. What is perceived by the senses is but an imperfect image of the eternal ideal, like shadows cast on a wall.

Perhaps the greatest gift Plato bestowed on science was his demand that philosophers take the study of nature very seriously because such studies were a means of revealing the intelligent design of the universe. Such a directive led to the search for the elegant mathematical relationships and abstract laws that lie beyond empirical observations. Eventually, the successes of this approach in astronomy and physics seemed to reflect badly on other fields and created what has been called the problem of "physics envy" among biologists. In terms of the immediate history of biology, the concept of species and genus and the beginnings of systematic taxonomy, which were established in the work of Aristotle, were inspired by Plato's search for the eternal ideal forms.

Aristotle, disciple of Plato and disappointed heir apparent, seems to have diverged more and more from the views of his mentor during a long and rather turbulent life. Just as Plato is the first of the philosophers whose writings have survived in bulk, Aristotle is the first great scientist to have this distinction. Many aspects of Aristotle's life provide insight into the complexities of the politics, customs, philosophy, and science of the Greek world in the fourth century B.C. Aristotle was born in Stagira, a town in Macedonia regarded as semi-barbaric by the more sophisticated, older Greek states. Macedonia, the largest of the Greek states at that time, was governed by Philip, father of Alexander the Great. Nicomachus, Aristotle's father, was a successful and wealthy physician at the royal court. As a youth, Aristotle must have learned the secrets of the Asclepiads (priest-physicians serving Asclepius the god of medicine), the theory and practice of Hippocratic medicine, and the advantages of royal patronage. Although a youth could be instructed in the rudiments of medicine in Macedonia, a properly educated physician also needed training in philosophy. Therefore, at 17 years of age Aristotle was sent to Plato's Academy, where he was immersed in a system of thought very different from that of the practicing physician, a system that valued the abstract over the practical. Apparently this new world suited Aristotle, for he remained at the Academy until the death of Plato, 20 years later.

During this time, Aristotle established himself as an independent philosopher and author, even expressing reservations about some aspects of his men-

Aristotle

tor's doctrines. Perhaps in response to such criticism, Plato appointed his nephew as his successor at the Academy instead of Aristotle. Embittered and disappointed, Aristotle abandoned the Academy, Athens, and mathematics. Biological researches, carried out mainly on the island of Lesbos, were his main interest during this period of his life. When a revolution deposed his patron, he fled to Macedonia where he served as tutor to Philip's son Alexander for 3 years. This was not the most successful student-teacher relationship in history, although it is one of the most famous. Certainly, Alexander was more interested in war and conquest than in philosophy, but he did remember his old tutor well enough to send back reports and specimens of peculiar animals found in exotic places during his triumphant military career.

In 334 B.C., at the age of 50, Aristotle returned to Athens and established his own school, called the Lyceum. Unfortunately, political turmoil soon ended this productive and relatively serene period. When Alexander died, Athens rebelled against Macedonian rule. Faced with an indictment for impiety, Aristotle fled the city to "prevent a fresh crime against philosophy." After a year spent in exile at Chalcis, the philosopher died. Although the bulk of Aristotle's works survived, there are gaps and uncertainties as to the order in which they were written. Some of the texts were apparently edited by his collaborators, and some might even be student notes reflecting a garbled version of what Aristotle really said and thought. A gift for organizing, arranging, and even appropriating other people's material and recognizing affinities enabled Aristotle to discuss almost every field of human knowledge. Very often all that is known of his predecessors comes from Aristotle's comments about them.

Aristotle's work falls into four major divisions: physics, ethics and politics, logic and metaphysics, and biology. It was Aristotle who formalized the earth-centered, closed, finite universe that dominated Western thought until the time of Nicolaus Copernicus (1473–1543). In clarifying difficult aspects of his views on ethics and politics, nature and society, Aristotle often used biological examples and analogies. While explicating his own principles of logic and metaphysics, Aristotle often called attention to deficiencies in the work of his predecessors, particularly Plato's theory of the ideal forms. Aristotle's biological writings reveal a remarkably acute observer of nature who, despite contemporary prejudice against manual labor, was willing to dirty his hands in dissections of even the lowliest creatures. Almost one quarter of Aristotle's surviving writings are essentially biological. The major biological works in this category include *Natural History of Animals*, *On the Parts of Animals*, and *On the Generation of Animals*.

Given the competing demands of abstract thought and the complexity of biological phenomena, it is not surprising that two Aristotles seem to emerge from his collected writings. One was the theorist and logician, who believed that the logical and the real were likely to be identical. The second Aristotle was the observer, who understood that he was working in the still uncharted territory of biological phenomena and advised his followers to rely on their own observations rather than his words. The extent to which Aristotle was influenced by the differing claims of abstract philosophy and scholarly medical tradition is unclear. But it should be remembered that long before Plato established a preference for pure thought over the evidence of the senses, the author of *Ancient Medicine* had argued that the only way to obtain clear knowledge about the nature of human beings is through medicine, not philosophy. Although not an unthinking empiricist, the Hippocratic physician could not rely on abstract theory. The successful physician had to confront the evidence of his senses in order to evaluate the symptoms presented by the each patient. In contrast, the physicist could search for the laws that govern the motion of all falling objects without considering their individuality. But in biology, as in medicine, close attention to details is essential.

Aristotle's biological investigations must have been very enjoyable and satisfying to him, but in terms of generating theories they were subordinate to his general philosophical system. Thus, although his descriptions of animals, even those as small and delicate as the eggs and embryos of fish and mollusks, are remarkably accurate, sometimes his powers of observation were distorted by his preconceived notions about the way things should be. As Einstein said, "It is the theory that decides what we can observe." Dissection is not invariably an instrument of understanding because function is not obvious in structure. For example, Aristotle believed that the function of the brain was to secrete mucus and cool the blood by balancing the heat of the heart, which served as the seat of the soul and the intellect. When the heat of the heart overpowers the cooling power of the brain, the blood is hot and man becomes passionate and excitable; when the brain predominates, the cooling of the blood results in sleep. A balanced state allows for rational life. The brain must be composed of earth and water to perform its cooling function because the element earth is cold and dry, while the element water is cold and wet. Aristotle used the vague metaphor of cooking or coction to explain nutrition as well as embryonic development. Some foods were cooked in the stomach and produced food vapors that ascended to the heart. This organ transformed them into blood and, through the movement of the blood, nutriments were transferred to the body and directly assimilated.

Both Plato and Aristotle could agree that the highest faculty of man was reason and the most sublime human activity was contemplation, but Aristotle also attached great significance to the collection of facts and data and the observation of general phenomena and particular examples. Aristotle took Plato's highly abstract theory of the Ideas or Forms that existed only in the world of pure thought and transformed it into a guide to the study of nature. Deliberate and systematic inquiries, Aristotle insisted, would reveal the causes of things because the Ideas or Forms exist in nature and not apart from it. Thus, although

Aristotle's world is certainly not one without a god, his inquiries into biological phenomena do not require the mythic apparatus that is at the core of Plato's *Timaeus*. Aristotle considered Plato's development of teleology (argument from design) an important innovation having its roots in the work of Anaxagoras and Empedocles. Aristotle's use of teleology, unlike that of Plato, does not rely on the intentional actions of a creator god. Still, Aristotle said in various passages that "Nature does nothing in vain" and "In all its workings Nature acts rationally."

Thinking about the *why* of things, that is, thinking about the primary cause of things and the cause of all kinds of physical changes, was, Aristotle asserted, the beginning of philosophy. His predecessors, from Thales to Plato, had speculated about the causes of events in nature in vague and often erroneous ways. Aristotelian philosophy incorporated four causes of changes in nature: the material, formal, efficient, and final causes. In other words, in order to explain an object or event, we must describe four essential factors: the matter of which it is made; the form or shape in which it exists; the moving or dynamic cause that is the active agent; and the final cause, which is its purpose or goal. In common modern usage only the efficient cause, and sometimes the final cause, are generally thought of as causes. Actually, Aristotle's "causes" are conditions necessary to account for the existence of a thing or event; he does not use the term cause in the sense of that which is both necessary and sufficient to produce a specific effect. His causes, therefore, include form and matter and can be applied to a description of the fabrication of a table by a carpenter as well as to the generation of a new human being.

By recognizing the value of studying organisms in and of themselves, Aristotle initiated an approach to philosophy and biology of great influence and fruitfulness. Moreover, the scope of his own biological research was so extensive that an account of the history of almost any branch of the life sciences must begin with Aristotle. The sources of Aristotle's biological knowledge included the works of his predecessors, information obtained from farmers, hunters, fishermen, and his own observations and dissections. The true philosopher, according to Aristotle, would find marvels in every realm of nature and would not recoil with "childish aversion from the examination of the humbler animals." The true philosopher would pursue the study of every kind of creature "without distaste," because each and every animal would reveal "something natural and something beautiful."

History of Animals, Parts of Animals, and *Generation of Animals* raised fundamental questions about reproduction, the nature of species, and relationships among organisms. Aristotle assumed that each species had its own nature, or essence. His species were those that had common names in Greek and whose nature could be discovered by investigating members of the species. Some scholars have attempted to recreate strict zoological hierarchies or construct a scale of nature from passages in Aristotle's work. But Aristotle recognized the difficulties entailed in establishing a complete classification system, especially when the information available to him was so limited; only about 500 species of animals are cited in Aristotle's works, and many of them seem to be subdivisions of common types. His own investigations probably included dissections of about 50 kinds of animals.

Broadly equivalent to genus and species, the terms *genos* and *eidos*, the only taxonomic categories Aristotle used, are the subject of much scholarly debate. Genus refers to family or kinship, and species, which comes from the Greek for "form," was used to deal with the common names of things. Many so-called primitive peoples could distinguish many more plant and animal species than Aristotle, but this does not mean that Aristotle was an incompetent classifier. Rather, it reflects that fact that he was a philosopher outlining methods and systems, not a taxonomist actually constructing a detailed system. Aristotle's goal was to find general principles that would allow scientists to organize many diverse organisms. He suggested the use of "essential traits" to form natural groups. It is the species rather than the individual that exhibits the essence that is eternal. Therefore, the philosopher must study the nature of the species and its specific characteristics. In other words, the philosopher must study the nature of man, rather than the accidents that make up a particular man such as Socrates or Aristotle.

A natural scheme of classification is one that gives attention to many characteristics and weighs them to determine the natural affinities beneath the variety of structures and functions. For example, it would be absurd to use habitat or means of locomotion alone to order animals. Natural affinities must be determined by investigating many traits, including structure, behavior, habitat, means of locomotion, and reproduction. Like Democritus, Aristotle began by dividing animals into groups with and without red blood, but he concluded that a more refined hierarchy could be constructed by arranging animals according to means of reproduction and level of development of birth. The group that we call mammals was at the top because the members of this group are warm and moist, not earthly. Their young were born alive and complete and were nursed by the females of the species. Next were creatures that were less warm, but still moist. These animals were born alive, but developed from eggs that grew inside the body of the mother. Lower down the scale were creatures like birds and reptiles that were warm and dry and laid complete eggs. Lower still were the cold and earthly animals like frogs that produced incomplete eggs. The lowest of all sexually reproducing forms were worms that produced eggs. Below these were the vermin that came into being by spontaneous generation. The quality of the soul or souls determined the habits and functions, as well as the structure and perfection, of each species. Plants have only a vegetative soul for growth and reproduction. In addition to the vegetative soul, animals have a sensitive soul for movement and sensation, and man has, in addition, a rational soul, whose seat is in the heart.

Three kinds of generation were known to Aristotle. Not surprisingly, sexual reproduction is due to the occurrence of the two sexes. In Aristotle's scheme the male is more complete and warmer than the female. The higher animals and some of the insects were produced by sexual reproduction. Animals such as the starfish, worms, and shellfish arose through asexual reproduction. Spontaneous generation routinely produced fleas, mosquitoes, and other vermin.

Plato found a solution to Parmenides' challenge concerning being, nonbeing, and change in his theory of the eternal and unchanging forms. The phenomenal world was a place of temporary associations and attributes. Aristotle objected to this solution, primarily on the grounds that it was obvious that the phenomenal world was composed of *things*, including men, animals, and other objects. For Aristotle, the *eidos* was the true unchanging essence, but it existed in and through the individuals residing in the phenomenal world. Aristotle's solution to the challenge of Parmenides was based on his persistent use of the distinction between *potentiality* and *actuality* for each variety of change. Perhaps the most intriguing aspects of the challenge of change, potentiality, and actuality are those raised by reproduction and the development of new organisms.

How could philosophy and science explain the generation of like by like, that is, the production of an animal by an animal, or a plant by a plant? Aristotle's *Generation of Animals* applies his concepts of actuality and potentiality, matter and form, to one of the great problems of biology and philosophy: understanding how a new life begins and develops. According to Aristotle, the embryo is not a preformed entity that expands during development, but a potentiality that is imposed on matter and expressed over time.

Aristotle believed that man is the most rational animal and therefore all other animals are inferior. When Aristotle described characteristics that are natural to particular species but different from those of human beings, such as lack of external ears in seals, he refers to them as deformities. Similarly, Aristotle asserts that the human female is a deformed male, even though the female is apparently necessary for the continued existence of the species. In the process of the generation of a new individual, the male provides the form principle (the moving cause of generation and development) via the semen; the female contribution is passive matter in the form of the menstrual blood. To explain the essential but subtle contribution of the male to generation, Aristotle offered the famous carpenter analogy: wood is passive and cannot transform itself into some useful entity, but the action of the carpenter can transform it without becoming part of it. That is, form is present in the soul of the carpenter. At first, this appears to be a compelling analogy, but analogies are often used to divert attention from the absence of a real explanation. Aristotle argued that during pregnancy the blood that was normally lost in menstruation was used to increase the size of the embryo. The Aristotelian theory of generation had the major virtue of accounting for the known facts, accommodating the prejudices of his time, and the minor defect of being completely wrong. Aristotle noted that men generally lived longer than women did, and that women who had many pregnancies had shorter life spans. Although Aristotle's form principle has been compared to DNA as a blueprint for development, his theory neglected the contribution of the female parent, a contribution that some of his predecessors had already acknowledged.

Although Aristotle had many disciples, he completely overshadows them. His ideas dominated biology as well as other aspects of human thought for generations. Although Aristotle's own approach to the study of nature called for close and accurate observations and the systematic collection of evidence, this aspect of his philosophy was generally neglected. Later scholars tended to become slavish followers of the words, rather than the spirit, of Aristotle. The disciples of Aristotle tended to take more modest portions of all possible knowledge for their investigations. Theophrastus (ca. 373 to ca. 285 B.C.), Aristotle's favorite pupil, friend, and chief assistant, and Strato (ca. 340 to ca. 270 B.C.) are the only followers of Aristotle at the Lyceum with a real claim to a substantive place in the history of science.

Inevitably overshadowed by his teacher, Theophrastus was a prodigious worker, voluminous writer, as well as an independent and original thinker, but only a fraction of his writings survived. *On the Causes of Plants* and *On the History of Plants* described the morphology, classification, and natural history of plants. In addition to his major studies of botany, Theophrastus also wrote about broader problems in metaphysics, biology, medicine, and the doctrine of the four elements. He was one of the first scientists to write a treatise on the classification of rocks and minerals. Some historians have found the fundamental concepts of ecology in his studies of the relationships between living things and their environment.

Theophrastus seems to have objected to the preoccupation of natural philosophers with the study of metaphysics (the study of first principles), the reckless use of teleological explanations, and their tendency to denigrate the direct study of nature. If natural phenomena were observed more closely, he argued, it would be easy to demonstrate that nature does not always act with universality of purpose and intelligent design. Theophrastus realized that the teleological explanation was often an oversimplification that masked and disguised our ignorance. Arguments from design provided instant answers to all questions but militated against asking hard questions and answering them in a meaningful way. Scientific inquiry into the universe, Theophrastus argued, required setting limits to the search for final causes and focusing on "the conditions of existence of real things and their relations with one another."

Theophrastus described philosophers as true citizens of the world, who could survive political upheavals and obtain employment in any realm. Strato, who abandoned the study of ethics for researches in pneumatics, apparently absorbed valuable lessons in survival and patronage as well as natural philosophy from his mentor and became the well-paid tutor of Ptolemy II of Alexandria. Strato was primarily interested in pneumatics and carried out many ingenious experiments to test the nature of air, motion, and the void. Unfortunately, with the exception of fragments apparently appropriated by later scientists, all of his writings were lost. Thus, some of his views on perception and the nature of the senses have been reconstructed. Unlike Plato, Strato believed that true knowledge was the result of experience. He rejected Plato's view of true knowledge as something possessed only by the soul and independent of experience. Through investigations of the relationship between the senses and the mind, Strato concluded that it is in the mind that a stimulus generated from outside the body is transformed into a sensation. According to Strato, perception and thought were the province of the same soul. Since animals have sense organs like those of human beings, they must also have minds. Unlike the Aristotelian view that animals are degenerate forms of man, Strato regarded man as merely a better kind of animal.

Strato succeeded Theophrastus as the head of the Lyceum in 287 B.C., and even though he realized that the institution was declining along with the numbers of students and the quality of their work, he retained that position until 269 B.C. By that time it was clear that the city of Athens was being eclipsed by Alexandria in Egypt as a vital center of intellectual activity.

While Aristotelian philosophy has always been an important component of Western thought, the grip of Aristotelianism has waxed and waned over the centuries since his death. After the Renaissance, the Reformation, and the Scientific Revolution, alternative ancient philosophies, such as those linked to the Stoics and Epicureans, and others known only through fragments preserved by Aristotle and his followers, attracted the attention of European intellectuals, philosophers, theologians, and scientists. In other words, the history of Western intellectual life can be seen in terms of the interaction of the whole heritage of ancient Greek thought with the body of thought characteristic of each succeeding era. The ideas associated with the ancient Greeks were so creative, intriguing, and fertile that they engaged the attention of early modern scientists. They remain challenging and fruitful today.

Stoicism and Epicureanism arose in late fourth century B.C. Athens. The founders of these schools, Epicurus (341–271 B.C.) and Zeno of Citium (334–262 B.C.) developed their ideas in reaction to the prevailing schools of Plato and Aristotle. Both Epicurus and Zeon were primarily concerned with ethical issues and insisted that understanding ethics required an understanding of the nature of the universe and human beings. Only a few fragments of their writings have survived, but the writings of their followers preserved and transmitted their basic concepts. Both the Stoics and the Epicureans believed that the good life led to a state of mental tranquility. Like the atomists, the Stoics and Epicureans expounded a philosophy of nature in which all natural phenomena, and even the human soul, were explained in terms of chance collisions among atoms moving through the void. The gods were not totally banished from their world picture but were made far enough removed from human affairs that fear of the gods and punishment after death would no longer disturb humans beings who accepted philosophy.

SUGGESTED READINGS

Africa, T. W. (1968). *Science and the State in Greece and Rome.* New York: Wiley.

Anton, J. P., ed. (1980). *Science and the Sciences in Plato.* New York: EIDOS.

Bag, A. K. (1985). *Science and Civilization in India.* Vol. 1: *Harappan Period.* New Delhi: Navrang.

Barnes, J. (1986). *The Presocratic Philosophers.* New York: Methuen.

Bodde, D. (1991). *Chinese Thought, Society, and Science: The Intellectual and Social Background of Science and Technology in Pre-modern China.* Honolulu: University of Hawaii Press.

Boylan, M. (1983). *Method and Practice in Aristotle's Biology.* Lanham, MD: University Press of America.

Cohen, M. N., and Armelagos, G. J., eds. (1983). *Paleopathology and the Origins of Agriculture.* New York: Academic Press.

Cohen, M. R., and Drabkin, I. E., eds. (1958). *A Source Book in Greek Science,* 2nd ed. Cambridge, MA: Harvard University Press.

Diogenes Laertius. (1969). *Lives of the Philosophers.* Trans. and ed. by A. R. Caponigri. Chicago: Regnery.

Farrington, B. (1961). *Greek Science, Its Meaning for Us.* New York: Penguin.

Ferejohn, M. T. (1991). *The Origins of Aristotelian Science.* New Haven, CT: Yale University Press.

Fortenbaugh, W. W., Huby, P. M., and Long, A. A., eds. (1985). *Theophrastus of Erusus: On His Life and Work.* New Brunswick, NJ: Rutgers University Press.

Guthrie, W. K. C. (1962–69). *A History of Greek Philosophy.* 6 vols. Cambridge: Cambridge University Press.

Huff, T. E. (1993). *The Rise of Early Modern Science: Islam, China, and the West.* New York: Cambridge University Press.

Hughes, J. D. (1975). *Ecology in Ancient Civilizations.* Albuquerque, NM: University of New Mexico Press.

Kirk, G. S., Raven, J. E., and Schofield, M. (1983). *The Presocratic Philosophers: A Critical History with a Selection of Texts,* 2nd ed. Cambridge: Cambridge University Press.

Lloyd, G. E. R. (1989). *The Revolutions of Wisdom: Studies in the Calims and Practice of Ancient Greek Science.* Berkeley, CA: University of California Press.

McKirahan, R. D. (1994). *Philosophy Before Socrates: An Introduction with Texts and Commentary*. Indianapolis, IN: Hackett Pub. Co.

Needham, J. (1954–2002). *Science and Civilization in China*. 7 volumes. Cambridge, England: Cambridge University Press.

O'Brien, D. (1969). *Empedocles' Cosmic Cycle: A Reconstruction from the Fragments and Secondary Sources*. London: Cambridge University Press.

Osler, M. J., ed. (1991). *Atoms, Pneuma, and Tranquillity. Epicurean and Stoic Themes in European Thought*. Cambridge: Cambridge University Press.

Plato (2000). *Timaeus*. Translated, with introduction, by Donald J. Zeyl. Indianapolis, IN: Hackett Pub. Co.

Rahman, A., ed. (1999). *History of Indian Science, Technology, and Culture, A.D. 1000–1800*. New Delhi: Oxford University Press.

Saggs, H. W. F. (1989). *Civilization Before Greece and Rome*. New Haven, CT: Yale University Press.

Sallares, R. (1991). *The Ecology of the Ancient Greek World*. Ithaca, NY: Cornell University Press.

Selin, H. (1992). *Science Across Cultures: An Annotated Bibliography of Books on Non-Western Science, Technology, and Medicine*. New York : Garland.

Selin, H., ed. (1997). *Encyclopaedia of the History of Science, Technology, and Medicine in Non-Western Cultures*. Boston: Kluwer Academic.

Woodbridge, F. J. E. (1965). *Aristotle's Vision of Nature*. New York: Columbia University Press.

2

THE GREEK LEGACY

THE ALEXANDRIAN ERA

As the vitality of the Lyceum declined, a new center of scholarship was established by the Ptolemies of Alexandria, heirs to the conquests of Alexander the Great (356–323 B.C.), the ruler of Macedonia who had once been Aristotle's student. The ascent of Macedonia, once a poor and backward province, began in earnest during the reign of Alexander's father Philip (393–336 B.C.). After conquering his barbarian neighbors and the city-states of Greece, Philip was killed by an assassin. Alexander, who was only 20 when his father died, efficiently disposed of rival claimants to the throne, defeated the Persians, conquered Egypt and Babylonia, and invaded India. Alexander's military forces were followed by a little army of scholars and historians, botanists and geographers, engineers and surveyors. His campaigns thus produced a wealth of information on the resources, natural history, and geography of the conquered areas. Before reaching his thirty-third birthday, Alexander fell ill with a fever and died. Immediately after his death his generals struggled for control of the army and the conquered territories. Ultimately, the Empire was divided among three successors: Antigonus Cyclops (382–301 B.C.) took Macedonia, Greece, and parts of Europe; Seleucus Nicator (356–281 B.C.) was awarded the Asiatic region; and Ptolemy Soter (367–283 B.C.) ruled over Egypt. The dynasty of the Ptolemies lasted for about 250 years and had the most lasting effect on the history of the sciences.

Alexander the Great and his physician

Ptolemy Soter, like Alexander, had studied with Aristotle. Ptolemy arranged for Aristotle's disciple Strato to tutor his son Philadelphus, who became Ptolemy II. The city of Alexandria in Egypt, symbolized by the lighthouse that was one of the seven wonders of the world, became the intellectual capital of the world. Scholars, poets, and philosophers came from Athens, as well as distant parts of the Empire. The confluence of cultures provided apparently created an intellectual climate that resulted in a burst of scientific and technical innovations.

The Ptolemies founded and supported the development of a magnificent Library and House of Muses, or Museum, a term that originally applied to a cult center dedicated to the cultivation and worship of the Muses, the nine goddesses who presided over literature and the arts and sciences. The Muses had been associated with philosophical studies since the time of Pythagoras. The evolution of the Museum in Alexandria into something like the ancient equivalent of a modern research university was, however, unique. Among the benefits that allegedly accrued to the scholars at the Museum were free meals and a tax-exempt salary generous enough to make the Greek poet Theocritus call Ptolemy "the best paymaster a free man can have." While the Museum was nominally under the direction of a high priest, its goals were more like those of a research institute. As a place for teaching, the Museum was modeled after the Lyceum, but it dwarfed that institution in terms of the size, if not the quality, of its faculty and student body. According to traditions based on later descriptions, the facilities and surroundings were magnificent, including a promenade, rooms for research, lectures, conferences, private and group study, an observatory, zoo, botanical garden, dissecting rooms, and a large dining room. Unfortunately, no contemporary accounts of the famous Museum have survived, and the evidence concerning the medical research presumably conducted at the Museum is ambiguous.

The Museum apparently served as an efficient and orderly way of producing engineers, physicians, geographers, astronomers, and mathematicians. Scientists and technicians were needed by the state, which was, after all, a Greek enclave in the ancient Egyptian world. Illustrative of the practical emphasis of Alexandrian research are many engineering advances in hydraulics and pneumatics and inventions such as the valve, pump, and screw. Most Alexandrian inventions, however, seem to have remained at the level of toys, except for weaponry, temple magic, and, to a lesser extent, medical instruments. The Ptolemies were, not surprisingly, interested in using religion, scholarship, science, and technology to enhance their power and prestige. Although Strato had boasted that he did not need the help of the gods to create a world, the science of pneumatics which he helped create allowed the rulers of Alexandria to make science the servant of the gods. By combining religious mysteries with the very practical side of pneumatic science, as developed by Hero of Alexandria, the temple priests were able to perform miracles on demand. Later writers satirized the Museum as a place where large numbers of scholars were kept like birds in a henhouse, endlessly squawking and bickering. In later Roman times, however, the Alexandrian Museum was essentially a secular teaching establishment whose fame and prestige inspired the development of similar centers.

Libraries were not unknown in Greece before the age of Alexander, but given the problems of producing books, large collections were rare. The Alexandrian Library may have housed half a million manuscript rolls. The Library seems to have exerted a major influence on the direction of research and teaching. Even the concept of the textbook can be traced to scholarship at the Library, where experts systematized their knowledge in a standardized format that surveyed general principles, well-established concepts, and ideas at the frontiers of research. In discussing the authenticity of various works in the Hippocratic collection, Galen of Pergamum (130–200) claimed that the rulers of Alexandria were so eager to enhance the Library that they confiscated all books found on ships entering the harbor. According to Galen, many forgeries were created to satisfy this passion for book collecting.

Clearly, the Alexandrian Era was one in which science and scholarship enjoyed unprecedented support. Yet little is known about the institutions and individuals who contributed to the intellectual life of this brief but brilliant period. Very little of the work carried out at the Library and Museum survived the destruction of the Library and the loss of its unique manuscripts. Even the exact sequence of events that led to the final destruction of the Library is unknown. During the reign of Cleopatra, the last of the Ptolemies, the Library is said to have contained 700,000 volumes. When Julius Caesar and his army visited Alexandria in 48 B.C., Cleopatra gave him many precious texts as gifts. During the riots against the Romans, fires destroyed tens of thousands of manuscripts. Later losses have been attributed to mob violence unleashed by early Christian leaders intent on destroying pagan institutions. According to traditional accounts of the lives of martyrs, the last scholar at the Museum was a woman mathematician named Hypatia, who was beaten to death in 415 by a frenzied Christian mob. Further attacks on the ancient pagan centers of learning were attributed to the Arab forces that conquered the city in 646 A.D.

THE ANATOMICAL RESEARCH OF HEROPHILUS AND ERASISTRATUS

Many sciences flourished at Alexandria, although research was primarily oriented towards specialized sciences and their practical applications. Attempts to encourage progress in medicine led to unprecedented advances in anatomical studies that, for the first time in history, included systematic dissection of human bodies. Perhaps the Egyptian tradition of cutting open the body and removing certain organs as part of the embalming ritual helped overcome traditional Greek antipathy to the mutilation of corpses. The ancient Egyptian practice of mummification was performed for religious reasons, and the embalmers were neither physicians nor scientists.

Despite the paucity of reliable primary sources, there is good evidence that sophisticated anatomical studies were carried out during the Alexandrian Era by at least two famous—or infamous—scientists: Herophilus (ca. 330/320–260/250 B.C.) and Erasistratus (ca. 310–250 B.C.). The exact relationship of these two anatomists to Alexandria and to each other is uncertain. When Christian theologians such as Tertullian and St. Augustine wanted evidence of the heinous acts committed by the pagans, they pointed to the notorious Herophilus and accused him of torturing 600 human beings to death. The Roman encyclopedist Celsus charged Herophilus and Erasistratus with performing vivisection on condemned criminals awarded to them by the rulers of Alexandria. While these accusers were obviously not eyewitnesses, some historians have accepted their allegations; others remain skeptical. The extent or existence of the practice of human vivisection during the Alexandrian Era remains controversial because the writings of Herophilus and Erasistratus have not survived and are known only through the diatribes of their enemies.

There seems to be little doubt that the Ptolemies gave anatomists access to the bodies of condemned criminals and that Herophilus and Erasistratus conducted many dissections. Performing public anatomies would have offended prevailing sentiments and perhaps gave rise to charges of even greater cruelties. Physician-scientists might also have used their patients and apprentices as experimental subjects in therapeutic research. Whatever conditions made the work of Herophilus and Erasistratus possible did not last long. Human dissection was probably discontinued by the end of the second century B.C. Some scholars, however, believe that human dissection continued until the first century A.D. In any case, anatomists might have used human skeletons for research and teaching even after they were forced to confine dissection and vivisection to other animal species.

Those who argue that it is inconceivable that Greek physicians would have performed human vivisections should reconsider the not uncommon references to the mutilation of bodies in Greek literature, as well as the public torture of slaves in law courts to extract evidence and the use of convicts in tests of poisons and their putative antidotes. Of course, the evidence from well-documented twentieth-century history suggests that there is no limit to possible human cruelty—individual or state organized. Herophilus, Erasistratus, and their contemporaries were engaged in constructing an entirely new science of human beings rather than an abstract philosophical system. The argument that Herophilus would not have made certain errors if he had done vivisections fails to allow for the influence of preexisting concepts on perception and interpretation and the intrinsic difficulty of novel anatomical studies.

Except for the possibility that Herophilus studied with Praxagoras of Cos, little is known of his life, although he is said to have come from Chalcedon. Among his students, according to a story that is often dismissed as a myth, was an Athenian woman named Agnodice, a female physician who disguised herself as a man in order to practice medicine. If the story of Agnodice really applies to Herophilus, it suggests that he must have spent some time in Athens. There is little doubt that Herophilus practiced medicine in Alexandria, but whether or not Herophilus performed his dissections and research at the Museum or received financial support from the state is unclear.

Later writers attributed many texts to Herophilus, including *On Anatomy*, *On the Eyes*, and a handbook for midwives. Some of the books written by Herophilus seem to have been in use for several centuries, but only a few fragments from his texts have survived. He apparently had a good overall picture of the nervous system, including the connection between the brain, spinal cord, and nerves. Rejecting Aristotle's claim that the heart was the seat of intelligence, Herophilus argued that the brain was the center of the nervous system. He also described the digestive system, called attention to the variability in the shape of the liver, and differentiated between tendons and nerves.

In his investigation of the circulatory system, Herophilus noted the difference between the arteries, with their strong pulsating walls, and the veins, with their relatively weak walls. Contrary to the prevailing assumption that the veins carried blood and the arteries carried air, Herophilus stated that both kinds of vessels carried blood. Intrigued by changes in the pulse that correlated with health and disease, Herophilus tried to measure the beat of the pulse using a water clock that had been developed in Alexandria.

Apparently known as a chronic skeptic, Herophilus regarded all physiological and pathological theories as hypothetical and provisional, including Hippocratic humoralism. While he probably did not totally reject humoral pathology, he seems to have preferred a theory of four life-guiding faculties that governed the body: a nourishing faculty in the liver and digestive organs, a warming power in the heart, a sensitive or perceptive faculty in the nerves, and a rational force in the brain. In clinical practice, Herophilus seems to have favored more active intervention than that recommended by Hippocrates, and he may have used the concept of hot-cold, moist-dry qualities as a guide to therapeutic decisions. Vigorous bloodletting and a system of complex pharmaceuticals became associated with Herophilean medicine, but the anatomist appears to have urged his students to familiarize themselves with dietetics, medicine, surgery, and obstetrics.

When scholars of the Renaissance sought to challenge the stifling authority of the ancients, especially that of Galen, the long neglected and much vilified Herophilus was lauded as the "Vesalius of antiquity." The title could also have applied to Erasistratus, another intriguing and rather shadowy figure, who was attacked by Galen for the unforgivable heresy of rejecting the Hippocratic philosophy of medicine. Galen wrote two books against Erasistratus and criticized his ideas whenever possible. The relationship between Erasistratus and Alexandrian anatomy is obscure. Galen claimed that Erasistratus and Herophilus were contemporaries, but Erasistratus may have been at least 30 years younger than Herophilus. Little is known of his life, except the tradition that having diagnosed his own illness as an incurable cancer, he committed suicide rather than suffer inexorable decline. He may have authored more than 50 books, but none of them has survived.

Some sources suggest that Erasistratus was a student or disciple of Strato, or even the nephew of Aristotle. Erasistratus was probably active in Antioch about 293 B.C. as physician to the court of Seleucus Nicator. If this is true, then human dissection must have been, at least briefly, accepted in both areas. There are striking similarities in the anatomical studies attributed to Herophilus and Erasistratus, but Erasistratus seems to have attempted to apply Alexandrian mechanics to his medical theories and to have been more concerned with physiology and experimentation than Herophilus. Both anatomists were associated with human dissection and vivisection and many anatomical and physiological discoveries. Both were interested in respiration, reproductive physiology, ophthalmology, the brain and nervous system, the heart and vascular system, the pulse, and fevers. In therapeutics, both generally favored treatment by opposites.

Galen accused Erasistratus of rejecting the Hippocratic philosophy of medicine and following the teachings of Aristotle. Like Herophilus, Erasistratus probably tried to replace humoral theory with a new doctrine. In the case of Erasistratus, this seems to have developed into a pathology of solids that perhaps did more to guide his anatomical research than his approach to therapeutics. Erasistratus is said to have been a gifted practitioner who rejected the idea that a general knowledge of the body and its functioning in health was necessary to the practicing physician. Many problems, he argued, could be prevented or treated with simple remedies and hygienic living. Nevertheless, he believed in studying pathological anatomy as a key to localized causes of disease. He was particularly interested in the possibility that disease and inflammation were caused by the accumulation of a localized plethora of blood that forced blood to pass from the veins into the arteries. At the time, naturalists believed that veins normally carried blood and arteries carried air. A local excess of blood formed from undigested food could damage surrounding tissues because blood in the overloaded veins could spill over into the arteries. When this occurred, the flow of pneuma, or vital spirit, which was supposed to be distributed by the arteries, would be obstructed.

One remarkable aspect of Erasistratus's theory was the assumption that invisible connections existed between arteries and veins. The direction of flow was, however, from veins to arteries, and the phenomenon occurred only in pathological states. If a living animal was injured so that an artery was cut, the air escaped and pulled some blood after it. Nature's horror of vacuums, as taught by the Aristotelians, was thus incorporated into the physiology of Erasistratus. The veins, which supposedly arose from the liver, and the arteries, which were thought to arise from the heart, were generally thought of as independent systems of dead-end canals through which blood and pneuma seeped slowly to the periphery of the body so that each part of the body could draw out its proper nourishment. Observations of engorged veins and collapsed arteries in the cadaver would appear to support these ideas. It logically followed that the goal of therapy was to reduce the plethora of blood. One way to accomplish this was to interrupt the production of blood at its point of origin by eliminating the supply of food. In addition to general starvation, a form of localized starvation could be accomplished by trapping blood in other parts of the body by tying ligatures around the roots of the limbs until the sick part had used up its plethora. The use of the ligature to stop the flow of blood from torn vessels was ascribed to Erasistratus.

Perhaps because of his theory that disease was caused by a local excess of blood, Erasistratus paid particular attention to the heart, veins, and arteries.

In his lost treatises, he apparently gave a detailed description of the heart, including the semilunar, tricuspid, and bicuspid valves. Mechanical analogies, dissections, and perhaps vivisection experiments apparently suggested to Erasistratus that the heart could be seen as a pump in which certain "membranes" served as the flap valves. Using a combination of logic, intuition, and imagination, Erasistratus traced the veins, arteries, and nerves to the finest subdivisions visible to the naked eye and speculated about further subdivisions beyond the limits of vision. He also gave a detailed description of the liver and gallbladder and initiated a study of the lacteals that was not improved upon until the work of Gasparo Aselli (1581–1626).

Although Erasistratus was sometimes called a materialist, atomist, or rationalist, he did not reject the concept of animating spirits. Apparently he believed that life processes were dependent on blood and pneuma, which was constantly replenished by respiration. Two kinds of pneuma were found in the body: the vital pneuma was carried in the arteries and regulated vegetative processes. Some of the vital pneuma got to the brain and was changed into animal spirits. Animal spirits were responsible for movement and sensation and were carried by the nerves, a system of hollow tubes. When animal spirits rushed into muscles, they caused distension that resulted in shortening of the muscle and thus movement. Perhaps inspired by Strato's experimental methods, Erasistratus is said to have attempted to provide quantitative solutions for physiological problems. In one experiment Erasistratus put a bird into a pot and kept a record of the weight of the bird and its excrement. He found a progressive weight loss between feedings, which led him to conclude that some invisible emanation was lost by vital processes.

The Alexandrian age provided unprecedented opportunities for scholarship and scientific research. The Library of Alexandria created the basic tools of scholarship: a storehouse of knowledge, a workplace for the duplication of manuscripts, a tradition of textual criticism, and reverence for the written word as a gift from the past and a legacy for the future. The Museum and Library flourished for barely two centuries, an oasis in the intellectual wastelands of much of Western history.

The later Ptolemies were not scholarly, enlightened, or wise and seem to have become increasingly divorced from Greek intellectual traditions. Greeks and foreign scholars were increasingly unwelcome in Alexandria, and many of them left to establish new centers of learning in other areas. Much of the deterioration can be blamed on political forces, but scientists and scholars also contributed to the uneasy intellectual climate. Fierce rivalries separated different schools of thought, with members of various factions impugning the work, methods, conclusions, and character of members of other groups. The tension that always exists in medicine between disinterested scientific research and the immediate needs of the sick grew and disrupted the ancient search for harmony and balance between the scientist and the healer. Critics of anatomical research charged that such pursuits distracted physicians from caring for patients. Finally, the worst of fates fell upon the Alexandrian scientists; they were persecuted, their grants were cut off, and they had to turn to teaching to eke out a living in new places. After Caesar conquered Egypt, Alexandria was reduced to the status of a provincial town in the great Roman Empire and the Museum and Library were fully eclipsed.

THE ROMAN WORLD

One of the major accomplishments of the Hellenistic era was the extensive dissemination of Greek culture. Nevertheless, despite the continuing influence of Greek literature, art, philosophy, and science, the Romans became the undisputed masters of political and military power. Like the Greeks, the Romans had the advantage of coming to civilization in the Iron Age. Narrowly defeated by Pyrrhus of Epirus in 279 B.C., 4 years later the Romans defeated Pyrrhus, the king who had expected to become the Alexander of the West. During the second century B.C. the Romans overcame the successors of Alexander in Macedonia and Syria. Soon the Greek cities of Asia Minor and the mainland succumbed to Roman rule.

Although the Romans were highly successful in mastering practical aspects of warfare, administration, agriculture, engineering, sanitation, and hygiene, Roman contributions to biology and medicine were modest. Despite considerable suspicion, hostility, and resistance, by the first century B.C. the Roman world had generally adopted and assimilated many aspects of Greek culture. Indeed it was not too difficult for an upper class Roman to acquire a veneer of Greek sophistication. Greek tutors could be purchased in the slave markets for less than the price of a good chef.

Roman writers were especially interesting in collecting and recording information on a grand scale. For some Roman authors the motive for writing was to show that Roman texts could be superior to those of the Greeks. This was the impulse that drove Cato the Censor (234–149 B.C.) to write on medicine and agriculture. Although his medicine was heavily imbued with the kinds of magical and superstitious formulas that had been rejected by Hippocrates, Cato boasted that Rome had long been healthy without doctors. On the other hand, the writings of the Roman poet Lucretius (99–44 B.C.), a follower of the philosophy of Epicurus (341–271 B.C.), indicate that some Roman authors were willing to explore philosophical and theoretical ideas. In his poem *De rerum natura* (*On the Nature of Things*), Lucretius used Epicurean teachings on ethics, atomic theory, the universe, and the nature of human life to combat the persistence of antiquated religious and superstitious ideas that he thought unworthy of intelligent beings. Politicians and demagogues, Lucretius argued, deliberately manipulated the superstitious fears that tormented men. In assuring his readers that a true philosophy of nature would allow them to defend themselves against state-sponsored mythology, Lucretius was making a deliberate political statement as well as defending Epicurean philosophy and atomist theory.

Unlike Aristotle, Lucretius did not believe that the world was the finite, eternal creation of a perfect God. Rather, he thought that the earth was fated to change and eventually perish. Like the Presocratics, Lucretius provided a natural explanation for the history of the universe and the nature of human society. Human beings were products of nature like all other creatures. Even dreams and the soul could be explained in terms of atoms. Lucretius advocated a return to a simpler mode of life and saw no promise in technological advances. He blamed religion and superstition for the decline of Roman civilization. Accustomed to consulting the will of the gods for all things, the people of Rome were easily manipulated by politicians who exploited the messages of the gods for their own purposes. Early Christian theologians disapproved of the materialistic concepts associated with the Epicureans, and few copies of *On the Nature of Things* survived the Middle Ages. In addition to the charge that they neglected the gods, the Epicureans were attacked for the idea that the goal of ethics was pleasure. Epicureanism, also known as Hedonism (from the Greek *hedone*, for pleasure), was misinterpreted as the pursuit of unrestrained sensuality. Actually, the Epicureans believed that the good life would result in a state of mental tranquility. During the Renaissance the work of Lucretius was highly regarded as a major source of Epicurean philosophy and a model for lucid scientific writing.

Unlike the systematic philosophical exposition so carefully crafted by Lucretius, the collection of facts and marvels complied by Pliny the Elder (23–79 A.D.), one of the most famous of the great Roman encyclopedists, was immensely popular and often imitated by medieval writers. Despite their obvious differences, Pliny and Lucretius shared the goal of freeing their contemporaries from the grip of superstition by providing information about the world and natural explanations of its phenomena. Until the sixteenth century, Pliny's much admired *Natural History* was regarded as a comprehensive and complete collection of information about the natural world. In 37 books, Pliny assembled facts and observations about all the natural sciences and human arts from some 2000 previous works written by 146 Roman authors and 326 Greeks.

In every way a model Roman citizen, Pliny was a member of a family of respected officials. Born in Novum Comum in northern Italy, Pliny practiced law before joining the military service. If travel is educational, Pliny was indeed well educated. He traveled through Germany, Gaul, Spain, and Africa, performing many official duties for the Roman emperors Vespasian and Titus. Curiosity and efficiency seem to have been Pliny's personal deities. To avoid wasting time that could be employed in reading and writing, he worked while traveling by carriage rather than horseback. Legend has it that his secretary read to him while he took his bath. It was curiosity that caused his death. When Mount Vesuvius erupted in 79 A.D., demolishing Pompeii and Herculaneum, Pliny insisted on studying the awesome phenomenon himself. He was overcome by the poisonous fumes of the volcano and died without sustaining any external wounds.

Enthusiastically collecting materials "as full of variety as nature itself," Pliny emphasized the Roman viewpoint that knowledge and science were useful either directly or as a means of teaching moral lessons. Unlike the abstract and abstruse works of the philosophers, Pliny's *Natural History* was a compendium of subjects of immediate importance for human life, presented in a style accessible to farmers and artisans, as well as men of leisure. Subjects such as mathematics, geometry, and the theory of music, which were so dear to the philosophers, were neglected, although Pliny included many dubious and bizarre stories. The *Natural History* is a rather like a random walk through observations about the heavens and the earth, geography and ethnography, natural history, stories of animals, fish, birds, and insects. The botanical sections deal with forestry, agriculture, and horticulture, as well as the manufacture of useful products from the plant world, such as wine, oil, and medicines.

Nature is so generous, Pliny assures us, that even the desert can provide the well-informed individual with a veritable drugstore. Following a wide-ranging discussion of the medicinal uses of plants, Pliny described drugs derived from human beings, as well as superstitions, magical practices, and physicians. Even his recommendations for pest control involved a mixture of religion, folk magic, and chemistry, including lizard's gall for protecting trees from caterpillars and soaking seeds in wine to prevent fungal diseases.

Beginning his discussion of animal life with humans, as appropriate for "the highest species in the order of creation," Pliny argued that all other creatures seem to have been produced by nature to serve man. Among all the human races, Pliny asserted, the Roman race was the most virtuous in the whole world. From other parts of the world came stories of marvelous races, strange births, and monsters. Among the more fortunate marvels was a race of sun-hardened people over 7 feet tall who never spit and never suffered from headache or toothache. Some "wonder people" were half male and half female, others had dogs' heads and communicated by barking, some had no mouths and derived all their nourishment from airs and odors. Some of the strange fur-covered Satyrs who alternately walked upright and ran about on all fours were probably monkeys or apes. Exotic animal species included the lion and the elephant, as well as the unicorn and the phoenix. The unicorn was a fierce chimera with the head of a stag, the feet of an elephant, the tail of a boar, the torso of a horse, and a 3-foot-long black horn projecting from the middle of its forehead. Pliny noted that the animal could not be captured alive; presumably, the creature has not been captured dead either.

Pliny was troubled by what he saw as the degeneration of Roman culture and the stagnation of the sciences, despite the enormous wealth and resources controlled by the empire. He thought it paradoxical that the sciences flourished during a period of constant tension and warfare among the Greek states but went into a state of decline after Rome imposed order and peace on the world. Complaining that his contemporaries were abusing the earth's resources in their pursuit of pleasure, wealth, and security, Pliny warned that the sad state of science was part of a general malaise overtaking the Roman world. Like the Stoic philosophers, and some twentieth-century ecologists, Pliny equated God with Nature, abhorred extravagance, and favored a return to his vision of a simpler but better life.

Major sections of Pliny's text dealt with drugs derived from plants, a subject of great practical importance. Pliny's contemporary, the Greek physician Dioscorides (ca. 40–90), composed a more systematic treatment of botany and herbal medicine. Little is known about Dioscorides, except that he probably studied at Alexandria before he became a *medicus* attached to the Roman forces in Asia under the emperor Nero. Military service gave him the opportunity to travel widely and study many exotic plant species. The text now known as *De materia medica* (*The Materials of Medicine*) refers to some 500 different plants. Dioscorides was an acute observer and keen naturalist, who recorded valuable information about medically useful plants, their place of origin, habitat, growth characteristics, and methods for preparing remedies. Drugs derived from animals and minerals were also described.

Some of the ingredients in the herbal remedies recommended by Dioscorides can be found among the vegetables and spices sold today in any grocery store, but the medicinal properties assigned to them would surprise modern cooks. Cinnamon and cassia, for example, were said to be useful in the treatment of internal inflammations, venomous bites, cough, diseases of the kidneys, and menstrual problems. Drinking a decoction of asparagus and wearing the stalk as an amulet purportedly induced sterility. Judging from the large number of remedies said to bring on menstruation and expel the fetus, menstrual disorders, contraception, and abortion must have been among the major concerns that patients brought to physicians. Even if earaches do not respond to remedies containing millipedes, cockroaches, or the lungs of a fox, recent studies of *De materia medica* suggest that Dioscorides organized plants according to a subtle and sophisticated drug affinity system, rather than simple criteria such as morphology and habitat. His classification scheme must have required close attention to the effects of drugs on a significant number of patients. Despite references to many bizarre ingredients, Dioscorides offered recipes for many effective drugs, including strong purgatives, mild emetics and laxatives, soporifics, analgesics, and antiseptics.

Dioscorides

In contrast to the popularity of *De materia medica*, the work of the Roman writer Aulus Cornelius Celsus (25 B.C.–A.D. 50) was essentially forgotten until the fifteenth century. Almost nothing is known about the life of Celsus, and his status as author of *De re medicina* (*On Medicine*) has been questioned. He has been honored as the creator of scientific Latin, but historians are divided as to whether he wrote or plagiarized one of the most famous Latin medical texts composed during the first century. Some say that Celsus was merely a compiler or translator, but the quality of the text and the display of critical judgment seem to militate against this assessment. Contemporaries, however, considered Celsus a man of quite modest talents. Given Roman traditions, it is unlikely that Celsus was a physician. Roman landlords were expected to assume responsibility for the medical care of the sick on their estates. Generally this simply meant knowing enough about medicine to supervise the women or slaves who carried out the unpleasant menial work of the healer.

While Celsus may have written a major encyclopedia, with sections on rhetoric, philosophy, jurisprudence, the art of warfare, agriculture, and medicine, only the section on medicine has survived. Because Celsus wrote in Latin during an era in which Greek was considered the language of scholars, the intellectual elite ignored his work. His manuscript remained in obscurity for some 13 centuries, but in the 1420s two copies of *De medicina* were discovered. The timing could not have been better, because *On Medicine* became the first medical work to be turned out on the new European printing press. To Renaissance scholars, Celsus represented a source of first-century Latin grammar and medical philosophy, uncorrupted by medieval copyists.

A master of organization, clarity, and style, Celsus established the historical context of Greek and Roman medicine. Celsus agreed with other critics of Greek influences that medicine and physicians had been unnecessary in ancient Rome. The Greek art of medicine became a necessity only after Roman society had become thoroughly infiltrated by indolence, luxury, and other Greek affectations that led to illness. While the early Greeks had refined the art of medicine more than any other people, after the death of Hippocrates Greek medicine had fragmented into various disputatious sects, such as the Methodists, Dogmatists, Pneumatists, and Empiricists. Without the analysis provided by Celsus, the origins and ideas peculiar to these sects would be totally obscure. The Dogmatists, for example, emphasized the study of anatomy and claimed Hippocrates, Herophilus, and Erasistratus as their founders. Members of other sects accused the Dogmatists of neglecting the practical aspects of medicine while pursing highly speculative theories. Emphasizing the role of experience in therapeutics, some members of the sect known as the Empiricists argued that the study of anatomy was totally useless to physicians.

Taking a balanced view of the rivalry between the various sects, Celsus rejected rigid approaches to medical science and called for moderation in all therapeutic interventions, as well as careful consideration of the needs of the individual patient. Although medical practice was primarily a matter of prescribing appropriate drugs, Celsus argued that physicians must also know anatomy and surgery, including operations such as amputation, couching the cataract, and the treatment of hernia and goiter. Critical of human experimentation, Celsus denounced Herophilus and Erasistratus for performing vivisections on criminals. Celsus argued that such cruel experiments were unethical and unnecessary because all too often accidents expose the internal organs of the body and provide windows of opportunity for physicians who could learn anatomy while engaging in acts of mercy.

On Medicine provides a valuable survey of the medical knowledge and practices of first-century Rome, but Celsus and all the medical writers of antiquity, with the exception of Hippocrates, have been overshadowed by the great physician, anatomist, philosopher, and writer, Galen of Pergamum (130 to ca. 210). According to some biographers, after a long and active life Galen died in Rome in 217, at the advanced age of 87.

Galen's writings were abstracted, rewritten, reorganized, and transmitted to medieval and Renaissance scholars in the form known as Galenism. As a medical authority, Galen was second only to Hippocrates; as anatomist and experimental physiologist, he had no rivals. Called the "First of Physicians and Philosophers" by the emperor Marcus Aurelius, Galen was referred to as "windbag" and "mulehead" by his critics and rivals. Galen wrote more than 300 treatises on subjects that included philosophy, religion, ethics, mathematics, grammar, law, anatomy, physiology, pathology, dietetics, hygiene, therapeutics, pharmacy, commentaries on Hippocrates, and autobiography. The only complete modern edition of his surviving works fills 20 large volumes; this German edition is said to have taken 12 years to complete and almost as long to read.

Fearing that his writings were being corrupted by careless copyists and that mediocre authors were trying to pass off their work as his, Galen composed a guide to the cautious reader called *On His Own Books*, which provided a list of the titles of his books, a description of his genuine works, and a reading program for aspiring physicians. He urged medical students and teachers to foster the love of truth in the young. This would inspire them to master the writings of the Ancients and devise ways to test and prove such knowledge. Medical practitioners, according to Galen, should maintain a scientific approach to anatomy, physiology, the nature of disease, and the nature of human beings. Galen portrayed himself as a scholar who understood that, despite his life-long search for truth, he would never be able to discover all that he so passionately wished to know.

Galen was born in Pergamum, in Asia Minor near the Aegean coast opposite the island of Lesbos. The city boasted a library and temple of Asclepius, the Greek god of medicine, and saw itself as a cultural rival of Alexandria. The Ptolemies tried to arrest the intellectual growth of Pergamum by forbidding the export of papyrus. In response to this threat, parchment was developed and the book as we now know it replaced the papyrus roll. Pergamum was also known as the site of intense Christian evangelism, and Galen did not fail to comment on the conflicts between pagans, Jews, and Christians.

Galen of Pergamum contemplating the rare opportunity of studying a human body

In autobiographical allusions, Galen refers to his father, Nikon, a wealthy architect, as a kind-hearted man, but he described his mother as a quarrelsome woman, so violent she sometimes bit the servants during temper tantrums. Despite his resolution that he would emulate the admirable qualities of his father and reject the disgraceful passions of his mother, Galen's writings reveal a penchant for acrimonious quarrels and bitter controversies that suggest he was unable to fully suppress his maternal inheritance. As a youth, Galen mastered mathematics and philosophy, but he began to study medicine after his father learned in a dream that his son was destined to become a physician. When Nikon died, Galen left Pergamum in order to study with famous teachers in Smyrna, Corinth, and Alexandria, which was still a distinguished center of anatomical instruction, despite the Roman prohibition of human dissection.

When he was about 28, Galen returned to Pergamum and became physician to the gladiators. Because of his success in treating the sick and the wounded, Galen also established a flourishing private practice. Restless and ambitious, in a few years Galen decided to move on to the Imperial City. In Rome, Galen was successful in treating several prominent individuals, won a reputation as a miracle worker, delivered public anatomical lectures, and composed some of his major anatomical and physiological texts. But 4 years later, complaining about the hostility of less successful physicians, Galen suddenly returned to Pergamum. His critics noted that his departure from Rome in 166 coincided with the appearance of a virulent epidemic disease. At the request of the emperor Marcus Aurelius, Galen eventually returned to Rome, where he resumed his busy schedule of imperial service, private practice, research, lecturing, and writing.

Like Plato and Aristotle, Galen believed that a form of divine intelligence had created the universe and living beings. Galen was tireless in his tirades against naturalists who failed to see evidence of design in the organization of the human body. Supporters of the atomic theory, which he considered materialistic and atheistic, were singled out for particular outrage. For Galen, the body was the instrument of the soul and proof of the existence and the wisdom of God. Anatomical research was the source of a "perfect theology" through which the anatomist demonstrated his reverence for the Creator by discovering his wisdom, power, and goodness through investigations of form and function.

Dissecting any animal, no matter how humble, revealed a little universe fashioned by the wisdom and skill of the Creator. Assuming that Nature acts with perfect wisdom and does nothing in vain, Galen argued that every part of the body was crafted for its proper function. Dissection might be a religious experience for Galen, but he understood that most practitioners needed anatomical knowledge for guidance in performing surgical operations and treating wounds. If the surgeon could choose the site of incision, knowledge of anatomy would allow him to do the least possible damage. Anatomical knowledge also allowed him to predict the kind of damage that would occur when muscles and nerves were injured, so that the physician could escape blame. Larger philosophical questions might also be answered by anatomy. For example, Aristotelians argued that the fact that the voice, which is the instrument of reason, came from the chest supported the contention that the seat of reason was in the heart, rather than the head. By demonstrating that the recurrent laryngeal nerves control the voice, Galen vindicated those who argued that reason resided in the brain.

Even though he could not carry out systematic dissections of human bodies, Galen was quite sure that his findings in pigs, goats, monkeys, and apes could be applied to human beings. Sometimes Galen was the beneficiary of fortuitous accidents that allowed him to inspect human cadavers and skeletons. In *On Bones*, Galen reported that he had often examined human bones exposed by the destruction of tombs and monuments. On one occasion a corpse had been washed out of its grave by a flood, swept downstream, and deposited on the bank of the river. The flesh had putrefied, but the bones were still closely attached to each other, leaving the skeleton perfectly prepared for an anatomy lesson. While some historians have suggested that Galen's patrons allowed him to carry out surreptitious human dissections, the enthusiasm with which Galen describes his encounters with cadavers and skeletons suggests their rarity. Indeed, while Galen was often critical of his predecessors, especially Herophilus and Erasistratus, he obviously envied their resources and privileges. Nevertheless, systematic dissections of other animals would have allowed Galen to take full advantage of the "windows" into the body provided by the injuries of his patients while he dressed their gruesome wounds.

The importance of Galen's work can hardly be exaggerated. Galen's word was the ultimate authority on medical and biological questions for hundreds of years. His anatomical studies were not seriously challenged until the sixteenth century, and his physiological concepts remained virtually unquestioned until the seventeenth century. Even in the nineteenth century, physicians fighting to reform medical practice complained about the continuing tyranny of Galenism. His errors became the focus of later attention, but it is essential to try to balance the merits and faults in Galen's work because both exerted a profound influence on the history of anatomy, physiology, and medicine. Galen's anatomy had the obvious defect that he had based his researches on animal models and not on the human body itself. Nevertheless, Galen was an acute observer whose writings could provide an excellent guide to readers who absorbed both the letter and the spirit of his teachings. Galen often told his readers that the only way to learn about the wonders of the human body was by studying the writings of the great physicians, such as himself and Hippocrates, and by actual experience. But, he warned the reader, even the writings of the great Hippocrates must be confirmed and tested by going directly to Nature and observing the structures and functions of animals.

Not satisfied with purely anatomical research, Galen sought ways to proceed from structure to function, from anatomical analysis to experimental physiology. Through dissection, vivisection, and speculation, Galen struggled to solve almost every possible problem in classical physiology. Given the magnitude of the task and the conceptual and technical difficulties inherent in such a program, it is not surprising that Galen's conclusions were sometimes mistaken. Ultimately his errors stimulated revolutions in anatomy and physiology, but for hundreds of years the Galenic synthesis was considered so complete and powerful that it fully satisfied the needs of scientists, scholars, and physicians.

Balancing the merits and defects of the Galenic synthesis is difficult for many reasons, not the least of which is the inherent complexity and obscurity of the system. While Galen may well have been misinterpreted and his writings corrupted by copyists, Galen made different assertions in various texts. Attempts to distinguish between what Galen actually said, or probably meant, and the layers of interpretation that eventually surrounded Galenic seed crystals are probably futile and counterproductive.

In essence, Galenic physiology rests on the doctrine of the threefold division of life into vital processes governed by vegetative, animal, and rational souls or spirits. Life ultimately depended on air, or *pneuma*, the breath of the cosmos. By integrating his anatomical and physiological knowledge with his philosophical system, Galen turned the human body into a place for the elaboration and distribution of three kinds of adaptations of pneuma. These were associated with the three principal organs of the body—the liver, heart, and brain, and the three types of vessels, the veins, arteries, and nerves. Pneuma was adapted by the liver to become the *natural spirits*, which were distributed by the veins and supported the vegetative functions of nutrition and growth. *Vital spirits*, which governed movement, were formed in the heart; the arteries distributed innate heat and pneuma or vital spirits to warm and vivify the parts of the body. Innate heat was produced as a result of the slow combustion process that occurred in the heart. The third adaptation occurred in the brain and resulted in *animal spirits*, which were needed for sensation as well as muscular movements; the nerves distributed animal spirits.

To work out the details of the elaboration and distribution of the three spirits required much ingenious research and a wealth of speculation concerning the functions of the parts of the body. Because theories of the motion of the heart and blood have played a central role in the history of biology, Galen's writings on these topics have been subject to particular scrutiny. Modern categories are, however, inappropriate in dealing with Galenic physiology; Galen's approach to the workings of the body made it impossible to separate questions about the circulatory system from the respiratory system and the generation and distribution of pneuma and spirits. Even Galen was not always sure about the roles, functions, and existence of parts of his system. Sometimes, for example, he suggests reservations about the nature and functions of the spirits. But when later commentators tried to clarify and simplify his arguments, they tended to transform his working hypotheses into dogmatic certainties.

Galen's ideas about the circulatory system illustrate both the strengths and weaknesses of his approach to physiological phenomena. By means of simple, direct experiments, Galen easily disproved the idea that the arteries of living beings contain air rather than blood. The artery of a living animal was exposed and tied at two points. When the vessel was cut between the two ligatures, blood, not air, escaped from the artery, indicating that the vessel normally contained blood. Similarly, the role of the arteries in nourishing the innate heat could be demonstrated by tying a ligature around a limb so that it was tight enough to cut off the arterial pulse, which was generated by the heart. The limb became cold and pale below the ligature because the arteries could no longer supply the innate heat.

In order to provide a complete explanation for the distribution of blood, air, and spirits, Galen asserted that the septum of the heart contained pores through which blood could pass from the right side of the heart to the left side. Blood was said to be continuously synthesized from ingested foods. The useful part of the food was transported as chyle from the intestines via the portal vein to the liver. The liver had the faculty of transforming chyle into dark venous blood. Since the liver was the seat of vegetative life, concerned with nourishment and growth, it imbued the blood with natural spirits. As the rich venous blood made its way through the veins, the parts of the body took up their proper nutriments. The useless part of the food was converted into black bile by the spleen.

Despite his detailed knowledge of the structure of the heart, its chambers, and valves, Galen's preconceived concept of its function led to ambiguities and obscurities. Galen thought that the right side of the heart was a main branch of the venous system and that the auricles were merely dilations of the greater veins, rather than important chambers of the heart itself. However, he demonstrated that when the chest of a living animal was opened and the heart examined, blood was found in the left ventricle of the heart. From the right side of the heart the blood could enter different pathways. Part of the blood discharged its impurities or sooty vapors into the lung by means of the artery-like vein (the pulmonary artery). Part of the blood passed through putative pores in the septum and entered the left heart, where it encountered the pneuma that had been brought in by way of the trachea and the vein-like artery (the pulmonary vein). The mixture of dark red blood and air resulted in the production of vital spirits and bright red blood; the vital spirits and refined blood were distributed by means of the arteries.

Completing the threefold system, some of the arteries led to a network of vessels at the base of the brain called the *rete mirabile*, where the vital spirits were transformed into animal spirits for distribution to the body via the system of hollow tubes called nerves. Clearly, the concept of continuous blood circulation is incompatible with a scheme in which blood is constantly synthesized from food by the liver and is then assimilated into the substance of the body as it ebbs and flows through the blood vessels. Of course, the nature of the Galenic system is so complex that the term "clearly" is hardly appropriate in any attempt at a brief and simplified reconstruction of Galenic physiology.

Because Galen used various animals in working out the details of his physiological system, he ascribed certain anatomical features found in other animals to human beings. For example, he described the liver as five-lobed, which is true in dogs but not in humans. Similarly, the *rete mirabile* is present in ruminants, but not in humans. Having denied that the heart is a muscle, he argued that the active phase of heart action was dilation rather than contraction. Ignoring discrepancies between the rate of respiration and the heart beat, Galen's system associated respiration with the movement of the heart; dilation of the heart would draw air into the body and contraction would drive it out. Galen's system seemed to provide a complete synthesis of medical and philosophical traditions. It combined the four humors of Hippocrates with Aristotle's threefold division of the spirits and incorporated the pneuma, or great cosmic spirit of the Stoics. Beyond these *desiderata*, it is deeply religious in the sense that it is wholly imbued with reverence and admiration for the work of the Creator.

It is difficult to imagine a physiological problem that Galen did not at least touch on. His mastery of experimental methods is demonstrated in his studies of the nervous system. Galen worked out the anatomy of the brain, spinal cord, and nerves and proved that the nerves originate in the brain and spinal cord and not, as Aristotle said, in the heart. In a remarkable series of experiments, Galen demonstrated that when the spinal cord was injured between the first and third vertebrae, death was instantaneous. Injury between the third and fourth vertebrae arrested respiration. Damage below the sixth vertebra caused paralysis of the thoracic muscles, while lesions lower in the cord produced paralysis of the lower limbs, bladder, and intestines. To prove that the kidneys and not the bladder made urine, Galen carried out experiments in which the ureters were ligated to prevent the flow of urine from the kidneys to the bladder. His study of the functions of the kidneys actually involved the quantitative falsification of the theories put forth by those who argued that urine was formed in the bladder and that the kidneys played no role in its formation. To study digestion, he put pigs on different kinds of diets and examined the contents of their stomachs after appropriate intervals.

In addition to his excellent reputation as a scientist and philosopher, Galen was regarded as an excellent clinician and brilliant diagnostician. By relying on the humoral theory of disease, Galen rationalized his approach to the prevention and treatment of disease. Because the humors were formed by the action of the innate heat on nutriments, relatively hot and cold foods and drugs could be prescribed as a means of altering the balance of the four humors. Despite his reverence for Hippocrates, Galen was unwilling to stand by doing no harm while waiting for Nature to provide a cure. Galen clearly favored active intervention, and his sophisticated individualized approach to the prevention and treatment of disease eventually degenerated into a system of active bleeding and purging, accompanied by large doses of complex drug mixtures. Such remedies were later called "Galenicals," and apothecary shops, or the School of Pharmacy at a modern university, might be identified by the sign of "Galen's Head" above the doorway.

In *On the Usefulness of the Parts*, Galen argued that human beings are the most perfect of all animals, but man is more perfect than woman because the male has more of Nature's primary instrument, which is excess heat. Reproductive anatomy is destiny because, according to Galen, the testicles performed the task of heating and perfecting the blood. It was a deficiency of heat that caused a fetus to become a female. Galen essentially explained the female reproductive organs as defective or inside-out versions of the perfect male parts; that is, because of the deficiency of heat, the reproductive parts formed inside the body of the fetus, but could not emerge and project themselves to the outside. Worse yet, brain development required greater heat and more perfected blood than that available to the female fetus. Evidence for such assertions, Galen argued, could be found by studying the effects of castration in humans and animals. Destruction of the testicles resulted in the loss of heat, virility, and manliness. It was the lack of heat in the female that resulted in the accumulation of fat; similar accumulations occurred in the body of a eunuch. As further proof that women were imperfect men, Galen offered ancient accounts of women who were suddenly transformed into men. The reverse transformation was impossible, Galen argued, because nature always strives towards perfection from the lower to the higher form.

On the other hand, in defense of his belief that bleeding was the appropriate treatment for almost every disorder that afflicted human beings, Galen asserted that women were immune to many of the diseases that attacked men because their superfluous blood was eliminated by menstruation or lactation. As long as their menstrual cycles were normal, women were supposedly spared the miseries of gout, arthritis, apoplexy, epilepsy, melancholy, and other diseases. Men fortunate enough to rid themselves of excess blood through hemorrhoids or nosebleeds could also enjoy protection against such diseases. In terms of humoral doctrine, bleeding seemed to accomplish the therapeutic goals shared by patients and physicians. Bleeding and purging seemed to rid the body of putrid, corrupt, and harmful materials. Indeed it is possible to explain the apparent therapeutic efficacy of bleeding in terms of the suppression of the clinical manifestations of certain diseases, such as malaria. Bleeding might inhibit the growth and multiplication of certain pathogens by reducing the availability of iron in the blood. Lowering the viscosity of the blood might also affect the body's response to disease by increasing the flow of blood through the capillary bed. Obviously, bleeding to the point of fainting, a tactic enthusiastically championed by Galen, would force a period of rest and recuperation on feverish and delirious patients. This would be of immense benefit to the caretaker, if not to the patient.

As skillful in the art of medicine as in the science, Galen was seen by his patients and admirers as a true miracle worker. Even his rivals acknowledged Galen's intelligence, productivity, and his passionate defense of his own ideas. Concerned with the dignity of the medical profession, Galen warned against displays of ambition, contentiousness, and greed, and urged physicians to distinguish themselves from the hordes of quacks and charlatans by means of skill, learning, and a disdainful attitude towards money. Clever students of medicine could also learn from Galen how to impress patients and observers by eliciting diagnostic clues through casual questions, making note of medicines already in use, observing the contents of all basins removed from the sickroom on the way to the dung heap, and secretly assessing the state of the pulse while conversing with the patient. When unsure of diagnosis and prognosis, the physician could take the case with a great show of reluctance, while warning the patient's friends and family to expect the worst. If the patient died, the physician's reputation was safe; if the patient survived, the physician would be considered a miracle worker.

Surprisingly, despite Galen's reputation and the fact that he gave public lectures and demonstrations in the Temple of Peace, he left no school or disciples. While his personality may have been admirably suited to dealing with Roman emperors, it seems to have intimidated or repulsed colleagues and potential students. Most of Galen's voluminous writings were unknown during the Middle Ages in Europe, but his work was not lost or entirely neglected. His medical texts were translated into Syriac and Arabic. Finally, the Arabic versions of his works were translated into Latin and inspired generations of European scholars. Galenic ideas, if not his experimental methods, became the foundation of Islamic and Christian medical studies. For many centuries the teachings of Galen were invested with great authority by theologians who were attracted by his religiosity, his concept of the three spirits, his denunciation of atheistic materialism, his support for a worldview that has been called *anthropomorphic teleology*, and his willingness to see proof of religious dogmas in scientific findings. It was unfortunate for science, and for Galen's reputation, that his most flagrant errors were treated as inviolable truths. When Galenic anatomy, pharmacology, and physiology were finally subjected to serious challenges in the sixteenth and seventeenth centuries, the scientists who revolutionized and reformed those fields took their instruction and inspiration from the study of the newly restored writings of Galen.

THE MIDDLE AGES

No physician-philosopher comparable to Galen appeared during the European Middle Ages, the period from about 500–1500, an era during which ideas about the nature of the physical universe, the nature of human beings, their proper place in the universe, and, above all, their relationship to God underwent profound changes and painful dislocations. During Galen's lifetime, the borders of the Roman Empire had encompassed the civilized Western world. By the time of his death, Greco-Roman culture and ancient science could well be characterized as essentially moribund. The medieval era was the outcome of the well-known story of the decline and fall of the Roman Empire after centuries of anarchy, turmoil, and warfare.

The Roman era, a period of relative peace and prosperity, might have been a time of great intellectual development, but long before Rome lost its place as undisputed political center of the Western world, many writers had complained about the malaise permeating Roman society. Pliny voiced the belief shared by many of his contemporaries that humanity had grown corrupt and degenerate. The decline of Rome seemed to owe more to internal decay than to barbarian invasions. Rome suffered the cumulative effects of misgovernment, corruption, war, endemic disease, a growing obsession with occultism, and a crisis in agriculture. The productivity of Italian lands diminished as a result of deforestation, erosion, and the neglect of irrigation canals; numerous farms and fields were abandoned. Mining enterprises failed because the richer veins of minerals had been exhausted. Observers in Italy noted significant declines in population and in the live birth rate.

Not long after 330 A.D., when the emperor Constantine established his capital in Byzantium (now Constantinople), the division of the Roman Empire between East and West became permanent. The Latin West entered into the era popularly known as the Dark Ages, during which Christians were warned to avoid all pagan books that might turn the young from the faith. In 529 A.D., by the order of the emperor Justinian, the Academy, the living heritage of Plato, was closed and pagans were forbidden to teach. Many pagan scholars were sent into exile in Persia, taking rare and precious manuscripts with them. Throughout antiquity the number of individuals engaging in inquiries that could be called scientific was always very small, but with the introduction of legal sanctions against pagan learning the few scholars who continued to engage in such activities faced increasing isolation as well as hostility and persecution.

Certainly, the climate of the European Middle Ages was not conducive to original researches in the biological and medical sciences. On the other hand, during the Middle Ages European society experienced significant changes in social organization, technological innovations, and the appearance of new institutions of learning. Many labor-saving devices, agricultural implements, techniques, and crops were either invented or successfully exploited for the first time. Significant improvements were made in metallurgy, construction, the production of textiles, and the use of watermills and windmills. By the twelfth century both food production and population were showing significant gains. As productivity and population grew, so did industry, the crafts, commerce, and urban centers. A revival of learning that gained momentum with the recovery and translation of ancient texts also contributed to the transformation of medieval European culture. By the thirteenth century, mathematics, astronomy, alchemy, and medicine joined theology as subjects worthy of serious attention. As new religious movements undermined the monolithic structure of faith, scholars turned to the philosophy and science of ancient Greece for answers to questions that faith alone could not answer.

Another factor in the transformation of European scholarship was the growth of the medieval university. Very different from the ancient centers of learning, the medieval university was also unlike modern institutions of higher learning, especially in terms of the relationships among students, faculty, and administrators. The community of scholars was replaced by the concept that those formally licensed to teach would instruct those seeking their own credentials. Informal discussions among independent scholars were replaced by formal lectures addressed to large groups of students. Free and open inquiry into areas of mutual interest was replaced by a standardized curriculum.

The exact origin of some of the major European universities is unknown, although ecclesiastical authorities seem to have dominated these nascent institutions. As the old monastic schools declined in prestige in the eleventh and twelfth centuries, cathedral schools and municipal schools began to supersede them. As the number of students increased, groups of teachers, called *magistri*, formed an organization called a *universitas magistorum*. The term *university* was originally applied to any corporate status or association of persons. The early universities were collections of colleges that sought the advantages of corporate status, rather like guilds. Indeed, during the eleventh century the words university and guild were essentially interchangeable. The term *studium generale* denoted a guild or university of scholars and students engaged in the pursuit of higher learning. Such institutions drew students from many countries to study with teachers known for particular areas of excellence. Eventually the term university was restricted to institutions of higher learning. In thirteenth-century Europe there were 5 significant universities in France, 2 in England, 11 in Italy, and 3 in Spain. By the end of the Middle Ages the number of European universities was about 80.

Even though the educated, or merely literate, were still a tiny minority of the population, some universities attracted large numbers of students, who, despite different national origins, had achieved competence in Latin, the language of learning. Students as young as 14 or 15 could enter the universities along with more mature scholars. The curriculum consisted of the seven liberal arts: grammar, rhetoric, logic, arithmetic, geometry, astronomy, and music.

Universities wielded considerable power and won large measures of self-government from civic and ecclesiastical authorities, but the relationship between "town and gown" was often quite tense. Some institutions, notably those in Paris and Oxford, were governed by guilds of instructors in the liberal arts. In contrast, Bologna, which specialized in legal studies, was dominated by a guild of students. Students had the reputation of being rowdy, armed, often drunk, and always dangerous. Given the lack of scholarships and student loans, it is not surprising that many were accused of making their living by a combination of begging and stealing. Despite their wild reputation, universities generally attempted to satisfy the eagerness of students to acquire knowledge, or at least access to careers that promised wealth, power, and prestige.

Medieval universities generally lacked the kind of campus grounds and buildings associated with the modern institution. Lectures were often held in the professors' apartments or in rented rooms. In the absence of standardized courses, groups of students might hire and pay teachers on a contractual basis. Students who passed an examination after 6 years of study became qualified to teach. Many rewarding and lucrative professions were open to the university graduate. He could take orders and become a church official, seek out the patronage of a wealthy noble, study medicine or law, or work as a copyist. But even in the thirteenth century there were students who rejected conventional careers and adopted the life of wandering scholars.

Subjects deemed appropriate for teaching were limited and regulated by religious dogma and convention. During the turmoil of the thirteenth century, secular authorities bowed to the ecclesiastical powers and orders of mendicant friars were established to combat alleged heresy at the universities. Nevertheless, the writings of Aristotle, as edited by Thomas Aquinas, became a source of authority and inspiration to philosophers and theologians. Aristotle's ideas about the earth as the center of the universe and the site of all change and imperfection, in contrast to the pure, incorruptible, and eternal heavens where the Creator abided, were very congenial to the Church. His view of human life as part of a rigid hierarchy created a welcome justification for the hierarchical and authoritarian arrangement of medieval society.

The structure and goals of the university encouraged a form of education called *scholasticism*, that is, the study of the ancient authorities as the repository of all that could be known. This approach essentially rejected the possibility of new discoveries and innovations. The increasingly formal organization of education and the professions also created rigid barriers that excluded all women and men lacking in wealth and status, who, of course, constituted the vast majority.

Editing the works of the ancients and providing interpretations of the works of Aristotle that were compatible with the teachings of the Church were the major tasks undertaken by medieval scholars such as Albertus Magnus (ca. 1200–1280), a member of the Dominican order, Bishop of Ratisbon, and a teacher at the University of Paris, known to his contemporaries as "Doctor Universalis." Although his writings reflect energy, originality, and encyclopedic knowledge, Albertus Magnus presented his prodigious scholarly efforts as commentaries on the work of Aristotle. His reputation as the foremost naturalist of medieval Europe is based on the detailed descriptions of plants, animals, and minerals that appear in his *De Mineralibus*, *De Vegatabilus et Plantis*, and *De Animalibus*. Albertus described several new plant species and may have been the first to prepare arsenic in a free form, but rather than attempt to form new theories, he forced his observations into a preexisting Aristotelian mold. Despite such caution, his reputation as a magician came to rival his fame as a scholar. Various works on astrology and the magical properties of plants, stones, and animals were falsely attributed to Albertus Magnus. *The Book of Secrets of Albertus Magnus*, one of the most popular texts of this kind, went through many editions well into the seventeenth century.

Roger Bacon (1214–1292), the medieval scholar known as "Doctor Mirabilis," has been called the first modern scientist because of the spirit in which he approached nature. After studying at Oxford and Paris, he joined the Franciscan order, but his views were so unorthodox that after 1257 he worked in isolation. For many years he was virtually imprisoned in Paris. Despite his contention that all branches of science were subordinate to theology, his studies of mathematics, astronomy, optics, and alchemy, combined with his quarrelsome nature, imaginative predictions, and emphasis on experimentation, led to his reputation as a magician as well as accusations of witchcraft and necromancy.

Other scholars, however, found the cloistered life congenial to intellectual and scientific pursuits. Hildegard of Bingen (1098–1179) was one of the very few medieval women to transcend the barriers that kept women out of the universities and the community of scholars. Widely respected during her lifetime as a writer, composer, naturalist, and healer, she was all but forgotten until nineteenth-century biblical scholars prepared new editions of her writings. Further studies brought Hildegard to the attention of historians of science, feminist scholars, musicians, poets, herbalists, and homeopathic practitioners. Although St. Hildegard has been called a mystic, a visionary, and a prophet, her writings suggest practical experience and boundless curiosity about the wonders of nature.

As the tenth and last child of a noble family, Hildegard was offered to God as a tithe and entered a life of religious seclusion at the age of 8. She took monastic vows in her teens and was chosen abbess of her Benedictine nunnery in 1136. At about 15 years of age, Hildegard began receiving revelations about the nature of the cosmos and humankind. The visions were explained to her in Latin by a voice from heaven. In 1141 a divine call commanded her to write down and explain her visions. When she began writing at the age of 43, Hildegard thought that she was the first woman to embark on such a mission; the names of other women writers were unknown to her. After a papal inquiry into the nature of her revelations, Hildegard became a veritable celebrity and was officially encouraged to continue her work.

Hildegard's *Liber simplicis medicinae* (*The Book of Simple Medicine*), generally referred to as the *Physica*, was probably the first book by a female author to discuss the elements and the therapeutic virtues of plants, animals, and metals. As the first book on natural history composed in Germany, it can be thought of as a pioneering effort in scientific botanical studies. The text includes much traditional medical lore concerning the therapeutic uses and toxic properties of many herbs, trees, mammals, reptiles, fishes, birds, minerals, gems, and metals. Some of the first recorded local plant names are those found in Hildegard's writings.

Hildegard's *Liber compositae medicinae* (*The Book of Compound Medicine*) included material on the nature, forms, causes, and treatments of disease, the elements, winds and stars, human physiology and sexuality, and so forth. Like Hippocrates, Hildegard attempted to find natural causes for mental and physical diseases. Some of her ideas were expressed in vivid comparisons of the workings of the microcosm and the macrocosm. For example, the movement of blood in the vessels was compared to the way the stars pulsate and move in the firmament of the heavens. Although Hildegard seemed to think of the heart as the seat of the rational soul, she considered the brain the regulator of the vital qualities and the center of life. In exploring mental as well as physical diseases, Hildegard argued that even the most bizarre mental states could have natural causes. Thus, people might think a man was possessed by a demon when the real problem might be a simultaneous attack of headache, migraine, and vertigo. Hildegard probably had a special interest in these disorders. Modern medical detectives have diagnosed her visions as classical examples of migraine.

ISLAMIC SCIENCE

The Middle Ages of European history roughly correspond to the Golden Age of Islam, the religion founded by the Prophet Muhammad (570–632). When Muhammad was about 40 years old, he received a series of visions in which the *Qur'an*, the sacred book of the Islamic religion, was revealed to him. By the time of his death, practically all of Arabia had accepted Islam, and a century later his followers had conquered half of Byzantine Asia, all of Persia, Egypt, North Africa, and Spain.

During this turbulent period, the Arabs adopted and disseminated some of the techniques essential to the expansion of literacy. At the beginning of the eighth century the Arabs captured Samarkand and learned the art of paper making from the Chinese, who had been using paper since about 100 A.D. A paper making plant was built in Baghdad in 794. When the Arabs conquered Sicily and Spain, they introduced paper to Europe. Mathematical concepts that originated in India, such as the zero and the so-called Arabic numeral system, were also introduced to Europe through the Arabs. Many words used in science and mathematics, such as alchemy, algebra, algorithm, alcohol, cipher, and nadir, are derived from the Arabic.

According to Muslim tradition, the Prophet respected and encouraged the pursuit of knowledge, although he was himself illiterate. "He who leaves his home in search of knowledge," said Muhammad, "walks in the path of God." The followers of Muhammad established elementary schools, usually in the mosques, where most boys and some girls received religious education. Secondary schools were eventually established to teach grammar, philology, rhetoric, literature, logic, mathematics, and astronomy. Just as Latin served as the common language of learning for students throughout Europe, Arabic was the language of learning throughout the Islamic world. Thus, Persians, Jews, and Christians, as well as Arabs, took part in the development of Arabic scientific literature.

From the ninth to the thirteenth century, Arab scholars could justly claim that Islam had captured the world of scholarship. Libraries were attached to most mosques, and books were regarded as great treasures. Scholars were encouraged to participate in the immense task of translating ancient manuscripts into Arabic and reconciling Greek philosophy with orthodox beliefs.

Pharmacology, optics, chemistry, and alchemy were of particular interest to Arab scientists. Islamic medicine and pharmacology drew on many sources. For example, the famous medical formulary of al-Kindi (Yaqub ibn-Ishaq al-Kindi, ca.800–870) included drugs from Mesopotamia, Egypt, Persia, Greece, and India, as well as traditional Arab remedies. al-Kindi's text served as a model for many Arabic treatises on pharmacology, botany, zoology, and mineralogy. Although the theory of vision may seem a rather esoteric branch of knowledge, al-Kindi argued that it would prove to be the key to the discovery of nature's most fundamental secrets. The Latin version of his work on optics, *De Aspectibus*, was very influential among Western scientists and philosophers, but it was only one of many Arabic works dealing with the anatomy of the eye, its role in vision, and the treatment of eye diseases. Islamic chemists invented and named the alembic, distinguished between alkalis and acids, accomplished the chemical analysis of many complex substances, and acknowledged the value of the exper-

imental approach to the preparation and evaluation of various drugs. Jabir ibn
Hayyan (ca. 721 to ca. 815), the most famous of the Arab alchemists, was
known in medieval Europe as Geber. More than 100 alchemical texts were
attributed to Geber, but many were later forgeries.

Discussions of Arab achievements in science, medicine, and philosophy
once focused on a single question: Were the Arabs merely the transmitters of
Greek achievements, or did they make any original contributions? Such a question
is inappropriate when applied to a period in which the very idea of a quest for
novel, empirical scientific knowledge was virtually unknown. Physicians, philosophers, and other scholars accepted the writings of the ancients as truth, example,
and authority, to be analyzed, explained, and preserved. Having no attachment to
the doctrine of the primacy of originality and progress, scholars saw tradition as
a treasure chest, rather than an obstacle. Like their counterparts in the Christian
West, scholars in the Islamic world had to find a means of peaceful coexistence
with powerful religious leaders who took the position that knowledge could come
only through the prophet Muhammad and his immediate followers. Despite the
competing claims of traditionalists, theologians, and skeptics, Greek philosophy,
science, and medicine eventually captivated Arab physicians and scholars, resulting in the formation of the modified Greco-Arabic medical system that is still
practiced by traditional healers in many parts of the world.

After the triumph of the Abbasid caliphs in 750 and the establishment of
Baghdad as the capital of the Islamic Empire, the scholars of the Age of Translations made it possible to assimilate the intellectual treasures of antiquity into
Islamic culture. The Persian city of Gondeshapur, conquered by the Arabs in
638, became a major center for the dissemination of ancient philosophical manuscripts. The community of scholars who had been drawn to the academy that
flourished in Gondeshapur were familiar with Greek, Aramaic, Hebrew, Syriac,
and Sanskrit. Christians, Jews, and Muslims were soon involved in an unprecedented effort to collect and translate philosophical, medical, and scientific manuscripts into Arabic.

A major challenge for translators was the need to create an appropriate
nomenclature for concepts never before expressed in Arabic. At the school for
translation established during the reign of the Caliph Al-Ma'mun (813–833),
many of Galen's medical and philosophical works were translated into Arabic
versions. One of the most important translators was Hunayn Ibn Ishaq (809–
875), a scholar whose work was a major factor in creating the great Galenic
synthesis. Hunayn and his disciples prepared Arabic editions of Galenic and
Hippocratic texts, the *Materia Medica* of Dioscorides, as well as abstracts, commentaries, and study guides for medical students. Although medieval physicians
and scholars, whether Muslims, Jews, or Christians, generally shared the assumption that Galenism was a complete and perfect system, the new versions
of the ancient texts inevitably reflected the perceptions of translators and editors.

The writings of Islamic physicians and philosophers, which were often presented as commentaries on the work of Galen, were eventually translated from Arabic into Latin and served as fundamental texts in European universities. But for many European scholars, medieval Arabic writers were significant only in terms of the role they played in preserving Greek philosophy during the European "Dark Ages." Much about the history of science and medicine, however, has been revealed through attempts to study the work of Islamic writers on their own terms. For example, Rhazes (865–923; Abu Bakr Muhammad Ibn Zakariya ar-Razi) was one of most scientifically minded physicians of the middle ages. The indefatigable Rhazes was the author of at least 200 medical and philosophical treatises, including the *Continens, or Comprehensive Book of Medicine*. In describing his own methods and goals, Rhazes explained that he had always been moderate in everything except acquiring knowledge and in writing books. By his own reckoning, he had written more than 20,000 pages in one year and had damaged his eyes and hands by working day and night for more than 15 years. This was the model he offered to those willing to commit themselves to the life of the mind. All biographical accounts agree that Rhazes became blind near the end of his life and that he refused treatment because he was weary of seeing the world and unwilling to undergo the ordeal of surgery. Eventually, biographical accounts adopted the spurious but irresistible story that Rhazes lost his vision when his patron al-Mansur had the physician beaten on the head with one of his own books for failing to provide proof of his alchemical theories. Rhazes' book *On Smallpox and Measles* provides valuable information about diagnosis, therapy, and concepts of diseases. Among the ancients, diseases were generally defined in terms of symptoms, such as fevers, fluxes, and skin lesions. By differentiating between smallpox and measles, Rhazes provided a paradigmatic case for thinking in terms of specific disease entities.

Islam's Prince of Physicians, known to the West as Avicenna (980–1037; Abu Ali Hysayn ibn Abdullah ibn Sina), was the first scholar to create a complete philosophical system in the Arabic language. His erudition was widely acknowledged, but his critics contended that he inhibited the development of medicine because no physician was willing to challenge the master of philosophy, natural science, and medicine. Avicenna's great medical treatise, the *Canon*, still serves as a guide to traditional practitioners.

Western scholars long maintained that the major historical contribution of Islamic medicine was the preservation of ancient Greek wisdom and that nothing original was produced by medieval Arabic writers. Because the Arabic manuscripts thought worthy of translation were those that most closely followed the Greek originals (all others being dismissed as corruptions), the original premise—lack of originality—was confirmed. The strange story of Ibn-an-Nafis (Al-Qurashi, d. 1288) and his discovery of the pulmonary circulation demonstrates the unsoundness of previous assumptions about the Arabic literature. The

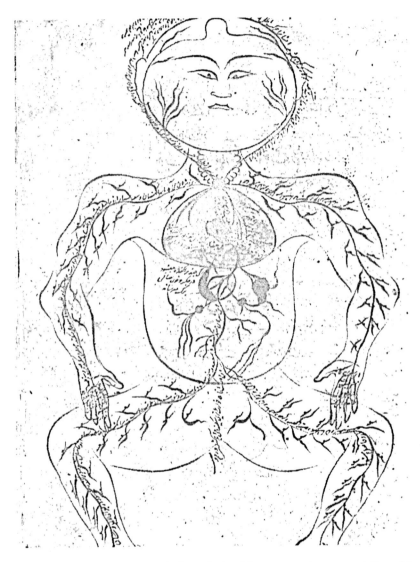

The internal organs of the human body as depicted in a manuscript representative of the
Golden Age of Islam

writings of Ibn-an-Nafis were essentially ignored until they came to the attention of twentieth-century historians of science.

Honored by his contemporaries as a learned physician and ingenious investigator, Ibn-an-Nafis was described as a tireless writer and a man of such great piety that when he was advised to take wine as a medicine for a dangerous illness he refused because he did not wish to meet his Creator with alcohol in his blood. In his writings, Ibn-an-Nafis invoked religious law and his own natural charity to explain why he did not perform dissections. Just how Ibn-an-Nafis was led to question Galenic physiology and discover the pulmonary circulation is unclear. Reading the works of other scholars and studying the medical illustrations typically found in medieval Arabic and Persian manuscripts were not likely to promote novel ideas.

After presenting a fairly conventional discussion of the structure and function of the heart, an-Nafis departed from Galenic orthodoxy and insisted that there were no visible or invisible passages between the two ventricles of the heart. Moreover, he argued that the septum between the two ventricles was thicker than other parts of the heart in order to prevent the harmful and inappropriate passage of blood or spirit between them. To account for the presence of blood in both sides of the heart, he reasoned that after the blood had been refined in the right ventricle, it was transmitted to the lung, where it was rarefied and mixed with air. The finest part of this blood was then clarified and transmitted from the lung to the left ventricle. In other words, the blood could only get into the left ventricle by way of the lungs.

Perhaps some still obscure Arabic, Persian, or Hebrew manuscript contains a commentary on the curious doctrines of Ibn-an-Nafis, but there is as yet no evidence that later Arab authors or European anatomists were interested in these anti-Galenic speculations. Thus, although Ibn-an-Nafis did not influence later writers, because what is hidden and unknown cannot be considered an influence, the fact that his concept was so boldly stated in the thirteenth century should lead us to question our assumptions about progress and originality in the history of science. Since only a small percentage of the pertinent manuscripts have been studied, edited, translated, and published, the question may go unanswered for quite some time.

SUGGESTED READINGS

Afnan, S. M. (1958). *Avicenna: His Life and Works*. London: G. Allen & Unwin.

Africa, T. W. (1968). *Science and the State in Greece and Rome*. New York: Wiley.

Amundsen, D. W. (1996). *Medicine, Society, and Faith in the Ancient and Medieval Worlds*. Baltimore: Johns Hopkins University Press.

Artz, F. B. (1980). *The Mind of the Middle Ages. An Historical Survey, A.D. 200–1500*. Chicago: University of Chicago Press.

Ascher, A., Halasi-Kun, T., and Kiraly, B. K., eds. (1978). *The Mutual Effects of the Islamic and Judeo-Christian Worlds*. Brooklyn, NY: Brooklyn College Press.

Avicenna (1930). *A Treatise on the Canon of Medicine of Avicenna Incorporating a Translation of the First Book*. Trans. by O. C. Gruner. London: Luzac.

Bender, T., ed. (1988). *The University and the City. From Medieval Origins to the Present*. New York: Oxford University Press.

Best, M. R., and Brightman, F., eds. (1973). *The Book of Secrets of Albertus Magnus of the Virtues of Herbs, Stones and Certain Beasts, also A Book of the Marvels of the World*. Oxford, England: Clarendon Press.

Bittar, E. E. (1955). A study of Ibn Nafis. *Bull. History Med.* 29:352–368, 429–447.

Brain, Peter (1986). *Galen on Bloodletting*. New York: Cambridge University Press.

Celsus (1960–61). *De medicina*. 3 vols. Trans. by W. G. Spencer. Loeb Classical Library. Cambridge, MA: Harvard University Press.

Chishti, G. M. (1988). *The Traditional Healer: A Comprehensive Guide to the Principles and Practice of Unani Herbal Medicine*. Rochester, VT: Healing Arts Press.

Dales, R. C. (1973). *The Scientific Achievement of the Middle Ages*. Philadelphia: University of Pennsylvania Press.

Dioscorides. *The Greek Herbal of Dioscorides*. (Illus. by a Byzantine in 512 A.D., Englished by John Goodyear, 1655 A.D.) Ed. by R. T. Gunther. Reprinted 1968. New York: Hafner.

Eamon, W. (1994). *Science and the Secrets of Nature: Books of Secrets in Medieval and Early Modern Culture*. Princeton, NJ: Princeton University Press.

French, R., and Greenaway, F., eds. (1986). *Science in the Early Roman Empire: Pliny the Elder, His Sources and His Influence*. Totowa, NJ: Barnes & Noble Books.

Friedman, J. B. (1981). *The Monstrous Races in Medieval Art and Thought*. Cambridge, MA: Harvard University Press.

Galen (1962). *On Anatomical Procedures. The Later Books*. Trans. by W. L. H. Duckworth. Cambridge: Cambridge University Press.

Galen (1968). *On the Usefulness of the Parts of the Body*. Trans., intro., and commentary by M. Tallmadge May. 2 vols. Ithaca, NY: Cornell University Press.

Galen (1984). *Galen: On Respiration and the Arteries*. Ed. and trans. by D. J. Furley and J. S. Wilkie. Princeton, NJ: Princeton University Press.

George, W. B., and Yapp, W. B. (1991). *The Naming of the Beasts: Natural History in the Medieval Bestiary*. London: Duckworth.

Gohlman, W. E. (1974). *The Life of Ibn Sina; A Critical Edition and Annotated Translation*. Albany: State University of New York Press.

Goichon, A. M. (1969). *The Philosophy of Avicenna and Its Influence on Medieval Europe*. Delhi: Motilal Banarsidass.

Goodman, L. E. (1992). *Avicenna*. New York: Routledge.

Grant, E. (1981). *Studies in Medieval Science and Natural Philosophy*. London: Variorum Reprints.

Grant, E. (1996). *The Foundations of Modern Science in the Middle Ages: Their Religious, Institutional, and Intellectual Contexts*. New York: Cambridge University Press.

Grmek, M. D., ed. (1998). *Western Medical Thought from Antiquity to the Middle Ages*. Cambridge, MA: Harvard University Press.

Grunebaum, G. D. (1963). *Lucretius and His Influence*. New York: Cooper Square.

Gruner, O. C. (1967). *Commentary on the Canon of Avicenna*. Karachi: Hamdard Foundation.

Hackett, J., ed. (1997). *Roger Bacon and the Sciences: Commemorative Essays*. New York: Brill.

Hahm, D. E., with Scarborough, J., and Toomer, G. J. (1984). *The History of Greek and Roman Science. A Selected, Annotated Bibliography*. New York: Garland Publishing, Inc.

Harris, C. R. S. (1973). *The Heart and the Vascular System in Ancient Greek Medicine from Alcmaeon to Galen*. Oxford: Clarendon.

Hildegard von Bingen (1994). *The Letters of Hildegard of Bingen*. Translated by J. L. Baird, and R. K. Ehrman. New York: Oxford University Press.

Hughes, J. D. (1994). *Pan's Travail: Environmental Problems of the Ancient Greeks and Romans*. Baltimore, MD: Johns Hopkins University Press.

Kibre, P. (1984). *Studies in Medieval Science: Alchemy, Astrology, Mathematics and Medicine*. London: Hambledon Press.

Kudlien, F., and Durling, R. J., eds. (1990). *Galen's Method of Healing*. New York: E. J. Brill.

Langermann, Y. T. (1999). *The Jews and the Sciences in the Middle Ages*. Brookfield, VT: Ashgate/Variorum.

Levey, M. (1973). *Early Arabic Pharmacology. An Introduction Based on Ancient and Medieval Sources*. Leiden: E. J. Brill.

Lindberg, D. C., ed. (1978). *Science in the Middle Ages*. Chicago: University of Chicago Press.

Lloyd, G. E. R. (1973). *Greek Science After Aristotle*. New York: Norton.

Luck, G. (1985). *Arcana mundi: Magic and the Occult in the Greek and Roman Worlds*. Baltimore: Johns Hopkins University Press.

Makdisi, G. (1981). *The Rise of Colleges. Institutions of Learning in Islam and the West*. Edinburgh: Edinburgh University Press.

Matheson, L. M., ed. (1994). *Popular and Practical Science of Medieval England*. East Lansing, MI: Colleagues Press.

Mayor, A. (2000). *The First Fossil Hunters. Paleontology in Greek and Roman Times*. Princeton, NJ: Princeton University Press.

Momigliano, A. D. (1975). *Alien Wisdom: The Limits of Hellenization*. Cambridge: Cambridge University Press.

Murdoch, J. E. (1984). *Antiquity and the Middle Ages*. New York: Scribner.

Nasr, S. H. (1968). *Science and Civilization in Islam*. Cambridge, MA: Harvard University Press.

Newman, B. (1987). *Sister of Wisdom: St. Hildegard's Theology of the Feminine*. Berkeley, CA: University of California Press.

Nutton, V., ed. (1981). *Galen: Problems and Prospects*. London: Wellcome Institute for the History of Medicine.

Parsons, E. A. (1952). *The Alexandrian Library: Glory of the Hellenic World*. New York: Elsevier.

Pliny. (1956–1966). Natural History. 10 vols. Trans. by H. Rackham, W. H. S. Jones, and D. E. Eichholz. Loeb Classical Library. Cambridge, MA: Harvard University Press.

Pojman, L. P., ed. (1998). *Classics of Philosophy*. New York: Oxford University Press.

Riddle, J. M. (1985). *Dioscorides on Pharmacy and Medicine*. Austin, TX: University Texas Press.

Riddle, J. M. (1992). *Quid pro quo: Studies in the History of Drugs*. Brookfield, VT: Ashgate Pub. Co.

Roberts, L. D., ed. (1982). *Approaches to Nature in the Middle Ages*. New York: Center for Medieval & Renaissance Studies.

Sarton, G. (1954). *Galen of Pergamon*. Lawrence, KS: University of Kansas Press.

Sassi, M. M. (2001). *The Science of Man in Ancient Greece*. Chicago: University of Chicago Press.

Schipperges, H. (1997). *Hildegard of Bingen: Healing and the Nature of the Cosmos*. Princeton, NJ: M. Wiener.

Shank, Michael H., ed. (1996). *The Scientific Enterprise in Antiquity and Middle Ages. Readings from Isis*. Chicago: University of Chicago Press.

Siegel, R. E. (1968). *Galen's System of Physiology and Medicine. An Analysis of his Doctrines and Observations on Bloodflow, Respiration, Humors and Internal Diseases*. Basel: Karger.

Siegel, R. E. (1970). *Galen on Sense Perception. His Doctrines, Observations and Experiments on Vision, Hearing, Smell, Taste, Touch and Pain, and Their Historical Sources*. Basel: Karger.

Siegel, R. E. (1973). *Galen on Psychology, Psychopathology, and Function and Diseases of the Nervous System; An Analysis of His Doctrines, Observations and Experiments*. New York: Karger.

Siraisi, N. G. (1973). *Arts and Sciences at Padua. The Studium of Padua Before 1350*. Toronto: Pontifical Institute of Mediaeval Studies.

Siraisi, N. G. (1990). *Medieval and Early Renaissance Medicine: An Introduction to Knowledge and Practice*. Chicago: University of Chicago Press.

Staden, H. von (1989). *Herophilus. The Art of Medicine in Early Alexandria*. New York: Cambridge University Press.

Stahl, W. H. (1962). *Roman Science: Origins, Development, and Influence to the Later Middle Ages*. Madison, WI: University of Wisconsin Press.

Steneck, N. H. (1976). *Science and Creation in the Middle Ages: Henry of Langenstein (d. 1397) on Genesis*. South Bend, IN: University of Notre Dame Press.

Strehlow, W., and Hertzka, G. (1988). *Hildegard of Bingen's Medicine*. Santa Fe, NM: Bear & Co.

Temkin, O. (1973). *Galenism: Rise and Decline of a Medical Philosophy*. Ithaca, NY: Cornell University Press.

Temkin, O. (1991). *Hippocrates In a World of Pagans and Christians*. Baltimore: Johns Hopkins University Press.

Turner, H. R. (1997). *Science in Medieval Islam: An Illustrated Introduction*. Austin, TX: University of Texas Press.

Wickens, G. M., ed. (1952). *Avicenna: Scientist and Philosopher. A Millenary Symposium*. London: Luzac.

3

THE RENAISSANCE AND THE SCIENTIFIC REVOLUTION

The European Renaissance has been called the age of discovery, but it was also the time of the rediscovery of Greek philosophy, art, and science. Thus, although the Renaissance seems in many ways unique and distinct, it might also be thought of as the natural culmination of medieval thought as well as a transitional period between the Middle Ages and modernity. Broadly speaking, the Renaissance, a term that literally means rebirth, comprises the period from about 1300 to 1650. Some historians, however, contend that focusing on the Renaissance as a separate historical period may cause us to misread the flow and meaning of ideas and events.The factors that led to and sustained the transformation of European intellectual, social, economic, and political life are complex and manifold. As medieval institutions disintegrated, the unifying principles of medieval society were challenged by new ideas and facts generated by the exploration of the word, the world, the heavens, and the human body.

Of course purely intellectual influences were not the only stimuli for change in Europe. The Renaissance was more than a rebirth of art and science; it was also a period of profound social and economic dislocation. Historians now see it as an era where social mobility, both upward and downward, became possible in a manner unthinkable during the Middle Ages. The newly emerging sense of individualism was also quite unlike the group identification characteristic of the Middle Ages. The new individualism was marked by an obsession with wealth, social status, and its symbols, a growing distance between the higher and lower social orders, a new economy whose inflationary tendencies often left peasants and artisans worse off than before, and struggles between the traditional feudal aristocracy and the new monarchies. The ravages of the great pandemic of bubonic plague of 1348, which may have killed one quarter to one half of the population in some regions, played a part in the broader transformation of society. Striking down peasants, clergymen, and nobles, the plague led to labor shortages, peasant uprisings, population shifts, and general confusion for church and state.

The expansion of commerce, mining, the growth of towns and cities, new inventions such as gunpowder weapons and the printing press, the development of a sea route to India, and the discovery of the New World were among the factors that encouraged cultural renewal and scientific advances. Natural philosophers, physicians, and surgeons were confronted with plants, animals, and diseases absolutely unknown to Hippocrates, Galen, and Pliny. Triumphantly exploring and encircling the earth, Europeans left in their wake a series of unprecedented ecological and demographic catastrophes, whose consequences are still unfolding.

One witness and participant in the great events of this era, philosopher of science Francis Bacon (1561–1639), said that of all the products of human ingenuity, the three most significant in terms of their "force, effect, and consequences" were the compass, gunpowder, and printing. Through their combined effects, Bacon said, these inventions had "changed the appearance and state of the whole world." At least in the West, the availability of printing and firearms at the end of the Middle Ages had an effect similar to that of the introduction of iron and the alphabet at the end of the Bronze Age. Gunpowder, a mixture of saltpeter, carbon, and sulfur, which was first reported in Europe in the thirteenth century, was probably invented much earlier in China. Documents from the 1320s depict cannons that were primitive but presumably deadly. Further refinements of gunpowder weapons meant that the mounted knight and the fortified castle were no longer invincible. Gunpowder weapons created injuries that could not be found in the writings of Galen and Hippocrates. Military power became concentrated in the hands of those who could control the production of gunpowder and cannons. The rise of absolute monarchies and the modern state was, therefore, given some impetus by the development of firearms.

While the printing press did not create the Renaissance, which had begun in Italy long before this invention appeared in Europe, it probably did sustain and accelerate the advance of the movement into northern Europe. Printing and paper making, like gunpowder, appeared first in China. According to Chinese chronicles, Ts'ai Lun, the god of paper makers, had presented the art of paper making to the Emperor of China in 105 A.D. Paper of many different kinds, colors, weights, and quality had been developed in China by the time the Arabs learned the art at Samarkand and brought it to Spain and Italy in the thirteenth century. The Chinese also invented various kinds of ink, including a special oily ink that was suitable for printing. At least in part because of the way in which

Chinese and European languages are written, the impact of new printing techniques was very different in China and Europe. For writing systems based on an alphabet, the invention of movable print was the key to the print revolution. The Chinese system, with many thousands of separate symbols or ideographs, was better served by block printing, a very ancient technique that was used to create seals, charms, and decorations, as well as books. Movable metal type was first invented in Korea, probably in the thirteenth century. The official Korean chronicles, however, set the date as 1392, when the Department of Books was established to take responsibility for the casting of type and printing of books. In China and the nations that used the Chinese writing system, printing was valued primarily as a means of preserving and authenticating the text, rather than a way of producing large numbers of relatively inexpensive books.

The earliest known document to be printed with movable type in Europe was a letter of indulgence printed by Johann Gutenberg (ca. 1398–1468) in 1454. Although Gutenberg's name is today almost synonymous with the early modern method of printing, it is likely that what he really did was improve techniques for cutting and casting metal type. By 1500, printing presses had been established throughout Europe, launching a communications revolution that accelerated the trend towards literacy, the diffusion of ideas and information, and the establishment of a popular vernacular literature. Despite some grumbling about the vulgarity of printing as compared to making copies by hand and the elitist fear than an excess of literacy might be subversive, books were soon being delivered to eager buyers by the wagonload. While most people remained illiterate, the print revolution initiated a transformation from a "scribal" and "image culture" to a "print culture." Reforms swept through the arena of elementary education as well as the university. Mass-produced books changed the relationship between teachers and students and made it possible for the young to study, and perhaps even learn, by reading on their own. It has been suggested that printing paved the way for the Enlightenment, the American and French Revolutions, and democratic movements. On the other hand, both sense and nonsense could be reproduced and distributed with unprecedented rapidity.

Scholars have raised questions about the uniqueness of the Renaissance and the Scientific Revolution, the fundamental change in the sciences that took place during the sixteenth and seventeenth centuries. Although the term *science* was not adopted as a substitute for natural philosophy until the nineteenth century, a form of inquiry that has become known as the scientific method was established during the seventeenth century. The exact relationship between the Renaissance and the Scientific Revolution is complex and controversial. Some scholars see the Scientific Revolution as a creation of the middle third of the twentieth century, when the term was introduced and popularized by scholars who believed that the establishment of the modern scientific attitude was probably more significant than the Renaissance and Reformation. The term Scientific

Revolution has been challenged and criticized for conjuring up the image of a sudden and definitive break with ancient ideas. Moreover, by the end of the twentieth century new approaches to the history and philosophy of science and changing attitudes towards science and modernity led some scholars to question the very existence of a scientific revolution. Despite these continuing disputes, the idea of the Scientific Revolution has been useful in exploring important aspects of the emergence of modern science. While pursuing innovative goals and methods, Renaissance artists, scholars, and scientists often appealed to classical authorities. The interplay between appeals to the past and aspirations to originality can be illuminated by considering the work of several key figures among the many remarkable artists, scientists, and philosophers of this era.

ART AND ANATOMY

Artists as well as anatomists are inextricably linked to the reform of anatomy during the Renaissance. Both medicine and art required accurate anatomical knowledge. Its special relationship with anatomy, optics, and mathematics, as well as a return to Greek ideals that encouraged the glorification and study of the human body, generated much of the distinctive character of Renaissance art. Dedicated to a more accurate portrayal of nature, Renaissance artists found that knowledge of the exterior of the body alone was insufficient to their artistic goals. Convinced that the study of the dead would make art more true to life, they wanted to study the working of muscles and bones and the organs of the interior of the body. Unlike their Greek predecessors, Renaissance artists were able to satisfy their demands for realism by observing and even performing human dissection. Many artists attended public anatomies and executions, studied intact and flayed bodies, and carried out dissections with their own hands, but probably none equaled Leonardo da Vinci (1452–1519) in terms of his insatiable curiosity and artistic and scientific imagination.

Leonardo was the illegitimate son of a Florentine lawyer and a peasant woman. Although his father married four times, Leonardo, who was brought up in his father's house, remained an only child until he was about 20. At 14 years of age, Leonardo was apprenticed to Andrea del Verrochio (1435–1488), the foremost teacher of art in Florence. Verrochio insisted that all his pupils learn anatomy. This included the study of surface anatomy and flayed bodies, attending public anatomies and executions, and carrying out animal dissections. Within 10 years Leonardo was recognized as a distinguished artist, but instead of settling down and turning out works pleasing to the wealthy and powerful Florentine patrons of the arts, he embarked on a restless and adventurous life, perhaps triggered by accusations of homosexuality.

Leonardo's genius and imagination have been universally applauded, but attempts to assess his influence on the sciences generally emphasize his penchant for secrecy and his failure to complete many of his ambitious projects. His legacy is quite modest when assessed in terms of his formidable gifts as painter, sculptor, architect, military engineer, and inventor. He did not leave a single complete statue, machine, or book despite sketches and plans for many exciting and innovative constructions. Thousands of pages of notes, full of ingenious plans and projects, sketches and hypotheses, went unpublished for centuries. Indeed, the secretive left-handed artist kept his notebooks in a kind of code. It is tempting to speculate that if Leonardo had published his preliminary investigations he might have revolutionized several scientific disciplines, but that which is unknown and incomplete cannot be considered a contribution to science. Nevertheless, it is worth remembering that Leonardo's unfinished work indicates the range of ideas and dreams possible in fifteenth-century Europe.

Observing the workings of nature in fine detail and grand design, Leonardo's mercurial mind darted from astronomy to anatomy, from music to machines. He thought about the movement of the earth, the nature of sound and light, the use of rings to determine the age of trees, and the nature of fossil shells. Studies of the superficial anatomy of the human body inexorably led to an exploration of general anatomy, comparative anatomy, and physiological experiments. Through dissection and experimentation, Leonardo believed he would uncover the mechanisms that governed movement and even life itself. Although many artists saw anatomical dissection only as a means to an end, Leonardo came to value anatomy as a science worth pursing. In addition to attending public anatomies, he dissected about 30 human bodies, including a 7-month fetus and a very elderly man, at a hospital in Florence.

Not all of Leonardo's methods are known, but he seems to have been the first anatomist to make wax casts of the ventricles of the brain by injecting hot wax into these hollow areas. When the wax solidified it was possible to do careful dissections of the delicate tissues. Leonardo used serial sections and developed other techniques for studies of soft tissues, such as the eye. To facilitate dissection, he placed the eye in the white of an egg and heated it until the albumen coagulated. He constructed models to study the mechanism of action of muscles and the heart valves and carried out vivisections to give him insight into the heartbeat. For example, he drilled through the thoracic wall of a pig and, keeping the incision open with pins, observed the motion of the heart. Although he realized that the heart was actually a very powerful muscle, he generally accepted Galen's views on the movement and distribution of the blood, including the imaginary pores in the septum. Leonardo's dissections of insects, fish, frogs, horses, dogs, cats, birds, and so forth, represent the first significant efforts since Aristotle and Galen to seriously exploit the advantages of comparative anatomy. Like so many of his projects, Leonardo's great book on human anatomy was left unfinished. When he died, his manuscripts were scattered among various libraries, and some were probably lost.

Renaissance artists and anatomists were not the first Europeans to revive the Alexandrian practice of human dissection. Anatomies had been a component of medical education in medieval Italian universities, and postmortems were occasionally conducted as part of the investigation of suspicious deaths and epidemic disease. Nevertheless, such events were quite rare. When anatomies were performed, the stylized ceremony was unlikely to produce new insights or inspire the typical student to challenge established authorities or engage in original research. The professor, trained in the scholastic pattern, did not perform the dissection himself. A barber-surgeon or technician did the actual autopsy, while the professor explicated passages from Galen. If the reader and technician were well synchronized, the students would see the organs as the professor described them, but even this was unlikely.

After a brief display of the mangled internal organs, the technician was supposed to demonstrate the muscles, nerves, and blood vessels. But this was generally too difficult for the technician and too boring for the onlookers. Presumably, Galen himself would have objected to this procedure, because he knew that he could not entrust even the task of skinning his Barbary apes to an assistant without losing valuable information. In any case, the audience generally preferred the scholastic disputations between the Galenists of the medical faculty and the Aristotelian professors of philosophy to the autopsy itself. Medical students knew that examinations and dissertations were based on authoritative texts, and they attended dissections primarily as a means of confirming accepted ideas and preparing for examinations. Like students in a modern laboratory course performing a typical "cookbook" experiment, medieval and Renaissance students were engaged in acquiring standard techniques rather than a search for novel observations.

Even if Renaissance anatomists dissected only to supplement their studies of Galenic texts, their efforts were confounded by many difficulties. In addition to the chronic shortage of bodies, anatomists lacked reliable guides to direct their efforts and tools with which to exploit the available materials. Furthermore, it would be difficult for them to express potentially interesting observations because, in the absence of a standardized nomenclature, the parts of the body were described in a jumble of Greek, Latin, Arabic, and vernacular terms. The same structure might be referred to by different names, or the same term might be applied to different structures. Illustrations appeared in some anatomical texts, but they were generally more distracting than informative. Formalized depictions of autopsies, "wound men," and astrological figures appeared in some texts, while others featured cheerful looking people holding up flaps of skin to display internal organs. Despite these deficiencies, by 1500 many of the most important Greek and Arabic manuscripts had been translated and anatomists were preparing Latin editions suitable for the use of medical students. Indeed, assisting Johannes Guinter (1505–1574) in preparing his text *Anatomical Insti-*

tutions According to the Opinion of Galen for Students of Medicine was an important step in the scientific awakening of Andreas Vesalius (1514–1564), the reformer of anatomy.

ANDREAS VESALIUS ON THE FABRIC OF THE HUMAN BODY

During the year 1543, two books appeared that challenged long-held convictions about the nature of the heavens and the human body. Just as *De revolutionibus orbium coelestium* (*The Revolutions of the Heavenly Spheres*), by Nicolaus Copernicus (1473–1543), transformed ideas about the movements of the earth and the heavens, *De humani corporis fabrica* (*On the Fabric of the Human Body*), by Andreas Vesalius, revolutionized ideas about the structure of the human body. Human dissection, the foundation of Vesalian reforms, became a central theme in medical education during the European Renaissance. Vesalius regarded his treatise as the first advance in anatomy since the time of Galen. Indeed, when the young Vesalius began his study of human anatomy, the authority of the ancient anatomist had grown so weighty that anatomists tended to explain away discrepancies between his descriptions and their own observations in terms of individual abnormalities or changes in the human body that had occurred since the time of Galen. For example, Jacobus Sylvius (1478–1555) defended Galen's description of the human sternum as made up of seven bones on the grounds that human beings in a more heroic age might have had more bones than their degenerate descendants.

According to a horoscope cast by Girolamo Cardano, Vesalius was born at 5:45 a.m. on December 31, 1514, in Brussels, Belgium. His father was imperial pharmacist to Charles V and often accompanied the emperor on his travels. When quite young, Vesalius began to teach himself anatomy by dissecting mice, moles, and other small animals. Although he studied at both Paris and Louvain, institutions noted for their extreme conservatism, his innate curiosity overcame the potentially stifling effect of education. As a medical student at the University of Paris, Vesalius served as assistant to Jacobus Sylvius, a dominant influence in Parisian anatomy and a staunch Galenist. To purge anatomy of the errors introduced by medieval copyists and translators, Sylvius was preparing a new anatomical textbook based on the work of Galen. While carrying out dissections directed by Sylvius, Vesalius became increasingly critical of professors who studied Galen's words and lectured about things they had never investigated for themselves. Later Vesalius described Sylvius as a bad-tempered man, more adept at using his knife at the banquet table than in the dissecting room. Pursuing his own interests, Vesalius became so skillful that he was called upon to perform public dissections. The outbreak of war forced him to leave Paris without a degree and return to Louvain in 1536.

After a brief stay in Louvain, Vesalius enrolled in the medical school of the University of Padua in Italy. In December 1537 he received his degree and an appointment as lecturer-demonstrator in anatomy and surgery. Because Padua was a relatively enlightened institution, Vesalius was able to institute many reforms in the teaching of anatomy. His lecture-demonstrations attracted hundreds of observers, although, in fact, dissection had little or no direct relevance for the medical practice of university-educated physicians. Surgeons and barber-surgeons, who actually performed bleeding and operations, were generally trained by apprenticeship rather than the formal university education that empowered physicians.

Never one to spare the feelings of his colleagues, Vesalius told students that they could learn more at a butcher shop than from arrogant professors who knew only the words of Galen, who had never had the opportunity to dissect human beings. Obtaining bodies for dissection, however, remained a chronic problem for anatomists and medical schools. Even though human dissection became a central theme in medical education, little progress was made in convincing the authorities to increase the supply of cadavers. Since medical students were required to observe human anatomies, teachers sometimes encouraged unorthodox means of obtaining the necessary material. Young anatomists often obtained at least part of their education in graveyards, snatching bones out of the teeth of savage dogs. In autobiographical passages Vesalius described the risks he had taken to obtain bodies. On one occasion he stole the bones of a criminal left on the gallows. He had to spend several nights outside the town, sneaking back each morning with bones hidden under his coat until finally he had a complete skeleton under his bed.

The anatomies conducted by Vesalius occupied him from morning to night for 3 weeks at a time. To minimize the problem of putrefaction, anatomies were done in the winter. A judge in Padua's criminal court became so interested in Vesalius's work that he thoughtfully set the time for the execution of convicted felons to suit the anatomist's schedule. Several bodies were dissected at the same time so that different parts of the body could be studied and their relationships examined. Large diagrams were prepared for the guidance of students. The anatomies began with the skeleton and then proceeded to the muscles, blood vessels, and nerves. Another cadaver would be used for the demonstration of the organs of the abdomen, chest, and brain. To improve the technical aspects of dissection, Vesalius introduced many new tools, some of his own design and some based on those used by various artisans he had consulted. Human dissection as performed by Vesalius in the crowded anatomical theater of Padua was a dramatic as well as enlightening event.

In 1540, as a mark of his independence from Galen, Vesalius conducted a dramatic demonstration in which he assembled the bones of an ape and a man in order to show that Galen had made hundreds of errors with respect to human

The title page of *On the Fabric of the Human Body* shows Andreas Vesalius, surrounded by excited observers, performing a dissection in the crowded anatomical theater at Padua.

anatomy. Finally, in 1543 Vesalius published his great anatomical treatise *On the Fabric of the Human Body* (*Fabrica*) and a shorter text, known as the *Epitome*, for students. The *Epitome* even contained figures rather like paper dolls and pictures of the internal organs that the reader could cut out and reassemble. About 250 blocks had been painstakingly prepared for incorporation into the text so that words and illustrations would complement and clarify each other. Both texts were extensively plagiarized and disseminated in editions so carelessly prepared that they caused Vesalius considerable pain and embarrassment. Of course the *Fabrica* infuriated conservative Galenists who denounced the text as "filth and sewage" and renamed its author "Vesanus" (madman). Conjuring up images of Vesalian anatomists as the "Lutherans of Physic," Galenists warned that the heresy of such medical innovators was as dangerous as Martin Luther's (1483–1546) effect on religion. Ironically, some critics attacked the *Fabrica* on the grounds that students would consider the remarkable illustrations a valid substitute for participation in dissections. Of course Vesalius, the supreme champion of anatomical research, constantly emphasized the importance of performing dissections.

Disgusted by the attacks of his conservative opponents, Vesalius decided to abandon research and the academic world. Following in the footsteps of his father, he became court physician to Emperor Charles V, Holy Roman Emperor and King of Spain, to whom he had dedicated the *Fabrica*. Once in service to the emperor, Vesalius had little opportunity for research or writing. The patronage of kings, popes, and wealthy noblemen was as important to Renaissance researchers as the patronage of government agencies and private foundations is to modern scientists, and imperial service could be almost as unpleasant as the stormy academic world. Charles V was a particular challenge to his physicians, with his bad combination of gout, asthma, gargantuan appetites, and predilection for quack remedies. When the University of Pisa offered Vesalius the Chair of Anatomy, Charles refused to release him. In 1556 Charles abdicated in order to retire to a monastery. Vesalius was transferred to the service of Philip II, the son of Charles V and successor to the throne. During this period Vesalius was sent to several other royal courts. When Henry II of France was injured while jousting, Vesalius and the famous French barber-surgeon Ambroise Paré (1517–1590) were among the medical consultants. After experiments on the heads of condemned criminals, Paré and Vesalius agreed that the wound was fatal. Unhappy with his role in service to Philip II, Vesalius hoped to return to the academic world and embarked on a pilgrimage to the Holy Land as a means of escaping from his employment. The fact of his death on October 15, 1564, during the return voyage from Jerusalem is known, but the exact cause and place of his death are uncertain.

Given the academic climate of his time, it is remarkable that Vesalius, a man steeped in conservative scholarly culture, confronted and rejected the

DECIMA SEPTIMI LIB
FIGVRA

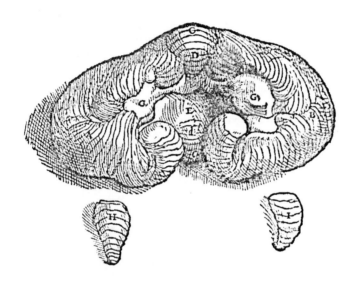

CIMAE FIGVRAE, EIVSDEMQVE C

Interior view of the cerebellum as depicted in *On the Fabric of the Human Body* (1543)

Galenic dogmas that generations of philosophers and physicians had accepted virtually without question. To understand the nature of the Vesalian reform of anatomy, we must abandon modern assumptions about the nature of progress and the evolution of science. Then we must ask what factors within the context of sixteenth-century science and culture forced Vesalius to demand that henceforth anatomists study only the "completely trustworthy books of man." After all, Vesalius was a scholar who knew the works of Galen better than most of his critics.

On several occasions Vesalius attributed his disillusionment with Galen to his "discovery" that Galen had never dissected the human body, but Galen had openly confessed to this deficiency. It is more likely that practical problems such as therapeutic venesection forced Vesalius to question Galenic dogma. As the texts of Hippocrates and Galen became available in their pristine form, medical humanists and physicians engaged in violent controversies about the proper means of performing phlebotomies. When Hippocratic and Galenic texts contradicted each other, the blame for obscurity and confusion could no longer be attributed solely to corrupt medieval influences. How then should physicians determine the proper site for venesection, the amount of blood that should be taken, and how often bleeding should be repeated? Which authorities should govern medical and surgical practice? When Vesalius began to ask himself whether such questions could be answered by means of facts established by anatomical investigations, the learned debates among his colleagues seemed to be nothing more than "horsefeathers and trifles."

While the *Fabrica* retained some errors reflecting the survival of Galenic traditions, Vesalius came close to this goal of describing the human body as it really is, without deference to ancient authorities. Vesalius opened the *Fabrica* with a defense of science and a warning about the dangers that result from obstacles to the arts and sciences. Along with a liberal dose of flattery for Charles V and the University of Padua, Vesalius praised the Senate of Venice for their liberality towards learning. Some of his arguments could well grace a modern research grant proposal. The first book of the *Fabrica* is devoted to the skeleton supporting the whole body and the second to the muscles. The third book describes the vascular system. The fourth book is devoted to the description of the nervous system. The abdominal viscera and generative organs are described in the fifth book; the sixth book describes the thoracic viscera. In book seven, Vesalius describes the brain, pituitary gland, and the eye. A concluding chapter on animal vivisection provided students with exercises to train their hands and senses for surgery and the closure of wounds.

In contrast to his success in anatomy, Vesalius did not establish any physiological concepts or methods that were fundamentally different from those of Aristotle and Galen. Although Vesalius gave an exhaustive description of the structure of the heart, arteries, and veins, he generally followed Galenic concepts

ıotum perdat,illumɋ; denuò occluſo uentriculo
ɛtionem quam me deſcripturum paulo antè pol
ı:quanquam uocis occaſione ſuem accipere ma-
utcunɋ illum afficias, ſubinde neque latrat,neɋ
n interdum expendere nequis. Primùm igitur
:dem,& liberum corporis truncum porrigat,aſ-
madmodum hîc modò interiecta tabella propo-

ʻgata,ac dein cuipiam aſſeris anulo, aut foramini,
:et collum exporrectum,caputɋ immotum ſit,&
     K k   interin

A depiction of a pig on a dissecting table

in explaining the function of these organs. He searched diligently for the Galenic pores in the septum of the heart but, unable to find them, concluded that they either did not exist or were invisible. In the second edition of the *Fabrica*, Vesalius denied that any Galenic channels could be found in the septum of the heart, but rather than using this as a way to question Galenic physiology, he merely called the reader's attention to this anatomical mystery.

Despite the rapid dissemination of the *Fabrica* and a plethora of unauthorized imitations, many anatomists continued to see the human body largely in terms of Galenic structures. Access to human cadavers remained limited, and anatomists often used the same animals Galen had dissected in order to clarify their studies of particular organs. Even Vesalius admitted that when he began performing public anatomies he always kept the head of an ox handy to demonstrate Galen's *rete mirabile* (wonderful net), a network of vessels that is found at the base of the brain in cattle but not in humans. Galenic theory required the presence of this "wonderful net" because it allegedly produced animal spirits. After he freed himself from Galen's influence, Vesalius openly declared this network was not present in humans and ridiculed his own youthful credulity. Nevertheless, although Vesalius and his contemporaries could accept originality in terms of direct observations and isolated facts, they found it more difficult to set themselves free of Aristotelian and Galenic theories. Thus, even if the *rete mirabile* was not found in humans, it was possible to suggest an alternative locus for the generation of the animal spirits.

Long after anatomical research was recognized as the cornerstone of Western medical science, anatomists were often forced into methods of obtaining bodies that were just as dangerous as those employed by the young Vesalius. Medical schools did not officially assist or encourage body snatching, but some were known to provide free tickets to the dissecting rooms for enterprising students. Rumors of bizarre experiments allegedly carried out by medical scientists reinforced widespread fear of those who carried out dissections and perhaps worse atrocities on human bodies. Even if anatomists were fortunate enough to receive the bodies of criminals who had earned the dreaded sentence of "death and dissection," they still faced grave dangers while conducting anatomies. Many anatomists apparently succumbed to the "putrid miasma" of the dissecting room, because the smallest cut acquired during dissection could become infected, resulting in a fatal blood poisoning.

THE RENAISSANCE, NATURAL MAGIC, AND ALCHEMY

Although the Renaissance has been characterized as the age of the rebirth of art and science, it was also an era in which superstition, mysticism, and the occult sciences flourished. Indeed, areas that today are unequivocally distinguished from the sciences were often inseparable aspects of the way in which Renaissance scholars and philosophers approached the mysteries of nature. Many varieties of mysticism, number magic, as well as neo-Platonism and Hermeticism challenged the authority of Aristotelian and Galenic doctrines. Natural magic and the occult sciences also affected medical philosophy and the nascent life sciences. While the relationship between science and mysticism had undergone profound changes since the Scientific Revolution, the persistent appeal of the occult sciences is reflected in the fact that many well-educated people who cannot describe the achievements of Andreas Vesalius, or even Copernicus, Galileo, and Newton, are likely to know their astrological sign and consult the horoscopes published in their daily newspaper.

In retrospect, dissection seems to be the dominant theme in the evolution of European medicine, but some Renaissance natural philosophers and physicians rejected anatomical research as irrelevant to understanding health and disease, life and death, and immersed themselves in alchemy, astrology, and natural magic. While chemistry as we know it today may be regarded as a young science, based on the atomic theory of John Dalton (1766–1844), chemistry can also be traced back to the technologies involved in working with metals, cements, glass, perfumes, medicines, and alcoholic beverages. Ancient descriptions of chemical manipulations suggest that elaborate rituals were used to standardize processes that were but vaguely understood. Methods of making bronze, brass, and iron and purifying gold and silver were of obvious economic importance, as well as a source of great power to those who controlled these techniques.

Alchemists occupy an ambiguous place in the history of science. During the European Renaissance, alchemists absorbed and transformed protochemical techniques and philosophy. They have alternatively been dismissed as mystics, quacks, and charlatans and praised as pioneers of modern chemistry. In terms of their role in the history of biology, alchemists can be thought of as the first natural philosophers to attempt chemical analyses of organic materials. They did not do so as proto-biochemists, but because they believed that the whole universe was alive. Thus, they did not limit their investigations to what we would now call inorganic chemistry. While their experiments may have been based on mystical religious and philosophical theories, their investigations led to more sophisticated techniques and improved laboratory instruments.

Alchemy did not originate as an *empirical* science, but as a *sacred* science concerned with manipulating the passions, marriages, growth, death, and transmutations of matter. Alchemical philosophy was shared by many ancient cultures, including Babylonia, China, India, Greece, and the Islamic world. According to alchemical doctrine, the *seeds* of noble metals are contained in the base metals. If the right conditions and methods were applied, these seeds could be encouraged to grow, just as the child grew in the womb. Alchemy was more than an attempt to transform base metals into gold. It encompassed a broad range of doctrines and practices, including the search for special elixirs of health, longevity, and immortality. Alchemists believed that beyond the known qualities of crude matter there must be other subtle and hidden qualities. In their search for these occult properties, the alchemists sometimes discovered or created new substances. For example, in trying to distill off the *essence* of wine, they discovered *aqua ardens* (strong liquors), which were transformed into medicinal liquors and cordials.

The search for the elixirs of life was not exclusively confined to Chinese alchemists. The European alchemist most closely associated with the quest for better living through chemistry was Philippus Aureolus Theophrastus Bombastus von Hohenheim (1493–1541). Fortunately, he is generally remembered as Paracelsus, which presumably indicated one who was greater than Celsus, The Paracelsians, his seventeenth-century followers, attributed the birth of a revolutionary new chemical or *spagyric* system of therapeutics to him. (Spagyric is derived from the Greek words meaning "to separate" and "to assemble.") Little is known of his early life or how he came to be called Paracelsus, but legends and speculation abound. His father, Wilhelm von Hohenheim, was a physician at Einsiedeln, a small town near Zurich. His mother, who was a nurse at the monastery hospital, apparently suffered from some form of insanity and eventually committed suicide.

After leaving home at the age of 14, and visiting various universities, Paracelsus rejected the stultifying atmosphere of the academic world and turned to the study of alchemy. He absorbed the philosophy, legends, and techniques of alchemy and metallurgy from the famous alchemist and abbot Johannes Trithemius (1462–1516). In search of the secrets of alchemists, astrologers, magicians, healers, and herbalists, he worked in the mines of the Tyrol and traveled through Germany, Spain, France, and possibly Russia and Egypt. His journeys were essential, he explained, because Nature was the divine book of Creation that the seeker of knowledge must tread with his feet. Although there is no evidence that he ever received a formal degree, he awarded himself the title of "Double Doctor," presumably for honors conferred on him by God and Nature, and he apparently enjoyed a successful, if always precarious private practice. Glorifying his own methods, Paracelsus ridiculed pharmacists and physicians as nothing more than "officially approved asses." Worse yet, he took away their business by undercutting their prices and curing their disillusioned patients.

In 1526 Paracelsus returned to Switzerland and, through the intervention of influential patients, was appointed Professor of Medicine and City Physician of Basel. At the University, Paracelsus demonstrated his talent for staging what would now be called media events by burning the works of Avicenna and Galen to show his contempt for ancient dogma and the curriculum favored by his outraged colleagues. Wearing the alchemist's leather apron rather than academic robes, he also broke with tradition by lecturing in the vernacular rather than

THEOPHRASTUS
PARACELSUS.

Ich hab gefunden, Was viele
Zu ihrem Unglück suchen.
Den Lapidem Philosophorum.

Geoffe Brock Gez del

Inveni, quem plurimi suo
cum Damno indagant.
Lapidem Philosophorum.

Ioh Georg Herd excud Aug Vind

Paracelsus surrounded by the symbols, texts, instruments, and apparatus appropriate to alchemy and astrology

Latin. Not surprisingly, he was forced to leave Basel only 2 years later. After many similar episodes, he was invited to settle in Salzburg by the Archbishop Duke Ernest of Bavaria, a man deeply interested in alchemy. Shortly afterwards, when only 48 years of age, Paracelsus died in a mysterious but certainly unnatural fashion. His friends said that he was the victim of a deliberate hostile attack, but his enemies said he met with an accident while drunk.

Despite his turbulent life, Paracelsus was a prolific writer. Very few of his books were published during his lifetime, but over 300 works, many of which were undoubtedly forgeries, were later attributed to him. Paracelsus claimed that through natural magic and alchemy human beings would one day see beyond the mountains, hear across the oceans, divine the future, make gold, cure all diseases, and even create life. Based on his writings and contemporary accounts of his eccentricities, Paracelsus has been called everything from genius to quack, from founder of medicinal chemistry to a bizarre footnote to the history of chemistry. Among other boasts, he apparently claimed that through natural magic and alchemy human beings would one day be able to make gold, cure disease, and even create life. Presumably, he was a combination of magician and scientist, healer and fraud.

For the Paracelsians, *natural magic* was the key to understanding Nature and her laws. While natural magic was essentially an occult science, allied to alchemy and astrology, it was also an experimental art with practical goals. Alchemists contended that life was a chemical process and thus the problems of physiology and pathology could be explained in chemical terms. That is, disease must be the result of a defect in body chemistry. Such concepts might be dismissed as too obvious for comment today, but they might be seen as enigmatic and profound in the sixteenth and seventeenth centuries. Paracelsus challenged the old Greek elements and humors, but his own chemical principles—*salt*, *sulfur*, and *mercury*—were equally ambiguous and did not lead to a theoretical system capable of replacing the ancient doctrines. Salt, according to Paracelsus, is the principle of solidity, sulfur is the principle of inflammability and malleability, and mercury is the principle of fluidity, density, and that which is metallic in nature.

Alchemy provided new chemical analogies for physiological functions. All physiological processes, according to Paracelsian doctrine, were fundamentally chemical transformations governed by the *archeus*, or internal alchemist of the body. Disease was the result of some malfunction of the archeus, and death was due to its final loss. Dispensing with traditional humoral explanations, Paracelsus argued that disease processes should be analyzed in accordance with the methods used to analyze chemicals. While obscure and inconsistent, some Paracelsian categories were peculiarly appropriate for metabolic diseases, dietary disorders, and certain occupational diseases. For example, gout and gouty arthritis were classified as "tartaric diseases" based on an analogy with the materials that precipitate out of wine casks. Because gout involves the abnormal local deposition of a metabolic product called uric acid, it is indeed a condition where body chemistry has gone wrong.

In terms of medical philosophy, Paracelsian alchemy could provide new chemical analogies for physiological functions; in therapeutics, alchemists could use the art to prepare new drugs. In contrast to the traditional preference for polypharmacy, which called for complex mixtures of numerous ingredients, Paracelsus argued that specific remedies should be designed for specific diseases in accordance with their specific causes. Orthodox physicians and pharmacists preferred expensive remedies that included exotic materials such as viper's flesh, mummy powder, and unicorn horn, in addition to many different medicinal herbs.

Specific Paracelsian or spagyric remedies might not have been any safer or more effective than orthodox prescriptions, but eventually better procedures for the purification of drugs and poisons made it possible to analyze the relationship between dosage and response. Alchemists often began their quest for remedies by focusing on poisons and venoms, reasoning that minute quantities of such materials were obviously very powerful. (Validation of this approach can be found in the twentieth-century successes in using toxins, venoms, and poisons, such as botulinus toxin, Chilean tarantula venom, and arsenic trioxide, in the treatment of muscle spasms, brain tumors, leukemia, and other disorders.) Alchemists also used astrological concepts, such as the correspondence between the seven planets, the seven metals, and the parts of the body as guides in the search for remedies. The ancient "doctrine of signatures" provided additional clues about potentially useful materials. According to this concept, natural objects, such as medicinal herbs, were stamped by the Creator with signs that suggested their usefulness against particular diseases.

Another guiding principle of Paracelsian medicine was the belief that violent diseases required violent remedies, especially those based on metals and their salts. Fortunately, preparations containing some of the more toxic minerals caused vomiting before the patient could consume a lethal dose. Laudanum (opium in wine) and ethyl ether, or "sweet vitriol," were among the most valuable drugs associated with Paracelsus. Although Paracelsus apparently recognized the soporific properties of ether, it was not used as a surgical anesthetic agent until the 1840s. Such examples can make the world of Paracelsus seem remarkably prescient, but, just as the products of his chemical manipulations were likely to be far from pure, safe, and effective, his writings were far from clear. It is easy to read too much into the alchemical literature and confuse obscurity with profundity. While scholars still argue about whether it is possible to find a structured system of thought in the writings of the Paracelsians, it is possible to say that the alchemists called attention to the fact that it was important to study vital functions in the human body, rather than maintain a narrow

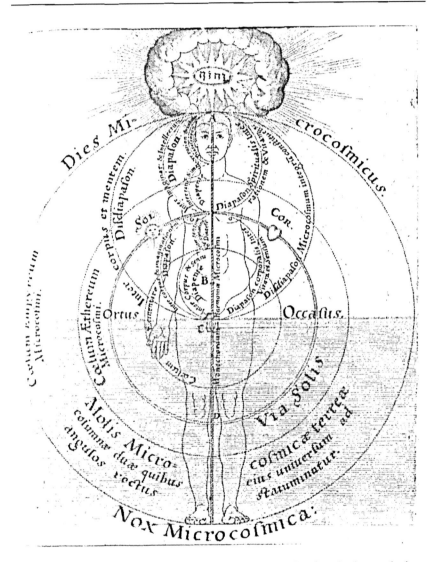

The microcosm (a seventeenth-century alchemical chart showing the human body as world soul)

focus on anatomical structure. Until the nineteenth century, however, attempts to deal with the chemistry of life were less successful than what might be called studies of animated anatomy.

ANATOMY AND PHYSIOLOGY

The Scientific Revolution is usually thought of in terms of the great transformation of the physical sciences that occurred during the sixteenth and seventeenth centuries. Nicolaus Copernicus, Galileo Galilei, and Isaac Newton challenged ancient systems of thought and removed the earth from its place at the center of the universe. Developments in anatomy and physiology during this period also revolutionized ideas about the nature of the microcosm, the little world of the human body. Renaissance anatomists and artists attempted to study the human body without deference to the ancients, but physiological investigations were still dominated by Aristotelian and Galenic doctrines. Questions about Galenic physiology were raised by some sixteenth-century anatomists, but it was not until the seventeenth century that William Harvey (1578–1657) provided a new way of thinking about the heartbeat, the pulse, and the movement of the blood.

Perhaps it is possible to imagine a people without an interest in astronomy (although some philosophers believe that looking up at the heavens was part of the process that transformed hominids into humans), but it is virtually impossible to imagine human beings who have no curiosity about their own life's blood. Ancient and modern humans have exhibited emotional responses to blood that have little to do with the specific physiological role of this remarkable liquid tissue. Blood has been used in religious rituals, charms, amulets, medicines, and, of course, in horror films. Many ancient cultures shared the belief that traits such as youth, courage, virility, character, and familial bonds could be transmitted through blood.

Scientists and philosophers also treated the blood as a most powerful fluid. Hippocrates and Aristotle shared the belief that the movement of the blood was fundamental to life. For Aristotle, the heart was the first organ of the body, the seat of intelligence, and the organ that added innate heat to the blood. Pulsations in the blood system were thought to be the result of a kind of boiling up that occurred when the blood in the heart met the pneuma drawn in by respiration. Galen demonstrated that both arteries and veins carry blood and attributed the bright red color of blood in the arteries and left ventricle to the presence of pneuma or vital spirits. Vesalius was unable to identify the pores in the septum that Galen's scheme required, but for the most part Vesalius was rather vague about the whole question of the distribution of the blood and spirits. Other sixteenth-century anatomists, however, were able to confront the issue of the invisible Galenic pores and suggest an alternative pathway from the right heart to the left heart, the pathway that is called the pulmonary or minor circulation.

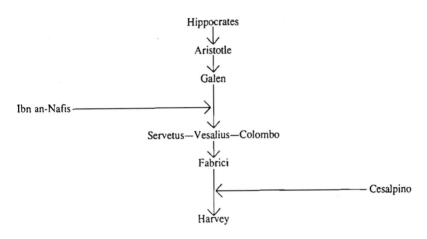

Discovery of the minor and major circulation

Michael Servetus (ca. 1511–1553), the first European to describe the pulmonary circulation, was a man who spent his whole life struggling against dog-
matism, conformity, and orthodox religion; he died a martyr to his principles.
Servetus so antagonized both Protestant and Catholic authorities that he was
burned alive by order of a Protestant court and in effigy by the Catholic Inquisition. In challenging theological doctrines about the soul, Servetus also challenged Galenic assumptions about the movement of the blood.

Born in Tudela, Spain, Servetus studied law at Toulouse and served for a
time as secretary to Charles V, Holy Roman Emperor. Perpetually restless, he
supported himself by writing, editing, and translating while he studied at various
universities. His youthful work, *On the Errors of the Trinity* (1531), convinced
both Protestant and Catholic censors that Servetus was a heretic who did not
deserve to live. To hide from his enemies, Servetus established a new identity
as Michael Villanovanus. By 1536 he was studying geography, astronomy, and
medicine at the University of Paris and, like Vesalius, serving as assistant to
Johann Guinter. At the time, both Servetus and Vesalius were anatomical nov-
ices and had barely begun to question Galenic dogma. Eventually, faced with
the threat of excommunication for indulging in the forbidden art of judicial
astrology (essentially fortune-telling), Servetus left Paris without a medical degree and resumed his travels.

To support himself, Servetus practiced medicine and wrote, edited, and
translated various books, including one about pharmaceutical products called
On Syrups and a new edition of Ptolemy's *Geographia*. Even when simply
editing classical texts, Servetus could not resist adding his own opinions. In a

description of France, Servetus called attention to the miraculous ceremony of the Royal Touch, in which the king was supposed to cure victims of scrofula (a form of tuberculosis). "I myself have seen the king touch many attacked by this ailment," Servetus wrote, "but I have never seen any cured."

While living in Vienna, Servetus took up a peculiar correspondence with John Calvin (1509–1564), the French Protestant reformer and champion of predestination. Ostensibly asking for information, Servetus sent his annotated copy of Calvin's *Institutiones* to Geneva along with an advance copy of his own radical treatise *On the Restoration of Christianity* (1553). Calvin responded by denouncing the author of the *Restoration* to the Catholic Inquisition and sending pages torn from the book as evidence. Servetus was arrested but managed to escape from prison before he was tried, convicted, and condemned to be burned in a "slow fire until reduced to ashes" along with his book. Within 4 months of his escape from death at the hands of the Catholic Inquisition, Servetus was captured while passing through Geneva. Once again he was condemned to be burned at the stake along with his book. Attempts to mitigate the sentence to death by the sword or immolation "with mercy" (meaning strangulation before the burning began) were unsuccessful. A monument to the unfortunate heretic was erected by the Calvinist congregation of Geneva in 1903.

In the *Restoration*, a 700-page theological treatise exploring the ramifications of his view of the relationship of God to the world and to human life, Servetus included a brief account of the pulmonary circulation. Although his primary inspiration was religious rather than scientific, Servetus drew on ideas from law, politics, astronomy, physics, and medicine to substantiate his theological arguments. To understand the relationship between God and humanity and to know the Holy Spirit, Servetus argued, one must understand the spirit of man. This required exact knowledge of the human body—its structures, functions, and its moving and nonmaterial components, especially the blood and the spirit—for as stated in Leviticus "the life of the flesh is in the blood."

In his discussions of human anatomy, Servetus mixed evidence from human dissection, passages from Galen, and his own theological concepts. Direct observation suggested that the septum did not have the pores required for Galen's scheme. If these pores did not exist, then little or no blood could pass from the right to the left heart. Simple considerations about the structure of the heart and the attached blood vessels also raised doubts about Galenic physiology. In particular, anatomists should ask why the pulmonary artery was so large and why so much blood was forcefully expelled from the heart to the lungs. Certainly more blood was sent to the lungs than was needed for their own nourishment. The blood must, therefore, go from the right side of the heart to the lungs for aeration as well as for the expulsion of sooty vapors. Then most of the blood should return to the left side of the heart via the pulmonary vein. According to Galen, aeration was the function of the left ventricle, but it was

Michael Servetus (in the background of this portrait, the ill-fated theologian is shown being burned at the stake)

during passage through the lungs that the color of the blood changed. Galen was led into error, Servetus explained, because he misunderstood the function of the lungs. Galen did not recognize the continuous flow of blood through the lungs, but assumed that if any blood returned to the heart from the lungs it was due to leakage.

Both biblical arguments and physiological observations, Servetus contended, proved that the soul is to be found in the tides of the blood rather than confined to the heart or liver or brain. Satisfied that he had reconciled physiology and theology as to the unity of the spirit, Servetus did not consider the systemic or major circulation of the blood, nor did he absolutely preclude the possibility that some blood might seep through the septum. Ibn-an-Nafis had been more forceful than Servetus in rejecting the existence of passages through the septum of the heart, but there is no compelling evidence that Servetus was aware of the work of this thirteenth-century Arab physician. Having established his main point, Servetus generally accepted other aspects of the Galenic system.

Since only three copies of the *Restoration* seem to have survived the flames, it is unlikely that Servetus's discovery of the pulmonary circulation was better known to anatomists and scholars than that of Ibn-an-Nafis. Even if anatomists were familiar with the work of Servetus, they were unlikely to admit any association with its ill-fated author. Still, it must be admitted that every age has its own invisible networks for the dissemination of ideas, especially ideas considered dangerous and subversive. If the story of Servetus was transmitted in whispers among his contemporaries, it might have been more instructive as a warning about the dangers of unorthodox thought than the errors of Galenic physiology. In any case, a more accessible account of the pulmonary circulation was also published in the 1550s by the well-known anatomist Realdo Colombo (Realdus Columbus; ca. 1510–1559).

Colombo, the son of an apothecary, served as apprentice to a surgeon for 7 years before studying medicine and anatomy at the University of Padua. According to university records for 1538, Colombo was an outstanding student of surgery. When Vesalius left Padua to supervise the publication of the *Fabrica*, Colombo was appointed as a temporary substitute. Eventually Colombo became professor of surgery and anatomy. In 1545 he left Padua for a professorship at Pisa. Three years later, at the invitation of Pope Pius IV, he accepted the chair of anatomy at the University at Rome.

Colombo and Vesalius seem to have had a cordial collegial relationship before Colombo assumed his professorship and became one of the most vociferous critics of the *Fabrica*. In response, Vesalius called his rival a scoundrel, an ignoramus, and an "uncultivated smatterer" whose general education was as deficient as his mastery of Latin. Later the anatomist Gabriele Fallopio (1523–1562) added to the attack on Colombo by accusing him of plagiarizing the discoveries of other anatomists. Throughout his career, Colombo boasted about his skills in dissection and his plans for an anatomical treatise that would supersede the *Fabrica*. This ambitious enterprise, which was to involve Michelangelo as the illustrator, never materialized. Columbo's legacy consists of one medical treatise, *De re anatomica*, which was published by his heirs in 1559. Lacking illustrations, this posthumous work was criticized as an obvious and inferior imitation of the *Fabrica*. While Colombo exaggerated his anatomical experience and originality, his description of the pulmonary circulation was the first unambiguous and accessible statement about the system made by a prominent European anatomist.

Unlike Servetus and Ibn-an-Nafis, Colombo made it clear that his ideas on the pulmonary circulation were based on clinical observations, dissections, and vivisection experiments on animals. In *De re anatomica* Colombo described the anatomy of the heart in detail and corrected many ancient errors. Praising his own resourcefulness and originality, he claimed to be the first to discover the role of the lungs in the preparation and generation of the vital spirits. Denouncing other anatomists for erroneously assuming the existence of a direct path from the right to the left ventricle, Colombo urged his readers to confirm his conclusions. The blood was carried from the right side of the heart by the "artery-like vein" (pulmonary artery) to the lungs, where it was rarefied and mixed with air. The "vein-like artery" (pulmonary vein) brought the mixture of blood and air back to the left ventricle of the heart to be distributed to the body via the arteries. Despite his glowing praise for his own originality and daring in revealing this error in the dogmas of Galen, Colombo's general views on the function of the heart, lungs, and blood were rather conservative. Even his claim to priority has been questioned on several accounts. In 1546 Servetus sent a manuscript copy of the *Restoration* to a doctor in Padua. If Colombo was not aware of this manuscript, he might have learned about the *Restoration* when he settled in Rome where the printed copies survived. Even if Colombo had begun his study of the pulmonary circulation as early as 1545, his ideas might have been stimulated or confirmed by the work of Servetus.

While the role of Ibn-an-Nafis and Michael Servetus in the series of events that led to the discovery of the circulation of the blood is ambiguous, some historians have argued that Andrea Cesalpino (1519–1603) should be honored as the discoverer of both the minor and major circulation. Cesalpino was a very learned man who combined a great reverence for the ancients, especially Aristotle, with an appreciation of more modern aspects of natural history and medicine. From 1555 to 1592 he was Professor of Medicine and Botany at Pisa. In 1592 he was called to Rome to serve as physician to Pope Clement VIII and as Professor at the Sapienza University.

Although his main scientific interest was botanical classification, Cesalpino also wrote a number of books pertaining to anatomy and practical medicine. As indicated by his *Quaestionum peripateticarum*, his scientific and medical work was firmly grounded on an Aristotelian foundation. Opposed to Galenic doctrines that conflicted with the views of Aristotle, Cesalpino was eager to support Aristotle's claim that the heart was the most important organ of the body.

For the Galenic system to work, Cesalpino noted, the lungs and the heart must expand and contract at the same time. Yet it was obvious that we can regulate our breath by our will, although we cannot control the heartbeat. Similarly, physicians knew that the pulses and the respiration might be fast or slow, strong or weak, and that changes in the respiration need not correspond to changes in the pulses. In a discussion of bloodletting, Cesalpino called attention to the well-known fact that veins swell on the far side of the ligature. This observation could have led him to an exploration of the idea that the direction of blood flow is from the arteries to the veins and from the veins to the heart. But preoccupied with Aristotelian ideas about the primacy of the heart and the nature of the innate heat, Cesalpino entered into a digression on Aristotle's concept of the nature of sleep and the movement of animal heat when sleeping and waking.

Like Aristotle, Cesalpino thought that the primary organ of the body was the heart, which he lovingly described as a fountain from which four great veins irrigated the body "like the four rivers that flow out from Paradise." Cesalpino had a remarkable gift for infusing ambiguous passages with words like *circulation* and *capillary vessels* that sound remarkably modern, at least in translation. If the scattered references to the movement of the heart and blood in his writings are brought together and arranged in appropriate patterns, it might appear that he had indeed discovered the circulation of the blood. On the other hand, such an exercise demonstrates the difficulty of establishing the relationship between observations, meaning and context, theory formation, and the nature of a scientific discovery.

In calling attention to the way in which observations made in the course of bloodletting could lead to insights into the workings of the body, Cesalpino was not alone. Several other sixteenth-century anatomists were led to the investigation of the valves of the veins from such considerations. When a ligature is tied around the arm in preparation for bleeding, little knots or swellings, which correspond to the valves of the veins, can be seen along the course of the veins. Studies of the venous valves play an important role in the history of the discovery of the circulation because William Harvey suggested that thinking about the purpose of these structures led him to wonder whether the blood might travel in a circle.

Girolamo Fabrici (Fabricius; ca. 1537–1619), Harvey's teacher at the University of Padua, probably began studying the structure, distribution, and function of the venous valves in 1574. Three years after the death of Gabriele Fallopio (1523–1562), Fabrici was appointed to the chair of anatomy and sur-

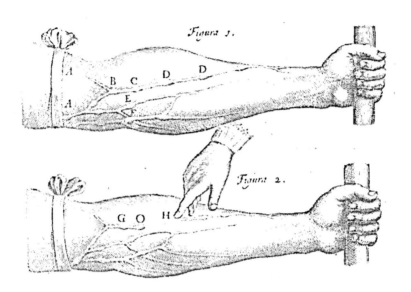

A simple demonstration of the valves on the veins when a ligature is tied around the arm in preparation for bloodletting inspired by William Harvey to think about the circulation of the blood

gery. Like his predecessor, Fabrici was much less interested in teaching than in pursuing his own research in anatomy and embryology. Sometimes he disappeared before completing his courses. When he was unable to avoid his teaching obligations, he angered students who had come to Padua from all over Europe by his obvious display of boredom and indifference. Presumably, then as now, students thought it more natural for the teacher to be attentive and enthusiastic and the students to be bored and apathetic. After serving the university for 50 years, Fabrici claimed that illness made his retirement necessary. Poor health did not prevent him from continuing to write; he published eight new works and revised several previous books during his retirement.

In 1603 Fabrici published a brief work called *On the Valves in the Veins*. Other anatomists had observed the peculiar membranous structures in the valves, but this was the first account of their structure, position, and distribution throughout the entire venous system. According to Fabrici, nature had fashioned the valves to oppose the flow of blood from the heart towards the periphery. Comparing these structures to the floodgates of a millpond, Fabrici suggested that their function was to regulate the volume of blood distributed to the parts of the body so that each part could obtain its proper nourishment. The arteries did not need such structures because of the pulsations of the thick arterial walls. In his anatomy courses and his public lectures, Fabrici frequently demonstrated the action of the venous valves. This is easily done by tying a ligature around the arm of a volunteer. Because the ligature blocks the flow of venous blood, little knots can soon be seen along the course of the engorged veins. Dissection of the veins in a cadaver suggested that the membranous structures distributed along the course of the veins corresponded to these swellings. A simple experiment involving an attempt to push the blood past these knots clearly demonstrates that the valves oppose the flow of blood. Intrigued by such experiments, Harvey realized that the venous valves did not retard the flow of blood from the heart to the periphery but directed the blood towards the heart.

William Harvey, a quiet and in many ways conservative man, was the eldest of seven sons of Thomas Harvey and the only member of this close-knit family of farmers and merchants to become a physician and scientist. After receiving the Bachelor of Arts from Caius College, Cambridge, Harvey studied medicine at the University of Padua. In 1602 Harvey received the degree of Doctor of Medicine and returned to England. He soon established a successful practice and married Elizabeth Browne, daughter of Lancelot Browne (d. 1605), physician to Queen Elizabeth and King James I. With these substantial credentials, Harvey was quickly elected Fellow of the College of Physicians (1607), appointed physician to St. Bartholomew's Hospital (1609), Lumleian Lecturer for the College of Physicians (1615–1656), and physician extraordinary to James I and Charles I. In 1631 Harvey was promoted to physician-in-ordinary to Charles I and became the king's senior physician-in-ordinary in 1639. (Peculiarities of nomenclature meant that the title *ordinary* ranked higher than *extraordinary* in the royal medical hierarchy.) Despite considerable personal risk and sacrifice, Harvey remained the faithful friend of Charles I throughout the Civil War in which the Parliamentary forces under Oliver Cromwell (1599–1658) defeated the followers of King Charles I.

When seen against the political and social turmoil of the age in which he lived, Harvey's achievements are particularly remarkable. As physician to the king, Harvey frequently traveled with the court to continental Europe, where he probably discussed his discoveries with other physicians. During the war, Harvey's house in London was attacked and many of his manuscripts and collections were destroyed. When the court moved from London to Oxford, Harvey obtained a professorship there. His remarkable temperament is illustrated by the story of how he sat under a hedge just beyond the battlefield during the battle of Edgehill reading a book. After Cromwell's followers prevailed and King Charles was publicly beheaded in 1649, Harvey retired to the countryside to live with his brothers. Suffering from gout, depressed, and often in pain, Harvey dosed himself with peculiar remedies and large quantities of opium. There is suggestive evidence that he made several suicide attempts. In his will, Harvey, who had no children, left his estate to Oxford University, which has an annual festival dedicated to him.

Despite the achievements of Andreas Vesalius, medical teaching and medical philosophy were still dominated by Galenic theories when Harvey first began to think about the circulation of the blood. Long before 1628, when he assembled his evidence and published *De motu cordis et sanguinis (Anatomical Treatise on the Movement of the Heart and Blood)*, Harvey seems to have arrived at an understanding of the motion of the heart and blood. The notes for his first Lumleian Lecture show that by 1616 he was performing demonstrations and conducting experiments to show that blood passes from the arteries into the veins. He concluded that the beat of the heart impels a continuous circular motion to the blood. Although it was possible to discuss innovative ideas as the Lumleian lecturer, Harvey said that he was reluctant to publish his novel and unprecedented ideas about the circulation for fear of having "mankind at large for my enemies."

In thinking about the function of the heart and the blood vessels, Harvey came to reject Galen's scheme while moving closer to Aristotle's idea that the heart is the most important organ in the body. Like Aristotle, whom he greatly admired, Harvey asked seemingly simple but truly profound questions about the nature and purpose of things in order to understand their final causes. Harvey wanted to know why the two structurally similar ventricles of the right and left heart should have such different functions as the control of the flow of nutritive venous blood and the distribution of the vital spirits in the arterial blood. If arterial blood and venous blood were fundamentally distinct, why did they differ only in color while they appeared to be identical in taste, odor, and other observable characteristics? Why should the artery-like vein nourish only the lungs, while the vein-like artery had to nourish the whole body? Why should the lungs need such an inordinate amount of nourishment? Why did the right ventricle have to move in addition to the movement of the lungs?

Harvey also posed a unique question that seems astonishingly simple, even child-like: How much blood is sent through the body with each beat of the heart? Unlike grand questions about final causes, this question could be answered by direct observations and experiments. Indeed, when reading *De motu cordis*, one can hardly avoid the thought that, at least in principle, the experiments by which Harvey answered this question could have been performed hundreds of years before. That is, Harvey's observations did not require the use of any materials or instruments that were not accessible to Aristotle, Galen, Vesalius, or the thousands of students who had dozed through lectures on the whole of anatomy.

In his first chapter, "Difficulties in Experiment," Harvey dismissed the work of anatomists who would study only one species, man, and that one only when dead. Despite differences between species, the proper use of comparative

EXERCITATIO
ANATOMICA DE
MOTV CORDIS ET SAN-
GVINIS IN ANIMALI-
BVS,
GVILIELMI HARVEI ANGLI,
*Medici Regii, & Professoris Anatomiæ in Col-
legio Medicorum Londinensi.*

FRANCOFVRTI,
Sumptibus GVILIELMI FITZERI.
ANNO M. DC. XXVIII.

The title page of William Harvey's *De motu cordis* (Courtesy of the National Library of Medicine)

anatomy and vivisection could provide valuable insights about vital functions in humans. Harvey admitted that, at first, the motion of the heart seemed so rapid and complex that he thought only God could comprehend it. Then he realized the advantages of using cold-blooded animals and mammals brought close to death by drugs or exsanguination. In these model systems the heartbeat was slower and easier to follow. Discovering the workings of the heart and arteries could be thought of as analogous to understanding the mechanism of a gun. Normally it was impossible to see the rapid action of trigger, flint, steel, spark, and powder, but one could logically separate out each step.

With arguments based on dissection, vivisection, clinical experience, and the works of Aristotle and Galen, Harvey proved that in the adult body all the blood must go through the lungs to get from the right to the left side of the heart. Acknowledging the work of his predecessors, Harvey clarified various aspects of the pulmonary circulation. He proved that the heart is muscular and that its important movement is the contraction. Previously, anatomists had taught that the heart was not a muscle and that dilation was the important function; movement had been ascribed to the chest and lungs rather than to the heart itself. Most of all, Harvey proved that it was the beat of the heart that caused a continuous circular motion of the blood: from the heart into the arteries and from the veins back to the heart.

In support of the novel idea of a continuous circulation of the blood, Harvey turned to quantitative considerations. Although he did experiments aimed at getting an accurate measurement of the quantity of blood put out by the heart with each beat, he emphasized that exact measurement was unnecessary. Even the most cursory calculation would prove that the amount of blood pumped out of the heart per hour was so great that it exceeded the weight of the entire body. That is, if the heart pumps out 2 ounces of blood with each beat and beats about 70 times per minute, the heart must expel about 600 pounds per hour. Such a quantity is three times the weight of even a very large man. Similar calculations were carried out for various animals, such as sheep and dogs. In all cases the amount of blood pumped out by the heart in even a half hour would be greater than the total amount of blood in the whole animal. Exsanguinating experimental animals, just as a butcher exsanguinates an ox, easily proved this predication. Such quantities of blood could not be continuously synthesized from food by the liver and distributed to all the parts of the body by the veins. Blood must, therefore, be impelled by the heart in order to move in a circle through the body.

Once this basic principle was grasped, many observations fell neatly into place. Every butcher knew, for example, that if he opened an artery while the heart was still beating he could exsanguinate an animal in less than half an hour. With Harvey's theory scientists could finally explain what every butcher already knew. Experience gained through phlebotomy and clinical observations, such as the throbbing of an arterial aneurysm and the pulses of the wrists, temples, and neck, similarly supported the central thesis of *De motu cordis*. Knowledge of the continuous circulation of the blood explained many puzzling clinical observations. The circulation of the blood explained why venom or poison or medicines introduced at a specific site could affect the whole system.

It is all too easy to take Harvey out of his seventeenth-century context and make him appear more modern in outlook and approach than is really appropriate. Moreover, his methods and conclusions appear so reasonable and compelling to his modern readers that it is easy to assume that his contemporaries must have been equally convinced. If that were true, *De motu cordis* should have dealt an immediate deathblow to the Galenic system and the practice of therapeutic phlebotomy. Instead, Harvey's methods and conclusions raised many questions and objections. Certainly, Galenism and therapeutic bloodletting were not immediately abandoned.

It is important to remember that the kind of evidence that appears unequivocal today did not necessarily appeal to Harvey's contemporaries. Moreover, although Harvey had discovered virtually all that could be known about the anatomical and mechanical aspects of the circulatory system, many troubling questions remained unanswered. In the absence of the microscope, Harvey could not see the capillaries that join the arteries and veins; he had to postulate hypothetical anastomoses or pores in the flesh to link the arteries and the veins. Without any information about the chemical nature of the difference between venous and arterial blood, the relationship between respiration and the circulation remained obscure. Harvey's theory of the circulation could not provide satisfactory substitutes for the Galenic spirits and the innate heat. Harvey acknowledged that his theory did not explain whether or not the heart added heat, spirits, perfection, or localized motion to the blood. Demonstrating a remarkable degree of modesty and restraint, he admitted that such questions were beyond the range of his experimental methods. A mechanical explanation of the movement of the heart and blood simply could not fully replace the all-encompassing Galenic system that had so completely, if incorrectly, explained the purpose of the heart, blood, lungs, liver, veins, arteries, and spirits.

Even Harvey's defenders realized that the new theory of the circulation raised more questions than it answered. How could the new system explain how the parts of the body secured their proper nourishment if the blood was not continuously consumed? If the venous blood was not continuously synthesized by the liver, what was the function of this large and bloody organ? If the blood moved in a continuous circle, how did the body distribute the vital spirits and the innate heat? If the vital spirit was not produced in the lungs or the left ventricle, what was the function of respiration? If the whole mass of the blood was constantly recirculated, what was the difference between the arterial and venous blood? Above all, what principles would guide medical practice if the great Galenic synthesis were sacrificed for the theory of the circulation?

Ironically, some defenders of Galen attacked Harvey for assuming that the results of experiments on animals could be applied to human anatomy and physiology. On the other hand, French botanists thought that the discovery of the circulation of the blood could be used to explain the circulation of sap in plants. This analogy was intriguing, but botanists soon realized that the facts of plant anatomy and physiology were not compatible with models based on the circulation of the blood in animals.

Several distinguished scientists accepted the general idea of the circulation but disagreed with particular aspects of Harvey's ideas, such as his estimate of the rate of the movement of the blood or the amount of blood driven into the arteries by the contraction of the ventricles. For example, the French physician Jean Riolan (1571–1657) suggested that the blood made the complete circuit through the body only two or three times a day. According to Riolan, only part of the blood returned to the heart, because the ebb and flow during this leisurely journey allowed the parts of the body to consume some of the blood. Rejecting the idea that the action of the heart drove the circulation, Riolan argued that the blood actually kept the heart in motion. Like Galen, Riolan believed that the liver was the most important organ of the body. The role of the heart, he argued, was to warm the blood and restore the spirits. Harvey took these arguments seriously enough to publish a refutation of Riolan's views in 1649.

In retrospect, it seems paradoxical that Harvey's discovery of the circulation did not lead to the rejection of venesection as a major therapeutic tool. Instead, ideas about the circulation provoked new controversies about the selection of sites for venesection. Even Harvey did not worry about the compatibility, or incompatibility, of venesection with the continuous circulation of the blood, nor does he seem to have considered the possibility of therapeutic blood transfusions. By 1660, however, his followers were engaged in attempts to transfuse and infuse many curious substances into animal and human veins. Seventeenth-century physiologists and physicians, quite unaware of the dangers of such experiments, had great hopes for the efficacy of blood transfusion. If the macrocosm and microcosm could be explained in terms of four elements and four humors, there was no compelling reason to expect major incompatibilities between the blood of different individuals, or even different species.

The first significant experiments on blood transfusion were performed by Christopher Wren (1632–1723), Richard Lower (1631–1691), and Robert Boyle (1627–1691) in England and by Jean-Baptiste Denis (1640–1704) in Paris. Wren and his colleagues began rather cautiously with a series of experiments on animals, but Denis proceeded immediately to experiments on humans. After the deaths of several patients, and a great deal of sordid publicity, Denis and his English counterparts abandoned this highly experimental form of therapy.

Sure of the truth of his own work, but well aware of its revolutionary nature, Harvey predicted that only the new generation of scientists would understand and accept it. Harvey was not the first natural philosopher to apply observation and experimentation to physiological problems, but he was one of the first to prove that experimentation was more powerful when used as a method of inquiry, rather than for its traditional use as a method of demonstration. Illness, the infirmities of age, and the loss of many of his precious manuscripts during the Civil War prevented Harvey from accomplishing all his goals, but he did live to see the beginnings of a new school of physiology. Ideas, methods, and questions inspired by the work of William Harvey provided the scientists who founded the Royal Society with a new research program for investigating the workings of the human body and developing a new experimental natural science.

SANTORIO SANTORIO AND THE QUANTITATIVE METHOD

It is easy to attribute much of Harvey's success to his brilliant use of the quantitative method, but not all seventeenth-century attempts to apply quantitative methods to the study of physiological problems were as successful. Although Santorio Santorio (Sanctorius; 1516–1636) has been called the founder of quantitative experimental physiology, his years of painstaking measurements primarily demonstrate that the methods available to him could not solve the fundamental problem he was trying to solve. After graduating from the University of Padua in 1582, Santorio practiced medicine in Venice. In 1611 he was appointed to the Chair of Theoretical Medicine at Padua, where he remained until his resignation in 1629, 5 years after his students charged him with negligence on the grounds that he was evading his teaching obligations in order to maintain his private practice. Freed from his duties at the University, he moved to Venice and devoted the rest of his life to his medical practice and research on the phenomenon known as *insensible perspiration*.

Many physicians and scientists thought that, in addition to the respiration that took place through the lungs, there was another form of respiration that involved the release of imperceptible vapors through the skin. This phenomenon was called insensible perspiration. Santorio believed that he could study the mechanism of insensible perspiration by exact measurements, using instruments of his own design. For much of his experimental career he sat on a special chair suspended from a steelyard, patiently weighing himself before and after eating, drinking, sleeping, resting, and exercising. In 1614 he published a book of aphorisms based on his work called *Ars de medicina statica aphorismi* (*Medical Statics*). The book went into at least 32 editions and was translated into several languages.

Santorio Santorio in his weighing chair

Although the *Medical Statics* provided only vague indications of experimental methods, Santorio claimed to have created a totally new field of exact measurement in medicine. Based on his measurement of the amount of food and drink and the sensible and insensible excreta of the body, Santorio provided deductions as to what changes had taken place inside the body under various conditions. In Santorio's opinion, these experiments had practical applications for health and medical practice. For example, a man could take his meals while sitting on a balance and know when he had eaten just the right amount of food. According to Santorio, physicians needed to understand the quantitative aspects of insensible perspiration in order to treat their patients. Critics responded that even if Santorio could provide quantitative measurements of the insensible perspiration, it was only the qualitative nature of the invisible vapors that was related to pathological phenomena.

Similar quantitative experiments carried out in the nineteenth century were subjected to scathing criticism by the great French physiologist Claude Bernard (1813–1878). For example, a study of nutrition reviewed by Bernard consisted of a balance sheet of all the substances taken into a cat's body and all of those excreted during 8 days of feeding and 19 days of fasting. Kittens born on the seventeenth day were included in the calculations as excreta. Bernard ridiculed such experiments as attempts to understand what occurred in a house by measuring who went in the door and what came out the chimney.

The *Medical Statics* is Santorio's most famous book, but it is so terse that it fails to reflect Santorio's motives for adopting the quantitative method and developing instruments such as the thermometer and pulse clock. Despite his interest in quantitative aspects of physiological phenomena, Santorio had no serious objections to Galenic medicine. While Galenic errors in anatomy could not be ignored, as long as a physician was guided by reason and experience, Santorio saw no reason to reject Galen as a medical authority. After all, it was Galen who taught anatomists to devote themselves to experimentation and observation.

SUGGESTED READINGS

Applebaum, W., ed. (2000). *Encyclopedia of the Scientific Revolution: From Copernicus to Newton.* New York: Garland Publications.

Bainton, R. H. (1960). *Hunted Heretic: The Life and Death of Michael Servetus, 1511–1553.* Boston: Beacon.

Bonelli, M. L. R., and Shea, W. R. (1975). *Reason, Experiment, and Mysticism in the Scientific Revolution.* New York: Science History.

Bylebyl, J. J., ed. (1978). *William Harvey and His Age. The Professional and Social Context of the Discovery of the Circulation.* Baltimore: Johns Hopkins University Press.

Cohen, H. F. (1994). *The Scientific Revolution: A Historiographical Inquiry*. Chicago: University of Chicago Press.

Cushing, H. (1962). *A Bio-bibliography of Andreas Vesalius*. 2nd ed. Hamden, CT: Archon Books.

Davis, A. B. (1973). *Circulation Physiology and Medical Chemistry in England 1650–80*. Lawrence, KS: Coronado Press.

Debus, A. G. (1978). *Man and Nature in the Renaissance*. New York: Cambridge University Press.

Debus, A. G. (1986). *Chemistry, Alchemy, and the New Philosophy, 1500–1700*. London: Variorum.

Eisenstein, E. (1983). *The Printing Revolution in Early Modern Europe*. New York: Cambridge University Press.

Field, J. V., and James, F. A. J. L., eds. (1993). *Renaissance and Revolution: Humanists, Scholars, Craftsmen, and Natural Philosophers in Early Modern Europe*. New York: Cambridge University Press.

Fishman, A. P., and Dickinson, W. R., eds. (1964). *Circulation of the Blood. Men and Ideas*. New York: Oxford University Press.

Frank, R. G., Jr. (1980). *Harvey and the Oxford Physiologists. Scientific Ideas and Social Interactions*. Berkeley, CA: University of California Press.

French, R. (1999). *Dissection and Vivisection in the European Renaissance*. Brookfield, VT: Ashgate.

Fulton, J. F., and Stanton, M. E. (1954). *Michael Servetus, Humanist and Martyr. With a Bibliography of His Works and Census of Known Copies*. New York: H. Reichner.

Hall, A. R. (1966). *The Scientific Revolution, 1500–1800: The Formation of the Modern Scientific Attitude*, 2nd ed. Boston: Beacon Books.

Harvey, W. (1976). *An Anatomical Disputation Concerning the Movement of the Heart and Blood in Living Creatures*. Trans. G. Whitteridge. Oxford: Blackwell.

Kaufmann, T. DaCosta (1993). *The Mastery of Nature: Aspects of Art, Science, and Humanism in the Renaissance*. Princeton, NJ: Princeton University Press.

Kerrigan, W., and Braden, G. (1989). *The Idea of the Renaissance*. Baltimore: Johns Hopkins University Press.

Leonardo da Vinci (1983). *Leonardo da Vinci on the Human Body*. Translations, text, and introduction by C. D. O'Malley and J. B. de C. M. Saunders. New York: Dover.

Lind, L. R. (1974). *Studies in Pre-Vesalian Anatomy: Biography, Translations, Documents*. Philadelphia: American Philosophical Society.

Lindberg, D. C., and Westman, R. S., eds. (1990). *Reappraisals of the Scientific Revolution*. New York: Cambridge University Press.

Lindeboom, G. A. (1975). *Andreas Vesalius and His Opus Magnum: A Biographical Sketch and an Introduction to the Fabrica*. Nieuwenkijk, Netherlands: de Forel.

McKnight, S. A., ed. (1992). *Science, Pseudo-Science, and Utopianism in Early Modern Thought*. Columbia, MO: University of Missouri Press.

Merchant, C. (1990). *The Death of Nature: Women, Ecology, and the Scientific Revolution*. San Francisco: Harper & Row.

O'Malley, C. D. (1964). *Andreas Vesalius of Brussels, 1514–1564*. Berkeley, CA: University of California Press.

Osler, M. J., ed. (2000). *Rethinking the Scientific Revolution*. New York: Cambridge University Press.

Pagel, W. (1984). *The Smiling Spleen. Paracelsianism in Storm and Stress*. Basel: Karger.

Porter, R., and Teich, M., eds. (1992). *The Scientific Revolution in National Context*. New York: Cambridge University Press.

Pumfrey, S., Rossi, P. L., and Slawinski, M., eds. (1994). *Science, Culture, and Popular Belief in Renaissance Europe*. New York: Manchester University Press.

Richardson, R. (2001). *Death, Dissection and the Destitute*, 2nd ed. Chicago: University of Chicago Press.

Saunders, J. B. de C. M., and O'Malley, C. D. (1947). *Andreas Vesalius Bruxellensis: The Bloodletting Letter of 1539*. New York: Henry Schuman.

Shapin, S. (1996). *The Scientific Revolution*. Chicago: University of Chicago Press.

Tambiah, S. (1990). *Magic, Science, Religion and the Scope of Rationality*. New York: Cambridge University Press.

Vesalius, A. (1969). *The Epitome of Andreas Vesalius*. Translated by L. R. Lind and C. W. Asling. Cambridge, MA: MIT Press.

Webster, C. (1982). *From Paracelsus to Newton: Magic and the Making of Modern Science*. New York: Cambridge University Press.

Whitteridge, G. (1971). *William Harvey and the Circulation of the Blood*. New York: American Elsevier.

4

THE FOUNDATIONS OF MODERN
SCIENCE: INSTITUTIONS
AND INSTRUMENTS

Looking back to the seventeenth century while preparing the second edition of his *Critique of Pure Reason* in 1787, the German philosopher Immanuel Kant saw it as the end of centuries of darkness. Suddenly, it seemed a new light had burst upon natural philosophers and illuminated the true methods of scientific inquiry. Certainly it is possible to see Kant's bright light reflected in the work of such giants of this age as Galileo, Newton, and Harvey. And yet certainly deeper, darker, and more fundamental forces were at work during the age that witnessed the construction of modern experimental science. Although it is important to remember that the terms "science" and "scientist" were not in general use until the nineteenth century, a discussion of the scientific revolution and the establishment of scientific societies that stringently excludes these terms because they are anachronistic tends to become awkward and confusing.

Never have the prevailing concepts of the earth, the heavens, and the nature of human beings been more radically transformed than during the seventeenth century, a time when European society was torn by the political, social, and spiritual turmoil that accompanied the shift from an agrarian society towards an urban and industrial world. The eternal and unchanging heavens had been radically rearranged by Galileo, and because of the work of William Harvey the blood was now known to travel in a circle in the body, just as the earth moved in an orbit around the sun. If the work of Galileo and Harvey raised questions that could not be answered by the authoritative texts that formed the basic curriculum of European universities, how could future natural philosophers be educated?

Given the essentially conservative role played by the university, it might not be surprising that few of the leading participants in the scientific revolution held university appointments. Before the sciences emerged as special disciplines, the individuals drawn to the study of natural philosophy were associated with various professions and institutions. The Presocratic philosophers, for example, were independent scholars, unattached to any formal educational institution. Plato and Aristotle, however, devoted some of their time to teaching and mentoring students and disciples. Scholarly activity in the Middle Ages was generally confined to the monasteries until the universities began to assume this role, but the sciences did not occupy a central role in their curriculum. Many European universities, moreover, were dominated by religious orders, and virtually all remained committed to Aristotelian thought. During the Renaissance, biological research was carried out by artists, scholars, and physicians, some of whom were associated with universities, while others earned their livelihood in private practice or in service to wealthy nobles and kings. The universities of the seventeenth century did not necessarily prevent the expression of new ideas, but the intellectual atmosphere was such that a scientist of Isaac Newton's stature could present his discoveries in a university lecture hall without creating a ripple of excitement or interest.

By the seventeenth century, membership in a novel institution called the *scientific society*, or *academy*, became the sign and symbol of a new scientific philosophy. In contrast to Plato's Academy and Aristotle's Lyceum, the members of the new scientific society were dedicated to practical and experimental goals rather than abstract and philosophical pursuits. Eminent philosophers of science, such as Francis Bacon (1561–1639) and René Descartes (1596–1650), provided the philosophical rationale for the new institution, while Harvey, Galileo, and others provided the experimental approach that inspired the academicians. New instruments, such as the telescope, thermometer, air pump, and microscope, made it possible for scientists to ask new questions and design new ways of finding empirical evidence.

The intimate association between experimental science and the academy raises an obvious question: Why did the university fail to serve as the incubator of the sciences? Clearly, the university and the scientific society differed fundamentally in structure, function, and philosophy. The pedagogical purposes of the university encouraged formal methods, based on teacher-student relationships that were not conducive to innovations. On the whole, universities continued to stress traditional subjects and authoritative texts. Universities generally regarded their mission as the preservation and transmission of the wisdom of the past rather than the search for new ideas and methods.

In contrast, the academy provided an informal setting where enthusiastic individuals could pursue common interests. Demonstrations and experiments brought together a critical mass of natural philosophers in a stimulating environment. The publication of journals, books, and the transactions of the societies enhanced the exchange of information. Academies that succeeded in obtaining government charters, official protection, and wealthy sponsors could serve as safe places for the discussion of new facts and ideas. Above all, the program of the scientific societies reflected the profound change in worldview that accompanied the fall of the Aristotelian system and the rise of the mechanical philosophy, rather than the impact of new facts in themselves. Those who directly participated in the early scientific societies constituted a very small elite, but they strongly believed that experimental science would ultimately transform society as a whole.

The earliest academies appeared in Italy, the heart of Renaissance art and culture. The short-lived *Accademia Secretorium Naturae*, established in the 1560s, met at the house of its leader, Giambattista della Porta (1538–1615) in Naples. Candidates for membership had to present a new fact in natural science as a condition for admission. A group called the *Accademia dei Lincei* (Academy of the Lynx Eye) met in Rome from 1603 to 1630 under the patronage of Duke Federigo Cesi. Distinguished members of the Academy included Galileo Galilei, della Porta, Francesco Stelluti, and Nicolas Fabri de Peiresc, a Frenchman interested in expediting the exchange of scientific information throughout Europe. Galileo's microscope was developed with the support of this academy, as were other important instruments and publications. The condemnation of the Copernican theory caused considerable anxiety among members of the Academy, and it disbanded shortly after the death of its patron.

Under the sponsorship of two members of the powerful Medici family, Grand Duke Ferdinand II and Prince Leopold, the Florentine *Accademia del Cimento* (Academy of Experiments) was formally established in 1657. Prince Leopold provided scientific instruments, as well as a place to meet, so that the small group of natural philosophers could conduct cooperative investigations. The Academy included pupils and disciples of Galileo, as well as Giovanni Borelli, a mathematician with a special interest in muscle physiology, and Francesco Redi, best known for his experiments on spontaneous generation. The Academy's 1667 *Report of Experiments* indicates that members had assembled and tested an outstanding collection of scientific instruments. Unfortunately, the Academy dissolved in that same year, when Leopold became a cardinal.

The early Italian societies emphasized experimental science and generally avoided speculative or controversial areas. This was not because the sciences had become well separated from philosophy and religion, but because natural philosophers, for their own protection, attempted to avoid potentially dangerous topics. While this may have stifled discussions of broad theoretical concerns, it did encourage experimentation, observation, and the refinement of scientific instruments, such as telescopes, thermometers, air pumps, and microscopes.

Academies developed in many European countries and in the United States, but none of them matched the prestige and stability of the Royal Society of London and the Royal Academy of Sciences in Paris. In England, the development of formal academies was somewhat inhibited by the Civil War. Charles II did not officially charter the Royal Society until 1662, but informal networks in London and Oxford, sometimes referred to as "invisible colleges," preceded it. During the 1640s groups of natural philosophers held meetings in private homes or friendly taverns. After a lecture by Christopher Wren at Gresham College in 1660, the assembled natural philosophers formally declared interest in establishing a college for the promotion of "Physico-Mathematical Experimental Learning." A chairman was elected, and, of the many persons interested in science, 41 men were chosen for the official project. The king gave his verbal approval, and 2 years later a formal charter was granted for the incorporation of the Royal Society for the Promotion of Natural Knowledge. Publication of the *Philosophical Transactions* of the Royal Society began in 1665. Thomas Sprat in his *History of the Royal Society* (1667) recorded these events in great detail. The invaluable secretary of the Society, Henry Oldenburg, another zealous advocate of the experimental method, carried out a massive correspondence that linked the Royal Society with scientists and academies all over Europe. Many of the fellows of the Royal Society regarded Frances Bacon as the guiding spirit of the Society.

FRANCIS BACON

Despite his status as prophet, critic, and philosopher of the new experimental science, Francis Bacon made no direct contributions to scientific knowledge, nor did he distinguish himself as a judge of facts and fancies. He failed, for example, to appreciate the role of mathematics in scientific theory and the importance of the work of Copernicus and Galileo. He did, however, influence the philosophy and institutions that directed the course of science. Moreover, he realized that in the future experimentation would serve a more powerful role as an organized method of investigation than as a method of demonstration. While defining the methodology of science, Bacon asserted that new scientific enterprises would increase human knowledge, power, and control over nature. He traced his distaste for Aristotelian philosophy to his student days at Trinity College, Cambridge.

Francis Bacon, son of Sir Nicholas Bacon, Keeper of the Great Seal, devoted much of his energy to the scramble for political power and wealth. His primary objective was to attain support for his great plan for the advancement of science and human learning. Under King James I, Bacon rose from Learned Counsel, Attorney General, Privy Councilor, Lord Keeper, to Lord High Chancellor, Baron Verulam, and Viscount St. Albans. In 1621, Bacon was accused of taking bribes. He admitted taking gifts, but he claimed that these bribes had not influenced his decisions as a judge. Nevertheless, he was severely punished.

In addition to his resignation, his sentence included a fine of £40,000, imprisonment in the Tower of London, and exclusion from Parliament and proximity to the king. Despite these disappointments, humiliations, and ill health, Bacon continued to write and offer his advice to king and country.

Bacon proposed grandiose schemes for the review and analysis of all branches of human knowledge, planned an encyclopedia of the crafts and experimental facts, and proposed plans for educational and institutional reforms that would transform and exploit the sciences in order to increase human welfare, comfort, and prosperity. Although William Harvey disparaged Bacon's scientific pretensions with the remark, "He writes philosophy like a Lord Chancellor," Bacon's *Advancement of Learning* (1605) became a landmark text in the history of science. Indeed, Thomas Jefferson's trinity of portraits of the world's greatest men consisted of Newton, Locke, and Bacon. Robert Hooke, Robert Boyle, Jean d'Alembert, Immanuel Kant, August Comte, and Charles Darwin all acknowledged Bacon's influence.

By the end of the twentieth century, Bacon, who so earnestly, if ineffectually, struggled to fulfill his vision of a life in science dedicated to the improvement of human welfare, became the target of attacks on rationalism, science, and technology. Moreover, feminist scholars and environmentalists have explored the deplorable consequences of his vision of the relationship between science and nature. The work of Bacon and Descartes can be seen as a historical turning point that led away from the possibility of an organic worldview. Once eulogized as the "father of modern science," Bacon has been vilified as the champion of the "new ethic" that justified and glorified the exploitation of nature. Critics urge a reassessment of the Baconian rationale for the mastery of nature from the vantagepoint of women, nature, and the poor.

Despite the failure of many of his own schemes for political and professional advancement, Bacon never lost his optimistic belief that through his methods nature could be understood and mastered, and that such knowledge would inevitably lead to unprecedented material and social progress. Comparing the investigation of nature to the interrogation and torture of a reluctant witness in the courtroom, Bacon contended that nature must "submit to the questions" posed by science. The secrets and resources that might still be hidden in the womb of nature were not passively discovered through observation. Nature should be treated as a "slave" who could be "constrained" and "molded" by technology and compelled to serve man. Through advances in the mechanical arts, man would conquer and subdue nature and "shake her to her foundations." Bacon wrote with apparent confidence that progress in the mechanical arts had reached the stage where man could successfully confront nature through "violence." But Bacon and his contemporaries knew that they lived in an uncertain world, where safety, stability, prosperity, and health were very much desired, but certainly not assured.

Mastery of language was also very important to Bacon. Between 1608 and 1620 he prepared at least 12 drafts of his most famous book, the *Novum Organum*. In this treatise Bacon described his program for the proper development of the sciences. In time, the "Baconian method" became virtually synonymous with the "scientific method." Bacon's method, also known as the *inductive method*, involves the exhaustive collection of particular instances or facts and the elimination of factors that do not invariably accompany the phenomenon under investigation. However, induction must work by elimination and not by simple enumeration. Rather than passively collect facts, the scientist must be actively involved in *putting the question to nature*. Recognizing the impossibility of absolute proof of inductive generalizations based on a finite number of observations, Bacon called attention to "the greater force of negative instances." In other words, experimental results that contradict a general theory, by falsifying an induction, may reveal more about nature than another bit of data that appears to support the theory. This principle, now known as *falsifiability*, is usually associated with the twentieth-century philosopher Karl Popper. Bacon also insisted that observations offered in support of a general theory be repeatable, rather than unique and unusual. Scientists would analyze experience "as if by a machine" to arrive at true conclusions by proceeding from less to more general propositions. The result of applying this scientific method, Bacon assured his readers, would be a great new synthesis of all human knowledge, a true and lawful marriage between the empirical and the rational faculty.

Using the picturesque metaphor of "idols," Bacon also discussed the logical and psychological causes of error in the pursuit of knowledge. The "idols of the tribe" are common intellectual errors, such as oversimplification. The "idols of the cave" represent the intellectual eccentricities of individuals. The "idols of the marketplace" are the kinds of error caused by the imperfections of language. Mistaken systems of philosophy, the "idols of the theatre," constitute the fourth group of idols.

The New Atlantis, Bacon's famous utopian novel, was probably written in 1614, but it was not published until 1626. Here Bacon presented a model scientific research institution called "Salomon's House." In this ideal institution, scientists collaborated in planning and performing their methodical investigations. Eventually, under the direction of such a community of experimental scientists, natural philosophy would be united with the craft traditions. This partnership would produce valuable knowledge and practical inventions that would improve the human condition. In order to prepare for a better future, the educational system should be based on science. Reform of the educational system was part of Bacon's plan. The ideal schools of the future would be based on the new experimental sciences rather than Aristotelian teachings. Unlike the universities, Baconian institutions would provide the utilitarian training and pragmatic focus needed to redirect science and technology.

Bacon planned to publish a comprehensive plan for the reorganization of the sciences under the title *Instauratio Magna* (The Great Instauration), but this work was never completed. The first part, *De Augmentis Scientiarum* (1623), was an enlarged version of the *Advancement of Learning*. It contains a detailed classification of the whole range of human knowledge. Only two of the six natural histories that Bacon had planned were completed: *Historia Ventorum* (*History of the Winds*, 1622) and *Historia Vitae et Mortis* (*History of Life and Death*, 1623).

Deeply suspicious of mathematics and pure *deductive* logic, which proceeds from the more to the less general by using intuitive thinking, Bacon was certain that valid hypotheses would be derived from the assembly and analysis of "Tables and Arrangements of Instances." Bacon rejected the scholasticism of the universities and launched open attacks on Aristotle, Plato, and scholasticism. Attacking "fruitless" scholasticism and Aristotelian science, Bacon argued that natural knowledge should continuously increase through the process of discovery. He ridiculed conservative philosophers who seemed to believe that the Aristotle and the Bible provided everything men needed to know. Fact gathering and experimentation, he insisted, must replace the sterile burden of deductive logic that could never generate new scientific knowledge. Instead of adding to human understanding, followers of the great thinkers had surrendered their intelligence to past masters of philosophy. Theology had to be separated from science and philosophy, but natural philosophy must be reunited with the craft traditions that had nourished it during the time of the Presocratics. Of all the ancient philosophers, Bacon had especially high regard for Democritus and Lucretius, the founders of atomic theory.

Financial aid for scientific research was a high priority item in Bacon's schemes. Properly supported, science could create a brotherhood of experts and an international government. Scientists and technologists would serve as the supervisors and reformers of a world made as beautiful as Eden was before the fall of man. For those who worried that science might subvert the authority of religion, Bacon dutifully portrayed science as the most faithful handmaid of religion and, except for the word of God, the "surest medicine" against superstition. When scientific knowledge had purged the mind of "fancies and vanities," it would be better able to submit to "the divine oracles" and "give to faith that which is faith's." Unfortunately, James I, who allegedly fell asleep while reading Bacon's *Novum Organum*, was unwilling to support Bacon's grand scheme.

Bacon's death has been attributed to his scientific curiosity and lack of judgment. On a cold day in March 1626, while driving through an area to the north of London, he began to wonder whether snow would delay the process of putrefaction. To test this hypothesis, he bought a chicken and attempted to stuff it with snow. Chilled by these exertions, Bacon contracted bronchitis and died.

RENÉ DESCARTES

Like Bacon, the French philosopher René Descartes believed that a new approach to natural philosophy would lead to knowledge that would promote human welfare. Unlike Bacon, Descartes was a gifted mathematician, honored as the inventor of analytic geometry, and the advocate of a deductive, mathematical approach to the sciences. Like Bacon, Descartes rejected the conceit that science should be purely theoretical and abstract, with pretensions to total separation from the world of practical problems. Rejecting the authority of the ancients, Descartes anticipated a rational philosophy through which human beings could master and possess nature's abundance and establish a new medical science capable of eliminating disease and the disabilities of old age.

Descartes was born in La Haye (now Descartes), France, and became interested in mathematics while a student at the Jesuit college at La Flèche. After earning a law degree, he worked as an engineer, dabbled in astrology, and traveled widely through Europe. Apparently restless and dissatisfied with these activities, Descartes enlisted in the service of the Prince of Orange. During a lull in the warfare in the Netherlands, Descartes went to Germany and joined the Bavarian army.

By 1920, Descartes had discovered a method of reasoning that he believed would be applicable to all the sciences. While sitting in a "stove" (actually a small, very warm room), he was overwhelmed by an emotional and intellectual experience that left him with a profound philosophical insight that the young philosopher immodestly described as the foundation of a "completely new science." Moving to Paris in 1622, Descartes associated with prominent writers, scholars, philosophers, mathematicians, and scientists. Worried by the increasingly intolerant climate in France, Descartes decided to move to the Netherlands, where he felt he could enjoy greater intellectual freedom.

Many eminent scientists, philosophers, and historians have agreed that Descartes's work constituted a major scientific revolution, despite the fact that no specific scientific principle or discovery is associated with his name. So pervasive was Descartes's influence on science and philosophy that many contemporaries of Newton thought of themselves as the disciples of Descartes. Unlike the revolutionary discoveries associated with Newton and Harvey, many of Descartes's theories about the nature of matter and space soon became mere historical curiosities. Still, the Cartesian spirit is so highly esteemed in France that French philosophers declare that "Descartes, c'est la France," even though Descartes spent 20 years in Holland because he thought his ideas were too unorthodox for France.

When the Inquisition censored Galileo for his support of the Copernican system in his *Dialogue Concerning the Two Chief World Systems* (1632), Descartes was about to publish a major work called *Le Monde* (*The World*). Despite his confidence that his own physics would eventually replace the flawed Aristotle system, Descartes suppressed this work and probably destroyed some of his manuscripts. Thereafter, fear of persecution for heresy influenced much of his behavior. His secrecy and habitual caution only delayed the inevitable. Finally, in 1649, as intolerance grew in both France and the Netherlands, Descartes sought refuge with Queen Christina of Sweden. Ridiculed as a "hermaphrodite" because of her interest in science and mathematics, Queen Christina expected Descartes to serve as a tutor. Five months after arriving in Stockholm, Descartes died of pneumonia. He probably contracted his final illness as a result of providing lessons for the Queen at 5 a.m. in harsh winter weather. The Roman Catholic Church put his works on the Index of Forbidden Books in 1667.

The *Discourse on Method*, Descartes's major work, was published in 1637. It was one of the first significant modern philosophical works written in French instead of Latin. Therefore, Descartes declared, all literate men and women could learn to use reason in the search for truth. As indicated by its subtitle "For the correct use of reason and for seeking truth in the sciences," the text provided simple rules for reasoning: (1) Accept nothing as true that is not self-evident; (2) divide complex problems into their simplest components; (3) solve problems by proceeding from simple to complex; (4) reevaluate the previous steps. According to Descartes, his own quest for certainly and scientific knowledge provided the foundation for these seemingly simple rules.

Descartes discovered his method for arriving at truth by doubting and then rejecting all knowledge based on authority, the senses, and reason. For Descartes, the only certainty was that he was thinking and he must, therefore, exist. It is impossible to leave Descartes without reference to his most famous axiom or first principle: "Cogito, ergo sum" (I think, therefore, I am.) A lesser known principle advanced by Descartes proclaimed: "Give me motion and extension, and I will construct the world." To escape the trap of solipsism (the belief that nothing exists but one's individual self and thoughts), Descartes argued that "clear and distinct ideas" that survived his intense analysis must be true. That is, by methodically confronting "clear and distinct ideas" with deep and profound skepticism, Descartes thought he could separate his search for truth from the errors of all past authorities. Applying his methods to science, philosophy, or any other rational inquiry, Descartes asserted, would not only resolve problems but would lead to the discovery of useful philosophical knowledge. As optimistically as Bacon, Descartes predicted "there can be nothing so remote that we cannot reach to it, or recondite we cannot discover it." Knowing the crafts of the artisans and the forces of bodies we could "render ourselves the masters and possessors of nature."

Unlike Bacon, whose work he had read and criticized, Descartes placed *a priori* principles first and subordinated his observations and experimental findings to them. Nevertheless, he too had a grand scheme and task for natural philosophy. An examination of method was a primary part of his plan to use the mathematical method in developing a general mechanical model of the workings of nature. Certainly, experimentation had a role in Descartes' system, but it was a subordinate one. In the Cartesian system, experiments should serve as illustrations of ideas that had been deduced from primary principles or should help decide between alternative possibilities when the consequences of deduction were ambiguous.

After developing his philosophical system and method of inquiry, Descartes devoted the rest of his life to the exploration of mechanics, medicine, and morals. His physiological studies were based on the belief that animals were purely mechanical, in contrast to humans who possessed a rational soul. Although Descartes portrayed man in terms of the dualism of mind and body, human beings could also be analyzed in terms of mechanical principles. These Cartesian principles had a profound impact on the development of physiology and psychology.

Neither Descartes nor Bacon alone could have served as a complete guide for the development of science and technology. The great mathematician and physicist Christiaan Huygens (1629–1687) recognized this when he remarked that Descartes had ignored the role of experimentation, while Bacon had failed to appreciate the role of mathematics in scientific method. A synthesis of the two approaches was needed, or the admission that there is no one scientific method sufficient for asking and solving all possible problems. Mechanical fact-finding, daydreams, tinkering, and flashes of intuition have played a role in science, no matter what formal method or fashionable doctrine scientists professed to follow.

Many early participants in the Royal Society of London and the French Academy of Sciences saw themselves as builders of Salomon's House and tried to carry out projects suggested by Bacon, such as inventories of the crafts and the compilation of catalogues and encyclopedias. Robert Hooke (1635–1703), Curator of Experiments for the Royal Society, reflected the Baconian principle that truth and utility are the same thing in his description of the "noble matters" pursued by the Society as a means of improving industry, agriculture, commerce, and navigation. One of the most versatile, hard-working, and underpaid scientists of all time, Hooke performed demonstrations for the Royal Society, worked as assistant to Robert Boyle, and designed and improved many experimental instruments such as the air pump, balance, and watch spring. While pursuing the useful arts, manufactures, and inventions, members of the Society pledged to avoid meddling with politics, morals, metaphysics, and divinity. To promote philosophical progress, members pledged that they would avoid "dogmatizing, and the espousal of any Hypothesis not sufficiently grounded and confirm'd by Experiments." Although the Royal Society did not emphasize biological studies, many of the inventions developed by members were fundamental to the life sciences. For example, the air pump was used in Boyle's studies of combustion and respiration. Many experiments were conducted on the intravenous injection of food and drugs, as well as blood transfusions between man and beast.

Early histories of the scientific revolution analyzed the age, class, professions, and religious affiliation of the members of the Royal Society in great detail. Until the rise of modern science was reexamined from a feminist perspective, it was considered highly significant that 62% of the early members were Puritan, but the fact that 100% were male was too obvious for comment. Excluded from the formal academies of England and France, women explored the possibilities of creating alternative academies, such as that proposed by Margaret Cavendish in her play "The Female Academy" (1662). In the 1690s, Mary Astell petitioned Queen Anne for funds to establish an educational institution for girls from noble families. Despite claims that the study of natural philosophy would enhance the moral virtue, modesty, and religious sensibilities of future wives and mothers, attempts to establish such institutions in England and France were generally unsuccessful.

Nevertheless, the exclusion of women from participation in the newly emerging experimental sciences was not necessarily inevitable while institutional arrangements remained uncertain and fluid. Although the rise of modern science is closely associated with the scientific academies of the seventeenth century, the informal salons of aristocrats, the courts of princes, the universities, and the workshops of artisans also played a part. In seventeenth-century Europe, natural philosophy was still a new and precarious enterprise. The sciences lacked a secure base of support within the existing power structure, and its practitioners were not assured of any particular means of earning a living. Those who were fortunate might find a wealthy and powerful patron willing to support artists, philosophers, or inventors. For ambitious young men, playing the role of intellectual ornament in the courts of princes or the salons of aristocrats was a rather precarious existence. On the other hand, such settings provided a place in which wealthy and aristocratic women could participate in the arts and sciences when all other opportunities were closed to them.

The medieval convent had provided a setting in which women such as Hildegard von Bingen (1098–1179) might pursue learning, but the medieval universities set the precedent for excluding women. European universities were generally closed to women until the late nineteenth century; some did not admit women until well into the twentieth century. Outside the privileged world of the convent and university, guilds, household workshops, and the practice of medicine served as possible pathways to science for some women. But by the end of the eighteenth century women generally found themselves legally barred from the practice of medicine, surgery, and pharmacology. Similar trends occurred in other fields and institutions. Thus, as the prestige of scientific academies increased during the seventeenth and eighteenth centuries and the trend towards the professionalization of science and medicine accelerated, the exclusion of women was also formalized. Women had been active participants in the Renaissance courts of Italy, aristocratic learned circles, and the salons that were the precursors of formal academies, but they were unwelcome in the new institutions, despite some controversy as to whether merit, regardless of sex, should be the primary criterion for admission.

Women could continue to engage in informal discussions of science at appropriate social gatherings and study with private tutors, but if they were too active in their pursuit of knowledge, even the most privileged women might be subjected to censure and ridicule. Generally, women who wrote about their scientific inquiries had to publish them anonymously. Women, who found themselves excluded from serious participation in science, warned that society would ultimately find it costly to lose the potential contributions that one half the human race might have made to the arts and sciences.

ACADÉMIE DES SCIENCES

Like the Royal Society of England, the Royal Academy of Sciences in France began with informal meetings of small groups of philosophers, but the two institutions developed quite differently thereafter. While the Royal Society of London developed without government funding, the French Academy became intimately linked to the state and so dependent on royal patronage that the academicians owed their appointments and salaries to the king. Up to this point in time, natural philosophers generally supported themselves as physicians, teachers, clerics, or government officials or sought the patronage of wealthy nobles. The French Academy helped establish a new concept of science; it was the first institution in Western Europe to turn scientific inquiry from an avocation into a profession financed by the state. The Academy served as a specialized niche in which amateurs could become professional scientists. In a society where rank and wealth ordinarily determined a person's worth, academicians ostensibly committed themselves to a community in which a member's prestige was based on experimental ingenuity and contributions to science.

Early in the seventeenth century, various groups of intellectuals met privately in and around Paris to explore the new scientific theories permeating Europe. Many individuals gravitated towards the fashionable salons that competed for the time and attention of the most amusing, learned, and witty intellectuals and scientists. Private academies attempted to gain the support of wealthy aristocrats in order to establish private laboratories and observatories. Wealthy patrons, however, were not reliable sources of support, because they often had a short attention span and quickly turned to other sources of amusement. When the disadvantages of instability and chronic financial stress clearly outweighed the benefits of independence, natural philosophers turned to the state for financial support.

On December 22, 1666, the Académie Royal des Sciences (Royal Academy of Sciences) held its first official meeting in the private library of Louis XIV. In 1699 the name of the society was changed to Académie Des Sciences. Although the king was portrayed as the patron and sponsor of the Academy, he had no personal interest in science. State support for the arts and sciences had become part of France's quest for fame, honor, international recognition, and economic advancement. The Academy owed its official status to Jean-Baptiste Colbert, the king's chief advisor and minister of finance. Colbert saw the Academy as a way to glorify the king and expand the wealth, power, and prestige of France. For the seed money granted to the Academy, Colbert expected to reap practical advances in industry, commerce, medicine, public health, and warfare.

In exchange for state support, the Academy accepted regulations dealing with its duties and relationship to king and state, membership and procedures. Luxuriously accommodated at the Louvre, holding meetings and conducting experiments in the Royal Library, the Academy became a creature of the state. Academicians enjoyed many advantages at first. The Academy had the resources to pay salaries, set national standards, establish regulatory power over industries and inventions, and serve the government as a "think tank" on technical matters. Colbert generally allowed the academicians wide latitude in choosing their projects, but he clearly expected the Academy to perform useful functions, such as analyzing the drinking water at Versailles, drawing up detailed maps of the tax district around Paris, improving navigational aids, studying the natural history of useful plants, and so forth.

The financial support granted by the king to the Academy of Sciences was not unprecedented in kind, but it was remarkable in its extent. The level of funding awarded to the seventeenth-century Academy of Sciences was equivalent to the annual income of a wealthy French monastery. Support was provided both for individuals, generally in the form of salaries or lodgings, and to the institution. The Academy enjoyed grants large enough for the construction of an observatory or chemical laboratory and funds for equipment, research assistants, expeditions, and publications. As academicians struggled to establish a professional identity, they learned how difficult it was to serve as advocates of the new experimental sciences while meeting the demands of the state. By the end of the seventeenth century, because of economic and political problems, pensions declined or even disappeared. Members argued that reliable subsidies were essential to provide the leisure, peace of mind, and good morale needed for a cooperative, creative scientific community, but politicians had their own agenda.

Many of the first projects adopted by the Academy, such as assembling catalogues, herbals, and encyclopedias, were guided by Baconian objectives. Interest in Baconian utilitarianism soon declined, and Cartesian philosophy became a major influence. Between 1666 and 1699 the Academy never had more than 34 or less than 19 members. During this period, about 60 men were awarded appointments as academicians, but no more than 20 actually worked on projects sponsored by the Academy at any one time. The seventeenth-century French Academy included a higher proportion of members trained as physicians, surgeons, or apothecaries than the Royal Society of England. This seemed to support predictions by Descartes and Bacon that scientific societies could initiate investigations that would improve medical practice and public health. Scientific societies were expected to encourage cooperation among scientists, but the Academy evolved as a closed, elite, and exclusive community. In the 1830s three eminent and learned women had been proposed as member of the Académie Française, the precursor of the Academy of Science. Despite considerable controversy, Mademoiselle de Scudéry, Madame des Moulières, and Madame Dacier were not admitted. In 1910, the year before Marie Curie became the first person ever to win a second Nobel Prize, the Academy of Science rejected her. Curie's opponents argued that other eminent women had been rejected by prestigious French academies and the tradition must be maintained.

Unfortunately for the academicians, Colbert was replaced by the Marquis de Louvois, a man who believed that scientific research should be devoted to useful ends in the service of the king and the state. The reorganization of the Academy in 1699 affected the way in which members conducted research. Academicians were ordered to undertake a variety of practical and even trivial projects, and political appointments led to increased control over the Academy by men unsympathetic to the sciences. The Academy was expanded to incorporate 70 members in hierarchically arranged positions that were paid for and controlled by the state.

Members of the Academy were envied and criticized; they were expected to referee disputes, solve technical problems, and serve as a symbol of progressive science. Tensions between the Academy and the increasingly restive French populace generated political pressures that impaired the institution. In 1793, after the French Revolution, the Academy was abolished. Some of the academicians perished in the Reign of Terror and others were forced to emigrate. Two years later the institution was reorganized as a branch of the National Institute of Arts and Sciences. The new Institute was expected to serve as proof of egalitarian French cultural achievements and make useful discoveries. All members of the Institute, salaried by the government, were expected to serve as government advisors and functionaries. Compared to the independent scientific and professional societies developing in other countries, the Institute was a conservative force that tended to undermine the early promise of French science. The former name of the Academy was restored in 1816, but the Académie des Sciences remained part of the National Institute of France.

SCIENTIFIC SOCIETIES IN THE UNITED STATES

Even during the early colonial period, science was a subject of intense interest in America. The English colonies were particularly influenced by the Royal Society, and several Americans, including John Winthrop the Younger (1606–1676), governor of colonial Connecticut, were elected fellows of the Society. The Reverend Increase Mather (1639–1723) established a scientific society in Boston in 1683 modeled on the Royal Society. This later became the Boston Philosophical Society, but disputes over religion and politics caused its premature demise.

Among the founding fathers of the new republic, Benjamin Franklin and Thomas Jefferson were particularly active in promoting the development of science. Franklin established a group called the Junto, which met weekly to discuss natural history and philosophy. The Junto was the precursor of the American Philosophical Society, the oldest surviving learned society in the United States. The American Academy of Arts and Sciences, incorporated in Boston in 1780, was the second major learned society established in the United States. George Washington, Benjamin Franklin, Thomas Jefferson, Alexander Hamilton, James Madison, and John Adams were among its first members. Various specialized and regional societies formed and dissolved, but only the American Philosophical Society and the American Academy of Arts and Sciences evolved into permanent national societies. Engendered in an age permeated by Baconian utilitarianism, the American societies also contributed to the spirit of cultural nationalism.

Like the Royal Society, American academies remained essentially independent of university or government ties, at least until World War I. Following the war, various societies were grouped into national councils or institutes: the American Council on Education, the National Academy of Sciences, the National Research Council, the American Council of Learned Societies, and the Social Science Research Council. The roots of the National Academy of Sciences go back to two societies founded in the 1840s: the National Institution for the Promotion of Science and the American Society for Geologists and Naturalists (which became the American Association for the Advancement of Science). In 1863 Congress chartered the National Academy of Science to act as the official adviser to the federal government on scientific and technical questions. According to its charter, the Academy must investigate and provide reports at the request of any department of the government. Except for reimbursement of the expenses incurred in carrying out such studies, the government was not expected to provide funding to the Academy. Every year the National Academy of Sciences, the nation's most exclusive club for researchers, elects about 60 new members. The process by which the candidates are selected is very secretive, but the Academy has repeatedly been attacked as an exclusive "old boys' club" that displayed marked resistance to the admission of women. Despite the increasing presence of women in scientific research in the second half of the twentieth century, the percentage of women in the Academy, generally no more than 5–10 percent of the total membership, hardly changed during the last 30 years of the twentieth century.

In his landmark report *Science, the Endless Frontier* (1945), Vannevar Bush, the director of the Office of Scientific Research and Development (OSRD) during World War II, urged the government of the United States to continue its tradition of opening new frontiers by recognizing that although the frontier represented by new lands for settlement had essentially disappeared, the frontiers of science remained. Like Bacon and Descartes, Bush expressed the optimistic belief that scientific progress would inevitably lead to improvements in human health, well-being, and security. *Science, the Endless Frontier* had been written in response to a letter from President Franklin D. Roosevelt designed to elicit a study of how OSRD's wartime experience might be used to create new enterprises, jobs, and a better national standard of living after the war.

The U.S. government agencies of special interest to twentieth-century researchers in the life sciences are the National Institutes of Health (NIH) and the National Science Foundation (NSF). The National Institute of Health was officially established in 1930 to promote research vital to the health and welfare of the nation. The precursor of this agency was created in 1887 when Dr. Joseph J. Kinyoun was awarded $300 to establish a bacteriological laboratory at the Marine Hospital on Staten Island as part of the United States Public Health Service. After World War II, the National Institute of Health engendered a series of institutes designed to stimulate disease-oriented research and the training of future biomedical scientists. The National Science Foundation, an institution originally dedicated to the support of pure research, was not created until 1950. Obviously quite different from the scientific societies generated by the first scientific revolution, these organizations illustrate the continuing interaction between science and society as negotiated in terms of institutional sponsorship and support. During the years between World War I and World War II, private foundations, such as those established by Andrew Carnegie and John D. Rockefeller, orchestrated and dominated the system of extramural grants that supported much of the scientific research carried out in the United States. It was only in the wake of World War II that "big science" first emerged and the federal government became the great patron of American science.

MICROSCOPES AND THE SMALL NEW WORLD

During the seventeenth century many instruments fundamental to modern experimental science were either invented or first put into use on a significant scale.
The projects and goals of academicians encouraged them to adopt new instruments, such as the telescope, microscope, air pump, and barometer, and improve
the precision and accuracy of older measuring devices, including clocks, rulers,
scales, and thermometers.

Instruments that apparently extended the range of the human senses also
created uncertainty about the relationship between the observer and the object
under investigation, changed the nature of the questions put to nature, and often
intensified old debates. The air pump, for example, was used to answer questions about the nature of air and the existence of the vacuum. All observers
could agree that pumping air out of various kinds of vessels allowed the experimenter to suffocate mice, put out candles, and so forth. But those who continued
to deny the existence of the vacuum could argue that removing useful air simply
allowed some unknown and subtle entity to enter the vessel and take its place.

Galileo might well assert that "anyone can see through my telescope," but
that did not ensure that all observers would interpret what they had seen in the
same way, or that skeptics would not continue to argue that the apparent images
were generated by the device itself and bore no relationship to real objects in the
distant heavens. In other words, instruments can be regarded as untrustworthy as
well as theory-laden entities. The telescope allowed scientists to see the immense and distant heavens more clearly, but the stars and planets had been the
object of study and speculation for thousands of years. The microscope gave
observers access to a small and secret new world on earth. Those who did accept
the value and reliability of information gathered with the new instruments had
to come to terms with the end of the Baconian ideal of assembling all observable
facts about the world, because it was possible that further advances in instrumentation would create unanticipated observations about unknown entities.
Even the most enthusiastic investigators had to admit that using the new instruments correctly was quite difficult and that sometimes the instruments exaggerated the faults and foibles of individual investigators.

Many hundreds of years elapsed between the time when the art of cutting
and polishing stones was invented and the development of lenses for eyeglasses,
telescopes, and microscopes. Lens-shaped gems were discovered in the ruins of
Nineveh, Pompeii, and Heraculaneum. Looking at the world through beautiful
gems might be very amusing, but these curiosities would probably not be of
much use to people with defective vision. The Roman Emperor Nero is said to
have watched performances in the arena with the aid of a jewel having curved
facets. Although it is possible that Nero's bauble contained concave facets to

correct myopia (nearsightedness), there are few reports suggesting the use of corrective lenses before the thirteenth century. The ancients knew that convex lenses or crystalline spheres could "concentrate" the rays of the sun and act as a burning glass. According to Pliny, doctors even used this technique to produce therapeutic burns on their patients. In other observations about the properties of gemstones, Pliny recommended looking at the world through emeralds to rest tired eyes. However, he did not mention using lenses to magnify images, and he thought that nearsightedness was incurable. The Roman philosopher Seneca said that small, indistinct writings appeared larger when seen through a globe filled with water, but he apparently assumed that water itself caused magnification.

An ingenious effort to analyze the way convex lenses produced a magnified image was made by Ibn-al-Haitham (962–1038), known to Europeans as Alhazen. His treatise was translated into Latin in the twelfth century as the *Opticae Thesaurus Alhazeni Arabis*. Alhazen dealt with experimental means of proving various propositions concerning the action of lenses. He also attempted to relate anatomical features of the eye to the physics of vision. Although Alhazen was interested in philosophical, mathematical, and medical aspects of optics, he did not suggest practical applications of this phenomenon. In his discussion of the work of Alhazen and his own studies of the properties of light, Roger Bacon (1214–1292) drew attention to the magnifying properties of lenses and the design of burning glasses, mirrors, and magnifiers. Some passages in his *Perspectiva* seem to describe telescopes, but modern scholars suggest Bacon was actually referring to objects that were inside glass domes and illusions created with sets of mirrors. Bacon did note that old men with weak eyes could use lenses curved on one side as an aid to reading.

Medieval scholars could have used lenses held in the hand just above a book or set directly on the page like some modern magnifiers, but this would make it very difficult to write and turn pages. This problem was solved towards the end of the thirteenth century with the invention of spectacles, a device that held lenses in front of the eyes. The inventor of the first spectacles might have been a glazier, that is, an artisan who created windows and glass ornaments, but the grave marker of Salvano d'Aramento degli Amati, a nobleman of Florence (d. 1317), states that he was the inventor of spectacles. Alessandro della Spina of Pisa (d. 1313) claimed that he had learned the secret of making spectacles and shared the discovery with others. Whatever the truth of such claims, by the beginning of the fourteenth century spectacles were mentioned quite frequently as a well-known, but still much admired invention. Spectacles were rare and expensive items that were listed in account books and wills as valuable property. Like literacy itself, eyeglasses became a mark of distinction and wisdom. Ancient sages and saints were depicted with quite anachronistic spectacles held in their hands or perched on their noses.

The use of spectacles must have occasioned a profound effect on attitudes towards human limitations and disabilities. Spectacles not only made it possible for scholars and copyists to continue their work, they accustomed people to the idea that certain physical limitations could be transcended by the use of human inventions. References to the use of concave lenses to correct nearsightedness first appeared in the sixteenth century. For example, Giambattista della Porta's *Natural Magic* (1589) contains a discussion of how "Strange Glasses" could be used to correct defects in human vision. Concave lenses were used to see things that were far away, while convex lenses were used to see things close to the eye. Boasting of his ingenious use of these techniques, della Porta claimed that he had designed special combinations of convex and concave lenses for friends with vision problem so that they could see things both far away and nearby.

By the sixteenth century the making of lenses was an established industry and spectacle makers had accumulated a good deal of empirical knowledge about optics. The invention of the telescope and the microscope was probably a somewhat accidental outcome of the art of making spectacles. Several texts and letters from the late sixteenth century seem to refer to the use of primitive telescopes. Once the first microscopes were constructed, the basic concept spread very quickly and many ingenious modifications were introduced. Credit for the invention has been attributed to three Dutch spectacle makers—Hans Janssen, his son Zacharias, and Hans Lippershey—but some admirers of Galileo claimed that he invented both the microscope and the telescope. Galileo's telescope could be used as a microscope by increasing the distance between its lenses. While Galileo did not actually invent the telescope, his astronomical discoveries revolutionized ideas about the heavens. With the telescope Galileo discovered the moons of Jupiter and the phases of Venus; under the microscope, he saw flies that seemed to be as big as lambs.

Perhaps as early as 1590, Zacharias or Hans Janssen created a crude compound microscope by combining a concave and a convex lens at the ends of a brass tube that was about an inch in diameter and 18 inches long. There is no evidence that the Janssens ever made any significant observations with their magnifying device. Cornelius Drebbel seems to have obtained one of the instruments constructed by the Janssens. After adding his own improvements, he attempted to bring the device to the attention of scientists and scholars. Improvements in the design of lens systems have also been attributed to Lippershey, who constructed an unusual binocular telescope.

Even with magnification no greater than 10-fold, the first microscopes provided exciting images; creatures just barely visible to the naked eye became complex and bizarre monsters. The ubiquitous flea was such a popular object of study that magnifying lenses were often referred to as "flea glasses." Some philosophers warned against believing in the images created by unnatural devices that misled the senses; less sophisticated skeptics considered use of such instruments a form of witchcraft. For example, George Stiernhielm, a Swedish poet who amused himself with scientific experiments, was denounced as a sorcerer and an atheist by a clergyman who had been persuaded to look at a flea through a magnifying glass.

With the new magnifying glasses, seventeenth-century scientists assembled a remarkable body of observations and experiments, beginning with the work of Francesco Stelluti in 1625. His strikingly detailed illustrations of the parts of the bee, obtained at magnifications of 5 and 10 diameters, were published by the Academy of the Lynx Eye. Giovanni Borelli (1608–1679), better known for his studies of muscle anatomy and physiology, investigated objects as varied as blood vessels, nematodes, textile fibers, and spider eggs. Athanasius Kircher (1602–1680) used the microscope in his studies of putrefaction and disease. Although many seventeenth-century scientists made some use of the microscope, Antoni van Leeuwenhoek, Marcello Malpighi, Robert Hooke, Jan Swammerdam, and Nehemiah Grew raised microscopy to a level not superseded until the middle of the nineteenth century.

Antoni van Leeuwenhoek (1632–1723), a man of immense energy and curiosity, was the most ingenious microscopist of the seventeenth century. The scope of his investigations was so extensive that well into the nineteenth century scientists were warned that they should take care to examine Leeuwenhoek's writings before claiming to have made an original microscopic observation. Leeuwenhoek's father, who died when Leeuwenhoek was only six, was a basket maker, and his mother was the daughter of a brewer. The family lived in Delft, Holland, a city famous for its beer, textiles, and chinaware. When his mother remarried, Leeuwenhoek was sent to grammar school at Warmond, a village near Leiden. At the age of 16 Leeuwenhoek was employed as a bookkeeper and cashier by a linen draper (cloth merchant) in Amsterdam. He returned to Delft 6 years later and bought a house and shop. In 1660 he began to add a variety of civil service positions to his résumé. Despite the demands of his business and municipal duties, he devoted so much of his time to his microscopes that he was accused of neglecting his family. Perhaps there was some merit in this charge; he survived two wives and all but one of his six children. Yet Leeuwenhoek himself lived, as recorded on his tomb, for 90 years, 10 months, and 2 days.

Leeuwenhoek's interest in lenses began when he was about 39 years old, but his inspiration for making microscopes is uncertain. One possibility is that he got the idea while working with special lenses that linen drapers used at the time to inspect the quality of cloth. Another possibility is that Leeuwenhoek learned about microscopes from someone with a real scientific interest in the device. Many possible mentors have been suggested, but it appears most likely that Robert Hooke's *Micrographia* (1665) was the model for Leeuwenhoek's early microscopic work. The preface of the *Micrographia* explains how to make

ANTONI VAN LEEUWENHOEK

A portrait of Antoni van Leeuwenhoek in 1686 by Jan Verkolje

and use simple magnifiers, also known as single microscopes. Hooke found these instruments very difficult to use and preferred the compound or multilens microscope. Some apparently random specimens that Leeuwenhoek sent to the Royal Society actually reproduced experiments described by Hooke. By 1674, however, Leeuwenhoek was publishing accounts of microscopic observations that were superior to Hooke's. Leeuwenhoek was interested in searching for truly novel entities, while Hooke was primarily interested in magnifying conventional objects.

Leeuwenhoek's first simple microscope or magnifier was a tiny lens, ground by hand from a globule of glass, clamped between perforated metal plates; the device had an attached specimen holder. As his curiosity and skills increased, Leeuwenhoek learned to make very small biconvex lenses of short focal length. Most of his lenses were made by grinding and polishing selected fragments of glass or even grains of sand, but some were probably made from molten glass. With these tiny, high-powered lenses, Leeuwenhoek's simple microscopes were superior to most of the seventeenth-century optical systems, because multiple lenses produced multiple aberrations and artifacts. During 50 years of research, Leeuwenhoek probably produced more than 500 lenses. When he died, a cabinet with 26 instruments and extra lenses was bequeathed to the Royal Society. Tests conducted by members of the Society demonstrated magnifying powers from 50- to 200-fold, but unfortunately these instruments were lost within 100 years of Leeuwenhoek's death. The magnification of Hooke's compound microscope was generally about 20–50 times.

Scientists and historians who were unsuccessful in imitating Leeuwenhoek's work generally concluded that he used special secret methods or compound lenses. Others have demonstrated that simple microscopes and appropriate methods of sample preparation and illumination can reproduce the observations reported by Leeuwenhoek, including cells, spermatozoa, nuclei, molds, and microbes. Various bacteria and pond infusoria would be visible with Leeuwenhoek's best microscopes. Compound microscopes became more ornate, complicated, and fashionable, but they were not really better than the best of the simple microscopes. Many distinguished nineteenth-century scientists, such as Robert Brown and Charles Darwin, continued to use simple microscopes.

In one of his early letters to the Royal Society, Leeuwenhoek described himself as lacking in style for written expression because he was brought up only "to business," not to languages or learning. Moreover, he frankly informed the learned gentlemen that he preferred to keep his methods to himself because he did not gladly suffer "contradiction or censure from others." Although Leeuwenhoek's inability to read the learned Latin treatises of his contemporaries may have kept him relatively free of prevailing dogmas and preconceptions, he did not work in total intellectual isolation. He read the work of Dutch authors and Dutch translations of standard works, studied the illustrations in books he could not read, and, when necessary, invoked the help of friends and professional translators. Perhaps he missed some useful information, but he maintained his independence of mind. Even the knowledge that ignorant people considered him a magician who showed people things that did not exist and that some learned men did not accept the truth of his writings did nothing to diminish his confidence in his own observations. Leeuwenhoek regarded speculation as something that belonged to an academic world that did not concern him. Generally, he presented his observations as facts, but he was certainly willing to suggest possible interpretations.

Leeuwenhoek's work was introduced to the Royal Society by his friend the physiologist Regnier de Graaf. Letters from Leeuwenhoek to the Royal Society were edited, translated into English or Latin, and published in the *Philosophical Transactions*. In 1680 Leeuwenhoek was elected a Fellow of the Royal Society. For his benefit, the certificate was written in Dutch. Although he treasured this honor, he often saw fame as a nuisance that brought many visitors in its wake, including heads of state and curiosity seekers. To cope with interruptions, Leeuwenhoek seems to have kept certain simple demonstrations on hand, but he kept his best microscopes to himself. Despite his reluctance to suffer interruptions, some visiting scholars and scientists presumably brought him useful information about the work of other microscopists.

The majority of Leeuwenhoek's 400 letters to the Royal Society have been preserved, but it was not until 1981 that Brian Ford discovered several packets of specimens still attached to a letter dated June 1, 1674. These specimens included fine sections of cork, sections of elder pith, bits of dust, spores, house mites, pollen grains, cottonseeds, sections of the optic nerve of a cow, and dried samples of pond infusoria. The finest parts of Leeuwenhoek's handcut section of cork were remarkably thin, and the cell walls were clearly visible when examined by scanning electron microscopy. Some specimens contained a few blood cells, which presumably came from Leeuwenhoek's razor. One packet contained a sample of the dried algal mats that were known as "paper from heaven." After studying this material, Leeuwenhoek concluded it was a sheet of dried plant material only superficially resembling paper. When these samples were reconstituted, some of the microscopic pond organisms discovered by Leeuwenhoek could be seen.

Unencumbered by a rigid philosophical system or a fashionable theoretical program, Leeuwenhoek allowed his insatiable curiosity to determine the course of his microscopical researches. Crystals, minerals, plants, animals, water from different sources, scrapings from his teeth, saliva, seminal fluid, even gunpowder were examined under his lenses. Unfortunately, Leeuwenhoek was a secretive man who could be quite contemptuous of those who lacked his own curiosity, dexterity, and persistence. Thus, when asked why he did not teach his methods to others, he responded that he saw no reason to train young people to grind lenses because most students lacked curiosity and sought instruction only as a means of making money or to earn a reputation. Such motives, he contended, did not inspire people to discover the world hidden from our sight.

While Leeuwenhoek obviously took great joy in his work, he also took pains to make it as accurate and quantitative as possible. Having no standard units of measurement, he used simple objects for making comparisons, such as strands of hair, or grains of sand, and stated his measurements in terms of fractions or multiples. For some objects he offered size estimates in terms of "biological units," such as the eye of a louse. His eminently practical scale allowed him to communicate with his contemporaries and provided good clues for historians and scientists attempting to recreate his experiments.

In the course of his long and vigorous career, Leeuwenhoek made so many discoveries that it is impossible to cover them all. His studies of microorganisms, sperm cells, and the capillaries were especially important, but he also contributed to pioneering studies of the life history of insects, structural studies of plant and animal tissues, parasitology, microbiology, and the battle against the doctrine of spontaneous generation. Leeuwenhoek studied the mating behavior and life cycle of eels, intestinal parasites, insects, and other creatures said to be spontaneously generated. Under the microscope, even mites, bees, and fleas were shown to be very complex creatures. Thus, Leeuwenhoek was convinced that even the smallest microscopic creatures were produced by parents like themselves. Although Leeuwenhoek also studied reproduction in plants, he apparently thought that the flowers were essentially decorative.

In 1661 Marcello Malpighi (1628–1694) had observed capillaries and the blood corpuscles within them, but he apparently thought that the blood cells were fat globules. Leeuwenhoek was probably unaware of Malpighi's observations when he rediscovered the blood corpuscles in 1674 and the capillaries in 1683. In addition to observing mammalian red blood cells, Leeuwenhoek attempted to determine the size of blood cells and their nuclei in fishes, frogs, and birds. Demonstrating that the capillaries link the arteries and veins, Leeuwenhoek helped to complete Harvey's work on the circulation of the blood. The microscope revealed that the passage of blood from the arteries into the veins took place in vessels so thin that only one corpuscle could be driven through at a time. From numerous observations of the capillary network in the tails of tadpoles and eels, Leeuwenhoek ventured the opinion that the same thing must occur in humans, but the process could not be seen because of the thickness of the skin.

In 1677 Johan Ham, a medical student, seems to have made the first observation of sperm in the seminal fluid of a man with gonorrhea. Ham, who told Leeuwenhoek about this discovery, assumed that the "animalcules" were products of putrefaction related to venereal disease. The idea that sperm were a sign of disease proved very persistent despite Leeuwenhoek's demonstration that these entities were found in the semen of healthy men and animals, including dogs, rabbits, birds, amphibians, and fishes. Leeuwenhoek concluded that the animalcules were normal constituents of semen throughout the animal kingdom. In a letter to the Royal Society he acknowledged that whole universities full of scholars refused to believe that there were living creatures in the male seed, but their opinions did not concern him because he knew that he was right. His studies of sperm and eggs from fish and frogs led him to reject the prevailing view that vapors from seminal fluid were responsible for fertilization. Although he was unable to demonstrate the actual union of sperm and egg, he believed that the sperm penetrated the egg. Assuming that motion was essentially equivalent to life, he was convinced that the sperm provided the origin of new animal life; the egg, in contrast, appeared to have no independent capacity for motion.

In 1674 Leeuwenhoek recorded his discovery of the true nature of microorganisms. Since he assumed that motility was characteristic of living beings, he concluded that the tiny moving things that he had observed under the microscope must be a form of animal life. About 30 of his letters to the Royal Society deal with microscopic organisms, including rotifers, protozoa, and bacteria. His reports generated a great deal of excitement and also some skepticism, to which Leeuwenhoek responded by sending testimonials from doctors, jurists, ministers, and other reliable witnesses.

Like Leeuwenhoek, Jan Swammerdam (1637–1680) was one of the great Dutch scientists produced in that country's golden age. In outlook, education, and methodology, Leeuwenhoek and Swammerdam differed in everything but their devotion to testing the limits of microscopy. Almost obsessively methodical in his approach to his work, Swammerdam lived a disorderly, restless, and tormented existence. His father was a prosperous apothecary who was interested in the natural sciences, but he insisted that his son should take religious orders rather than study natural history. As a compromise, Swammerdam was sent to the medical school at Leiden. He graduated in 1667 but never practiced medicine. Absorbed in his research, Swammerdam was constantly at odds with his father, who seems to have been a stingy, cantankerous man obsessed with the idea that his adult son should support himself rather than waste time studying vermin.

Through his research on the natural history and anatomy of insects, Swammerdam established himself as one of the founders of modern comparative anatomy, entomology, and microscopy. Using the microscope as a tool rather than an end in itself, Swammerdam undertook systematic studies of entomology and comparative anatomy. He investigated the fine structure of insects, plants, animals, and humans and discovered the minute "seeds" of ferns. Fascinated by the activities of insects, Swammerdam willingly let lice and other insects bite him so that he could observe the operation of their mouthparts. Through human dissection and microscopic examinations, he made important discoveries about the uterus, lymphatic system, spinal medulla, and the organs of respiration. His proof that the lungs of a newborn mammal would float in water if the animal had begun to breathe on its own but would sink if respiration had not begun had important implications for forensic science. To examine minute insects, Swammerdam had constructed special instruments that allowed him to refine the art of dissection. Like Leeuwenhoek, Swammerdam had many different microscopes for specialized purposes. To examine minute insects, Swammerdam constructed special instruments that allowed him to refine the art of dissection. His dissecting tools included scalpels carefully ground under a magnifying glass and fine glass tubes that could be used to inject air, colored liquids, ink, or wax into his specimens.

After publishing a general history of insects in 1669, the impoverished and sickly entomologist contracted malaria, a mosquito-borne disease that was to trouble him for the rest of his life. When his father sent him to the countryside to recuperate, Swammerdam used the time for his classic study of the morphology and natural history of the mayfly. For Swammerdam, the study of the mayfly was not just a scientific account of a rather insignificant insect, but a deeply felt reflection of his religious commitment. Knowledge of the brief life of the mayfly might, he hoped, give human beings a vivid image of the shortness of earthly existence and so inspire them to a better life. Chronically ill and desperate for money, he tried to sell his collections of specimens and even his microscopes. Eventually he became involved in religious mysticism and abandoned science as a vain and unworthy enterprise.

The bulk of Swammerdam's writings were not published until 1737. After a precarious existence in various private collections, his manuscripts were purchased by the great chemist and teacher Herman Boerhaave (1668–1738), who published them in Dutch and Latin with plates engraved from Swammerdam's own drawings. Swammerdam's *Bible of Nature* was the first major study of insect microanatomy, metamorphoses, and classification. The work also contains observations of marine invertebrates and amphibian metamorphosis.

During the eighteenth century, Swammerdam's work on insects was used in support of the theory of embryological development known as *preformation*. His studies of the stages of insect metamorphosis indicated that it was a precisely predetermined sequence of events during which later stages grew from tiny parts already present at earlier stages. Comparing the development of humans, insects, and amphibians, Swammerdam suggested that a process like molting occurred in each case; that is, the new parts developed under the old skin, which was then shed.

Second only to Leeuwenhoek for the breadth of his interests and his enthusiasm for microscopic studies, Marcello Malpighi (1628–1694) was a pioneer in physiology, embryology, plant anatomy, comparative anatomy, and histology. In particular, Malpighi used the instrument to extend and refine knowledge of the form and function of the human body. Malpighi was described by contemporaries as "modest, quiet and of a pacific disposition." His parents died when he was 21, and, as the oldest child, he was responsible for his seven siblings. Nevertheless, he earned the degree of Doctor of Medicine from the university of Bologna by the time he was 25. At age 38 he decided to devote his free time to anatomical and microscopical research.

While studying philosophy at Bologna, Malpighi became interested in the new experimental approach to natural philosophy. Bartolomo Massaari, professor of anatomy, invited Malpighi to use his private library and become a member of his scientific academy. After completing his studies of medicine and philosophy, Malpighi established a private practice and obtained a position as lecturer in logic at the University of Bologna, where he became notorious for his anti-Galenist views. In 1656 a chair of theoretical medicine was created for him at the University of Pisa. Malpighi found the intellectual climate of Pisa a decided improvement, especially after he became a member of the Accademia del Cimento (Academy of Experiments). The international nature of the seventeenth-century scientific community is reflected in Malpighi's correspondence with Henry Oldenburg, Secretary of the Royal Society, and his election as a Fellow of the Royal Society.

Especially interested in the physiology of respiration, Malpighi conducted painstaking investigations of the capillaries and the fine structure of the lungs. He also studied the kidneys, cerebral cortex, plant microanatomy, and the natural history of invertebrates. Malpighi, who was born in the year *De Motu Cordis* was published, demonstrated the physical existence of Harvey's hypothetical anastomoses. When Malpighi began his studies in 1660, the lung was thought of as a fleshy organ consisting of a porous parenchyma. Without the microscope, the smallest divisions of the blood vessels and the ramifications of the windpipe were, of course, invisible. Anatomists, therefore, assumed that blood and air simply mixed within the parenchyma of the lung, but Malpighi demonstrated that the lung was composed of membranous vesicles filled with air. By washing out the blood, inflating the lungs through the windpipe, and allowing the tissues to dry, Malpighi was able to trace the ramifications of the trachea or windpipe to a network of thin-walled bladders. Further investigation revealed that blood did not leak out of its proper vessels into the air spaces, but made its way from artery to vein through the minute structures known as capillaries.

Despite the crudity of his instruments and what he called his own clumsiness, through reason, observation, and experiment, Malpighi linked his structural studies to new theories about the function of the lungs and kidneys. Urging others to follow his example, he explained that he did not come to his conclusions about the structure of the body by means of books, but through long, patient, and varied use of the microscope. While his instruments and techniques were limited, Malpighi carefully described his use of mercury and wax injections, methods for staining microscopic preparations with ink and other readily available liquids, and the use of various kinds of lenses and means of illuminations. Malpighi explored the fine structure of many organs, always trying to define their smallest components and their physiological significance. The single best known microscopic observation of the seventeenth century, however, is probably the description of the cellular structures that Robert Hooke (1635–1703) found in cork.

Overshadowed by his great rival Isaac Newton, Robert Hooke was one of the seventeenth century's leading inventors and designers of scientific instruments. Hooke's contributions to astronomical instrumentation were particularly valuable, but he also introduced improvements in the microscope and methods of illumination. Indeed, the study of light was one of his main theoretical preoccupations. As Curator of the Royal Society, Hooke was responsible for the demonstrations performed at the weekly meetings. In 1663 he was assigned the task of undertaking an extensive program of microscopical investigations so that he could present at least one novel observation at each meeting. Despite an eye disorder that made microscopic work difficult and even painful, Hooke presented studies of various insects, such as fleas, lice, and gnats, and demonstrated different types of hair, various textiles, the point of a needle, the edge of a razor, molds, moss, and the cells seen in sections of freshly cut cork.

These observations were published in Hooke's *Micrographia* (1665), which greatly stimulated interest in microscopy. The text was a popular treasury of information about microscopic technique and illustrations of the wonders that appeared when familiar objects were magnified. Although writing for a general audience, Hooke used the *Micrographia* as a vehicle for his theoretical work and speculations about the nature of light, the relationship between respiration and combustion, and the origin of fossils. Hooke described the streaming juices in live plant cells, but his colleague Nehemiah Grew (1628–1712) carried out more extensive microscopic studies of plant and animal anatomy.

Nehemiah Grew was the only son of the Reverend Obadiah Grew, an anti-Royalist clergyman who lost his position when Charles II was restored to the throne. After graduating from Cambridge, Grew left England to study medicine at the University of Leiden. Having established a successful private practice, Grew devoted much of his energy to studies of the anatomy of plants and animals. As he explained in an essay presented to the Fellows of the Royal Society in 1671, his research was guided by the idea that animals and plants must possess fundamentally similar structures because both were designed by "the same Wisdom." When Henry Oldenburg died in 1677, Grew became Joint Secretary of the Royal Society with Robert Hooke. Grew's treatise, *The Comparative Anatomy of Stomachs and Guts begun*, appeared as an appendix to a catalog of specimens in the museum of the Royal Society. Modest about his work, Grew similarly entitled his pioneering work on plants *The Anatomy of Vegetables begun* (1682). This study was the first part of *The Anatomy of Plants*, which included *The Anatomy of Roots*, *The Comparative Anatomy of Trunks*, and finally *The Anatomy of Leaves, Flowers, Fruits and Seeds*.

The Anatomy of Plants helped revive the science of botany, which had stagnated, except for the production of herbals, since the time of Theophrastus. Grew studied the vascular tissue of plants, noted its tubular nature, and discussed its different distribution in roots and stems. Although his discussion was more detailed than that of Malpighi, Grew acknowledged Malpighi's priority in the discovery of the vascular system of plants. Grew speculated about the possibility that flowering plants might undergo sexual reproduction and suggested that the flowers contained the sexual organs. He recognized the pistil as the female part but was uncertain about the purpose of the stamens. Calling attention to the fact that the pistil and stamens were sometimes present in different flowers and sometimes appeared in the same flower, he suggested that some flowers were hermaphroditic, like snails. In 1694, 12 years after the publication of Grew's *Anatomy of Plants*, Rudolph Camerarius (1665–1721) described plant sexuality based on anatomical and experimental studies.

Comparative Anatomy of Stomachs and Guts begun was the first zoological book with the phrase "comparative anatomy" in the title and the first book to deal with one system of organs by the comparative method. Viewing human beings as the highest type, Grew compared all other animals to the human standard. Approaching more physiological ideas, Grew argued that the morphological significance of any organ could not be grasped until its function had been considered. This point was illustrated by a discussion of the foods eaten by each animal and the type of gut required to process different foods. By studying freshly killed animals, Grew was able to see the peristaltic action of the gut and study the villi of the mucosa. Patient and thorough in his research, Grew dissected the guts of 40 different species of birds alone. Many species of fish were also examined, though clearly Grew found the mammals most worthy of study. The strong religious influence that pervaded Grew's life appeared most clearly in his last work, *Cosmologica Sacra* (1701), in which he attempted to demonstrate the wisdom of God in the excellence of His creations.

Unfortunately, none of the great seventeenth-century microscopists seems to have been willing or able to train pupils or attract disciples capable of continuing their work. There were, on the other hand, critics who regarded the instrument and its use as somewhat sacrilegious, that is, an attempt to pry into matters God had hidden from human eyes. Some of the most respected physicians and philosophers of the late seventeenth century, such as Thomas Sydenham (1624–1689) and John Locke (1632–1704), condemned microscopy as a distraction that was irrelevant to understanding the nature of human life, health, and disease.

The lack of interest in microscopy in the eighteenth century is quite surprising, given the remarkable range of observations made in the seventeenth century. Physicians in particular were hostile to the instrument, because the findings of microscopists had not yet improved the art of diagnosing and treating disease. During the eighteenth century, the microscope became, for the most part, a toy to entertain wealthy amateurs. The instrument that can be considered a system of "brass and glass" became one in which the "brass" aspects were the subject of much elaboration, sometimes in the form of silver and gold ornamentation, while the lenses, which determine resolution and magnification, were often neglected. Nevertheless, some self-styled virtuosi developed considerable skill, if not critical judgment, in their attempts to amuse themselves and impress their friends. Their willingness to pay for quality instruments stimulated instrument makers to improve the mechanical design and convenience of the instrument and to provide better forms of sample preparation and manipulation.

Indeed, the remarkable accomplishments of the seventeenth-century microscopists were essentially unsurpassed until the middle of the nineteenth century, when serious problems with the lens systems were finally solved. Chromatic and spherical aberrations were the most serious problems, although cloudy lenses and distracting inclusions also plagued microscopists. As more convex lenses were fashioned to increase magnification, lenses became more like prisms, separating light into different wavelengths. Although some physicists thought that the problem of chromatic aberration was insoluble, not all lens makers were willing to accept this declaration as the final word on the subject. Chromatic and spherical aberration were less of a problem with single lenses than for compound microscopes, but most microscopists preferred the compound microscope because of its size and shape. The length of the body of the instrument allowed it to stand on a table at a convenient height; thus, the microscopist could easily take notes while using the instrument. Because the simple microscope had to be held in the hand and brought to the eye, note taking involved frequent changes of position.

While it is true that chromatic aberration is inevitable when single lenses are used with ordinary light, the problem could be avoided by using monochromatic light or by constructing doublet or triplet lenses with different indices of refraction. The construction of achromatic lenses is generally attributed to John Dollond (1706–1761), but he may have learned about previous attempts to create such lenses by combining a convex lens of crown glass with a concave lens of flint glass. Dollond was a silk weaver and an amateur scientist with a special interest in theoretical optics, including the work of Isaac Newton and the mathematician Leonhard Euler (1707–1783). After joining his son's optical business in 1752, Dollond took up the challenge of making achromatic lenses. Using a combination of crown glass and flint glass, Dollond created a successful lens system for a telescope and took out a patent. Other inventors proposed similar solutions at about the same time, but it was Dollond who won the Copley Medal of the Royal Society for a telescope using his achromatic lenses. In 1761 the

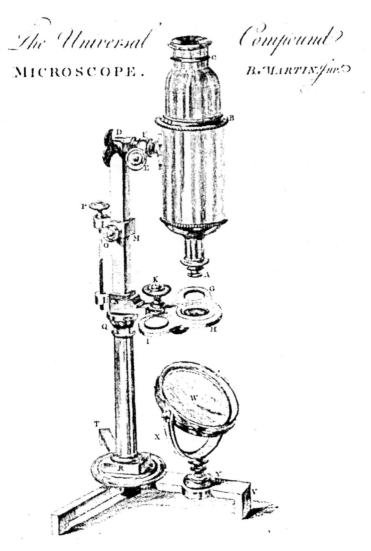

An eighteenth-century universal compound microscope

former silk weaver was elected a Fellow of the Royal Society and appointed optician to the king.

In the 1820s Giovanni Battista Amici (1786–1863), astronomer and mathematician, initiated a series of improvements in the construction of the microscope. After serving as Professor of Mathematics at the University of Modena from 1815 to 1825, Amici became astronomer to the grand Duke of Tuscany and Director of the Observatory at the Royal Museum in Florence. Using his improved microscope, Amici made some of the earliest observations on the growth of the pollen tube. The nature of fertilization in flowers was not, however, completely clarified until the 1880s. In 1830, Joseph Jackson Lister (1786–1869) provided a solution to the problem of the accumulation of spherical errors in lens combinations. Lister's design was considered especially significant because it was based on optical theory rather than trial and error.

Microscopists are, of course, concerned with both resolution and magnification. Resolution is the ability to distinguish closely spaced points into separate entities. Since in practice it is resolution that determines the limits of the details that can be studied, resolution is more significant than magnification. Several levels of magnification are possible with simple magnifiers and ordinary light microscopes. At about 10-fold magnification familiar objects can be seen more clearly. Hooke usually used low-power microscopy, which involves magnifying objects about 50 times. This level of magnification reveals interesting anatomical details in plant and animal sections and is useful in visualizing the "animalcules" found in pond water. A magnification of about 300-fold reveals the interior components of cells, including the nucleus and mitochondria, and various microorganisms, including bacteria. Even higher magnification and maximal resolution can be achieved with oil immersion microscopy.

At least in principle, Hooke had anticipated immersion microscopy for use with the simple microscope. Amici is often credited with priority in the introduction of immersion lenses, but Sir David Brewster (1781–1868) had adopted immersion microscopy as early as 1812. Attempting to develop achromatic lenses, Brewster suggested that the front element of a microscope's objective lens could be immersed in the liquid in which the object of study was mounted. Brewster may have been the first microscopist to use filters to produce monochromatic light. These ideas were elaborated in his work *A New Treatise on Philosophical Instruments* (1813). Working independently, Amici had developed immersion microscopes as a means of diminishing the loss of light in high-power lens systems in order to improve definition. Oil immersion microscopes can provide magnification up to 600-fold and resolution near the limits imposed by the nature of light.

Working together, the physicist Ernst Abbe (1840–1905) and the instrument maker Carl Zeiss (1816–1888) produced further advances in microscopy. One of Abbe's goals was to remind microscopists that increased magnification without improved resolution was futile. In their skills and education, Abbe and Zeiss nicely complemented each other: one had expertise in theoretical science and mathematics, and the other had essential practical experience. Their attempts to develop achromatic compound lenses were frustrated by the limited range of optical glasses then available, but their work stimulated the search for new kinds of glass. Abbe developed the apochromatic objective in 1886, improved the oil-immersion microscope, and developed oil-immersion objectives with at least eight component lenses. The substage condenser was developed in the 1870s to complement his high-power objectives.

Phase contrast microscopy, a modification of the light microscope, was developed by Fritz Zernike (1888–1966), Professor of Physics at the University of Groningen and winner of the 1953 Nobel Prize. Structures that differ in refractive index from the surrounding matrix appear as differences in intensity under phase contrast microscopy. This approach was particularly suited to the study of biological specimens, including living cells, which are often colorless and transparent. In favorable cases, the phase contrast microscope avoids the need for dehydration, fixation, staining, and other aspects of sample preparation that are likely to introduce artifacts. Although the production of phase contrast microscopes began during the early years of World War II, such instruments were not generally available until after the war.

Improvements in the light microscope, as well as new methods of sample preparation, made it possible to explore the fine structure of complex organisms and the nature of microorganisms. After many of the technical difficulties of optical microscopy had been solved, however, one fundamental problem—the finite size of light rays—put a limit on the usefulness of the instrument. The theoretical limits of light microscopes, about 500-fold magnification and a resolution of 0.2 micrometers, were reached in the early 1930s. To obtain higher magnification and resolving power, microscopists had to find ways of working with rays having a shorter wavelength than ordinary light. In theory, ultraviolet microscopes could extend the limits of resolution. Practical problems, including the fact that the human eye is not sensitive to light in the ultraviolet range, make the use of ultraviolet microscopy complicated and expensive. Despite the difficulties, during the 1930s ultraviolet microscopy was used to locate nucleic acids within the cell and to study the behavior of chromosomes during cell division.

To analyze the details of subcellular components, such as the nucleus, chloroplasts, and mitochondria, biologists needed much greater magnifications. Using a focused beam of electrons instead of light, the electron microscope made it possible to explore the fine structure of biological specimens. The basic discoveries that made the electron microscope possible occurred in the nineteenth century, but the first primitive instruments were not produced until the 1930s. While the theoretical possibilities of the instrument might have been obvious, most physicists thought that the device could not be practical because almost everything put under the electron beam would burn to a cinder. Many fundamental problems had to be resolved before the electron microscope became a useful tool, including several patent battles involving industrial firms, independent scientists, research laboratories, and universities in Germany, Brussels, and the United States. In 1931 the German physicists Max Knoll (1897–1969) and Ernst Ruska (1906–1988) built a model they called a supermicroscope. They were able to obtain magnifications of about 400-fold, but most of their poorly focused images were of no practical use. By 1933 Ruska had built the instrument that might be regarded as the ancestor of the first generation of successful electron microscopes. Within 2 years an improved instrument produced remarkable photographs of the legs and wings of a housefly. In 1934 Ladislaus Laszlo Marton (1901–1979) published an electron micrograph of a biological specimen: a thin section of a sundew leaf impregnated with osmium salts. Marton published what he considered his first successful bacterial micrographs in 1937. When Marton gave a lecture about the electron microscope to an audience of medical and biological scientists, immunologist and Nobel Laureate Jules Bordet (1870–1961) protested, more or less facetiously, that biologists were having enough trouble interpreting the images produced by the light microscope.

As scientists and instrument makers became increasingly convinced that the electron microscope had major potential, several teams entered the competition to produce a practical scientific instrument, and instrument manufacturers decided to explore its commercial possibilities. It was soon apparent that one of the major problems for electron microscopists would be finding appropriate ways to prepare specimens. Research on the electron microscope was somewhat impeded by World War II, but early versions of the instrument were used in the development of war materials such as metal alloys, synthetic rubber, catalysts, and cements. After the war, a new wave of research, theoretical insights, technical improvements, and commercially successful instruments appeared. Success led to bitter priority battles, but most scientists agreed that even though the winner of the patent wars could be considered the inventor as a matter of law, the electron microscope was clearly the result of work of many people. In 1986 Ruska shared the Nobel Prize in Physics with Gerd Binnig and Heinrich Rohrer, who were honored for the invention of the scanning tunneling microscope.

After high-resolution electron microscopy had become routine, new forms of microscopy, including scanning tunneling microscopy, atomic force microscopy, near-field optical microscopy, and acoustic microscopy, made it possible to "see" in entirely new ways. Basically, these innovations are variations on the basic wonders of microscopy. As noted by microscopists and historians of science, the microscope is different from the telescope in that the objects to be studied by the telescope are obvious and the desire to study the objects in the heavens was very ancient. In contrast, once microscopists got beyond the stage of admiring magnified fleas, they were able to cross a threshold into an entirely new world in which the nature of the once invisible objects was even more mysterious than that of the heavenly bodies. Potentially, the microscope offered not just a new way of seeing, but a revolutionary change in the world of thought and imagination.

SUGGESTED READINGS

Amato, J. A. (2000). *Dust: A History of the Small and the Invisible.* Chicago: University of Chicago Press.

Ben-David, J. (1991). *Scientific Growth: Essays on the Social Organization and Ethos of Science.* Berkeley, CA: University of California Press.

Benjamin, M., ed. (1994). *Science and Sensibility. Gender and Scientific Enquiry, 1780–1945.* Cambridge, MA: Blackwell.

Boffey, P. (1975). *The Brain Bank of America.* New York: McGraw-Hill.

Bradbury, S., and Turner, G. L. E., eds. (1967). *Historical Aspects of Microscopy.* Cambridge, England: W. Heffer & Sons Ltd.

Crosby, A. W. (1997). *The Measure of Reality: Quantification and Western Society, 1250–1600.* New York: Cambridge University Press.

Descartes, R. (1972). *Treatise of Man.* Translated by T. S. Hall. Cambridge, MA: Harvard University Press.

Dobell, D. (1932). *Antony van Leeuwenhoek and His "Little Animals."* New York: Harcourt, Brace. Reprinted, Dover Publications, 1960.

Dupree, A. H. (1986). *Science in the Federal Government: A History of Policies and Activities.* Cambridge, MA: Belknap Press of Harvard University.

Eisenhard, M. A, and Finkel, E. (1998). *Women's Science. Learning and Succeeding from the Margins.* Chicago: University of Chicago Press.

Ford, B. J. (1985). *Single Lens. The Story of the Simple Microscope.* New York: Harper & Row.

Frangsmyr, T., ed. (1990). *Salomon's House Revisited. The Organization and Institutionalization of Science.* Canton, MA: Science History Publications.

Hahn, R. (1971). *The Anatomy of a Scientific Institution: The Paris Academy of Sciences, 1666–1803.* Berkeley, CA: University of California Press.

Hawkes, P. W., ed. (1985). *The Beginnings of Electron Microscopy.* Orlando, FL: Academic Press.

Hooke, R. (1665). *Micrographia.* New York: Dover Publications, reprinted 1961.

Hunter, L., and Hutton, S., eds. (1997). *Women, Science and Medicine 1500–1700: Mothers and Sisters of the Royal Society.* Thrupp, Stroud, Gloucestershire: Sutton Pub.

Kohlstedt, S. G., Sokal, M., and Lewenstein, B. (1999). *The Establishment of Science in America.* New Brunswick, NJ: Rutgers University Press.

Leeuwenhoek, A. van (1977). *The Select Works of Antony van Leeuwenhoek (1632–1723).* Translated by S. Hoole. Facsimile edition. New York: Arno Press.

McClellan, J. E. (1985). *Science Reorganized: Scientific Societies in the Eighteenth Century.* New York: Columbia University Press.

Oleson, A., and Brown, S. C., eds. (1986). *The Pursuit of Knowledge in the Early American Republic: American Scientific and Learned Society from Colonial Times to the Civil War.* Baltimore: Johns Hopkins University Press.

Peltonen, M., ed. (1996). *The Cambridge Companion to Bacon.* New York: Cambridge University Press.

Rasmussen, N. (1999). *Picture Control: The Electron Microscope and the Transformation of Biology in America, 1940–1960.* Stanford, CA: Stanford University Press.

Ruska, E. (1980). *The Early Development of Electron Lenses and Electron Microscopy.* Translated by T. Malvey. Stuttgart, Germany: S. Hirzel.

Schiebinger, L. (1989). *The Mind Has No Sex? Women in the Origins of Modern Science.* Cambridge, MA: Harvard University Press.

Shea, W. R. (1991). *The Magic of Numbers and Motion: The Scientific Career of René Descartes.* Canton, MA: Science History Publications.

Stroup, A. (1990). *A Company of Scientists. Botany, Patronage, and Community at the Seventeenth-Century Parisian Royal Academy of Sciences.* Berkeley, CA: University of California Press.

Wilson, C. (1995). *The Invisible World: Early Modern Philosophy and the Invention of the Microscope.* Princeton, NJ: Princeton University Press.

Zagorin, P. (1999). *Francis Bacon.* Princeton, NJ: Princeton University Press.

5

PROBLEMS IN GENERATION:
PREFORMATION AND EPIGENESIS

Although the term generation has become obsolete, it originally encompassed some of the most exciting aspects of modern biology—reproduction, embryology, development and differentiation, regeneration of parts, and genetics. Long before philosophers began debating basic questions about the ways in which new organisms were produced, human beings had been tending crops, pollinating plants, selecting desirable traits in domesticated animals, and castrating both humans and animals. Thus, early theories of generation were built on a rich but confusing collection of facts, folklore, and myths.

In classical usage, generation referred to the coming into existence of new animals and plants, regardless of the methods involved. Although the higher animals were known to have sexual methods of reproduction, lower forms of life seemed to arise through spontaneous generation from mud or slime, while plants arose from seeds or through vegetative methods. Early Greek philosophers posed fundamental questions about generation. Why were offspring similar to but usually not identical to their parents? What caused births that deviated from the normal range, and what produced strange and monstrous births? In sexual reproduction, what contributions were made by the two sexes? What caused organisms to grow from simple, even unrecognizable forms into adults like their parents? How was it possible for some species to regenerate lost parts while others were permanently mutilated by even trivial accidents? Some fragments from the writings of Alcmaeon refer to the developing chick egg, but the first proposal for systematic embryological studies appeared in the Hippocratic treatise *De natura pueri*. If about 20 eggs were selected and incubated, one egg could be taken each day and opened for examination. There is little evidence, however, that this systematic program was actually carried out during the time of Hippocrates or even by Aristotle.

One of the great debates that runs through the study of generation was initiated by Aristotle, who analyzed the alternative models of development known as *preformation* and *epigenesis*. Preformationist theories assume that a miniature individual exists in either the egg or the semen and begins to grow into its adult form as the result of the proper stimulus. Many preformationists believed that God had formed all the individuals of each species during creation and encased them within the first organism of that species. The theory of epigenesis, which Aristotle favored, is based on the belief that each embryo or organism is gradually produced from an undifferentiated mass by a series of steps and stages during which new parts are added. Various explanations for the means by which development occurred were proposed, but all advocates of epigenesis were opposed to the concept of preformation or preexistence.

For almost 2000 years, Aristotle's ideas about *form, essence*, and *potential*, the nature of semen, the primacy of the heart, the roles of the two sexes, the nature of epigenetic development, and fetal nutrition formed the core of embryological thought. Despite the ambiguity of Aristotle's use of the terms form and essence, most philosophers believe that he did not mean to include accidental, physical features of specific organisms. According to Aristotle, the female parent contributed only unorganized *matter* to the new individual, while the male provided the important principle of *form* (which might also be called soul or essence). All the parts were not formed simultaneously, but the *potential* for the whole organism existed in the semen. The heart, which was necessarily the first part to be formed, possessed the principle of growth. If development proceeded properly, the embryo developed towards "parental likeness." When the *movements* proper to form could not gain control over matter, a deformity or monstrosity developed. Given the fact that Aristotle regarded females as "damaged males," characterized by a deficiency of heat, the semen presumably failed to live up to its potential in about half of all births.

Ideas about generation underwent little fundamental change from the time of Aristotle to that of William Harvey, although Galen did disagree with some of Aristotle's conclusions. Based on animal dissections, Galen described the fetal membranes, the blood vessels of the uterus, the fetal organs, and the foramen ovale, including the way it closes after birth. In *On the Semen*, Galen disputed Aristotle's idea that the female does not form any counterpart to the male semen. The female testes (ovaries), Galen argued, secrete semen into the horns of the uterus, where the male and female sexual products mix with blood furnished by the mother to form the fetus. As the fetus developed, the liver, heart, and brain appeared. After the formation of the liver, fetal life was rather like that of a plant, because it was nourished under the influence of the vegetative soul. With the formation of the heart, fetal life was like that of an animal because the heart served as the source of heat. The brain was the last of the important organs to be formed because the fetus did not hear, taste, smell, or engage in any voluntary action, nor did it have any intellectual faculties. Thus, in accordance with Galen's teleological principle, it did not need a brain until it was ready to be born. In *On the Usefullness of the Parts*, Galen argued that human beings are the most perfect of all animals, but man is more perfect than woman because the male has more innate heat. Indeed, it was a deficiency of innate heat that caused a fetus to become a female. As proof of this theory, Galen cited the effects of castration. By removing heat, castration caused the body of a eunuch to become more like that of a woman.

Although some medieval scholars began to suspect that Aristotle's writings had not exhausted the potential limits of wisdom, most sixteenth-century anatomists generally accepted his theories of generation. Even Vesalius based his discussion of embryology on dissections of animals and perpetuated many Aristotelian and Galenic ideas about the human fetus. Although he had few opportunities to study the human fetus, Realdo Colombo challenged Aristotle's doctrine of conception. Colombo boasted that he alone had discovered a thick white semen produced by the female testes. In the uterus, this material supposedly united with the semen provided by the male. According to Colombo, the sex of the fetus depended on whether male or female semen predominated. Rejecting the Aristotelian notion that the fetus was nourished by menstrual blood, Colombo suggested that the umbilical vein supplied a very pure and perfect blood.

Girolamo Fabrici, who is generally remembered for his studies of the venous valves, might also be regarded as the founder of modern embryology. During his anatomical lectures of 1576, Fabrici dissected three cadavers, one of which was a woman who had died in labor. He demonstrated the foramen ovale, the anatomy of the fetal horse and sheep, and closed a long discussion of development with the vivisection of a pregnant ewe. In addition to his anatomical writings, Fabrici published two important embryological works: *On the Formed Fetus* and *On the Development of the Egg and the Chick*. The illustrations in these texts reveal remarkable details of the development of the chick, especially considering that Fabrici worked without the benefit of magnifying lenses. The text, however, reflects the tension between original observation and deference to authority that still inhibited sixteenth-century anatomists. While trying to preserve the framework provided by Aristotle and Galen, Fabrici described his own observations of the ovary, oviduct, and the developing chick egg. Although nineteenth-century naturalists rejected much of Fabrici's work as "sheer Galenism," William Harvey said that his own embryological investigations had been guided primarily by Aristotle and Fabrici.

More than any other aspect of his work, Harvey's investigations of generation demonstrate the ingenuity of his experimental approach and the difficulties posed by the lack of reliable instrumental and theoretical guides. *On the Genera-tion of Animals* did not appear in print until 1651, some 23 years after the publication of Harvey's treatise on the circulation of the blood. Nevertheless, *On the Generation of Animals* represented the outcome of many years of research concerning a problem of profound importance to Harvey as an experimental scientist and natural philosopher. Having set out to provide experimental proof for the venerable Aristotelian theory of epigenetic development, Harvey came close to demonstrating that all of Aristotle's fundamental assumptions were wrong. Aristotle assumed that the embryo was a combination of matter from the female and the form-building principle contributed by the male. Immediately after mating, the process of coagulation or precipitation led to the appearance of a *conceptual mass* in the uterus. With pregnancy established, blood from the female was supposed to nourish the embryo and provide additional matter for growth. This nutritional function explained the cessation of the menses during pregnancy.

For his experiments on generation, Harvey generally used chick eggs and deer from the Royal Park. Searching for the conceptual mass from which the embryo was supposed to develop, Harvey examined uteri from numerous does during and after the mating season. A careful series of dissections revealed that until about 6 or 7 weeks after mating no conceptual mass could be found in the uterus. Like Aristotle and Fabrici, Harvey concluded that the male made no material contribution to the developing embryo and that it was anatomically impossible for the semen to reach the eggs in the uterus. Nonetheless, the semen was essential because its fertilizing power affected not only the eggs, but some-how made the uterus and the entire female body fertile. The semen, according to Fabrici, exerted its fertilizing power by emitting some "spirituous substance." Rejecting this idea as unnecessary, unproved, and excessively obscure, Harvey suggested that fertilization was similar to the spread of a *contagion*, a vague term that denoted something transferred by contact. The seed of the male, Har-vey argued, affected the female with a contagion that transmitted the semen's prolific quality so that the eggs and uterus became fruitful. The power of this contagion was obvious when human females became pregnant. While the em-bryo developed, the whole system of the female was affected, as indicated by the enlargement of the breasts, the swelling of the uterus, and other well-known symptoms of pregnancy. While Harvey admitted that he could not explain the way in which this putative contagion accomplished its mission, he was sure that the concept was useful. It has been said that a good theory need not be true, but must be fruitful. In this sense Harvey's pioneering investigations, like his hypothetical fertilizing contagion, were ultimately fruitful.

As in the study of the heart and blood, although he began with respect for Aristotle and ancient philosophy, Harvey's investigations resulted in the wreck-age of the ancient foundations of prevailing theory. In the case of his studies of

generation, however, Harvey was unable to construct a new theoretical framework or provide a satisfactory solution to ancient puzzles. Based on his studies
of the developing chick egg, Harvey agreed with Aristotle that the embryo developed by epigenesis. As epitomized in his dictum *Omne vivum ex ovo*, Harvey
believed that all animals arose from eggs. The "egg" that Harvey referred to
was actually the fetus at a stage large enough to be seen without a magnifying
lens. While the true egg was just as invisible as Harvey's hypothetical anastomoses between the arteries and veins, it was just as logically necessary in view of
the evidence for its existence.

As in the case of the circulation of the blood, Harvey's work on generation
inspired natural philosophers to follow his methods and find experimental approaches to the problems he had identified. Thus, Harvey became one of the
last major advocates of epigenesis during the late seventeenth century and for
most of the eighteenth century. The mechanistic view of the world that his
work had stimulated led to the growing influence of preformationist models of
development. During this period, preformationists were generally defenders of
mechanism, while epigenesis seemed to require mysterious outside forces. The
idea of cell theory or a set of instructions in the egg was, of course, unknown.
The rapid acceptance of the preformationist theories suggests that this concept
must have resonated with the deepest philosophical needs of many eighteenth-
century naturalists. The theory of preexistence eliminated the atheistic and materialist implications of mechanical epigenesis, which seemed to imply that unorganized matter could produce life without the intervention of the Divine Creator.
Preformationist theories, in contrast, were in harmony with the theological doctrine that all human beings were literally the descendants of Adam and Eve.

Although preformationist theories enjoyed a prominent place in eighteenth-century science, the philosophical roots of this concept can be traced back
to Democritus, Empedocles, Lucretius, Seneca, and the Church Fathers. During
the seventeenth century, several naturalists claimed to have seen chick embryos
in unincubated eggs and plants in ungerminated seeds. In overcoming the authority of Aristotle and Harvey, preformationists drew on the increasing acceptance of the mechanical philosophy, based on the work of Galileo, Newton, and
Descartes, and microscopic observations that allowed naturalists to see details
of embryological development at earlier stages. Enlightenment philosophers believed that the rational and objective mind could eventually explain the universe
in terms of matter and motion. The development of a complete organism by the
gradual addition of parts, however, seemed to elude all attempts to formulate a
mechanical explanation. One possible solution to this problem was to deny the
idea of individual development altogether and reformulate the doctrine of preformation in terms of mechanical theories. New individuals might grow according to mechanical laws from preformed miniatures originally created by God.
The idea that the new individual grew from a preformed miniature created by

God seemed to fit Descartes's ideas about the infinite divisibility of matter as well as theological doctrines. Before cell theory there was no logically defined lower limit to the size of living or "preformed" beings. If life on earth was only a few thousand years old, the number of past generations would be fairly small, but it was plausible that the microscope could not really "see" the tiny individuals or their parts.

Microscopic observations reported by Marcello Malpighi and Jan Swammerdam, two of the great pioneers of microscopy, helped to stimulate interest in preformation. On many occasions Malpighi complained about the poor quality of his microscopes, but he was able to reveal details of earlier stages of the embryo than had ever been seen before. He described the development of the heart and gill arches, the dorsal folds, brain, mesoblastic somites, the amnion and the allantois. Although Malpighi thought that the development of the chick egg generally proceeded gradually, the microscope seemed to reveal the existence of embryos in eggs that had been fertilized but not yet incubated. In his *Dissertation on the development of the chick egg* (1673), Malpighi speculated that the heat of summer might have affected development. Although Malpighi generally avoided making dogmatic statements about the mechanism of generation, later preformationists considered him one of the pioneers of seventeenth-century preformationism.

Studies of the life cycle of insects and the metamorphosis of amphibians convinced Swammerdam that the parts of the adult were already present in caterpillars and tadpoles. By carefully dissecting a silkworm, he could reveal the presence of the parts of the future adult, such as wings, antennae, and proboscis. Such demonstrations convinced Swammerdam that no new parts were formed during insect metamorphosis. He concluded that the four stages of insect life—egg, larva, pupa, and adult—were packaged within each other as a series of boxes. Similarly, the limbs of the frog could be dissected out from under the skin of the tadpole some time before they would normally emerge. Calling attention to similarities between his study of the metamorphosis of the frog and Malpighi's description of the developing chick, Swammerdam suggested that this work had broader implications.

The Cartesian philosopher Nicolas Malebranche (1638–1715) reformulated Swammerdam's rather vague ideas into a philosophical scheme involving a series of embryos preexisting within each other like a nest of boxes. In a popular series of books, translated into English as *Search after Truth*, Malebranche introduced the expression *emboîtement*, which could be translated as "encasement." The term inevitably conjures up the image of the nest of Russian dolls that has become a common metaphor for preformation, but it also suggests the myths found in many cultures about boxes within boxes that contain some precious gift or key. According to Malebranche, an infinite series of plants and animals were contained within the seed or the egg, but only naturalists with sufficient skill and experience could detect their presence. Microscopists could demonstrate the existence of tiny chicks in fresh, unincubated eggs and tadpoles within the eggs of the frog. Based on this evidence, Malebranche speculated that the bodies of all the men and animals that would be born until the end of time might have been created at the creation of the world. If so, the first females of each species might have contained all future generations of their kind. Similar speculations had appeared long before the invention of the microscope. The Roman writer Seneca (3 B.C. to 65 A.D.), for example, suggested that "all the parts of the body of the man," including the roots of his beard and hair, were present in the tiny speck growing in his mother's womb. Some seventeenth-century writers suggested that the whole human race, as well as all human parasites, might have been placed within the ovary of Eve. As might be expected in an age fascinated by the new infinitesimal calculus, such speculations stimulated the kinds of calculations implied by preformationist theory, such as the size of the primordial ancestor and the number of generations since the Creation.

While preformationist theory appeared to offer philosophical solutions to the puzzle of generation, the discovery of spermatozoa created a new enigma. Logically, the boxes-within-boxes doctrine allows only one parent to serve as the source of the preformed individual. The egg was well known for many species and seemed the obvious repository for the ancestral line. Within this theoretical system, the semen of the male was accorded the critical role of providing the generative principle needed to initiate the growth and development of the preformed organism. When the microscope revealed great numbers of active "little animals" or "worms" in the semen, some naturalists rejected ovist preformation and contended that the male parent might carry the germs of the preexisting individuals. The debate between ovists and spermists became more central to speculations about generation than the old controversy of preformation and epigenesis.

Spermists assumed the existence of a "little man within the sperm," but they did not use the term "homunculus," which was actually used as a form of ridicule at the time. The term was associated with the artificial creation of life, such as the secret recipe Paracelsus allegedly boasted about for creating a "homunculus" that looked quite human. Legends about the production of a homunculus usually referred to some kind of monster, like the legendary Golem. Karl Ernst van Baer coined the term "spermatozoon" in 1827; it means "animals in the sperm" or "seed animals."

Just which naturalist was the first to discover sperm is unclear, because seventeenth-century microscopists enthusiastically examined almost every conceivable fluid with their lenses. Before the microscope revealed the little animals in the semen, many natural philosophers accepted Galen's belief that semen was related to blood, while others argued that semen was a form of milk, sweat, or saliva. Many early microscopists assumed that the "spermatic worms" were probably parasites, or signs of putrefaction, but others thought they were insignificant fibers and filaments. Indeed, the term "worm" implied a very lowly form of life. Although some microscopists suggested that the newly discovered creatures were associated with disease or were themselves permanent parasites, Antoni van Leeuwenhoek believed that they were a normal constituent of male semen. According to one version of the discovery, in 1677 Johan Ham, a medical student at Leiden, observed little animals in semen from a man with gonorrhea. Seeking advice, Ham apparently brought a sample of this material to Leeuwenhoek.

After confirming Ham's observations and extending his investigations to the semen of healthy men and other animals, Leeuwenhoek studied the appearance, motility, survival, and other characteristics of these animalcules. He explained that the observations of human semen were not obtained by any "sinful contrivance," but were made with "the excess with which Nature provided me in my conjugal relations." In fresh human semen, examined before "six beats of the pulse had intervened" between ejaculation and microscopic analysis, Leeuwenhoek found a great number of living animals. Noting that these animalcules were smaller than the red corpuscles of the blood, he concluded that a million

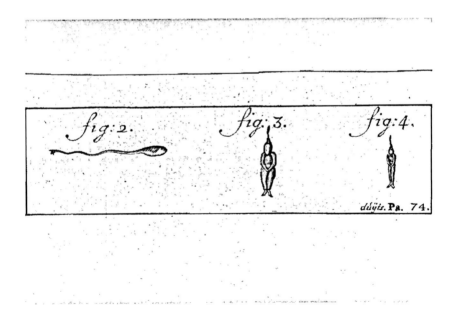

An engraving of three sperm accompanied a letter sent by Antoni van Leeuwenhoek to Nehemiah Grew in March 1678. Two show tiny people in the head of the sperm.

of them would not equal the size of a large grain of sand. The little animals, which had a round body and a long thin tail, swam about with the snake-like motion of eels. Convinced that he had detected two sorts of animalcules in semen from humans and dogs, Leeuwenhoek suggested that they might give rise to male and females offspring. François de Plantade (aka Dalenpatius) sent Leeuwenhoek drawings of little men, with hats and beards, inside the body of the little animals in the semen. However, Leeuwenhoek did not believe that it would be possible to see the whole shape of the future adult in the sperm any more than it was possible to see a tree in an apple seed.

When he first described his observations in a letter to the Royal Society, Leeuwenhoek suggested that if the Secretary found them disgusting or scandalous he should feel free to destroy them. Fortunately, the Fellows of the Royal Society were intrigued rather than scandalized, and these observations were published in a Latin translation in the *Philosophical Transactions* of 1679. In all, Leeuwenhoek studied the spermatozoa of about 30 different kinds of animals, including mammals, birds, amphibians, fish, mollusks, and arthropods. He also investigated the ovaries of cows and sheep to study the Graafian follicles. In opposition to Regnier de Graaf (1641–1673), Leeuwenhoek argued that these putative eggs could not possibly get to the uterus because they would have to pass through the fallopian tubes, which were no larger in diameter than a pin, whereas the alleged eggs were as large as peas. Although Leeuwenhoek had considerable respect for learned authorities, he had no hesitation in disagreeing with them on the basis of his own observations. It was obvious, Leeuwenhoek noted in his letter to the Royal Society, that all of the learned authorities who disagreed with him were mistaken. Some other curious discoveries reported by Leeuwenhoek, including parthenogenesis in aphids, the budding of hydra, and the formation of daughter colonies within *Volvox*, were used by other scientists in support of alternative theories of preformation.

One of Leeuwenhoek's chief rivals, Nicolas Hartsoeker (1656–1725), claimed that he had discovered sperm long before Leeuwenhoek. This discovery supposedly occurred in 1674 when Hartsoeker was only 18 years old. At the time, however, the young naturalist was apparently too shocked to make further observations or to discuss his findings with others. About 3 years later, with help from a teacher and a friend, he conducted microscopic studies of sperm from a dog, a pigeon, and a rooster.

The French physician Nicolas Andry de Bois-Regard (1658–1742) was an advocate of spermist preformation and the concept that the germs of life were encased within each other. In *An Account of the Breeding of Worms in Human Bodies* (1700), Andry described the differences between spermatic worms and ordinary parasites. Andry believed that spermatic worms entered eggs in order to be nourished while the parts of the fetus developed and grew. Although Andry engaged in some creative speculations about the characteristics of sperm, the descriptions of battles between spermatic animals that left some looking like wounded warriors were actually created by later authors. The early spermists did not adopt this concept, even though it could be used to account for birth defects and the sudden appearance of monsters.

The microscopic techniques of the eighteenth century could not answer basic questions about the nature of the little animals in male semen, but they provided observations that stimulated endless disputes between ovists and animalculists. Spermist preformationists called themselves "animalculists," because they assumed that motion was a fundamental characteristic of animal life. Classifying these little creatures as animals raised many questions: Did they have organs of digestion, circulation, and copulation? Did they exist as male and female entities that mated and produced another generation of little animals?

Moral and religious issues were major problems for spermists. Although spermists argued that the testes of Adam served as the repository of the human race, rather than the ovary of Eve, some theologians cited arguments against this concept. For example, St. Augustine had said that God gave men testicles as a reminder of original sin. The idea that God had put all future generations in Adam's testicles was thus highly offensive. Moreover, this arrangement raised the issue of the "spilling of seed" and the wastage of the vast numbers of sperm. Some spermists objected that excess was common in nature; after all, trees produced many flowers and fruits although only one seed was needed to produce a new tree. The Dutch insect anatomist Pierre Lyonet (1706–1789) argued that the enormity of wastage proved that the spermatic animals could not be the "seeds" of human life. Dismissing arguments about wastage in the plant world, he pointed out that excess fruits and seeds were meant to serve as foods for animals and humans.

One ingenious response to arguments about the wastage of sperm was the hypothesis that those that did not succeed in entering an egg might serve some other purpose. One such possibility, the doctrine known as "panspermism," was supported by the mathematician-philosopher Gottfried Wilhelm Leibniz (1646–1716). Excess sperm, and other primitive animals, might become the "seeds of life" that were carried on the wind until they found a suitable "host" in which to develop. Panspermism was subject to considerable ridicule by the end of the seventeenth century, but the concept reappeared several times, even in the twentieth century. Francis Crick and Leslie Orgel explored panspermism theory in a book called *Life Itself: Its Origin and Nature* (1981).

Preformationist doctrines won the support of many competent and cautious scientists and philosophers, but they also attracted extremists who saw tiny people stuffed into spermatozoa. Some animalculists were quite convinced that their microscopes allowed them to see sperm displaying behaviors characteristic of the types of animals to which they would give rise. Other eighteenth-century naturalists rejected both ovist and spermist preformationist views. Many eighteenth-century philosophers were appalled by seventeenth-century spermists. In particular, they rejected the idea that God would put so many potential human beings into entities destined for mass destruction. Nevertheless, spermism did not disappear in the eighteenth century, although arguments in support of the concept seemed more abstract. For example, although Jean Astruc (1684–1766), physician to King Auguste II of France and author of a landmark work on sexually transmitted diseases, has been called the last of the animalculists, he actually believed that both sexes contributed to the new organism. Puzzled by the origin and subsequent wastage of the vast majority of spermatic animals, he concluded that it was not possible for either the ovaries of Eve or the testes of Adam to contain the seeds of the whole human race. Spermists and ovists failed to explain the mixture of parental traits found in offspring. Attempting to account for the contributions of both parents, Astruc suggested that the father's semen was the source of the embryo, but the spermatic animal was molded and transformed as it made its way through the pore or "passage" of the mother's egg. Other eighteenth-century naturalists agreed that theories of generation should accommodate contributions from both parents.

Attempting to find a Newtonian explanation for how the world came to its present state, the French naturalist Georges Louis Leclerc, Comte de Buffon (1707–1788), suggested that the embryo was formed from "organic particles" present in the male and female semen. An "internal mold" directed the particles to their proper places. The mathematician Pierre Louis Moreau de Maupertuis (1698–1759) proposed a similar scheme. His interest in problems of inheritance and development was stimulated when he saw an "albino Negro" and concluded that the existence of albinos raised doubts about preformationism and the existence of distinct human races. Furthermore, if matter was not infinitely divisible, Maupertuis reasoned, it was impossible to believe that a fetus formed today could be as distinct as its primordial ancestor. In his *Venus physique* (1745), Maupertuis argued that both parents contributed to the "semen" that mixed in the womb and served as the basis of development. Particles from each part of the adult body found their way into the seminal fluid. Some obscure force, vaguely reminiscent of gravity or electrical attraction, drew particles to the embryo, although Maupertuis could not say how they recognized their proper place. There might be more seminal particles than the embryo needed, but the selective attraction between the particles determined the combination found in the new embryo.

Despite doubts raised by critics, some of the most respected scientists of the eighteenth century, namely, Albrecht von Haller (1708–1777), Charles Bonnet (1720–1793), Lazzaro Spallanzani (1729–1799), and René Antoine Ferchault de Réaumur (1683–1757), were proponents of ovist preformation. In terms of the range and significance of the philosophical questions posed, the most important of the debates between preformationists and epigenesists during this time period involved Haller and Caspar Friedrich Wolff (1733–1794). Considered a Newtonian mechanist, Haller is best known for his work as a physiologist, but he had a deep interest in embryology. Thus, the debate between Haller and Caspar Wolff concerning the nature of embryological development provides an especially valuable case study of the role of observation, explanation, and metaphysical assumptions in establishing the position taken by scientists with opposing viewpoints.

At different points in his life, Haller supported three different theories of embryological development. Under the influence of his mentor Hermann Boerhaave (1668–1738), Haller accepted spermist preformation. Then, during the 1740s, after reflecting on the implications of Abraham Trembley's (1710–1784) discovery that the hydra, or freshwater polyp, could produce complete new animals when it was cut in half, Haller converted to epigenesis. Finally, in the 1750s Haller adopted ovist preformation based on observations, experience, and his religious conviction that God rather than chance must be at work in embryological development. Both spontaneous generation and epigenesis seemed to promote the blasphemous assumption that matter, without the action of God, could produce living beings.

"Trembley's polyp" fascinated and confounded many eighteenth-century naturalists, even though Leeuwenhoek's description of the microscopic freshwater hydra in the *Philosophical Transactions of the Royal Society* (1703) had attracted little attention. Leeuwenhoek thought the tiny creature was an animal, and he said that it reproduced by budding. Réaumur called the creature a "polyp" because it was similar to the octopus. At first Abraham Trembley could not decide whether the polyp was a plant or an animal, but when he saw it using its arms or tentacles to catch food and put its prey into its central cavity or stomach he decided that it was an animal. Trembley told Réaumur that the creature was carnivorous, that it reacted to touch, light, and warmth, and appeared to walk with a primitive foot. Yet in the morphological sense the creature had plant-like features, because not all of the polyps had the same number of arms. Reasoning that animals should die when cut up, Trembley cut some polyps in half. He was shocked to find that both halves regenerated into complete polyps. Further experiments proved that even when cut into several pieces, or turned inside out, the creatures were able to survive and become complete animals. The fact that lizards and salamanders could regenerate their tails and starfish their arms was well known, but Trembley's experiments were regarded as new and exciting. A preliminary report of his observations was published in the *Histoire de l'Académie des Sciences* (1741). After further experiments to settle his own doubts, Trembley published a complete account of his work in 1744. Some naturalists expected to find a macroscopic version of Trembley's polyp that would allow them to conduct more systematic dissections and experiments. Based on the assumption that there was a correspondence between the macroscopic and the microscopic world, naturalists predicted that God must have used the same blueprints for both.

Studies of regeneration challenged preformation theories, but not all naturalists came to the same conclusion. For example, Hartsoeker permanently abandoned spermism in 1722, because of his reflections on regeneration in the crayfish. Haller temporarily abandoned spermist preformation because he thought that if an animal like the hydra could form two new individuals when cut in half, the preformed germ of a new individual could not preexist in either the egg or the sperm. Other preformationists, however, did not regard regeneration in Trembley's polyp as a fatal challenge to their theory. Réaumur and Bonnet concluded that their own experiments on regeneration indicated that the "soul" of the polyp was somehow diffused throughout its body. Bonnet argued that his concept of the "germ" could accommodate parthenogenesis as well as regeneration. In contrast, some very radical materialists concluded that these observations might mean that there was no soul. The hydra, which is a complex colony of individual cells, eventually assumed a special role in developmental biology as a model system for studies of pattern formation and differentiation.

In the third phase of Haller's work on embryology, he returned to preformation. This time his own studies of the developing chick egg convinced him that the embryo was already present in the unfertilized egg. Development began when the male semen stimulated the egg and caused the heart to start beating. As fluids began to move through the transparent embryo, its parts solidified and became more opaque, which made the fetus visible. Another impetus to Haller's conversion to ovism was his anxiety about the implications of contemporary theories of mechanist epigenesis. For example, Haller objected to Buffon's notion of an "interior mold" in which organic particles from the semen were somehow drawn together and put into their proper places. In other words, mechanist epigenesis suggested that innate forces and matter could create life without the need for a Divine Creator. Even though purely physical forces could apparently produce crystals and snowflakes, Haller could not accept the idea that inert matter and Buffon's "penetrating force" could create the "wonderful plan of the human body." Not only did Haller reject Buffon's concept with respect to embryology, he argued that forces like gravity, magnetism, and irritability were not inherent in matter. According to Haller, God was the source of all the forces allegedly possessed by matter.

Given such strong motives for rejecting mechanist epigenesis, Haller eagerly accepted Charles Bonnet's studies of parthenogenesis (virgin births) in aphids as compelling evidence in support of ovist preformationism. Not all naturalists considered parthenogenesis rigorous proof for ovism, as demonstrated by Leeuwenhoek's attempts to reconcile spermism with his discovery of parthenogenesis in aphids. Like Haller, Bonnet was extremely pious and also hoped to find harmony between natural philosophy and religious doctrine. After an initial loose attachment to epigenetic views of development, Bonnet became convinced of the merits of preformationist theory and worked out the details of a more sophisticated encapsulation theory. In this he was influenced by his own research and his correspondence with Haller, his friend and colleague. Plagued by poor health throughout his life, Bonnet lost his hearing and later his eyesight. Although he had to give up experimental work, his philosophical studies gave him hope that reason would achieve a true victory over the senses.

While studying the reproduction of aphids, Bonnet found that females that hatched during the summer gave birth to live offspring without fertilization. In the autumn, the new generation of males and females mated and the females laid eggs. By carefully segregating young females from the males, Bonnet was able to raise 30 generations of aphids by parthenogenesis. According to Bonnet, the first female of every species contained within her ovaries the miniature precursors of all future members of that species. Male semen was needed to initiate growth, but the female parent provided the nutrition needed for development. In order to cope with troublesome cases and apparent exceptions, preformationist theory had to become ever more complex and convoluted. For example, while the germs in higher animals might be confined to the reproductive organs, in lower animals the germs had to be scattered throughout the body. Such reasoning was essential in explaining the behavior of hydra and similar polyps that reproduced by budding and the ancient observation that certain creatures could regenerate lost parts or form complete new individuals from segments.

There was some doubt in Bonnet's mind as to whether the germ was actually a particular individual or merely a generalized member of the species. Eventually he adopted the idea that the germ must carry the imprint of the species rather than that of the individual. That is, the original germ determined the species characteristics, while natural processes and factors encountered during gestation, such as the mother's health, size, and nutritional status, determined its particular fate. Once initiated, the unfolding and growth of the germ could be regarded as a purely mechanical process in which nutriments were absorbed into the appropriate parts of the expanding embryo. By allowing for the effects of accidents, disease, and poor nutrition, this preformationist theory exonerated God in the matter of "monstrosities," even those that appeared over several generations. Recognizing the many difficulties posed by this theory, Bonnet acknowledged that preformation theory had to be understood as a great triumph of rational thought over direct observation.

Further support for the ovist preformationist model was provided by Lazzaro Spallanzani, one of the great figures in experimental physiology and the eighteenth century's foremost authority on semen. Originally a student of law, Spallanzani was drawn to science by his cousin Laura Bassi (1711–1778), a remarkable woman who had earned a doctorate in philosophy at Bologna and was the first woman to hold a chair of physics at any European university. In addition to her academic work and studies in mechanics, Bassi was the mother of eight children. Another woman who was important to Spallanzani was his sister Marianna, who helped him with experimental work. Spallanzani took religious orders, but he also held various university appointments and traveled widely. Versatile and energetic, Spallanzani taught mathematics, physics, philosophy, and Greek. Contemporaries familiar with his research on digestion, blood circulation, respiration, regeneration, spontaneous generation, reproduction in plants and animals, the senses of bats, the nature of sponges, and the electrical activity of the torpedo (electric ray or crampfish), referred to him as "Magnifico."

A series of studies on the nature of regeneration led Spallanzani to choose amphibians, such as frogs, toads, and newts, as a model system. Spallanzani believed that, under the right conditions, such animals could regenerate severed tails, limbs, and even heads. Experiments on regeneration led to an interest in the life history of amphibia and their point of origin, which he concluded was undoubtedly the egg. In his *Prospectus Concerning Animal Reproduction* (1768), Spallanzani claimed to have discovered the preformed germ of the tadpole in the eggs of the frog. Later, in his *Dissertations Relative to the Natural History of Animals and Vegetables* (1789), he assured his readers that further studies of additional species had substantiated his previous claim that the germ preexisted in the eggs of the female before fertilization occurred. Moreover, he convinced himself that the preexistence of the germ was a general law of nature. Studying the development of amphibian eggs, he found that frog eggs increased in size while still inside the mother. He interpreted this observation as proof of ovism, because he assumed that growth could only be explained if the frog were already present in the egg. In other words, eggs could be thought of as tiny tadpoles ready to unfold themselves and begin developing when exposed to the fecundating semen of the male. Amphibia, an order that naturalists considered oviparous, therefore, should be understood as actually viviparous. Observations of the development of the eggs of insects, birds, crocodiles, and other reptiles were also integrated into Spallanzani's arguments in favor of ovist preformation.

Since the mammalian ovum was not discovered until 1827 and the union of sperm and egg nuclei was not seen until the end of the nineteenth century, eighteenth-century discourse about the nature of the process known as *fecundation* or *fertilization* was necessarily cast in very different terms. Many naturalists argued that fertilization of the egg always occurred inside the body of the mother. Some believed that the egg had nothing to do with fertilization, but existed only as a means of nourishing the new germ of life. While upholding the importance of the male semen, some naturalists assumed that the spermatic animals were some kind of parasite, perhaps spontaneously generated like other infusoria. Thus, microscopic observations forced naturalists to confront questions about which component of semen represented the male fertilizing power:

spermatic animals, liquid, or the *aura seminalis*, which could have been a vaporous entity or an invisible force vaguely analogous to gravity or magnetism.

An ingenious experimentalist, Spallanzani provided rigorous proofs that fertilization in frogs and toads was external and that it was not caused by some mysterious vapor. Having observed the normal mating behavior of frogs, he performed experiments in which female frogs were killed while the eggs were being emitted. Eggs that had been discharged and came in contact with the semen developed normally, but eggs dissected from within the body of the female did not develop. To confirm his hypothesis that fertilization was external and that contact with semen was necessary for development, Spallanzani designed special tight-fitting taffeta overalls for his frogs. Despite these extraordinary costumes, the frogs still attempted to mate. The females discharged eggs, but none of them developed. When some of these eggs were mixed with semen that had been retained by the little trousers, normal development took place. When Bonnet learned about these experiments, he suggested Spallanzani could use his remarkable method of artificial fertilization to create exotic new hybrids or even human beings. Indeed, not long afterwards, John Hunter (1728–1793), physician and anatomist, very discreetly supervised artificial insemination on behalf of a patient with a rather peculiar physical problem, thus allowing him to father a child.

Encouraged by the success of his experiments on artificial fertilization, Spallanzani decided to use the amphibian system to determine which portion of the semen caused fecundation. To test the possibility that the *aura seminalis* stimulated development, Spallanzani placed a small quantity of semen in a watch glass and attached a number of eggs to a smaller watch glass. The small watch glass was then inverted over the vessel containing the semen. Although the eggs were kept quite moist and almost one tenth of the semen evaporated during the experiments, the eggs failed to develop. To test the activity of the remaining semen, Spallanzani applied some of it to other eggs, which did undergo development. In another test of the possible influence of "vapors," Spallanzani tied eggs to a string and suspend it above a watch glass containing frog semen. Again, the eggs failed to develop. Physical contact between eggs and semen was, therefore, essential. Further tests showed that other liquids, such as blood, urine, wine, and vinegar, could not substitute for semen. Moreover, various physical and chemical agents, such as heat, drying, wine, and vinegar could destroy the fecundating power of semen.

Despite his ingenious experiments, in his assessment of the role of the spermatic worms, Spallanzani reached conclusions that appear to be illogical within the context of his own work and incorrect in terms of modern science. Having ruled out a role for the vaporous portion of the semen, Spallanzani attempted to separate the seminal worms from the liquid component by filtering semen through paper, cotton, and chiffon. The material trapped in the filter teemed with seminal worms and caused fertilization, but Spallanzani concluded that the liquid portion of the semen—not the *vermicelli spermatici* (little spermatic worms)—accounted for fertilization. Still convinced of the validity of ovism, Spallanzani suggested that the spermatic worms were needed to stimulate the fetal heart and promote growth. Spallanzani thought that his experiments with supposedly sperm-free fluids proved that the spermatic worms were not responsible for fertilization. Some scholars have suggested that when Spallanzani painted allegedly sperm-free fluids onto eggs, he triggered artificial parthenogenesis, but others argue that touching frog eggs with a brush or needle cannot cause them to divide, because frog eggs do not contain centrosomes. If the needle carried other kinds of cells, the egg might have begun to develop after incorporating their centrosomes. Another possible explanation is that Spallanzani's "sperm-free" fluid still contained active sperm.

Convinced that the sperm were real animals like other microscopic creatures he had studied, Spallanzani concluded that the spermatic worms were parasites passed on from one generation to the next during intercourse or within the uterus during pregnancy and that the seminal fluid was their natural habitat. He observed that sperm could not survive very long once they were exposed to air. This led to the conclusion that they were true parasites of the sexual organs. If the parasites came from inside the body, they could have been circulating in the blood before they entered the testes. Assuming that the seminal fluids were necessary for the development of the egg, these parasites might even penetrate the egg, find their way into the tiny preformed genital organs, and remain there until puberty. Although he remained an ovist, Spallanzani noted that the resemblance of children to both parents raised some questions concerning the preexistence of the new individual. Unable to resolve these puzzles, or further define the function of the semen, Spallanzani concluded that such questions transcended the sphere of human knowledge.

Despite the authority of Haller, Bonnet, and Spallanzani, preformationist theory and the mechanical philosophy did not go unchallenged even in the eighteenth century. One of the first biologists to adopt a new mode of thought known as *nature philosophy* was the embryologist Caspar Friedrich Wolff (1733–1794). While studying medicine at Berlin and later at Halle, Wolff absorbed large doses of philosophy from Christian Wolff (1679–1754), professor of mathematics and philosophy, and author of *Thoughts on God, the Soul, and the World*. Caspar Wolff was particularly affected by his mentor's philosophical principle that everything that happens must have an adequate reason for doing so, for it would be absurd for something to come from nothing. In 1759, Wolff completed a dissertation entitled "Theoria generationis" ("Theory of Generation"), which was later recognized as a landmark in the history of embryology. A German version of his thesis was published at the same time.

Wolff has been called the only significant academic defender of epigenesis and opponent of Haller in this period; however, during his lifetime his work brought him little recognition. Denied a position at the Medical College of Berlin, Wolff emigrated to St. Petersburg, Russia, where he became Academician for Anatomy and Physiology. Although Wolff is generally considered one of the German embryologists, Russian historians have argued that his work was part of a distinctly Russian tradition that shaped and nourished late eighteenth- and early nineteenth-century embryological theory. While working at the Petersburg Academy of Science, Wolff published his major work on the formation of the intestines (*On the Formation of the Intestinal Canal in the Incubated Chicken Egg*) and reports on his studies of "monsters." A museum at the tsar's Summer Palace in St. Petersburg, known as the Kunstkamera, housed a major teratological collection of anatomical and embryological materials.

The leading preformationists Haller, Bonnet, and Spallanzani saw life and science from a profoundly religious perspective. Consciously or subconsciously, they wanted to preserve and enhance evidence for the biblical account of creation, but as natural philosophers they wanted to place generation within a generally mechanistic physiology. Wolff, in contrast, believed that studies of generation could only be purely descriptive since it was impossible to determine the actual mechanism of development. His theory of generation was founded on the philosophical assumption that development necessarily occurred by epigenesis. When Haller rejected Wolff's theory of generation on largely religious grounds, Wolff retorted that a scientist must search only for truth and could not prejudge biological questions on religious rather than scientific grounds. Yet Wolff's own theoretical preconceptions led him to seek out observations favorable to his epigenetic theory of development. For Wolff, nature philosophy provided a view of a world constructed in conformance with a preestablished harmony. Within this philosophical view, it was assumed that life could not be explained in mechanical or physicochemical terms. Understanding life according to the precepts of nature philosophy meant attaining full descriptive knowledge of the complete natural history of different life forms. In the marvelously coordinated development of the embryo, nature philosophy found its most typical and often most lyrical expression.

In terms that reflect the basic assumptions of nature philosophy, Wolff asserted his conclusion that the organs of the body did not exist at the beginning of gestation, but were formed successively. Naturalists need not explain exactly how the addition of parts took place, only demonstrate that it did indeed occur by means of a process in which each step in development supported the next step. Wolff defined generation as the formation of a body by the gradual creation of its parts. Because the actual mechanism of generation was unknowable, a vague descriptive term, the *vis essentialis*, or essential force, was introduced as the agent that brought about change. In an ingenious attempt to save epigenesis from relying on archaic vital forces such as Wolff's *vis essentialis*, Immanuel Kant (1724–1804) and his colleague J. F. Blumenbach (1752–1840) postulated a goal-directed "developing drive" (*Bildungstrieb*). Kant argued that the developing drive was an inherent property of the organism and was inherited through the germ cells.

In addition to observations of the chick egg, Wolff studied the metamorphosis of plants. Basic to all of Wolff's work and conclusions was the assumption that plants and animals are essentially the same in terms of their original undifferentiated materials. This assumption was essential to Wolff's theory, but beyond that, as a broad statement of principles, it eventually provided the essential connection between embryology and cell theory. In terms of Wolff's program, plant metamorphosis was a valuable model system because he was not interested in fertilization or the origins of the primary undifferentiated material, but only in the pattern of differentiation once the process had been initiated. While this model system suited his philosophy, it also helped him circumvent the technological limitations of microscopy. Given the poor quality of microscopes and the lack of staining techniques, plant materials were preferable objects of study because more details could be seen in plants than in unstained animal tissues. Thus, much of what Wolff said about embryogenesis was an extrapolation of what he had seen in developing plant parts.

According to Wolff, the inner life force of the plant causes liquid to be drawn up from the soil. The liquid collects at the growing point of the stem and becomes a kind of thin jelly. As evaporation of the liquid occurs, small sacs or vesicles are formed. When a bud was first transformed into the rudiment of a leaf, or as growth began at the apex of the stem, Wolff could easily count the number of vesicles. As growth proceeded, the size of the individual vesicles remained the same, but they soon became too numerous to count. Further development resulted in a hollowing out of tracks in the jellied mass of vesicles, which allowed the ducts of the plant vascular system to develop.

Having demonstrated the lack of preexisting structures in the rudiments of the parts of the plant, at least to his own satisfaction, Wolff criticized preformationists for describing unreal structures that were merely the products of their "rich imaginations." Similarly, Wolff asserted, the chick was formed from an undifferentiated mass of little globules that contained no bodily structures or parts. Eventually, the vascular system of the chick formed out of this material just as it did in plants. Since the globules that preceded the appearance of the heart, vessels, and other parts of the body were visible under the microscope, Wolff disputed the hypothesis that the parts themselves could be present but too small to be seen. Objecting to claims that the embryo was present but transparent and thus invisible, Wolff argued that even if alcohol was used to make structures more opaque, it was impossible to find blood vessels in the chick embryo at the onset of incubation. To lend more credence to his plant-animal analogies, Wolff concentrated on the development of intestines, blood vessels, and kidneys, because their final structures are essentially tubular. The little vesicles Wolff described might have been cells, but many contemporary reports of globules in biological preparations have been dismissed as artifacts or aberrations produced by defective lenses. While Wolff thought that these little sacs could change into plant or animal organs, he regarded them as secondary structures in a mass of undifferentiated jelly rather than the primary unit of origin. Thus, it is not appropriate to interpret Wolff's descriptions of "vesicles" as cells, or his references to "layers" in the early embryo as the "germs layers" described by nineteenth-century embryologists.

After publishing his treatise on the development of the intestines, Wolff's major area of interest was teratology, the study of the structure of "monsters" and malformations. He planned to write a major treatise on theoretical aspects of teratology, based on materials in Russian museums, but he was unable to complete this project. He did, however, publish several descriptive studies of animal monsters, including a double-headed chicken, a chicken with four legs and four wings, and a calf with two heads and two necks. Such monsters had been explained as the accidental collision and union of two normal embryos, or the development of an embryo that was predestined to be a monster at the creation of the first ancestor. Rejecting the idea that monsters were directly created by God, Wolff argued that monsters were works of nature that must be investigated in the same manner as normal embryos. Although from a modern perspective, monsters would seem to represent a challenge to both preformation and epigenesis, most eighteenth-century naturalists did not consider this a significant problem for their theories.

Although Wolff had great respect for Haller, he had no sympathy for the theory of preformation, which he called a pitiful chimera. Those who adopt a system of preformation, Wolff warned, do not explain development but simply say that it does not occur. To Wolff, the philosophical limitations of preformation rendered it useless as a guide to scientific research. Indeed, preformation might be thought of as a theory that would ultimately discourage research on the embryo, for if development is nothing but the expansion of a miniature individual, knowledge of the fully formed organism would be formally equivalent to knowledge of the embryo at its earliest stage of development. A descriptive epigenetic approach provided a compelling reason to study the course of development, because it asserts that development entails change as well as growth.

A change in the climate of opinion late in the eighteenth century made Wolff's philosophy more congenial to naturalists because, with the growing influence of nature philosophy, many investigators rejected attempts to apply the mechanistic philosophy to living systems. Embryology, in the form of developmentalism, was especially attractive to advocates of nature philosophy as a means of demonstrating the unity of nature and the spiritual or immaterial forces that guided development in the material world. By the end of the nineteenth century, the conflict between preformation and epigenesis was essentially put aside as embryologists adjusted to fundamental changes in their worldview. First, cell theory provided a framework for a new understanding of egg, sperm, and developing embryo. Second, scientists essentially abandoned mechanistic models and nature philosophy and realized that they need not treat living organisms as machines, nor give up all hope of ever explaining the mechanisms that governed living beings.

CELL THEORY

Cell theory is a fundamental aspect of modern biology, prerequisite to and implicit in our concept of the structure of the body, the mechanism of inheritance, fertilization, development and differentiation, the unity of life from simple to complex organisms, and evolutionary theory. To say that the cell is the fundamental unit of life is to provide a tremendously powerful generalization integrating the study of structure and function, reproduction and inheritance, growth and differentiation. By presenting the fabric of the body in terms of cells and cell products, cell theory provides one of many possible answers to the question: What is an organism? Ancient philosophers and anatomists speculated about the nature of the constituents of the human body that might exist below the level of human vision, but even after the introduction of the microscope, investigators differed as to the level of resolution that might be attainable and appropriate to studies of the human body in health and disease. By the eighteenth century, many anatomists had abandoned humoral pathology and hoped to discover correlations between localized lesions and the process of disease through the analysis of the structure and function of organs and systems of organs.

Tissue doctrine, elaborated by the great French anatomist Marie François Xavier Bichat (1771–1802), represents an ambitious and influential attempt to systematize the study of the composition and organization of the body. The indefatigable Bichat seemed to live in the anatomical theater and dissection room of the Hôtel Dieu. A prodigious worker, he performed at least 600 autopsies in the space of a year. Such a regimen was obviously not conducive to health and longevity; Bichat's death at 31 years of age was probably caused by tuberculosis. Working in the autopsy rooms and wards of the hospitals of Paris, Bichat and his colleagues proclaimed their conviction that medicine could only become a true science if physician-researchers adopted the method of philosophical analysis used in the other natural sciences in order to transform observations of general and complex phenomena into precise and distinct categories. This approach and the movement to forge a linkage between postmortem observations and clinical studies of disease was largely inspired by the work of another French physician, Philippe Pinel (1755–1826), author of *Philocophical Nosography* (1798). Pinel argued that diseases must be understood not by reference to humoral pathology, but by tracing them back to the organic lesions that were their sources. Because organs were composed of different elements, research must be directed towards revealing these constituents. Organs that manifested analogous traits in health or disease must share some common structural and/or functional components. Failing to find this analogy at the organ level, Bichat conceived the idea that there might be some such analogy at a deeper level.

To reach a finer level of resolution, Bichat attempted to study the body in terms of organs, by "decomposing" them and analyzing their fundamental structural and vital elements, which he referred to as *tissues*. Organs had to be teased apart by dissection, maceration, cooking, drying, and exposure to chemical agents such as acids, alkalis, and alcohol. According to Bichat, the human body could be resolved into 21 different kinds of tissue, such as the nervous, vascular, connective, fibrous, and cellular tissues. Organs, which were made up of assemblages of tissues, were in turn components of more complex entities known as the respiratory, nervous, and digestive systems. The actions of tissues were explained in terms of *irritability* (the ability to react to stimuli), *sensibility* (the ability to perceive stimuli), and *sympathy* (the mutual effects parts of the body exert on each other in sickness and health).

Obviously, Bichat's simple tissues were themselves complex and compound, except in the sense that they were simpler than organs, organ systems, or the body as a whole. Tissues, as Bichat himself acknowledged, consisted of combinations of interlaced vessels and fibers. Thus, unlike atomic theory or cell theory, Bichat's tissue theory of general anatomy provides no actual unit of basic structure or material to carry the basic properties of life. Embryology was essentially outside the boundaries of Bichat's own research program. His investigations of the arrangement of animal tissues generally ignored the problem of tracing the embryological origins of specific organs and tissues. Bichat's goals were thus very different from those that motivated the founders of cell theory. In formulating tissue doctrine, Bichat's objective was not merely to extend knowledge of descriptive anatomy but to provide a scientific language with which to describe pathological changes. Through an understanding of the specific sites of disease, better therapeutic methods and the means of assessing the efficacy of such treatments should eventually emerge. Although Bichat's work is often regarded as the foundation of the new science of histology, the word *histology* was actually coined about 20 years after his death to denote a new approach to exploring the body in terms of the developmental history of its tissues.

Having no theoretical objectives for the pursuit of microanatomy beyond the tissues, many of Bichat's followers regarded the tissue as the body's natural unit of structure and function, even after cell theory was well established. One reason for this was Bichat's well-known, and often justifiable, skepticism concerning reports of microscopic observations. The microscope was not a trustworthy tool for exploring the structure of the body, Bichat warned, because every person using it gazed into the same darkness and saw a different vision.

Ever since microscopic investigations began in the seventeenth century, observations of various cells, vesicles, and globules had been reported. In 1665 the word *cell* and illustrations of cellular structures appeared in Robert Hooke's *Micrographia*. Studying sections of freshly cut cork under his microscope, Hooke could see that it was perforated and porous, rather like a honeycomb. Counting the number of little boxes or cells at the surface, Hooke found more than one thousand along a linear inch and estimated that a cubic inch must contain more than one billion. While the cells in cork apparently contained nothing but air, Hooke noted that living plants contained cells full of streaming green juices. Nehemiah Grew, Hooke's colleague at the Royal Society, noted that the cells in the younger parts of plants seemed to be closely packed and very full of juices. Similarly, Malpighi found plants to be made up of little bodies, rather like bags, and Leeuwenhoek reported seeing blood corpuscles, sperm, infusoria, and bacteria.

During the nineteenth century, as instruments and staining techniques improved, microscopists reported an avalanche of discoveries unprecedented since the seventeenth century. Cell theory, however, was established before the major wave of technical improvements in the microscope and sample preparation occurred. Since plant and even animal cells are generally large enough to be visible with about 100-fold magnification, it seems likely that the "failure" of the first wave of microscopy to produce cell theory had more to do with the intellectual and conceptual limits, attitudes, and expectations of the observers than the technical limitations of their instruments and methods of sample preparation.

One of the keys to bringing order to confusing and diverse observations of "globules" in biological materials was the study of the cell nucleus conducted by Robert Brown (1773–1858), a naturalist best known for his studies of the random thermal motion of small particles. It was not until 1905 that Albert Einstein proposed an explanation for the phenomenon now known as Brownian motion and thus provided the first proof for the physical existence of atoms. Just as Brown was not the first to see "Brownian movement," he was not the first to see the nucleus, but in both cases his work was more fruitful than that of his predecessors. Some of Leeuwenhoek's studies of blood cells from fish show that he had seen nuclei, but it was Brown who first realized that these entities could be found in the tissues of many different organisms.

While studying pollen grains in 1827, Brown saw minute specks, which seemed to be in constant motion. He called these particles *active molecules*, but rather than attributing their motion to some innate vital force, Brown decided to look for this phenomenon in nonliving entities of similar size, such as particles of coal dust, powdered glass, and minerals. Four years later, in a report on the reproductive organs of orchids and other plants, Brown described his observations of the germination of pollen grains on the stigma of the flower and the growth of the pollen tube into the ovary. Describing the cell nucleus in his paper on "Fecundation in Orchideae" (1832), Brown noted that each cell in orchids and other kinds of plants contained a dark "circular areola."

Through the discovery of the cell nucleus, studies of plant cells began to assume a more coherent pattern in the 1830s. It was more difficult to generalize about the nature of animal cells, but by the end of that decade zoologists as well as botanists had constructed an explanatory framework for many seemingly diverse phenomena. Many of the naturalists involved in establishing the doctrine known as cell theory were influenced by Romanticism, a form of nature philosophy that emphasized the subjective contemplation of nature.

Naturalists inspired by Romantic doctrines hoped to establish a new science of life that would uncover the fundamental unity underlying the superficial but bewildering diversity of the biological realm. Their goal was to discover the basic elements of life and describe the organization and historical development of organisms. The universe envisioned by the Romantics was not a great machine, but a living entity in which spirit, will, or God was constantly present. All phenomena in nature were manifestations of this spirit and, thus, could not be reduced to matter and motion. Even Immanuel Kant was convinced that there would never be a Newtonian revolution in biology because knowledge of purely mechanical principles could never explain organized living creatures. The reproduction of living beings seemed to offer the greatest challenge to the old mechanistic philosophy. Cartesian physiologists could argue that animals were machines, but they could not explain how such "machines" could produce little machines like themselves, whereas mechanical contrivances could not.

Champions of nature philosophy, or the Romantic approach to science, saw nature as a mysterious world of hidden forces, irreducible complexity, and continuous but nonrandom change. The fundamental unity of the natural world, however, implied that studies of the life history of the individual, or the species, should ultimately provide an understanding of life itself. The poet Johann Wolfgang von Goethe (1749–1832) helped popularize ideas that intrigued and inspired many German scientists. In essence, Goethe advocated a new pattern of thought that glorified a poetic, dynamic, developmental ideology of nature. In his *Metamorphosis of Plants* (1790), Goethe expressed ideas about embryology that were similar to those of Caspar Wolff. Critics who later reevaluated Goethe's contributions to science objected that all that he thought he had discovered could be considered "wrong, silly, or already known."

The German naturalist Lorenz Oken (1779–1851) was another influential champion of nature philosophy. While visiting a remote island in the North Sea, Oken distracted himself with the study of marine animals, much as Aristotle had done many centuries before. During this period, Oken received his inspiration about the relationship between all forms of animal life and the *archetype*, that is, the generalized form that reflected nature's basic plan. Understanding the archetype, Oken declared, would provide the key to understanding all living organisms. Combining microscopic observations with philosophical speculations, Oken postulated the existence of a primitive, undifferentiated mucus-like fluid called *Urschleim*. Spherical vesicles were said to arise from this jelly-like material to produce infusoria, the simplest living things. According to Oken, complex organisms were actually aggregates of these simple entities. Every animal or plant was, therefore, a colony of infusoria that had given up their independence to subordinate their life to that of the organism as a whole. Those who rejected the mechanical philosophies, which as Goethe complained had tried to extort Nature's secrets by means of "levers and screws," tended to find such ideas congenial.

Oken's *Urschleim* was eventually replaced by *protoplasm*, an equally vague term for the contents of the cells, used by Johannes Evangelista Purkinje (1787–1869) in 1839 to describe the cell substance. Theologians had used the word *protoplast* for Adam, the first formed, but for Purkinje the term designated whatever was first produced in the development of the individual plant or animal cell. According to Purkinje, the body was composed of fluids such as blood, plasma, and lymph, fibers such as those found in the tendons, and cells. Devoted to teaching as well as research, Purkinje developed innovative teaching techniques, a knife that was a precursor of the microtome, established the use of balsam-sealed preparations, and adapted Louis J. M. Daguerre's (1789–1851) methods to produce the first photographs of microscopic materials. Because university officials were unwilling to meet his demands for space and equipment, much of his research and teaching was carried out in his own home laboratory, which became known as the cradle of histology. Purkinje's observations of nerve cells led to a detailed description of the large flask-shaped cells in the cerebellar cortex that are named for him.

Apparently independently of Purkinje, Hugo von Mohl (1805–1872) used the word *protoplasm* to describe the part of the plant cell within the cell membrane. In a series of influential articles, von Mohl helped make the term protoplasm a general part of the vocabulary of biology. In 1861 Max Schultze (1825–1874), professor of anatomy at Bonn, inelegantly, but succinctly defined the cell as a lump of nucleated protoplasm. According to Schultze, protoplasm was the physical basis of life. In all forms of life, plants and animals, higher and lower forms, protoplasm provided unity of structure and function. Thomas Henry Huxley (1825–1895) introduced the term protoplasm to the general public at Edinburgh in 1868 in an address on "The Physical Basis of Life." As used by Huxley, the term was thoroughly divorced from its previous religious associations. "All vital action," Huxley announced, "may be said to be the result of the molecular forces of the protoplasm which display it."

Johannes Peter Müller (1801–1858), an eminent physiologist and comparative anatomist, was the mentor of many famous biologists, including Jacob Henle, Robert Remak, Hermann von Helmholtz, Rudolf Virchow, and Theodor Schwann. Müller served as professor of anatomy and physiology and director of the Museum of Comparative Anatomy at the University of Berlin. When the revolution of 1848 broke out, Müller, who was then serving as the rector of the University of Berlin, was charged with controlling the disturbances that broke out among students and staff. Student demands included free education and an end to examinations. After the student rebellion was put down by the military, dealing with the continuing hostility and sabotage that became part of campus life wrecked Müller's physical and mental health. Nevertheless, he retained his position at the University until his death.

A man of broad interests, Müller taught human and comparative anatomy, embryology, physiology, and pathological anatomy and made original contributions to each field. When he died, probably from overwork, three people had to be appointed to replace him. Stimulated by the work of Purkinje, Müller became one of the first to use the new microscopic approach in studies of pathological phenomena. While Müller's work was significant in stimulating interest in a finer level of resolution in pathological anatomy, credit for the formal statement of cell theory is generally attributed to his student Theodor Schwann (1810–1882) and the botanist Matthias Jacob Schleiden (1804–1881).

After studying law at the University of Heidelberg, Schleiden tried to establish a practice as a barrister in Hamburg, but he was such a failure that he decided to commit suicide. Even with a gun aimed at his own forehead, Schleiden was unsuccessful. When he recovered from the self-inflicted but superficial wound, he decided to switch from law to natural science. After earning doctorates in both medicine and philosophy, he was appointed professor of botany at the University of Jena. Despite his success in research and teaching, Schleiden resigned 12 years later in order to travel and rest his nerves. Contemporaries characterized Schleiden as arrogant, temperamental, and unsparing in his attacks on his rivals and predecessors. Schleiden did, however, respect the work of Charles Brisseau-Mirbel (1776–1854), an eminent French botanist and microscopist who suggested that cells were found in all parts of the plant. New cells, according to Brisseau-Mirbel, were formed in a primitive fluid in a manner analogous to the formation of a network of bubbles in the foam of a fermenting liquid.

Confronting the practitioners of what he considered archaic systematic botany, Schleiden redefined botany as an inductive science encompassing all aspects of the study of the laws and forms of the vegetable kingdom. Because botany had been under the control of pedantic systematists, it had established few facts and had discovered no fundamental principles and ideas that could lead to the establishment of its natural laws. Only the chemistry and physiology of plants were truly significant, Schleiden argued. Work on the systematic ar-

Matthias Jacob Schleiden

rangement of plants was dismissed as a waste of time. In order to transform botany into a true science, botanists must study the plant world in all possible ways, including microscopic examination of parts invisible to the naked eye.

In 1838 Schleiden published "Contributions to Phytogenesis" in Müller's Archives for Anatomy and Physiology. Recognizing the importance of Robert Brown's work on the nucleus, Schleiden saw it as the key to understanding the growth and development of plants. Soon he came to regard the nucleus, which he renamed the *cytoblast*, as a universal elementary organ in the plant world. For Schleiden, all plants of any complexity were aggregates of cells, which he characterized as fully individualized, independent, separate entities. Within the plant, each cell led a double life: one pertaining to its own independent development and another that allowed it to serve as an integral part of a plant. Thus, all aspects of plant physiology were fundamentally manifestations of the vital activity of cells.

Beyond their function as structural components of plants, cells were of interest because their origin might reveal the origin of life itself. Some naturalists believed that the origin of life occurred in a kind of fluid "blastema" out of which organic solids precipitated. The first structures to emerge from this blastema might be the elementary units of life. Most early nineteenth-century biologists identified these elements as fibers, while others saw them as globules. Early versions of cell theory were very similar in concept to globulist ideas.

Schleiden described several possible methods of cell formation in "Contributions to Phytogenesis" and later in *Principles of Botany*, but he favored the hypothesis known as "free-cell formation." According to this doctrine, cell growth was analogous to the process of crystallization: a nucleolus grew by accumulation of minute granules out of the cytoblastema, a fluid rich in sugars and mucus. As mucus particles aggregated, part of the fluid was transformed into a relatively insoluble substance and formed the cytoblast around the nucleolus. When the cytoblast attained its full size, the young cell began to develop as a delicate transparent vesicle that gradually expanded, and finally formed a complete cell within a rigid cell wall. Schleiden thought that plants could also grow by the formation of cells within cells in the presence of the cytoblastema. In this case, the entire contents of a cell were divided into two or more parts and a gelatinous membrane immediately formed around each part.

Although cell theory is generally associated with Schleiden, he was not the first botanist to describe cells and speculate about their origin. Franz Julius Ferdinand Meyen (1804–1840) described investigations of cells and cell division in his three-volume treatise *On Plant Physiology* (1837–39). Unlike Schleiden, who thought that the nucleus formed de novo from fluids, Meyen believed that the division of previous cells formed new cells. In response, Schleiden claimed that Meyen's ideas suggested a return to preformationism.

Unfortunately, Meyen died too young to contribute to or challenge the establishment of early cell theory.

Despite Schleiden's attachment to an oversimplified model of cell formation based on the growth of inorganic crystals, he categorically rejected the idea that organized life forms could arise through spontaneous generation. Even the lowly algae, lichens, and fungi reproduced their own kind. Confining his work and speculations to the plant world, Schleiden observed that many eminent men had struggled to establish an analogy between the animal and vegetable kingdoms. Failure had been generally acknowledged, although few understood that the precise reason was the impossibility of applying the idea of individuality as used in the animal kingdom to the plant world. Only the very lowest orders of plants, which consisted of single cells, could be called individuals. By clearly enunciating the idea that the plant is a community of individual cells, Schleiden provided the key principle unifying plant and animal life. Indeed, by sharing his ideas about the plant as a community of cells with Theodor Schwann in 1837, Schleiden helped to bring about the extension of cell theory from the plant world to the animal kingdom.

Quite unlike the abrasive, heterodox Schleiden, Theodor Schwann seems to have been a timid, introspective, and extremely pious person. Schwann received his early education at the Jesuits' College in Cologne and then studied medicine at the Universities of Bonn, Würzburg, and Berlin. After graduating in 1834 he became one of Müller's favorite disciples. During the time Schwann had Müller's encouragement, energy, and willpower to keep him at work, he made numerous contributions to histology, physiology, and microbiology, in addition to his famous work on cell theory. While making histological preparations for Müller, Schwann discovered the sheath surrounding nerve fibers that has been named for him, and, in studying the process of digestion, he discovered the ferment (enzyme) named pepsin. Schwann also studied the respiration of the chick embryo and its need for oxygen. A vigorous program of experiments on fermentation carried out by Schwann challenged the theory of spontaneous generation and suggested that microorganisms are responsible for the chemical changes involved in putrefaction and fermentation. Unable to cope with the violent quarrels that were so much a part of nineteenth-century academic life, Schwann eventually retreated into scientific exile rather than compete for a professorship in a German university.

When Schwann began to think about cells, some affinities between plant and animal structures had been suggested, but the great variety of forms found in the animal kingdom seemed more significant than any similarities. Even when animal cells, fibers, corpuscles, and so forth were seen, Schwann explained, they remained merely descriptive aspects of natural history. Without a unifying theory, studies of the mode of development of one kind of cell could

Theodor Schwann

not be related to that of any other kind. Animals appeared to be more diversified than plants in both their internal and external forms. Even with the best microscopes, it was difficult to see any detail in animal cells because they were generally transparent and lacked the cell walls found in plant tissues. Schwann had noticed nucleated entities in certain animal tissues, but until he talked with Schleiden he did not recognize the implications of these observations. After Schleiden described the role that the nucleus played in the development of plant cells, Schwann suddenly realized how important it would be to show that the nucleus found in animal cells performed the same role as the nucleus of plant cells.

Like Schleiden, Schwann saw the nucleus as the key to elucidating the relationship between the plant and animal world and the composition of animal tissues. Beneath the myriad forms that animal tissues assumed, forms as diverse as muscles, nerves, and blood corpuscles, there was the unifying factor of the cell nucleus. Schwann's purpose, as set forth in *Microscopical Researches into the Accordance in the Structure and Growth of Animals and Plants* (1839), was to prove the basic unity of the two kingdoms of organic nature. In this endeavor Schwann was quite successful; moreover, he also provided evidence that even the most physiologically diverse parts of an animal developed according to the same basic principles.

In the first section of *Microscopical Researches*, Schwann described the structure and growth of the notochord of the tadpole and cartilage from various sources. Section two provided evidence that cells are the basis of all animal tissues, no matter how specialized. And finally, in the third section, Schwann explicated his theory of cells. Close examination of various animal tissues indicated that they did indeed originate from cells that were analogous to the cells of plants in all important respects. Such findings removed the great barrier thought to separate the animal and plant kingdoms. Animal tissues, like plant tissues, contained cells, cell membranes, cytoplasm, nuclei, and nucleoli. Building on the work of Brown and Schleiden, Schwann argued that the presence of the nucleus was the criterion that characterized true cells.

In much the same manner as Schleiden described the double life of plant cells, Schwann proposed that animal cells similarly possessed an independent individual life and a life subordinated to the functioning of the organism as a whole. Two major modifications of cell life were defined. Independent cells were those in which the cell membrane remained clearly distinguishable from those of neighboring structures. Coalesced cells were those in which the cell membranes blended, partially or entirely, with neighboring cells, or intercellular materials, to form a homogeneous substance. Thus, tissues could be classified in terms of the degree of development that cells had to undergo in order to form a particular tissue. This analysis was extended to all of the body's most highly differentiated tissues, including muscles, nerves, bones, teeth, hooves, and feathers. No matter how unique and noncellular any body part might appear, if traced backwards in terms of embryonic development, all the most complex and specialized tissues and parts of the animals were derived from cells.

In some cases, Schwann thought that cells appeared to be formed within previously existing cells, but he also accepted Schleiden's idea that cells could be formed from a structureless fluid, or cytoblastema, by a process analogous to crystal growth. While Schwann took pains to explain that cell growth was only figuratively similar to crystallization, he did seem to find the metaphor of crystallization from the "mother-liquor of life" very powerful, and he urged other scientists to pursue this analogy in their research. Schwann's work has been interpreted as the culmination of the Romantic search for the common origin of all living forms, but Schwann chose to present his conclusions as the result of pure empirical investigations. In any case, Schwann's theory of the cells proclaimed the fundamental unity of anatomical and physiological principles in plants and animals and provided a new framework for the investigation of the origin and development of the embryo.

In the third section of *Microscopical Researches*, Schwann summarized all of his research and explicitly elaborated the generalization known as cell theory. His first major proposition states that there is one universal principle of development for all the diverse parts of organisms and that this principle is the formation of cells. His second proposition described the generation of cells from a structureless substance that was present either around or inside existing cells. In a section called "Theory of the Cells," Schwann attempted to deal with the most fundamental aspects of cellular phenomena and the question of whether they should be considered mechanical or vital. Actions carried out altogether blindly in accord with the laws of necessity, Schwann argued, may seem to be adaptive to some higher purpose. Thus, organized bodies that seem to possess powers not found in inorganic nature might actually be acting in terms of physical and chemical laws. Understanding organized bodies in a scientific sense resolved itself into a question of studying the fundamental powers of individual cells. Cellular phenomena were then divided into two natural groups. First, *plastic* phenomena were those related to the combination of molecules to form a cell. Second, chemical changes in the component particles of the cell itself, or in the surrounding cytoblastema, could be called *metabolic* phenomena. The word "metabolic" was coined from the Greek to describe "that which is liable to occasion or to suffer change."

Schwann recognized metabolic power as a universal property of cells and, therefore, of life. The ancient enigma of animal heat could, therefore, be explained as the product of cell metabolism. Furthermore, because the so-called fermentation-granules or yeasts were actually cells, fermentation provided a practical model system for investigating the metabolic activities of cells. Schwann's theory of intracellular fermentation and his broader view of the cell as the unit of metabolism were accepted only after a fierce battle between the German chemist Justus von Liebig and the French chemist and microbiologist Louis Pasteur. Liebig and his associate Friedrich Wöhler published an anonymous satire of Schwann's theory in which yeast in solution gave rise to eggs that hatched into animals shaped like alembics (special glass vessels used for distillation). These creatures devoured the sugars in solution, digested their meal, and then rudely belched forth carbon dioxide and excreted alcohol. Temperamentally unable to respond to such attacks, Schwann left the battlefield to be defended by Louis Pasteur, a man who was described as never one to put up his sword until all enemies had been conquered or killed.

While Schwann and Schleiden had provided a powerful new framework for understanding the structure, development, and physiology of plants and animals, their theory was different from modern cell theory in several important respects, primarily in terms of their concept of free-cell formation and the notion of the cytoblastema. In subsequent years, various botanists, zoologists, and microscopists including Karl Nägeli, Hugo Von Mohl, and Rudolf Virchow attacked this concept. Karl Nägeli (1817–1891) realized that certain algae were useful models for studies of cell division and for observing the movement and behavior of protoplasm. At first Nägeli, coeditor with Schleiden of a short-lived journal, defended Schleiden's theories, but his comparative studies of the production of cells in various groups of plants eventually convinced him that new cells arose from the division of preexisting parental cells.

Swammerdam and Spallanzani had described the cleavage of eggs. Yet when Jean Louis Prévost (1790–1850) and Jean Baptiste Dumas (1800–1884) saw the segmentation of frog eggs in 1824, they still could not explain the meaning of the phenomenon. In 1854 Martin Barry (1802–1855) published illustrations of rabbit eggs undergoing cleavage. But Barry believed that cells of later stages were derived directly from nuclei. Albert Kolliker (1817–1905), who studied the development of the eggs of the cuttlefish, was probably the first to lay great stress on the division of the nucleus during the process of segmentation of the egg. Kolliker applied Schwann's theory to embryonic development in a great number of animal species and tissues.

It was primarily through the work of Franz Leydig (1821–1908) and Robert Remak (1815–1865) that the behavior of the nucleus during cell division was clarified. Studies of the developing frog embryo in 1852 convinced Remak that every cell was produced by the division of a preexisting cell. Moreover, division of the nucleus preceded cell division. Later, Remak observed the division of embryonic blood corpuscles in the developing chick. He suggested that under normal growth conditions cells increase in number by the division of one cell into two new cells. Rudolf Virchow (1821–1902) was one of the first to accept Remak's evidence as a refutation of free-cell formation. Nevertheless, even after Virchow's declaration *omnis cellula e cellula* (all cells from preexist-

ing cells), many biologists continued to accept the possibility of free-cell formation. About 30 years later, Walther Flemming (1843–1905) provided proof of
the division of the nucleus as a fundamental aspect of cell division and declared
omnis nucleus e nucleo (all nuclei from nuclei).

Cell theory in its modern form was, therefore, elaborated and incorporated
into scientific medicine by Rudolf Virchow, a man of many talents and accomplishments. Virchow was a prominent member of the reformist social and intellectual movements of the nineteenth century. Indeed, Virchow and his like-
minded colleagues were quite outspoken about the direct relationship between
their scientific doctrines and the positions they took in opposing the repressive
Prussian state. In 1847, Virchow participated in an official investigation of a
typhus fever outbreak in an industrial district of Silesia. Convinced that the
fundamental causes of the epidemic could be found in the abysmal social and
sanitary conditions prevailing in the district, Virchow prepared a report that
blamed the government for the misery he had observed. This led to his immediate dismissal from his official position and established his reputation as a radical
socialist and medical materialist. Following the revolution of 1848, his known
sympathy for the opposition made it expedient for him to leave Berlin. Fortunately, he obtained a position at the University of Würzburg, the first chair of
pathological anatomy in Germany. Virchow spent 7 very productive years there,
until the chair at Berlin became vacant in 1856. After obtaining a commitment
from the University to establish an institute of pathology for his research, Virchow accepted the position and remained at the University of Berlin until his
death.

In 1847 Virchow and Benno Reinhardt (1819–1852) founded the *Archives
for Pathological Anatomy*. Editing the journal gave Virchow the opportunity to
encourage original scholarship in a variety of fields. Despite the title of the
journal, Virchow included articles on comparative anatomy and physiology,
anthropology, oriental languages, translations of medieval Greek and Arabic
manuscripts, as well as the expected papers on infectious diseases and the histology of tumors. In 1850 Virchow was elected to the Berlin City Council and in
1861 to the Prussian Diet, where he actively opposed the policies of Otto von
Bismarck (1815–1898). He remained on the City Council for the rest of his life
and initiated many social, sanitary, and medical reforms. During the wars of
1866 and 1870, Virchow was responsible for military hospitals and the development of the first hospital trains. He served as a member of the Reichstag from
1880 to 1893. In 1869 he founded the Berlin Society of Anthropology, Ethnology and Prehistory, serving as president until his death.

In the first volume of the *Archives for Pathological Anatomy*, Virchow
reviewed prevailing ideas on the organization and growth of tissues. Reflecting
the influence of Schleiden and Schwann, he described the origin of cells in a
formless blastema, a fluid exuded from vessels. During his studies of the process

of inflammation, Virchow had seen white corpuscles from the blood enter the area of a wound in great numbers and become macrophages. This process seemed to fit Schleiden's theory. On the other hand, while investigating the healing of the cornea, he had seen phenomena inconsistent with free-cell formation. Further microscopic studies of pathological processes led Virchow to the same conclusion that Remak had come to from embryological studies. By 1854 Virchow was convinced that "there is no life but through direct succession"; that is, all cells were derived from preexisting cells. In 1855 Virchow published a paper on "Cellular Pathology" that included the famous motto *omnis cellula e cellula*. Resistance to this fundamental insight, Virchow suggested, could be attributed to the fact that microscopy was still new to medicine and was not deeply ingrained in the thinking of conservative physicians. Furthermore, overly enthusiastic and uncritical use of the instrument by naïve investigators generated justifiable skepticism. Virchow's experiences might have been similar to those of an Oxford University student attending newly established courses in microscopic anatomy in the 1840s. When one of the older faculty members was persuaded to look at the microscopic preparations that illustrated the lectures, the elderly professor proclaimed that he did not believe in such evidence, but even if it were true he did not think that God meant for human beings to know such things.

For Virchow, the cell was the fundamental link in the great chain that formed the hierarchy of tissues, organs, systems, and, ultimately, the complete organism. In a series of lectures later published under the title *Cellular Pathology* (1858), Virchow reviewed previous ideas about plant and animal structure, compared plant and animal cells, and analyzed the structure of the tissues that made up human organs in health and disease. Humoral pathology, the ancient concept of general disease, was set aside as Virchow demanded that medical and scientific inquiry focus on a new question: Where in terms of the body's cells is the disease? All disease, according to Virchow, is simply modified life; there is no essential difference between normal and pathological states. Thus, the study of pathology must be linked to the study of physiology in order to describe the subtle changes that take place in a pathological state. For example, research on leukemia convinced Virchow that cancer cells differ from normal cells primarily in their behavior rather than their structure. While other scientists had advanced some of these ideas, none had stated them as convincingly as Rudolf Virchow, the man who was often called Germany's "Pope of Pathology."

In his influential treatise *The Cell in Development and Heredity* (1925), Edmund B. Wilson (1856–1939) outlined three stages in the development of cell theory after its enunciation by Schleiden and Schwann. Between 1840 and 1870, scientists labored at the foundations of the theory, marking out the fundamental outlines and principles of genetic continuity. The second period, from

1870 to 1900, witnessed the maturation of cytology and cellular embryology, as well as the development of new concepts of the physical basis of heredity and the mechanism of development. Wilson wrote from the perspective of a scientist in the third stage, after the rediscovery of Mendel's laws when the focus of attention had shifted towards studies of heredity. By this stage, improvements in microscopes and, more particularly, improved methods of preparing and staining biological materials made it possible to analyze the fine structure of the cell.

Nineteenth-century biologists usually prepared their materials by teasing out or squashing fresh material into a thin layer and studying this directly under the microscope. The structure of relatively uniform tissues might be adequately studied this way, but the interrelationships among different types of cells in complex organs would be essentially obliterated. Fixatives were first used as preservatives for gross specimens. The eminent seventeenth-century chemist Robert Boyle had suggested many preservatives, but it is uncertain whether he actually tested them. Microscopists used various chemicals, such as alcohol, acetic acid, and chromic acid and its salts, primarily as hardening agents, and soaked tissue samples until they became rigid enough to be sliced with razors and knives. Many of the early pioneers of histological technique were English amateurs who were more interested in making beautiful microscopic prepara-tions for display than in learning new facts about nature. English microscopists probably developed the precursor of the microscopic slide and the first micro-tomes. In Germany, in contrast, microscopes were seen as serious tools for the use of scientists, who had little patience with merely cosmetic, time-wasting techniques. Until German scientists became interested in subtle details of tissue and cell structure, they continued to use free-hand sections and ridiculed the conceits of British amateurs.

The availability of a whole panoply of synthetic dyes in the late nineteenth century vastly expanded the ability of microscopists to see structural details within biological preparations. Natural dyes had, of course, been available for thousands of years. Dyes imparted color to fabrics, paper, leather, wood, oils, fats, waxes, and to plant and animal tissues prepared for microscopic examina-tion. Staining methods in the first half of the nineteenth century were generally quite crude, and the origins of many common procedures are obscure. Early microscopists used natural dyes such as blueberry juice, red cabbage juice, in-digo, madder, carmine, and logwood extract. Carmine is derived from *Coccus cacti*, an insect that had been cultivated and used for dyeing in Mexico long before the Spanish conquest. The female insects, which contain a purple-colored sap, were harvested, killed, and dried just before egg laying, and the pulverized insect preparation was sold as cochineal. Hematoxylin, a very important reagent in histology, comes from logwood.

Crude staining methods, coupled with overactive imaginations and specu-lative tendencies, led to reports of various phantom structures, such as the internal organs of Christian Gottfried Ehrenberg's (1795–1876) amazing microscopic animals. Convinced that the infusoria were complex creatures with internal organs, Ehrenberg interpreted staining patterns as evidence of their digestive system. He named these organisms *Polygastrica* in honor of the multiple stomachs revealed by their consumption of dyes. Later investigators would reinterpret such observations as evidence of differential staining, that is, the inner portions of the cells Ehrenberg had observed absorbed dyes while the outer parts did not.

Joseph von Gerlach (1820–1896), professor of anatomy and physiology at Erlangen, Germany, and author of the landmark text *Handbook of General and Special Histology* (1848), has been called the founder of modern staining technique. Although he was not the first to use stains in research, Gerlach was one of Europe's best known anatomists and most influential microscopists. Unlike some of the other pioneers of staining, Gerlach appears to have carefully controlled his experiments so that they could be described in detail and reproduced by others. In the 1850s he introduced one of the most successful of the early histological stains, a transparent solution of carmine, ammonia, and gelatin, generally called "Gerlach's stain." Use of this stain produced a noticeable difference in the degree of staining of the nucleus and the intercellular substance. These studies suggested that the uptake of stain indicated that specific chemical reactions had occurred and that specific cell components differed in their ability to combine with dyes. The work of Paul Mayer (1848–1923) was especially valuable in creating a coordinated and systematic approach to histological techniques. In the 1870s, when Mayer began his research, histologists were already using chemicals as fixatives and stains to improve their microscopic preparations. Chemicals such as glycerin, calcium chloride, chromic acid, osmium tetroxide, carmine, and hematoxylin were used, but the work was crude and unsystematic. Mayer systematized staining methods and provided logical guides to combining and comparing existing methods.

William Henry Perkin (1848–1907) discovered mauve, the first of the aniline dyes, while trying—very unsuccessfully—to synthesize quinine (a valuable anti-malarial drug). Within a few decades of Perkin's first patent, a rainbow of dyes, including safranin, methyl violet, aniline blue, methyl green, fuchsin, and crystal violet, had been synthesized. By the 1860s, many aniline dyes were being used as biological stains, but even in the late nineteenth century commercially available dyes were often impure mixtures, and using them systematically and reproducibly was difficult. Further confusion resulted from the fact that the names used for specific dyes varied in different countries.

Generally, the period from 1875 to 1895 was rich in discoveries concerning fundamental cytological phenomena, such as mitosis, meiosis, and fertilization, and important cellular organelles, such as mitochondria, chloroplasts, and the Golgi apparatus. These methods made possible the discoveries and theories that linked cytology to inheritance and development. Eduard Strasburger (1844–1912), professor of botany at Bonn, helped to unify the field with his monumental *Cell-Formation and Cell-Division* (1875). His descriptions of the complex processes taking place in the division of plant cells, however, are not as well known as Walther Flemming's (1843–1905) studies of cell division in animals. In 1876 Flemming became professor of anatomy and director of the Anatomical Institute at the University of Kiel, where he remained until his retirement in 1901. His first papers on the cell and the nucleus were published in 1877. Flemming described the chromosomes in the late 1870s, but the term was first used in 1888 by Heinrich W. G. Waldeyer (1836–1921), who introduced the use of hematoxylin as a histological stain. Flemming had used the term *chromatin* for the nuclear substance and he gave the name *mitosis* to cell division. Flemming's *Cell Substance, Nucleus, and Cell Division* (1882) established a basic framework for further exploration of the stages of cell division. Thus, cell theory came to include two more generalizations: cells in animals and plants are formed by the equal division of existing cells, and division of the nucleus precedes division of the cell. Further studies of the cells of higher organisms revealed a microcosm full of minute inclusions and organelles, some of which contain their own genetic machinery.

EMBRYOLOGY AND NATURE PHILOSOPHY

Studies of embryology might be summarized in terms of changes in the level of resolution at which scientists were able to work. Embryos could be thought of as minute but complete organisms or as assemblages of germ layers, globules, vesicles, and fibers. Once cell theory had been established, embryos could be analyzed in terms of cells and components of cells, such as nucleus, cytoplasm, membranes, chromosomes, intracellular organelles, and macromolecules. In thinking about developmental phenomena, biologists were also affected by shifts in fundamental philosophical systems and the controversies surrounding ideologies such as preformationism and epigenesis, mechanism and vitalism, reductionism and organicism. Nineteenth-century nature philosophy generated an intense interest in comparative anatomy and morphology, as well as the historical description of growth and change in organisms, cultures, and institutions. Embryological studies were particularly valued because they could provide the most detailed histories of organisms. Advocates of a new approach to embryology, which has been called *developmentalism*, saw their work as a branch of morphology primarily concerned with explaining how the fertilized egg gave rise to the structure, form, and apparently goal-directed behavior of the organism.

Many of the scientists involved in the development of cell theory and embryology shared an interest in nature philosophy, sometimes in its most extreme forms, and some seem to have suffered from severe mental and emotional problems. Karl Ernst von Baer (1792–1876) might be seen as the prototype of this species. Originally, von Baer planned to follow a military career, but he began medical studies at Dorpat in 1910. In his *Autobiography*, von Baer recalled that the lectures on general anatomy at Dorpat ignored microscopy and followed Bichat's teachings. Worse yet, some of the professors "suffered from a surfeit of rather useless scholarship," as well as a "lack of spirit." Occasionally, students were warned about something called nature philosophy, but they were not told what harm such ideas might cause. Naturally, this only made them more curious. Perhaps his problems with depression as a student made von Baer particularly sensitive to pernicious educational practices. He warned that it was wrong and counterproductive to overburden young minds, because students would feel crushed by their assignments rather than absorbed by the work itself.

When Napoleon invaded Russia in 1812, von Baer volunteered for service at an army hospital to care for the wounded and the victims of epidemic typhus fever. Almost all of the student volunteers contracted typhus fever. Seriously ill, von Baer received the standard "wait-and-see" treatment. The experience provided very little medical knowledge, but it taught him about the horrors of war. By the time he earned his degree, he had little confidence in himself or medicine. Further experience in the great hospitals of Vienna left him with little enthusiasm for a life spent as a practicing physician. The prospect of making his way in the world of science was more exciting, and he turned to the study of anatomy, embryology, physiology, and comparative anatomy. After years of peripatetic poverty he obtained a position at Königsberg, where he carried out most of his embryological work. His major treatise *On the Developmental History of Animals* appeared in 1828. In 1834 he moved to Saint Petersburg, but peace of mind still eluded him and his obsession with scientific projects threatened to destroy his own nervous system. Convinced that only travel could help him overcome illness and depression, he embarked on various journeys through Europe, Russia, and Lapland. When his sight and hearing failed he was forced to retire, but he continued to pursue some aspects of his researches until his death at 84 years of age.

Finding the mammalian ovum was the goal of von Baer's early embryological research, but in his *Autobiography* he claimed that at first he lacked the courage to search for this elusive entity. After all, even Albrecht von Haller, a man characterized by unsurpassed erudition and prodigious diligence, had been unsuccessful in his attempts to investigate the origin of the mammalian ovum. Eminent naturalists such as Jan Swammerdam and Regnier de Graaf had become engaged in a priority battle about finding the mammalian ovum, but both had been mistaken. In 1672 Swammerdam published an account of his study of the "female testes," including his claim that he had discovered the human ovum. During the same year, De Graaf published a treatise called *De mulierum organis generationi inservientibus* (*The Generative Organs of Women*), which was primarily a study of development in the rabbit. When de Graaf discovered large, round swellings on the ovaries of rabbits, ewes, and women, he assumed they were mammalian eggs. De Graaf also described the *corpus luteum* (follicles from which the mature egg had already escaped). Leeuwenhoek argued that the structures now known as Graafian follicles could not be eggs, but Haller suggested that the egg might be formed by the coagulation of the fluid in the Graafian follicle. De Graaf noted that the "egg" did not contain a tiny embryo, but he thought it did contain the "germ" of the future organism. Although the rabbit embryo was not visible until about 10 days of gestation, about one third of the total time of pregnancy, de Graaf's work seemed to provide microscopic evidence that viviparous animals arise from eggs formed by the ovaries, and ovist preformationists assumed that his observations supported their doctrine.

Haller and his associates had carefully dissected about 40 ewes, but they could not find an embryo in the uterus until about 17–19 days after mating. After examining intact Graafian follicles, burst follicles, the corpus luteum, the oviducts, and the uterus, Haller concluded that a fluid must have been sent to the uterus, where it became more mucus-like, and then produced the embryo by coagulation or crystallization. Haller's theory of the formation of the mammalian and human embryo was still being taught at Dorpat University when von Baer was a student. But despite Haller, many scientists continued to believe that the Graafian follicle was the actual mammalian ovum and that it had to enter the oviducts and travel to the uterus. Von Baer's 1827 treatise *On the Origin of the Mammalian and Human Ovum* finally clarified the relationship between the Graafian follicle and the mammalian ovum.

During a comparative study of the ovaries of various animals, von Baer discovered the ovum in the Graafian follicle of a dog that had been sacrificed for this study only a few days after coming into heat. While examining this dog, von Baer noticed that he could find a small yellow entity within follicles that were still intact, but close to bursting. Opening a follicle, he removed the tiny spot and examined it under the microscope. On seeing a small, well-defined yellow yolk mass, von Baer realized that he had previously noticed similar entities making their way through the oviducts to the uterus. Based on these findings, von Baer set about observing, analyzing, and comparing the course of development in a wide range of organisms, including the human female.

Despite the importance of this discovery, von Baer felt that his work was not properly appreciated until the 1840s after the establishment of cell theory. At that point, his work was finally honored as the "first building block" of the new embryology. When later critics accused him of being an opponent of cell theory, von Baer recalled he had seen cell division in some eggs, but he had assumed the eggs were spoiled or defective. Indeed, before Schwann's cell theory had been elaborated, terms such as "cell" and "cell division" were completely foreign to him. In his early publications, von Baer called the components of the embryo "histological elements" or "morphological elements." Reflecting on the significance of Schwann's theory, von Baer acknowledged that it certainly had contributed to the development of histology. Nevertheless, he thought that Schwann had overestimated the importance of the "so-called life of the cell." How could the cells build the animal organism on their own, von Baer asked, unless they had "a great deal of morphogenetic intelligence"?

Reflecting on the status of embryology in the 1860s, von Baer noted that it was still impossible to explain how the male contribution rendered the egg capable of developing and how the characteristics of the father passed into the new individual. Von Baer thought that his proof that the mammalian embryo did not coagulate out of some mucoid fluid, but developed by a series of transformations from a previously organized corpuscle, was an important factor in diminishing support for the doctrine of spontaneous generation. While he regarded spontaneous generation as "highly problematic," he did not think that the question had been unequivocally settled.

Although von Baer has generally been considered the founder of a new form of epigenesis, he actually presented a more subtle interpretation of embryological development than a distinction between epigenesis and preformation would suggest. Aware of the dispute between Haller and Caspar Wolff, von Baer explained that Wolff had seen his work in opposition to the preformation theory favored by Haller. Contrasting his own work to that of Wolff, von Baer argued that Wolff's concept of epigenesis, as the truly new formation of all parts and of the entire embryo, had gone too far. While it was certainly true that no limbs or specific parts were present in the ovum, von Baer argued that the parts "do not come into existence by truly new formation, but by a transformation of something already existing." The term *evolution*, in the sense of metamorphosis, von Baer suggested, was more appropriate for embryological development than Wolff's terminology, because, while nothing corporeal was actually preformed, the course of development was in some sense preformed.

Before embryological development was linked to the theory of the cell, several investigators suggested that development does not proceed directly from egg to organ formation, but involves intermediate embryonic structures. The idea that the early embryo consists of leaf-like layers from which new structures develop can be found in the work of Caspar Wolff, but the establishment of the so-called germ layer concept is primarily associated with Heinrich Christian Pander (1794–1865) and von Baer. In 1817, Pander published a well-illustrated treatise in which he described three primordial or germ layers in the chick embryo. As a descriptive guide, rather than an attempt to explain the mechanism of development, the germ layer theory simply states that, despite the great differences between adult vertebrates of different species, similar organs are derived from comparable germ layers. In 1855 Robert Remak refined the concept and called the three germ layers *ectoderm, mesoderm*, and *endoderm*. Ectoderm (outside skin) gives rise to skin and the nervous system; mesoderm (middle skin) produces muscles, skeleton, and the excretory system; and endoderm differentiates to form the notochord, digestive system, and associated glands. The notochord, or chorda dorsalis, a transient embryonic structure discovered by von Baer, eventually develops into the backbone. Von Baer thought the term dorsal cord, or vertebral cord, preferable to notochord. While the structure cannot be seen in adult vertebrates, except for certain fish, it serves as an important tool for determining the vertebrate nature of questionable organisms.

Another important generalization established by von Baer was known as the law of corresponding stages. Discussing the progressive specialization that occurred during development, von Baer outlined four descriptive propositions associated with this precept. First, during development general characters appear before specialized ones. Second, the most general characteristics gradually develop towards the less general and then to the most specific. For example, limb buds become recognizable as limbs that later differentiate into hands, wings, or flippers. Third, during development, the embryo of a given species continuously diverges from those of other species. Fourth, the embryo of a higher species goes through stages that resemble the stages of development of lower animals.

Scientists less cautious than von Baer transformed his general principles into the "law of recapitulation," also known as the "biogenetic law." Johann Friedrich Meckel (1781–1833), an enthusiastic proponent of the recapitulation doctrine, introduced his "law of parallelism" in 1821 to describe the growth of the human embryo. During development, he suggested, the human embryo essentially climbs the hierarchy of animal forms from lowest to highest: fish, reptile, mammal, human. Von Baer challenged Meckel's concept and argued that the human embryo never assumed forms equivalent to the adult forms of lower animals.

Ernst Haeckel's (1834–1919) dogmatic and oversimplified version of the biogenetic law is epitomized in the phrase "ontogeny recapitulates phylogeny." One of the first German biologists to adopt Charles Darwin's (1809–1882) ideas about evolution, Haeckel argued that *ontogeny*, the embryological development of the individual, repeats *phylogeny*, the evolutionary development of the species. Phylogeny, Haeckel insisted, is the mechanical cause of ontogeny. Interpreting embryological history as evidence that all species evolved from common ancestors, Haeckel assumed that during development embryos progressed though stages that were virtually identical to the adult forms of their ancestors. In some sense, Haeckel's biogenetic law shared a deep affinity with the metaphor at the core of nature philosophy. If the universe is thought of as a great living being, then the "gestation of nature" has established a profound fundamental unity among organic beings. Defining nature philosophy as the study of the "generative history of the world," Lorenz Oken (1779–1851) saw the history of the individual as, in essence, the history of the universe. All other animals, according to Oken, were persistent fetal stages in the production of man, the creature that was the prototype and model of all existence. The individual development of man, therefore, replicated the development of life on earth because the production of man was the goal of the great universal developmental tendency. For Oken and other Romantics, the history of development was a reflection of the establishment of archetypes rather than a series of transmutation of types in the evolutionary sense. However, many evolutionists accepted and exploited the biogenetic law as a means of incorporating embryology into the body of evolutionary theory. Charles Darwin, realizing the value of recapitulation theory, called the field of embryology "second in importance to none in natural history." While Thomas Henry Huxley was willing to argue that "in the womb, we climb the ladder of our family tree," despite his admiration for the "excellent Huxley," von Baer never accepted this interpretation of his work.

Embryological researches and the use of comparative methods were stimulated by von Baer's work. Having shown how complex and absorbing embryological development really was, he also provided the guidelines for others to follow. In the 1870s the great Swiss anatomist Wilhelm His (1863–1934) urged embryologists to exploit the vast and untapped potential of von Baer's analytical approach to embryology. Von Baer's studies of the mammalian ovum made it possible to see cell multiplication as the basis of embryonic development, and a new focus on the cell made it possible to think about how cell theory might apply to the process of inheritance as well as the study of reproduction and development. Such shifts in conceptual categories and the establishment of new scientific specialties and institutions in the late nineteenth century were reflected in the excitement generated by the research program known as experimental embryology.

EXPERIMENTAL EMBRYOLOGY

The founders of experimental embryology, Wilhelm Roux (1850–1924) and Hans Driesch (1867–1941), were primarily interested in the question of how factors intrinsic or extrinsic to an egg or its parts could govern the development of the embryo. Convinced that descriptive and comparative studies of embryonic development were inadequate, Roux demanded a new experimental approach and saw himself as the founder of a new discipline, which he called *developmental mechanics*. Some of his followers preferred the terms *causal analytical embryology* or *developmental physiology*.

When celebrating the founding of a new discipline, the appropriate nineteenth-century ritual was the establishment of a new journal, in this case the *Archive for Developmental Mechanics of Organisms*, the first volume of which appeared in 1894. "After sufficient observation and description," Roux exhorted his colleagues, "it is time to take the further step towards knowledge of the processes that produce them." The protocol required for this task involved the resolution of developmental processes into simple though still complex functional processes. These functional processes would then be reduced to their components, so that eventually the truly simple processes could be resolved into their physicochemical factors. Unencumbered by modesty, Roux predicted that several centuries later students would read his work with the same intensity of interest with which he had studied Descartes. The terminology of the new experimental embryology owed a debt to Darwin and Haeckel as well as Descartes, as evident in Roux's 1881 paper "The Struggle of the Parts in the Organism: A Contribution to the Completion of a Mechanical Theory of Teleology."

Embryology, Roux argued, must be investigated by means of the tools so successfully developed by the founders of mechanistic physiology. Indeed, Roux's place in the history of embryology must be understood as part of the complex interactions between mechanistic and vitalistic biological theories. In order to analyze the immediate causes of development and resolve or transcend fruitless debates about preformation and epigenesis, embryologists had to adopt experimental methods. Rather than deal with the old historical impasse by traditional means, Roux recast embryological questions in terms of new analytical concepts.

The primary question Roux posed was whether development proceeded by means of *self-differentiation* or *correlative dependent differentiation*. Self-differentiation, or *mosaic development*, was defined as the capacity of the egg or of any part of the embryo to undergo further differentiation independently of extraneous factors or of neighboring parts in the embryo. Correlative dependent differentiation was defined as being dependent on extraneous stimuli or on other parts of the embryo. These were operational definitions, but, Roux insisted, they were powerful because the alternatives could be subjected to experimental tests: transplantation and isolation. Establishing a new conceptual dichotomy had important implications: above all, it removed the adversarial burden of the old terms preformation and epigenesis. Now one could ask: To what extent is the differentiation of a given part of the embryo, at a given point in time, self-differentiation, and to what extent is it dependent differentiation? By defining development as the production of perceptible manifoldness, Roux ignored the ancient conflict between preformation and epigenesis. His concept of development encompassed two principles: first, a true increase in manifoldness, which could be called *neoepigenesis*; second, the transformation of imperceptible manifoldness, which could be called *neopreformation*.

On theoretical grounds, and in keeping with his most famous but seriously flawed experiment, Roux believed that self-differentiation served as the mechanism of development. In other words, Roux tended to visualize the fertilized egg as a complex machine and development as a process that involved the distribution of parts of the machine to the appropriate daughter cells. A series of experiments demonstrated that, although some extreme conditions could damage the embryo, more subtle changes in environment had little or no effect on development. Confident that external forces could be neglected, as predicted by the mosaic model, Roux set out to study the "formative forces" within the egg. The first question was whether all the parts of the egg must collaborate to cause normal development or if the separate parts could develop independently. To answer this question, Roux conducted the famous "pricking experiment," in which he destroyed one of the cells of a frog embryo at the two-cell stage. When he pricked one cell with a hot needle, the undamaged cell developed into a half-embryo.

These experiments seemed to confirm Roux's belief that each cell normally develops independently of its neighbors and that total development is the sum of the separate differentiation of each part. While Roux thought that experimental embryology proved the validity of the mosaic model of development, when other scientists modified his procedures they rapidly destroyed the experimental base of his theory. Nevertheless, long after embryologists adopted other theories of development, Roux was still cited as the scientist who had formulated the core questions of modern embryology.

Hans Adolf Eduard Driesch (1867–1941), one of the first to follow Roux's protocol for experimental embryology, provided definitive evidence against Roux's model of mosaic development. According to Driesch, the embryo seemed to develop epigenetically as a harmonious equipotential system. Driesch and Roux differed greatly in philosophical viewpoints and technical approach, but both had studied with Ernst Haeckel. Driesch's interests were very broad, encompassing mathematics, physics, and philosophy. Even as a doctoral candidate, Driesch had questioned the wisdom of his mentors, and his work presented a direct challenge to Haeckel as well as Roux. Relations between Driesch and Haeckel deteriorated to the point where Haeckel advised his former student to take some time off and spend it in a mental hospital. Although Driesch did not take Haeckel's advice, he did eventually abandon experimental embryology and devoted himself to philosophical inquiries, including parapsychology and occultism. He traveled widely and supported himself by giving lectures on philosophy and science.

Several important differences in experimental conditions led to the rebuttal of Roux's dogmatic extrapolation of the results obtained with his injured, but unseparated frog cells. First, Driesch used sea urchin eggs instead of frog embryos. Second, having discovered that sea urchin embryos at the two-cell stage could be separated into individual cells merely by shaking, he was able to separate the embryonic cells rather than kill one of them. The separated cells formed advanced embryos that were normal in configuration, but smaller than their natural counterparts. This proved that both of the first two cells of the embryo contained all the components needed for full development. Roux's concept of mosaic development was, therefore, false. With improved techniques, Driesch was able to extend his experiments to the four-, eight-, and even later cell stages.

The implications of Driesch's experiments were momentous, but Driesch was not particularly happy about disproving a nice hypothesis. On seeing his half-embryos developing into typical whole gastrulas, Driesch lamented that the experiment had turned out as it must, not as he had expected. Still, his experiment seemed a step backward from Roux's initial success in explaining development. Driesch concluded that since a new organism could be generated from parts of the embryo, it could not be regarded as a machine. The parts of a machine, being necessarily simpler than the original machine, cannot reproduce the whole. The process of development is harmonious because the parts normally work together to form one individual even though each could form an independent individual. In contrast to Roux, Driesch emphasized the epigenetic nature of early development and compared the *presumptive significance* of an embryonic part (what it would form under normal circumstances) and its *prospective potency* (what it might form under altered conditions). The results of experimental manipulations demonstrated that the prospective potency of an embryonic cell was much greater than its presumptive significance. Therefore, using the terminology of analytical geometry, the fate of a given cell was a function of its relative position in the whole.

Instead of finding that more experimental work rewarded him with a clearer picture of embryological development, Driesch had apparently reached a more profound level of confusion, which seemed to end all hope of finding a mechanistic explanation for development. There was no way Driesch could picture a machine that could develop into two whole machines, identical to the original, when divided into its component parts. Embryonic development appeared to involve a harmonious-equipotential system. That is, each part had equal capacity or potentiality to substitute for any other part, and the formation of a harmonious whole was the end result of the interplay of parts. This seemed to imply that at some fundamental level all the parts of the early embryo were identical to each other. Such a system, Driesch logically argued, would be devoid of a structural or material cause for change or differentiation. Some nonmaterial causal agent, therefore, had to be involved in setting differentiation into motion. At this point, Driesch turned to Aristotle for inspiration and revived that venerable old mechanic, the *entelechy*, an internal perfecting principle (that which carries the end in itself). To add to the confusion, other embryologists observed very different patterns of development when they manipulated the embryos of different species. Some species produced a "Roux-like" half larva, others produced "Driesch-like" complete entities, and some displayed a mixture of the two kinds of development.

Roux and Driesch were frustrated by the results of their own experiments, but their work established the foundations of a rigorous analytical approach to embryonic development. Their contemporaries Hans Spemann (1869–1941) in Germany and Ross G. Harrison (1870–1959) in the United States were especially important for their role in refining the basic techniques of experimental embryology: isolation, transplantation, and tissue culture. The work of Spemann and Harrison transformed embryological theories and made it possible for embryologists to establish links with biochemistry and molecular genetics.

In the 1890s, Harrison began using grafting as a way to study growth and regeneration in amphibians. He later described experiments in which he had grafted the head of the frog larva of one species onto the body of a larva of a different species. By using species of different colors, Harrison was able to study the development of the transplanted tissues, including limbs, neural crest, optic nerve, eye, and inner ear. Between 1905 and 1907, Harrison adapted methods used in embryology to study the growth of isolated tissues, especially nerve fibers, under controlled conditions. Previous attempts to grow tissues and cells outside the living body had met with very little success. Tissue culture and cell culture became a valuable research technique in genetics, virology, and oncology, as well as embryology. In addition to his experimental work, Harrison was involved in the construction of an approach to embryology known as modern synthetic organicism, a more holistic way to understand the structure and growth of organisms, and an alternative to vitalism and mechanism.

While still a medical student at the University of Heidelberg, Hans Spemann, who had fallen under the spell of Goethe and Haeckel, decided he would prefer research in comparative anatomy and embryology to clinical medicine. At the Zoological Institute of the University of Würzburg, Spemann worked with the cytologist Theodor Boveri (1862–1915) and graduated in 1895. While undergoing a rest cure for tuberculosis, Spemann read August Weismann's (1834–1914) book *The Germ Plasm: A Theory of Heredity* (1892) and became intrigued by the puzzle of heredity and the relationship between the reproductive cells and embryological development.

On returning to the laboratory, Spemann began a series of experiments based on those of Roux. Instead of using frogs, Spemann selected the salamander as his experimental system and developed methods of separating and rearranging the cells of the early embryo. These studies led to a series of papers entitled "Developmental Physiological Studies on the Triton Egg" (1901–1903), which introduced the technique of manipulating and constricting the egg with a loop of fine baby's hair. Spemann found that if he constricted salamander eggs but did not completely separate the developing cells, he could produce animals with two heads and one trunk and tail. In his autobiography he recalled his fascination with watching the behavior of such "twin embryos."

For Spemann, embryonic development was the study of the "physiology of development." His primary experimental goal was discovering the precise moment when a particular embryonic structure became irrevocably determined in its path towards differentiation. During the 1890s, Spemann generated tadpoles with abnormally large, medially located eyes. These "Cyclopean eyes" contained a single, central lens. Following up these observations, Spemann spent 12 years investigating the mechanism of lens induction, thereby refining his techniques for removing tissue from a donor embryo and introducing it into a host embryo. Performing these operations without standardized, sterile culture media was extremely difficult, but Spemann was able to demonstrate the process now known as embryonic induction. Attempts by other scientists to extend Spemann's work on the induction of the lens demonstrated the importance of using a maker system that distinguishes between the cells of the donor and the host in order to detect inadvertent contamination by unwanted donor cells. Finding sophisticated methods for marking and tracing the fate of individual cells remains an important factor in developmental biology.

Refining the art of microdissection, Spemann and his associates carried out a series of experiments in which selected parts of one embryo were transferred to a specific region of another. Embryos of different species were used so that color differences would indicate the identity of the transplanted area during development. In 1921, the task of analyzing the effect of transplanting a region known as the dorsal lip was assigned to Hilde Pröscholdt (1898–1924). During that same year, Pröscholdt, who had joined the Spemann laboratory in 1920, married Otto Mangold, one of Spemann's first students. Hilde Mangold appears as the second author of Spemann's landmark paper "Induction of Embryonic Primordia by Implantation of Organizers from a Different Species" (1924). While Spemann's other students had been allowed to publish their thesis work as sole authors, Spemann insisted on adding his name to Hilde Mangold's thesis publication. Moreover, he insisted on having his name precede hers. After the discovery of the organizer region, Hilde Mangold abruptly disappears from accounts of the history of embryology as if she gave up scientific work for marriage and motherhood. Actually, Mangold died in 1924, at the age of 26, from severe burns caused by the explosion of a gasoline heater in her home, at about the time the organizer paper was published.

Mangold's experiments involved grafting the region known as the dorsal lip from an unpigmented species onto the flank of a host embryo, which belonged to a pigmented species. Three days after the operation an almost complete secondary embryo formed on the host embryo; the secondary embryo was composed of a mosaic of host and donor cells, which indicated that both the transplanted tissue and the host embryo participated in its formation. Because the tiny bit of tissue from the donor embryo could cause the formation of a new embryo, it was called the *organizer region*. During this time period, researchers were using pond water as their medium, and keeping experimental embryos alive was very difficult.

Only a few of the hundreds of embryos that Mangold operated on managed to survive. The longest-lived embryo in her experiments survived to the tailbud stage. Mangold described five experimental cases in detail and briefly noted several others. Despite the small number of successful experiments, many developmental biologists considered this demonstration of the organizer effect one of the most important experiments in modern embryology.

In many ways the discovery of the organizer represents the culmination of the approach to embryology initiated by Roux and Driesch. In 1935 Spemann was awarded the Nobel Prize in Medicine or Physiology. One year later he published his final account of his ideas and experiments in *Embryonic Development and Induction*. Given the state of German society in the 1930s and Spemann's own authoritarian proclivities, it is not surprising that historians have pointed out connections between the concept of the organizer as the mastermind of development and the desire for a dictator to bring order out of the chaos afflicting the social organism.

In 1928, Johannes Holtfreter (1901–1992) joined Otto Mangold's laboratory, where he attempted to determine how the organizer influenced the developmental path taken by the cells of the host. Using a special salt solution and sterile conditions, Holtfreter successfully recreated Hilde Mangold's experiments. The balanced salt solution used in these experiments is now known as Holtfreter's medium. The secondary embryos in Holtfreter's experiments were so complete that the products of the experiment looked like twins fused at the flank or belly. Holtfreter also carried out experiments to determine whether the organizer was still effective after it was "devitalized" by heat, alcohol, drying, or freezing. In contrast to Spemann's belief that the organizer must be alive, Holtfreter found that dead organizer material still induced development. Spemann had previously discovered that the organizer was active even if the cells had been crushed, but his philosophical approach remained largely that of an organicist seeking holistic explanations, rather than biochemical theories. Eventually, Holtfreter became convinced that the term "organizer" was misleading. Testing various kinds of embryonic and adult tissue, Holtfreter discovered that many different tissues caused neural induction. His work led to important insights into cell competence, inductive signals, and the ways in which signals from the organizer provoked developmental phenomena in sensitized cells. Other researchers confirmed Holtfreter's discovery and found that many other bits of animal tissues, alive or dead, could induce complex structures, including heads, internal organs, and tails on host embryos.

Despite Spemann's antipathy to chemical explanations, the idea that dead embryonic tissues could retain their capacity to induce organized structures had a profound impact on experimental embryology. Other embryologists drew the logical conclusion that chemical signals control the pattern of development and differentiation. In attempting to identify those chemical signals, researchers linked biochemistry to developmental biology. This new era in embryological research was somewhat disconcerting to the old pioneers, as evident in Spemann's protest that "a dead 'organizer' is a contradiction in itself."

Spemann's approach to embryology was built on an organismic, holistic view of embryos and their development. Such attempts to understand developmental processes in terms of the potentialities and interactions of supracellular embryonic parts were largely displaced by the growing force of reductionist trends in post World War II biology. As the focus of biological theory and technique shifted from the organism as a whole to the cellular, subcellular, and molecular levels, classical experimental embryology was superceded by developmental biology.

Experimental embryology left a legacy of unfinished business, which offered intriguing problems for researchers willing to address old problems with the theories and techniques of molecular biology and new model systems such as fruit flies, nematodes, and slime molds. Thus, by the end of the twentieth century, fundamental questions about development and differentiation were of special interest to geneticists and molecular biologists. Developmental biology became an interdisciplinary field, representing a synthesis of embryology, morphology, cell biology, biochemistry, genetics, molecular biology, and genomics. In 1995 the Nobel Prize for Physiology or Medicine was awarded to three developmental biologists: Edward B. Lewis (1918-), Eric F. Wieschaus (1947-), and Christiane Nüsslein-Volhard (1942-). It was the first Nobel Prize for research in developmental biology since Hans Spemann was awarded the prize in 1935. Their work led to the discovery of a family of genes that control the development of individual body segments in *Drosophila*, but similar genes have been found in many different species, including humans. Some scientists believe that these genes will serve as a "Rosetta stone" for the study of development, phylogeny, and evolution. Insights into developmental biology have many practical implications, such as prenatal screening, cloning, regeneration and organogenesis, stem cell therapy, and the identification of new targets for therapeutic drugs. Following the history of embryology further, therefore, requires a thorough immersion in biochemistry, genetics, molecular biology, and genomics.

SUGGESTED READINGS

Adelmann, H. B. (1966). *Marcello Malpighi and the Evolution of Embryology*. 5 vols. Ithaca, NY: Cornell University Press.

Amrine, F., Zucker, F. J., and Wheeler, H. (1987). *Goethe and the Sciences: A Reappraisal*. Dordrecht: Reidel.

Anderson, L. (1982). *Charles Bonnet and the Order of the Known*. Boston: Reidel.

Baer, E. von (1986). *Autobiography of Dr. Karl Ernst von Baer*. Edited and with a Preface by Jane M. Oppenheimer. Translated from the 1886 German edition. Canton, MA: Science History Publications.

Blyakher, L. Y. (Blacher, L. I.) (1982). *History of Embryology in Russia from the Middle of the Eighteenth to the Middle of the Nineteenth Century*. With an Introduction by Jane Maienschein. Translated from the Russian edition of 1955 by Hosni Ibrahim Youssef and Boulos Abdel Malek. Edited by G. A. Schmidt. Washington, DC: The Smithsonian Institution.

Bonner, J. (2001). *First Signals: The Evolution of Multicellular Development*. Princeton, NJ: Princeton University Press.

Cunningham, A., and Jardine, N., eds. (1990). *Romanticism and the Sciences*. New York: Cambridge University Press.

Dawson, V. P. (1987). *Nature's Enigma: The Problem of the Polyp and the Letters of Bonnet, Trembley, and Réaumur*. Philadelphia, PA: American Philosophical Society.

Gaskings, E. (1967). *Investigations into Generation 1651–1828*. Baltimore, MD: Johns Hopkins University Press.

Gehring, W. J. (1998). *Master Control Genes in Development and Evolution: The Homeobox Story*. New Haven, CT: Yale University Press.

Gilbert, S. F., ed. (1991). *A Conceptual History of Modern Embryology*. New York: Plenum Press.

Hamburger, V. (1988). *The Heritage of Experimental Embryology: Hans Spemann and the Organizer*. Oxford: Oxford University Press.

Harvey, W. (1847). *The Works of William Harvey, M.D.* Translated with a life of the author by Robert Willis. London: The Sydenham Society. New York: Johnson Reprint, 1965.

Horder, T. J., Witkowski, J. A., and Wylie, C. C., eds. (1986). *A History of Embryology*. New York: Cambridge University Press.

Nakamura, O., and Toivonen, S., eds. (1978). *Organizer: A Milestone of a Half Century from Spemann*. New York: Elsevier.

Pinto-Correia, C. (1997). *The Ovary of Eve. Egg and Sperm and Preformation*. Chicago, IL: University of Chicago Press.

Rather, L. J., Rather, P., and Freriches, J. B. (1986). *Johannes Muller and the Nineteenth Century Origins of Cell Theory*. Canton, MA: Science History Publications.

Roe, S. A. (1981). *Matter, Life, and Generation. Eighteenth Century Embryology and the Haller-Wolff Debate*. New York: Cambridge University Press.

Schleiden, M. J. (1849). *Principles of Scientific Botany; or, Botany as an Inductive Science*. Translated by E. Lankester with a new introduction by J. Lorch. Reprinted 1968. No. 40: The Sources of Science. New York: Johnson Reprint, 1968.

Schwann, T. (1847). *Microscopical Researches into the Accordance in the Structure and Growth of Animals and Plants*. Translated by Henry Smith. New York: Kraus Reprint, 1969.

Spemann, H. (1938). *Embryonic Development and Induction*. New Haven, CT: Yale University Press.

Virchow, R. (1860). *Cellular Pathology*. New York: Dover (reprint 1971).

Willier, B. H., and Oppenheimer, J. M., eds. (1974). *Foundations of Experimental Embryology*. 2nd ed. New York: Hafner.

Wilson, E. B. (1925). *The Cell in Development and Heredity*. New York: MacMillan.

6

PHYSIOLOGY

In ancient philosophical systems the study of anatomy and physiology were virtually inseparable. As used by the ancient Greek philosophers, the term physiology originally encompassed inquiries into the nature of living and nonliving things. Eventually, physiological studies were those most closely associated with studies of the vital activities of normal, healthy human beings. Although the activities that distinguish living from nonliving things might be considered more exciting than their morphology, anatomical inquiries could be pursued with only the naked eye and a few simple tools. An understanding of physiological phenomena, in contrast, depends on subtle inferences drawn from chemistry and physics. Nevertheless, many fundamental physiological concepts were formulated by the authors of the Hippocratic texts, the ancient Greek philosophers, and Renaissance scientists. Indeed, for hundreds of years, humoral theory, based on the writings of Hippocrates, Aristotle, and Galen, served as the basic explanation for human health and disease.

Implicitly or explicitly, a philosophical framework that has been either mechanistic or vitalistic has guided physiological studies from the most ancient times to the present. A mechanical philosophy asserts that all life phenomena can be completely explained in terms of the physical-chemical laws that govern the inanimate world. Vitalists claim that the real entity of life is the soul or vital force and that the body exists for and through the soul, an entity that is incomprehensible in strictly scientific terms. Teleology (arguments from design) provided another fundamental guiding principle for understanding the vital functions of living beings. According to the teleological principles established by

Aristotle and Galen, every part of the body is formed for a purpose. If Nature acts with perfect wisdom and does nothing in vain, every part of the body must have been designed for its proper function. Function can, therefore, be deduced from the study of structure.

It was not until the seventeenth century that new ways of dealing with the dynamic functions of the body were realized. By the end of that remarkable century, a new concept of the cosmos was well established. Through the work of Galileo, Kepler, and Newton, the universe emerged as a great law-bound machine with the earth merely one of the planets circling the sun. Gravity, not crystalline spheres and angels, kept the heavenly bodies moving in their orbits. Although the scientific revolution may not have transformed the life sciences as profoundly as it did physics, it was not without effect. Two different ways of exploring vital phenomena emerged with new clarity in this period. These might be called the *metric experimental* approach, best exemplified by the work of William Harvey (1578–1657), and the *rational-philosophical* approach of the philosopher and mathematician René Descartes (1596–1650).

The method exploited so well by Harvey could be seen as the culmination of the work of Aristotle, Galen, and Vesalius. Harvey said that through vivisec-

The private laboratory of a professor of physiology at Atlanta Medical College in 1913

tion he was led to truth. By investigating the structures and activities of the living body, Harvey found ways to explain vital phenomena in a new way without any special knowledge drawn from chemistry or physics. Indeed, Harvey had essentially no tools or instruments that had not been available to his famous predecessors. On the other hand, the career of Santorio Santorio (1561–1636), who sat so patiently on his balance and measured what could be measured, proves that measurement and patient experimentation are not enough to answer fundamental physiological questions. Quantitation alone gives little insight into vital phenomena without an appropriate theoretical framework for guidance.

In the seventeenth century physiologists dealt most successfully with problems of a mechanical nature and problems that could at least be conceptualized in mechanical terms. Recognizing the power of mechanical explanation in Harvey's work, other physiologists tried to force vital phenomena to fit mechanical analogies. The jaws were explained as pincers, the stomach as a mill, the veins and arteries as hydraulic tubes, the heart as a pump or spring, the muscles and bones as a system of cords and pulleys, the lungs as bellows, and the viscera and kidneys as sieves and filters. Digestion and metabolism, muscle and nerve action, however, are problems that require an understanding of chemical and electrical phenomena. In contrast to the physiologists known as *iatromechanists*, who believed that all functions of the living body could be explained in terms of physical and mathematical principles, the *iatrochemists* and Paracelsians attempted to explain vital phenomena as chemical events.

The writings of René Descartes provided a philosophical framework for a mechanistic approach to physiology that influenced many generations of scientists. While Descartes made no original physiological discoveries and dissected only to support his preconceived ideas, he played a critical role in assimilating and integrating the work of Galileo, Kepler, and Harvey into a new system of philosophy and provided his followers with an apparently complete and satisfying mechanistic system. His writings on physiology, which include *On Man* and *On the Formation of the Fetus*, were part of a general system of philosophy. The fundamental platform of the mechanical philosophy was set forth in a bold and comprehensive manner in his *Principles of Philosophy* (1644). According to Cartesian philosophy, all natural phenomena could be explained in terms of matter and motion.

Tragedy and illness marked Descartes's childhood, although his family enjoyed both wealth and prestige. His mother died when he was born, leaving a sickly baby whose pulmonary weakness seemed to doom him to an early grave. During years spent at a Jesuit school, Descartes was drawn to mathematics as the only area of truth and stability to be found in a world where Aristotelian philosophy was under attack and the Copernican theory was being widely adopted by enlightened thinkers. Doubt may have been the major product of his Jesuit education, but the profound impact of this religious indoctrination was at the core of a system that has been described as the metaphysics of a Roman Catholic mathematician. His universe seemed to be a great mechanical system in which God was the first cause of all motion. Critics, however, claimed that Descartes protected himself with a veneer of piety while he created a thoroughly materialistic and mechanistic science. Although convinced that he could erect a philosophical framework that would unite all the physical and natural sciences, when Descartes heard of the persecution of Galileo, he suppressed his own *Treatise on the World* to avoid the appearance of heresy. Descartes destroyed some of his manuscripts and turned his attention to safer subjects. Parts of the *Treatise on the World* were finally published after his death.

Even a superficial introduction to philosophy must include Descartes's system and his famous phrase, *cogito, ergo sum* (I think, therefore I am). This was the one idea that Descartes was sure of even as he went about systematically doubting everything else; that is, mind must exist. Although Descartes acknowledged the importance of observations, his approach was a form of rationalism that subordinated the crude facts of observation and experiment to the test of reason and a priori principles. Because it was inconceivable that God would deceive us, clear and distinct ideas about matter must also be true. The world, therefore, must be composed of two distinct entities: mind and matter. Ideas that were true and self-evident became the premises by which other truths were established. Matter had the essential qualities of extension in space, divisibility, and motion. Rejecting the atomic theory of Democritus and Lucretius, Descartes ruled out the existence of their great void. "Give me motion and extension," Descartes said, "and I will construct the world."

Cartesian doctrine treated animals as automata whose activities were explained in purely mechanical terms as the motions of material corpuscles and the heat generated by the heart. The Galenic doctrine of the three spirits was replaced by one where only the animal spirits remained, albeit transformed into a subtle fluid utilized in the brain and nerves. Even human beings emerged from Descartes's exposition as another kind of earthly machine. In his *Treatise of Man*, Descartes presented his analysis of the mind and body of human beings. To do this, he sifted through the anatomical knowledge of his time, carried out his own dissections, added certain self-evident ideas, and concocted a theory of the body as a machine working in accordance with physical laws. Descartes challenged scientists to treat the physical and mental aspects of human beings in the same manner as all other scientific problems. Yet humans differed from animal automata by virtue of the rational soul that served as the agent of thought, will, memory, imagination, and reason. Nevertheless, except for ideas, all physiological functions of the human body were as mechanical as the workings of a clock. The Cartesian mind-matter dualism was reflected in a strict mind-body dualism. Essentially inaccessible to science, the mind was defined in terms of its faculty of knowing. Serving as the agent of thought, governing volition, conscious perception, memory, imagination, and reason, the rational soul was the only entity exempted from a purely mechanical explanation.

While Harvey's work on the motion of the heart and the blood seemed to present the perfect example of a mechanical system, Descartes imposed his own interpretation on fundamental aspects of the workings and purposes of the circulatory system. According to Descartes, the heart was a heat machine rather than a pump. Thus, in contrast to Harvey, Descartes endorsed the ancient idea that the action stroke of the heart was the expansion phase, rather than the contraction. As soon as blood entered the fiery chambers of the heart it was warmed, dilated, and vaporized. The purpose of the fire in the heart was, therefore, to heat and vaporize the blood so that it continually fell drop by drop through a passage from the vena cava into the right chamber, from which it was driven into the lungs. In contrast to the fiery heart, the lung was described as a delicate, soft organ in which the hot vapors of the blood were cooled by fresh air so that they condensed and fell drop by drop into the left cavity of the heart. It might have been possible to obtain estimates of the temperature of the heart and other

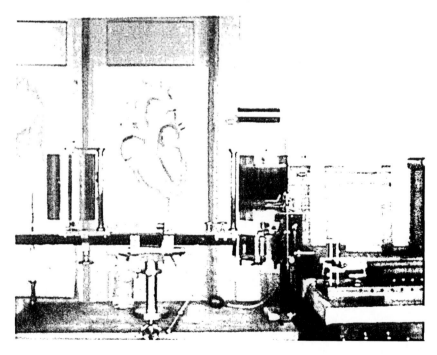

Demonstrating the action of the heart in a physiology laboratory

internal organs with the thermometers available in the seventeenth century, but Cartesian physiology did not call for testing self-evident ideas by crude experimental verification.

Similar mechanical explanations were proposed for other physiological phenomena, but the nervous system presented a formidable challenge. According to Descartes, all other systems were subservient to the nervous system, which carried out the commands of the rational soul. The strongest and most subtle parts of the blood were carried to the cavities of the brain by arteries taking the most direct route from the heart. These special parts of the blood nourished and sustained the substance of the brain and produced the animal spirits, which Descartes compared to a subtle wind or very pure flame. Through conduits in the brain the animals spirits were able to enter the nerves and affect the muscles and cause movement. The animal spirits of the Galenic system were transformed into a special subtle fluid that could be described in purely mechanical terms. The nerves were hollow tubes in which purely hypothetical Cartesian valves governed the direction of flow of the nervous fluid, just as the venous valves governed the flow of the blood.

In the Cartesian system, the pineal gland served as the seat of the soul, the reservoir for the animal spirits, and the only site in which a direct interaction between the immaterial soul and the corporeal machine took place. Descartes selected the pineal gland for this crucial role primarily because it is a single rather than a paired organ located at the base of the brain. Moreover, Descartes erroneously believed that the pineal gland was a very soft and mobile organ present in humans but absent in lower animals. In addition to purely hypothetical valves, Cartesian nerves were endowed with delicate threads along the length of their cavities that connected the brain to the sense organs. The tiniest motion along the thread exerted a pull upon the site of the brain where the thread originated, opening up the orifices of certain pores on the internal surface of the brain. This allowed the animal spirits to flow into the muscles and caused the machinery to move. Movements of the body were, therefore, the result of a reflex arc triggered by external stimuli that led to internal responses. The concept is reminiscent of Plato's description of men as the puppets of the gods. Further mechanical analogies equated the nerves, muscles, and tendons of the body and the engines and springs that moved the devices in the elaborate fountains and grottoes of the French royal gardens. These pleasure gardens were rather like an early version of Disneyland, where mechanical gods, goddesses, and monsters moved when hidden triggers were activated. Indeed, in the Cartesian system all physiological functions might be as mechanistic as the workings of clocks, mills, and other machines.

Although Descartes was not always consistent as to whether ideas were differentiations impressed upon matter or aspects of the soul alone, he attempted to explain thoughts and emotions in mechanical terms. Ideas were impressions received by the animal spirits as they left the pineal gland. Memories were apparently physically recorded, rather like permanent press, into the fabric of the nerves. Since human beings were a combination of soul and body, emotion was experienced both in the soul and in the machinery of the body. To establish a mechanical basis for passions and emotions, Descartes suggested that stimulation of the sense organs caused changes in the flow of the subtle fluid in the nerves. These currents in the nervous fluid caused distinctive vibrations in the pineal gland, which were interpreted as emotions and passions. For example, the joy experienced in the soul had its counterpart in the body because in this state the blood would become finer, expanding more readily in the heart, and thus facilitating the excitement of the nerves. The mind could not change physical reactions to external stimuli directly, but it could modulate the pineal vibrations and thus the outcome of a specific stimulus. That is, an external stimulus could cause fear, but the mind could determine whether the bodily reaction would be flight or fight.

Understanding the activities and functions of the nervous system remains one of the great challenges in physiology, but many scientists have adopted Descartes's belief that even mental phenomena could be explained in terms of physiological principles. By the 1920s, Sir Charles Sherrington (1857–1952) and others had elucidated the electrical aspects of the nerve impulse, as well as the role of the synapse and the transmission of nerve impulses between muscles and the central nervous system. Epitomizing the results of this phase of research, Sherrington proposed the concept known as the "integrative action of the central nervous system." Further researchers found evidence that chemical transmitters, such as acetylcholine and noradrenaline (nonepinephrine), were involved in the action of the nervous system. By the end of the twentieth century, physiologists had discovered the close association between hormonal regulation and the regulatory effects exerted by the nervous system.

Descartes's work was widely read, imitated, and honored. The Cartesian system challenged scientists to treat the physical and mental aspects of human beings in the same manner as all other scientific problems. Disciples of Descartes saw him as the first philosopher to dare to explain all the functions of human beings, even the brain, in a purely mechanical manner. Like many other systems of thought, Descartes's physiology began as heresy and ended as dogma. The mechanistic approach was adopted by philosophers and experimentalists, especially those like Giovanni Alfonso Borelli (1608–1697), who had a special interest in the problem of muscle action.

Like Descartes, Borelli favored the mechanical mode of explanation for physiological processes, but if Descartes may be regarded as the founder of iatromechanism as a philosophy, Borelli was the founder of iatrophysics as an experimental science. Little is known about Borelli's early life other than his precocious mastery of mathematics and his enthusiastic commitment to the legacy of Galileo. In 1656 Borelli was awarded a professorship at the University of Pisa. As one of the founders of the Academia del Cimento, Borelli helped make Pisa a respected center of research in mathematics, medical science, and the new experimental approach to natural science. After 12 years at Pisa and many quarrels with his colleagues, Borelli left the university to seek quieter and healthier surroundings. Returning to the University of Messina, Borelli continued his work on the problem of animal motion. In 1674 he was accused of participating in a political conspiracy to free Sicily from Spain. Forced into exile, he fled to Rome. Overwhelmed by financial problems and illness, Borelli died before his great work *De motu animalium* (*On Motion in Animals*) was published. An introduction was added by an ecclesiastic dignitary, who commended Borelli for upholding the authority of the Church in his lectures on astronomy. In physiological matters, too, Borelli remained a faithful son of the Church by acknowledging that all the mechanical phenomena described in his book were ultimately governed by the soul.

De motu animalium was an attempt to apply mathematical and mechanical principles to the study of muscle action. The text was a sustained analysis of the mechanics of muscle contraction, which dealt with the movements of individual muscles and groups of muscles treated geometrically in terms of mechanical principles. Animal movements were divided into external movements, such as those carried out by the skeletal muscles, and internal movements, such as those of the heart and viscera. Borelli's method of study involved a progression from the simplest element of the motor system, the independent muscles, up to the more complicated organs and organ systems, and finally the power of movement of the organism considered as a whole. According to Borelli and his contemporary Nicolaus Steno (1638–1686), a pioneer in the microscopic study of muscle structure, the fleshy muscular fibers played a fundamental role in muscle contraction; the fibers of the tendons, in contrast, were merely passive agents that did not take part in contraction.

The action of the heart particularly intrigued Borelli. Unlike Descartes, he recognized that the heart was a muscular pump rather than a heat engine and confirmed this by simple experiments. Measuring the temperature of the heart and other internal organs in a vivisection experiment on a deer, he proved that the temperature of the heart was not significantly different from that of other parts of the body. During the eighteenth century, several scientists took up the question of the regulation of body temperature. Of course the ancients had noted that body temperature was apparently constant in healthy people but varied in pathological states. It was also apparent that no matter what the ambient temperature, members of certain species are warm to the touch when alive and become cold when dead. Other species were relatively cold or varied in temperature and activity depending on ambient temperature. With the clinical thermometer, experimenters could demonstrate that human body temperature remained quite constant, except during attacks of fever. Charles Blagden (1748–1820) and John Hunter (1728–1793) performed an ingenious series of experiments that proved that body temperature was constant at a broad range of ambient temperatures. Small rooms were maintained at temperatures of 110–120, 180–190, and even 260°F, while outdoor temperature was about 32°F. Blagden and Hunter worked in the hot rooms, noted their own reactions, and measured body temperature. Nonliving materials and foods, such as eggs, meats, wines, and water, served as controls. From these experiments, Blagden concluded that temperature regulation was a fundamental characteristic of life.

While measuring the weight that different muscles could support, Borelli found that the muscles on the two sides of the jaw could support a weight of more than 300 pounds when acting together. He assumed that the force of contraction of all healthy muscles would be the same for a given unit of bulk. Thus, he extrapolated from his measurement of some exterior muscle to the forces that presumably could be generated by the heart and other internal muscles. The strong muscles of the stomach and their ability to crush foods and other objects particularly impressed him. To test the grinding capacity of the stomach, he introduced hollow glass spheres, hollow lead cubes, wooden pyramids, and other objects into the stomachs of turkeys. The next day the objects were crushed, eroded, and pulverized. Borelli admitted that chemical reactions also played a role in digestion, but some iatromechanists refused to admit that digestive juices played any significant physiological role.

In explaining the action of the heart, Borelli compared the ventricles to a winepress or piston and noted that during systole (contraction), the walls of the ventricles obliterated the cavities from which the blood had been driven. He concluded that when heart muscles contracted, they increased in bulk and that this was a general phenomenon in muscle action. The hardening and tension apparent during muscular contraction must be caused by inflation of the muscular substance by something flowing into the muscle, or a sudden fermentation in the muscle itself, triggered by animal spirits traveling from the brain through the nerve and into the muscle. Trying to formulate a strictly mechanical explanation, Borelli suggested that muscular fibers are chains of rhombuses and that contraction was due to inflation caused by the sudden insertion of a number of wedges.

Basically, Borelli believed that muscles increased in bulk during contraction and ascribed this apparent inflation to a sudden fermentation triggered by animal spirits traveling from the brain through the nerve and into the muscle. In his attempts to analyze the interaction between nerves and muscles, Borelli tested theories involving the movement of an incorporeal influence, vapor, or air into the muscles. A simple experiment in which the muscles of a living animal were divided length-wise while the animal was held under water appeared to refute this hypothesis. If some spirituous gas had entered the muscles, it should burst out of the wound and bubble up through the water. Inflation of the muscles, therefore, must result from something that occurred in the muscles themselves rather than the movement of an air or vapor transmitted by the nerves. In response to some influence from the brain, a form of fermentation must have occurred within the muscle itself to cause its sudden inflation. This seemed an admirable explanation, but Nicolaus Steno, Jan Swammerdam, Francis Glisson, and other seventeenth-century microscopists and physiologists were able to demonstrate that muscles do not increase in volume when they contract.

Defending the value of science in medicine, Steno challenged critics to think about the basis of their glib explanation of diseases that affect "animal movements," such as paralysis, convulsions, apoplexy, and syncope. All too often, erroneous theories were used to justify therapeutic interventions that were more likely to do away with the patient, rather than cure the disease. Steno believed that Galileo's methods and philosophy could be applied to biological problems, such as muscle contraction. His experiments demonstrated that the apparent swelling of a working muscle was due to shortening of the fibers and did not involve an increase in volume. Although Steno's approach was primarily mechanical and geometrical, his ideas about the microscopic structure of various tissues were influenced by his collaboration with Jan Swammerdam. Working together, Steno and Swammerdam demonstrated that a frog continued to move for some time after its heart was removed. Another experiment proved that an isolated heart continued to beat, even though blood and spirits could no longer enter the organ. Therefore, according to Steno, all previous writings on the relationship between spirits or subtle fluids and muscular action were merely attempts to disguise a profound ignorance. Under the microscope, muscles appeared to be collections of motor fibers. Each motor fiber was in turn a complex of very minute fibrils arranged lengthways, with a middle part that differed from the ends in its consistency, thickness, and color. Only the fleshy part of the motor fiber was involved in muscle contraction; the tendons did not change. Such observations clarified the differences between tendons and nerves and refuted the ancient assumption that tendons caused motion and that muscles were merely passive, fleshy material.

Further proof that muscle contraction did not involve an increase in muscle volume was established by Francis Glisson (1597–1677), a member of the remarkable group of natural philosophers who met informally at Gresham College and later founded the Royal Society. An accomplished anatomist and pathologist, Glisson investigated a broad range of medical and physiological problem, such as rickets, the fine structure of the liver, and the physiological property referred to as irritability (contractility). To refute Thomas Willis's (1621–1675) argument that muscle contraction was the result of the entry of some gaseous spirit into the passive fibers of the muscle, Glisson carried out a classical experiment reminiscent of Archimedes (ca. 287–212 B.C.) in the bath,

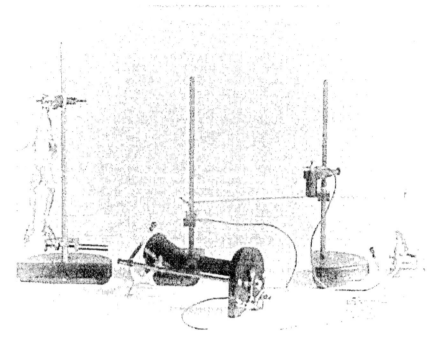

Apparatus for measuring the contraction of the muscles of a frog

contemplating the amount of gold in Hiero's crown. The arm of a muscular man was totally submerged in a glass container so that the water level could be measured during muscular contractions. This experiment proved that muscles do not increase in volume during contraction. While studying the anatomy of the liver, Glisson tried to explain why the bile was not discharged into the intestines continuously, but only when needed. He discovered that the gall bladder and biliary duct discharged more bile when they were irritated. This occured, he argued, because they have the capacity to be irritated. That is, Glisson thought of irritability as a general vital property. The doctrine of irritability became very influential in physiology, largely through the work of Albrecht von Haller (1708–1777).

PHYSIOLOGY IN THE EIGHTEENTH CENTURY

Having assimilated Newtonian physics and Cartesian philosophy, many eighteenth-century naturalists, especially the French *philosophes*, attempted to explain physiological functions in terms of mechanistic materialism. As adapted by Enlightenment philosophers, however, the mechanical philosophy was not necessarily atheistic. Pious naturalists could investigate matter and motion, anatomy and physiology while assuming that God was not directly involved in the ordinary motions of the universe or the normal activities of living beings.

In the eighteenth century, as in the seventeenth, physiology remained a mixture of speculation and experimentation, but the revolution in chemistry provided new ways of understanding digestion, respiration, and the role of the circulation of the blood. Teachers at Europe's leading medical schools were actively involved in integrating chemical concepts into physiology and medicine and disseminating these new systems of thought as a guide to medical practice. One of the best known educators of this period was the indefatigable Hermann Boerhaave (1668–1738), who taught chemistry, physics, botany, ophthalmology, and clinical medicine at the University of Leiden. Although Boerhaave is not associated with any particularly striking discovery or theory, he was so famous in his day that a letter addressed to "The Greatest Physician in the World" was supposedly delivered directly to Boerhaave. Almost a century after the publication of his *Elements of Chemistry*, Boerhaave was still lauded as the author of the "most learned and luminous treatise on chemistry" ever written. Boerhaave's approach to anatomy, physics, and chemistry was absorbed by legions of devoted students and disciples, including Albrecht von Haller.

Physiology, like embryology, owes a great debt to Albrecht von Haller, who carried out enormous numbers of experiments despite almost paralyzing melancholy, bad health, and profound revulsion for the pain caused by the vivisection of animals. Haller's landmark treatise, *Elements of Physiology* (1757–66), helped stimulate the development of modern physiology. Although Haller considered this textbook a concise summary of the field, the range of this work was so extensive that the great French physiologist François Magendie (1783–1855) complained that whenever he thought he had performed a new experiment, he always found it had already been attempted or described by Haller.

In his studies of the form and function of various organs and organ systems, Haller attempted to link anatomical knowledge to physiology by means of experiment. Indeed, he defined physiology as animated anatomy. Reviving Glisson's concept of irritability, Haller refined the concept and contrasted the *irritability* of muscle to the *sensibility* of nerves. Irritable tissues were those in which the special force, the *vis insita* (inherent force), resided. The task of the physiologist was to determine which parts of the body were irritable and which were not. The irritable parts were defined as those that contract when touched, while the sensible parts were those that conveyed a message to the mind when they were stimulated. While Glisson had used the term irritability to describe a broad range of phenomena, Haller tried to confine the concept to the special contractile force in muscles that caused them to contract in response to external stimuli. Haller concluded that irritability is a property of muscle fibers and sensibility is a property of nerves. Some physiologists called the *vis insita* the vital force, but Haller rejected this doctrine because he found that the inherent force in muscles survived for some time after death. Another force that was carried to the muscles from the brain by means of the nerves was called the *vis nervosa*. This force initiated muscular contractions, but it could not be called the vital force either because it also remained for a time in dead animals. Thus, the two special forces, the *vis insita* and the *vis nervosa*, were associated with muscles and nerves, respectively.

Because the nerves served as the instruments of sensation, only those parts of the body served by nerves experienced sensations. The nerves also elicited the power of contraction in the muscles that served as the instruments of movement. While using evidence from pathological lesions and vivisection experiments to determine whether specific parts of the brain had particular properties and functions, Haller concluded that the question was too complex to be answered satisfactorily at the time. He rejected the idea that the nerves act as solid bodies like elastic strings conveying vibrations and concluded that the material substrate of the nerves must be a subtle and unique fluid known to us only by its effects. The nerves and fibers of the brain, therefore, must be hollow in order to transmit the nervous fluid that was responsible for sensation, movement, and the preservation of life. While denying the possibility that the soul was diffused over the whole body, Haller argued that the soul had nothing in common with the body, except for sensation and movement. Because both sensation and movement seemed to have their source in the medulla, it must be the seat of the soul.

As early as the seventeenth century, physiologists had gained some insights into muscle action, but the physiology of the nervous system proved to be much more complicated. Following the path established by Haller, physiologists attempted to study the action of the nervous system by systematically stimulating or destroying specific parts. One of the most famous series of experiments on the relationship between nerves, muscles, and electricity was carried out by the Italian physician Luigi Galvani (1737–1798). Galvani's theory of "animal electricity" originated in his accidental discovery that when frog legs were suspended from copper wires on an iron railing, the legs twitched when they came in contact with the iron bar. Further studies, using a crude electrostatic machine and a Leyden jar, convinced Galvani that electrical impulses in the nerves caused muscular contractions. His results and conclusions were published as *De Viribus Electricitatis in Motu Musculari* (On Electrical Powers in the Movement of Muscles) in 1792. Quacks, charlatans, and assorted healers quickly added the remarkable concept of "animal magnetism" to their assortment of miracle cures. Following up Galvani's experiments, Alessandro Volta (1745–1827) invented the Voltaic pile, the first electrical battery, and proved that electrical stimulation could force a muscle to contract continuously.

The work of Julien de La Mettrie (1709–1751) offers a sharp contrast to that of the pious Haller. Indeed, the concept of the body as a machine reached its ultimate expression in his most infamous treatise, *Man, the Machine* (1748). La Mettrie was quite willing to let his materialistic scientific theories conflict with or contradict Christian dogma, despite the fact that he was himself a priest. The priesthood was a career chosen by his father, a rich merchant who belittled his son's desire to be a poet. Advised by a friend that it would be more rewarding to be a mediocre physician than a good priest, La Mettrie became one of Boerhaave's students. In addition to translating the works of Boerhaave into French, La Mettrie published a series of pamphlets satirizing the conservative Parisian medical community. His unorthodox ideas about the nature of vital phenomena were revealed in a controversial book, *A Natural History of the Soul*. This treatise on the soul traced the evolution of an active principle in matter through plants and animals to humans. According to La Mettrie, we cannot know what the soul truly is, but neither do we know what matter is. Since we never find a soul without a body, to study the properties of the soul we must study the body. To do this we must investigate the laws of matter. Using personal experience to support his thesis, La Mettrie explained that while suffering from a fever, he realized that the clearness of his thinking varied with the severity of his illness. This proved to him that thought is a function of the brain, dependent on physical conditions. After further annoying his conservative adversaries by publishing a book called *The Politics of Physicians*, La Mettrie sought refuge in Holland, where he amused himself by writing another attack on physicians.

Feeling relatively safe in Leiden, La Mettrie composed the unorthodox work *Man, the Machine* (1748), but he discreetly published it anonymously. Without Descartes's veneer of piety, La Mettrie discarded Descartes's "rational soul" and proposed a fully materialistic theory of human physiology. In contrast to Descartes, La Mettrie rejected the idea that humans were essentially different from animals. For La Mettrie, the human being was rather like a special variety of monkey, superior mainly by virtue of the power of language. Indeed, he argued that animals were not merely machines, although their ability to reason was less developed than that of humans. Experiments on animals indicated that peristalsis continued after death and that isolated muscles could be stimulated to contract. Presumably, if this occurred in animals, it must also be true for humans because both were essentially the same in terms of bodily composition. Dismissing Cartesian mind-body dualism, La Mettrie argued that all human actions, even those attributed to the mind, were entirely dependent on physical and chemical factors. Substances such as opium, coffee, and alcohol obviously affected both the body and the mind, as demonstrated by changes in heart beat, respiration, wakefulness, mood, imagination, thoughts, and volition. Many diseases obviously attacked both the mind and the body. Pleased by the controversy provoked by his book, La Mettrie played a rather sardonic joke on his critics by publishing a satire called *Man, More than a Machine*. Eventually La Mettrie found refuge at the court of Frederick II of Prussia, where he was able to write, lecture, and practice medicine. The physician who had called death the conclusion of a farce provided a demonstration of the truth of this witticism at a feast given in his honor by a grateful patient. Immediately after eating an enormous quantity of a truffle pastry, La Mettrie fell ill and died. The cause of death was unclear, but Voltaire pronounced it a great occasion since, for once, the patient had killed the doctor.

By the middle of the eighteenth century, optimism about the explanatory power of the mechanical philosophy began to decline. Vital phenomena such as reproduction, embryological development, digestion, nutrition, and growth were difficult, if not impossible, to reconcile with the mechanical philosophy. Physiologists attempted to solve intractable problems by dividing complex vital phenomena into simpler components that could be analyzed, or at least described and linked to each other. At the same time, new ideas and methods from chemistry and physics helped transform physiological research.

THE CHEMICAL APPROACH TO LIFE

Despite the general preference for iatromechanism in the eighteenth century, some natural philosophers were attracted to chemical explanations of life. Like the Renaissance alchemist Paracelsus, Johannes Baptista van Helmont (1579–1644) believed that the workings of the universe could be explained in chemical terms. All physiological processes, therefore, were chemical transformations governed by the *archeus*, the body's own internal alchemist. Unlike Paracelsus, who has been labeled a quack and a mystic, van Helmont has a secure place in the history of chemistry for his pioneering studies of gases and fermentation. It was van Helmont who introduced the term *gas* to replace the Paracelsian word *chaos* (which indicated what typically happened in the alchemist's laboratory when substances being heated in closed vessels released great quantities of gas). Nevertheless, alchemy provided vivid chemical analogies for many physiological functions as well as new approaches to therapeutics.

As a student of philosophy and theology at the University of Louvain, van Helmont found so-called higher education entirely empty and unsatisfying. He referred to the Master of Arts degree as a meaningless symbol of scholasticism rather than a sign of learning. Rejecting philosophy, he turned to botany, law, and finally medicine. At the age of 22 he earned the degree of Doctor of Medicine and then spent 10 years traveling through Europe. Eventually, he settled near Brussels, where he practiced medicine as a form of charity while devoting most of his time to chemical research. Although van Helmont was a devout Catholic, he came into conflict with Church authorities, who thought that his naturalistic exposition of magnetic cures in *De magnetica vulnerum curatione* (1621) conflicted with orthodox interpretations of cures as miracles. Van Helmont published very little after this experience, but just before his death he entrusted his manuscripts to his son for posthumous publication.

While striving to reconcile the chemical view of life with a vitalist philosophical outlook, van Helmont conducted chemical experiments of an exact and quantitative nature in which the idea of the indestructibility of matter was implicit. For example, he proved that a metal could be dissolved in acid and then recovered without loss of weight. Intrigued by the transformation of liquid water into air and the way in which this vapor could again be transformed into liquid water, van Helmont planned and executed an experiment to prove that water is the source of all things. In a large earthenware vessel he placed 200 pounds of dried earth and a willow tree that weighed 5 pounds. After watering the tree with nothing but pure rainwater for 5 years, he found that the tree weighed almost 170 pounds, but the weight of the soil was virtually unchanged. Van Helmont was satisfied that his experiment proved that water was the primary element. This experiment certainly proved something fundamental—that meticulous measurements will not illuminate the secrets of nature where theory throws a false light.

To confirm his belief that earth could be formed from burning vegetation, which in turn had come from water, van Helmont carried out an experiment in which he burned 62 pounds of charcoal. Having determined that only one pound of ash remained, he assumed that the rest of the material had been drive off as *gas sylvestre*, the same gas that was released when organic materials were burned, during fermentation, and from shells and limestone treated with acids. While van Helmont apparently realized that different gases existed, he lacked the special apparatus required to collect them for further study and characterization. However, his research on gases led him to reject the Paracelsian concept of the three elements—sulfur, mercury, and salt. Instead, van Helmont proposed that there are only two elements—air (the natural atmosphere) and water (everything that is not air).

According to van Helmont, all physiological phenomena could be explained in terms of chemical processes that were governed by a hierarchical series of *archaei* in various organs. Subordinate to the *archaei* was an entity that he named the *blas*, which performed specific functions. A more prosaic way of explaining these concepts is to say that all changes in the body were due to the action of ferments; that is, the *blas* or *archaeus* acted on matter through ferments. For example, digestion involved a series of conversions that transformed food into living flesh. The first stage was said to occur in the stomach by means of an acid ferment produced by the spleen. The acidic chyle prepared in the stomach then passed into the duodenum, where very complicated ferments from the bile were at work. Another stage of digestion was caused by the liver as a prelude to the conversion of chyle into crude blood. After the nutritious chyle was absorbed, the useless refuse passed into the large intestine where another ferment converted it into feces. Another digestion supposedly took place in the heart and arteries to convert the darker and thicker blood of the vena cava into blood that was lighter in color and more volatile. In explaining the work of the heart, von Helmont suggested that there were minute pores in the septum that allowed the passage of *vital spirit* from the left side to the right side. The pores allowed blood to pass from the right side of the heart to the left side but prevented movement in the opposite direction. Finally, each part of the body took the nutrients it needed from the blood and transformed them into its own special components. Although much of this theory was obscure and confusing, von Helmont's emphasis on the relationship between fermentation and physiological processes was of considerable value to other iatrochemists, such as Franciscus Sylvius (1614–1672), who established a vigorous program of chemical and experimental approaches to physiology at the University of Leiden. Fascinated by the properties of acids, bases, and salts, Sylvius transformed his observations into a theory that explained disease as the result of an excess of acridity, which could be either acidic or basic. If physicians were guided by this theory, Sylvius asserted, they would be able to design rational therapies.

A very important and bold insight expounded in Sylvius's work is the idea that the chemistry of living things is the same as the chemistry of nonliving things. Thus, at least in theory, it should ultimately be possible to reproduce in the laboratory chemical events supposedly peculiar to the living body. Digestion, for example, could be explained as a fermentation involving the saliva, the bile, and the pancreatic and gastric juices. At least in iatrochemical theory, all mystical *archaei* could be banished from the digestive process. Like Borelli, Sylvius was the founder of a distinct school of thought. While Borelli insisted that physiological phenomena could be explained in purely mechanical terms (iatromechanism), Sylvius rested his case on the science of chemistry (iatrochemistry). Both Sylvius and Borelli had great confidence in their methods and theories, and both thought that they had the exclusive key to the explanation of physiological phenomena. But the question of digestion was one where the advocates of iatrochemistry had a significant advantage.

Based on studies of the pancreatic and gastric juices carried out by Regnier de Graaf (1641–1673) and René Antoine Ferchault Réaumur (1683–1757), the intrepid Lazzaro Spallanzani (1729–1799) used himself and his students as guinea pigs in attempts to understand the mechanism of digestion. Spallanzani began his investigation by testing the action of saliva on various foods. He went on to verify Réaumur's work on the power of the gastric juices of birds of prey to digest food and prevent putrefaction. To determine whether observations made with other animals were applicable to humans, Spallanzani valiantly swallowed tubes and bags containing selected foods, despite the possibility that such materials could cause a fatal obstruction of the digestive system.

Independently of Spallanzani, Edward Stevens published a thesis in 1777 that contained one of the first descriptions of the isolation of human gastric juice and in vitro studies of its properties. Little is known about Stevens other than the date that he published his thesis. A translation of part of his thesis was appended to the English edition of Spallanzani's *Dissertation Relative to the Natural History of Animals*. Stevens conducted his experiments with the cooperation of a man who made his living by swallowing and regurgitating stones for the amusement of spectators. After 20 years of swallowing stones, this human stone-swallowing regurgitator might have found Stevens's perforated silver spheres a pleasant delicacy. Various foods were placed inside these spheres in order to study the action of the digestive juices. Knowledge of human digestion was not substantially improved until 1833 when the American surgeon William Beaumont (1785–1853) published the results of studies conducted through the gastric fistula of Alexis St. Martin, a hardy Canadian who had survived a shotgun blast in the abdomen.

A very different approach to chemical phenomena eventually led to an understanding of the mechanism of combustion and the physiology of respiration. This was the doctrine of phlogiston, derived from the alchemical theories of Johann Joachim Becher (1635–1681). Despite his lack of formal education, Becher was able to attain great academic and financial success, alternating with periods of exile and infamy. In 1666 he was appointed professor of medicine at the University of Mainz, but his interest in economic issues led to a position at the Commercial College of Vienna. By 1678 he was forced to seek refuge in Holland. Two years later, after the failure of his alchemical enterprises, he fled to England. According to Becher, bodies consist of air, water, and three kinds of earth: *terra pinguis* (fatty earth), *terra mercurialis* (metallic earth), and *terra lapidia* (stony earth). During combustion, terra pinguis was released as fire.

Through the work of George Ernst Stahl (1660–1734), *terra pinguis* was transformed into *phlogiston*, the key to a new chemical system that seemed to explain many bewildering phenomena, while guiding chemical research for at least a century. After studying medicine at the University of Jena, Stahl held a teaching position there until 1693, when he was invited to teach at the new University of Halle. His colleague Friedrich Hoffmann (1660–1742), a distinguished physician and founder of one of the most famous medical systems of the eighteenth century, taught chemistry, physics, anatomy, surgery, and medical practice. Stahl's responsibilities included botany, physiology, dietetics, pathology, and materia medica. In 1703 Stahl published a new edition of Becher's *Physica suberranea* and appended his own exposition of the mechanism of combustion. Phlogiston was defined as a material principle, a component of combustible materials, and both the material and the principle of fire. Phlogiston, however, was not fire itself. Phlogiston was found in the vegetable, animal, and mineral kingdoms, but was more abundant in the first two, which accordingly left little residue when they burned and released their phlogiston.

One problem for phlogiston theory was the apparent gain in weight that occurred when metals were burned. But, because phlogiston could not be trapped and analyzed, there was no compelling reason to assume that it was an ordinary substance with weight. Like heat, light, magnetism, or electricity, it might be a weightless subtle fluid. Indeed, instead of being attracted to the earth, like other elements, phlogiston might have a tendency to rise. This would explain the increase in the weight of metal oxides and the weight loss that occurred when the oxides were reduced, but it had little relevance to Stahl's physiological theories. For Stahl, all chemical changes in the living body were fundamentally different in nature from ordinary chemical events. In the living body, he asserted, all chemical changes were controlled by the *anima sensitiva* (sensitive soul). Rejecting Cartesian dualism and claims that the body was a machine, Stahl contended that physiological phenomena obeyed laws entirely different from those governing the inanimate world. Although phlogiston theory implicitly accepted the alchemical concept that substances were composed of matter plus intangible spirits, essences, or principles, it retained the loyalty of some of the most distinguished eighteenth-century chemists.

While a connection between respiration and combustion had been suspected long before, the chemical basis of this relationship was not demonstrated until the end of the eighteenth century. Solving the riddle of respiration required the confluence of studies of the circulation of the blood, the exchange of gases in respiration, the microanatomy of the lungs, and the chemistry of the gases. It was primarily the invention of the pneumatic trough by Stephen Hales (1677–1761) that made the isolation and characterization of various gases possible. Reverend Hales, the Perpetual Curate of Teddington, investigated the hydraulics of the vascular system, analyzed the "water economy" of plants, the chemistry of gases, respiration in plants and animals, and blood pressure in animals and invented several practical ventilation devices. Much of this work was reported in *Vegetable Staticks* (1727) and *Statistical Essays, containing Haemastaticks* (1733). As a Fellow of the Royal Society, Hales often served as a member of various commissions investigating matters of public health and alleged wonder cures. This led to a special interest in practical problems such as ventilation and ways of introducing fresh air into confined quarters, such as ships, prisons, and hospitals.

About 100 years after Harvey published his theory of the circulation of the blood, Hales provided rigorous experimental confirmation through his analysis of the hydrodynamics of the vascular system. Hales's scientific work was memorialized in a poem by Thomas Twining, who noted that in his serene retreat in Teddington the good pastor searched for Nature's secrets by methods that ranged from weighing "moisture in a pair of scales" to stripping the skins from living frogs. In one series of experiments Hales decapitated frogs in order to observe the reflexes triggered by pricking the skin at various points. Through these experiments Hales discovered the connection between the stimulated nerves and the spinal cord. He came to the conclusion that muscle movements were related to electrical activity rather than the pressure of the blood.

Another series of experiments established the general outlines of the physiology of plants, including the circulation of the sap, the interactions between the plant and its environment, the uptake of water by the roots, the transport of water to the leaves, the transpiration of water by the leaves, the growth of the parts of plants, and the proportional aspects of plant growth. Hales also extended previous work on the relationship between respiration, or combustion, and air.

Microscopic examinations of plants in the seventeenth century, especially those of Nehemiah Grew, revealed that plants contained ramifying systems of tubes that appeared to run from the roots, through the stem and branches, and into the leaves. Since some of the tubes appeared to be filled with liquid and others with air, Grew and other naturalists thought that the movement of sap in plants might be analogous to the circulation of the blood in animals. Advocates of this theory generally assumed that sap moved up via the inner part of the stem and down through the outer parts. Hales proved that such a hydraulic cycle or circulation does not occur in plants. His experiments indicated that it was the evaporation of water from the leaves rather than pressure of water in the roots that caused upward movement of the sap. Having clarified the water economy of plants, Hales realized that plant nutrition and its relationship to the exchange of gases required further study. By separating his reaction vessels from the collection vessel to create a device called a pneumatic trough, Hales was able to collect and store several different gases, including carbon dioxide and oxygen. The ancient Greek element "air" could now be resolved into several "airs" or gases with very different properties. Applying more rigorous qualitative methods to the study of plants and animals, Hales returned to van Helmont's plant growth experiment. Rather than accepting the assumption that water alone entered into new plant material, Hales measured the amount of water taken up by roots and given off by leaves. This led to the conclusion that in addition to water, plants must take part of their nourishment—some "secret food of life"—from the air.

Hales's invention and his ideas inspired revolutionary work on the chemistry of gases by a group of scientists known as the pneumatic chemists, which included Joseph Black (1728–1799), Henry Cavendish (1731–1810), Carl Wilhelm Scheele (1742–1786), Joseph Priestley (1733–1804), and Antoine-Laurent Lavoisier (1743–1794). In 1754 Joseph Black received the doctorate of medicine for a dissertation entitled "On the acid humour arising from food and on magnesia carbonate alba." Stones formed in the bladder and gravel passed in the urine were of considerable interest to medical researchers at the time. A heated controversy was raging over the use of caustic agents as solvents for such stones. To express his satisfaction with Mrs. Joanna Stephens's secret remedy for the stone, Prime Minister Walpole paid £5000 for her recipe and had it published in the *London Gazette* (1739). When Black discovered carbon dioxide, the gas he called "fixed air," he was testing the chemistry of calcined snails, which happened to be a major component of Mrs. Stephens's powders and pills. According to phlogiston theory, limestone became caustic quick lime by taking up phlogiston; when quick lime was slaked, it supposedly gave off phlogiston. Black discovered that when chalk was calcined or burned, it gave off a gas he called "fixed air." Driving this gas through a clear solution of limewater caused the formation of "mild lime," which precipitated out of solution. Fixed air was found in expired air. Burning charcoal also produced it, and it was released during fermentation. This gas was deadly to animals and would extinguish a flame. At first, Black thought that all of the atmospheric air that was unsuited for respiration must be fixed air, but other chemists discovered another noxious gas, later identified as nitrogen.

Another remarkable species of gas was discovered by Henry Cavendish, a man said to be as eccentric as he was wealthy. Even as a youth, Cavendish refused to submit to tests of orthodoxy. Therefore, Cavendish left Cambridge without a degree because he objected to the strict religious tests applied to candidates. After studying physics and mathematics in Paris, he settled in London where he used his large inheritance to purchase the finest instruments and reagents for his scientific experiments. In 1766 Cavendish published an account of his experiments on "factitious airs" in the *Philosophical Transactions of Royal Society*. By dissolving zinc, iron, or tin in vitriolic acid, Cavendish produced hydrogen, which he called "inflammable air." At first he thought that inflammable air might be phlogiston itself. In one of his most significant experiments, Cavendish exploded a mixture of hydrogen and oxygen and thus proved that water was not an element, but a compound of two gases.

Carl Scheele and Joseph Priestley discovered oxygen almost simultaneously, but both were staunch advocates of phlogiston theory who thought that this new gas was "dephlogisticated air." Conducting a series of experiments to prove that air is made up of two components, "foul air" and "fire air," Scheele determined the ratio to be one part of fire air to three parts of foul air on a volume/volume basis. Scheele believed that the function of fire air was to absorb the phlogiston given off by burning substances. When air became saturated with phlogiston, it could no longer support combustion. Scheele and Priestley knew that fire air was used up in combustion and that the remaining air had less weight and volume than the original mixture. This was ingeniously explained by postulating that the combination of air and phlogiston produced a compound so subtle that it passed through the pores of the glass and disappeared into the atmosphere.

Although mainly remembered as an experimental chemist, Joseph Priestley was also a minister, theologian, author, and educator. During his lifetime,
he was more famous as a radical theologian and political thinker than as a
scientist. His experiments on gases began in the public brewery next to his
house in Leeds. The gas that bubbled out of the beermaking vats, now known
as carbon dioxide, was identical to Joseph Black's fixed air. Although the gas
was largely insoluble in water, Priestley discovered that it produced a very
pleasant effervescent beverage, comparable to the very best mineral waters. The
College of Physicians suggested that Priestley's soda water might be used as a
cure for scurvy. In 1791, during a series of attacks on dissenters and radicals, a
mob burned down Priestley's house. Priestley barely escaped the riots with his
life, but his home, library, laboratory, and many manuscripts were destroyed.
Reluctantly, he decided to emigrate to America where he settled in Northumber-
land, Pennsylvania.

An enthusiastic and independent experimentalist, Priestley was curious
about everything, but unsystematic in his researches. Indeed, he often said that
if he had known any chemistry he would never have made any discoveries.
Chance and careful observation, he argued, were more significant in making
discoveries than preconceived theories. Among the gases Priestley discovered
were ammonia, sulfur dioxide, carbon monoxide, nitric oxide, and hydrogen
sulfide. As Sir Humphry Davy (1778–1829) said: "No single person ever discovered so many new and curious substances." Priestley did not think that it
was possible to prepare air that was purer than the best common air, but this
was what appeared to have happened in 1774 when he extracted a new air from
mercuric oxide. The new air caused a candle to burn with a very vigorous flame
and allowed mice to survive for a longer period than a similar limited quantity
of ordinary air. Priestley called this gas *dephlogisticated air*. After breathing
this air himself, Priestley predicted that it would be useful in medicine, although
it might be dangerous as well.

During a series of experiments on the relationship between various gases
and the nature of respiration, Priestley discovered that air that had been "injured" by animal respiration or by the burning of candles could be "revivified"
by plants. This observation puzzled him at first because he had assumed that
both plants and animals should affect air in the same manner. Working out the
puzzle in terms of phlogiston theory, Priestley reasoned that ordinary air became
saturated with phlogiston from combustion, animal respiration, and other chemical processes. Plants, in contrast, took up phlogiston and thus purified the air.
Dark venous blood, laden with phlogiston, released phlogiston in the expired air.
Bright red dephlogisticated arterial blood was, therefore, capable of absorbing
phlogiston from body tissues. Blood exposed in vitro to dephlogisticated air
became as bright as arterial blood. Although Priestley acknowledged that no
one had been able to ascertain the weight of phlogiston, he did not see this as a

reason for rejecting the theory. No one had weighed light or heat, he argued, but scientists did not doubt their existence.

Intrigued by Priestley's studies of the effects of plants and animals on the various gases, Jan Ingenhousz (also known as Ingen-Housz, 1730–1799) undertook a series of experiments on plant nutrition that led to the discovery of photosynthesis, the process by which green plants absorb carbon dioxide in the presence of sunlight and release oxygen. In 1771 Priestley had reported that plants restore air that had been spoiled by burning candles or the death of an animal through suffocation. Some chemists, however, were unable to duplicate these results. Carefully repeating Priestley's experiments, Ingenhousz demonstrated that only the green parts of plants could improve spoiled air and that, most importantly, they could do this only in the presence of sunlight. As he reported in his book *Experiments Upon Vegetables, Discovering Their Great Power of Purifying the Common Air in the Sunshine and of Injuring it in the Shade and at Night* (1779), in the dark, all parts of the plant perform respiration, and produced "fixed air" (carbon dioxide), just like animals. But, when green plants received the visible part of sunlight, they released "dephlogisticated air" (oxygen). Plants and animals, therefore, mutually supported each other. Animals consumed oxygen and produced carbon dioxide, whereas plants exposed to sunlight consumed less oxygen by respiration than they produced by photosynthesis. Thus, through his experiments on photosynthesis Ingenhousz established the fundamental similarities and significant differences between plants and animals.

Unlike Priestley, Antoine-Laurent Lavoisier approached chemistry as a scientist who had mapped out a logical and systematic protocol for the reformation of chemistry at the outset of his career. While studying law, Lavoisier became interested in the natural sciences. At the age of 21 he submitted the first of many articles to the Royal Academy of Sciences. Four years later, when he became a member of the Academy, he decided to devote himself to science. To support himself in a manner that would not interfere with his research, Lavoisier became a member of France's network of private, and much hated, tax collectors. In 1771, he married 14-year-old Marie Anne Pierrette Paulze. Marie illustrated his scientific works, assisted him in the laboratory, kept his notes, translated the works of English chemists into French for him, and entertained his famous visitors. By the time Lavoisier published his *Treatise on the Elements of Chemistry* (1789), he was quite satisfied that he had established a revolution in chemistry. Astute as he was in science, Lavoisier did not seem to recognize the danger of the social and political changes fermenting all around him. Along with many other tax-farmers, Lavoisier was arrested, tried, and condemned to the guillotine. Madame Lavoisier married another chemist, Benjamin Thompson (1753–1814), later known as Count Rumford.

In contrast to Priestley, Lavoisier's work was not especially remarkable for experimental originality or the discovery of new substances, but it was memorable for its precision, planning, and explanatory power. On November 1, 1772, Lavoisier deposited a sealed letter with the Secretary of the Academy of Science. This was to serve as his claim to priority against future publications, providing all went as he expected and his investigations confirmed his hypotheses. His plan was to bring about a revolution in physics and chemistry by repeating previous work with new precautions. The revolution began to take shape in 1772 when Lavoisier met Priestley and learned about his dephlogisticated air. After repeating Priestley's experiments, Lavoisier renamed the new gas *oxygen*. Further experiments demonstrated that "eminently respirable air" was converted into "fixed air" by both combustion and respiration. To exorcise the spirit of phlogiston from the body of chemical theory, in 1777 the Lavoisiers ceremoniously burned the writings of George Ernst Stahl while chanting an appropriate requiem.

In "Experiments on the respiration of animals and on the changes which the air undergoes in passing through the lungs" (1777), Lavoisier explained the process of respiration as a slow combustion or oxidation that used oxygen and released carbon dioxide. Unlike his predecessors, Lavoisier performed quantitative measurements of oxygen consumption in living beings. To provide quantitative measurements of the production of animal heat by the slow combustion that occurred within the animal body, Lavoisier and Pierre Simon Laplace (1749–1827) designed a calorimeter. Using this device, Lavoisier proved that both combustion and respiration consume oxygen and produce carbon dioxide. Lavoisier's experimental program made it possible to link the chemical revolution with modern physiological research. Adopting the principle that the same natural laws apply to both living and nonliving entities, Lavoisier initiated the experimental analysis of the energetics of living systems. Calorimetry was later used to determine the relationship between the composition of foodstuffs and their value as sources of metabolic energy. Lavoisier and his followers introduced a new chemical language that was incorporated into the science of different European states with varying degrees of satisfaction and comprehension. Tragically, the violence of the French Revolution truncated Lavoisier's career just at the point where he was learning to apply the new chemistry to physiological phenomena. The idea that physiological phenomena could be analyzed and explained in chemical terms had, however, been well established.

GENERAL PHYSIOLOGY

Through the work of eighteenth-century chemists and early nineteenth century cell biologists, physiology developed into a mature discipline utilizing chemical and instrumental techniques that distinguished it from its origins in anatomy. Although physiologists still wrestled with fundamental philosophical questions, by the second half of the nineteenth century general physiology had become a science based on experimental methods and concepts often derived from chemistry and physics. The founders of general physiology include Claude Bernard (1813–1878) in France; Johannes Müller (1801–1858), Justus von Liebig (1803–1873), and Carl Ludwig (1816–1895) in Germany; and Sir Michael Foster (1836–1907) in England. The philosophical approach and intellectual milieu that nurtured nineteenth-century pioneers in physiology differed significantly from country to country. Foster introduced educational reforms and established a model physiological research laboratory. Although his own research did not produce outstanding discoveries, his laboratory trained many eminent scientists and his writings on physiology and the history of physiology were highly regarded. In response to growing antivivisectionist activity, Foster helped establish the Physiological Society, the first organization of professional physiologists, and the *Journal of Physiology*.

Johannes Müller, whose name is immortalized in several of his anatomical discoveries, was particularly interested in the effect of stimuli on the sense organs. His experiments on the direction of nerve impulses in spinal nerves confirmed the so-called Bell-Magendie law and provided new insights into reflex action. While Müller was primarily interested in the comparative aspects of animal function and anatomy, Liebig and Ludwig are most associated with the innovative application of chemical and physical methods to physiology. Their work provided many useful techniques, such as more precise measurement of respiration, muscular action, and blood pressure, and methods for analyzing the nature of body fluids. Ludwig was a founder of the physicochemical school of physiology. In addition to inventing the kymograph, a blood pump for sampling gases in the blood, and other instruments, Ludwig developed methods of keeping animal organs alive in vitro by perfusing them with a solution that mimicked the composition of blood plasma. In collaboration with Henry Bowditch, he formulated the "all-or-none law" of cardiac muscle action, which states that in response to any stimulus the heart muscle contracts to the fullest extent or not at all. Ludwig also studied urine production by the kidney, the relationship between nitrogen in the urine and protein metabolism in the whole animal, and the effect of secretory nerves on the human digestive glands.

FROM ANIMAL CHEMISTRY TO BIOCHEMISTRY

Studies of the animal economy or animal chemistry became a prominent aspect of the research program of early nineteenth-century scientists such as Jöns Jacob Berzelius (1779–1848), Friedrich Wöhler (1800–1882), and Justus von Liebig (1803–1873). After completing his medical studies, Berzelius discovered that he preferred the study of chemistry to the practice of medicine. Building on the work of Lavoisier and John Dalton (1766–1844), the founder of modern atomic theory, Berzelius carried out the tedious analytical work involved in preparing a table of the 50 elements then known and determining their atomic weights. He also discovered many elements, established a journal, developed new analytical methods, discovered pyruvic acid, developed the concepts of isomerism and catalysis, and trained so many chemists that almost all of the major nineteenth-century chemists were either his students or students of his students. A prolific writer, Berzelius unleashed a torrent of articles and books that helped unify and direct chemical studies for almost a century. Since Berzelius was trained in medicine, it was not surprising that he would eventually turn to the analysis of complex organic substances such as bile, blood, and feces.

Beyond his specific discoveries, Berzelius was important in bringing the chemicals of life within the purview of atomic theory. In some respects, however, Berzelius retained a vitalist's view of the origin of natural products. Not surprisingly, Friedrich Wöhler, the first man to synthesize an organic chemical in the laboratory, was one of Berzelius's disciples. As a student, Wöhler was, according to his own admission, distinguished neither by special zeal nor broad learning, but at Heidelberg University he was allowed to abandon routine course work in order to devote himself to research. Wöhler's most famous experiment involved the production of urea from ammonium cyanate, a demonstration that helped blur the distinction between organic and inorganic compounds. For the first time, a chemical produced by living beings had been synthesized from materials that were available, at least in principle, from nonliving matter. With great excitement, Wöhler announced that he could "make urea without kidney of man or dog." Although the preparation of urea is often seen as proof of the theoretical equivalence of inorganic and organic chemicals, many other demonstrations were needed to abolish the conceptual abyss separating "animal chemistry" from inorganic chemistry. Hermann Kolbe (1818–1884) is generally considered the first chemist to perform the complete in vitro synthesis of an organic compound, i.e., acetic acid. While Kolbe may have excelled as a chemist, he appears to have had little tolerance for imagination as a stimulus to scientific thought. Attempts to think about the geometrical arrangement of the atoms within molecules struck Kolbe as the worst kind of pseudoscientific nonsense.

Prior to 1750 nutrients were regarded as necessary for the animal as lubricants for the muscles and joints and as replacements for parts worn out by the wear and tear of daily life. Chemists assumed that these components were present in foodstuffs and were directly assimilated into the tissues. Organic chemicals seemed to be overwhelmingly complicated, but by the nineteenth century chemists were confident that knowledge of chemical composition would provide a new scientific basis for agriculture and nutrition. Believing that empirical formulas adequately characterized organic chemicals, chemists still found themselves unable to deal with all the chemical species found in foodstuffs. They turned instead to an analysis of bulk foods, body fluids, solids, and excrements and created simplicity from complexity by seeing all foods as mixtures of saccharine, oleaginous, and albuminous materials (that is, carbohydrates, fats, and proteins). Chemists and physiologists differed as to the proper approach to elucidating the vital phenomena of the body. Many physiologists saw vivisection, or animated anatomy, as the only way to unravel the complex phenomena of life. Test tube chemistry, in their view, could contribute nothing to understanding life. The work of the German chemist and educator Justus von Liebig and the French physiologist Claude Bernard illustrates significant aspects of this dispute.

Struggling to create a community of chemists in Germany, Liebig saw it as his mission to prove that the science of chemistry would revolutionize agriculture, industry, nutrition, medicine, and sanitation. Science could even find ways to make meat available to the lower classes. During the first phase of his career, Liebig investigated classical organic chemistry, developing and improving methods of analysis and identification. He later turned to more complex problems of agricultural chemistry and the chemistry of living things. Convinced that he had established a true chemical theory of fermentation, Liebig rejected evidence that yeast is a living organism and ridiculed the work of Schwann and Pasteur. Yeast, Liebig insisted, was the product of fermentation rather than the cause. Studies of animal chemistry and nutrition contributed to Liebig's outstanding reputation and influence, but they also revealed his tendency to formulate grandiose theories from a shaky and meager base of data. Critics later said that by the end of the century, Liebig was Germany's most famous chemist, but no one could remember why.

According to Liebig, only green plants could build up complicated organic substances from the simple inorganic elements they took from the air and the soil. In contrast to plants, which were synthetic chemical factories, animals were degradative chemical processors that ultimately took the components they needed for their tissues from plants and used the rest as fuel. In simplified forms, Liebig's ideas about nutrition stimulated much food faddism, including that of Sylvester Graham, the American health reformer whose name is immortalized in the graham cracker. While Liebig assumed that the chemical transformations that took place in living beings were not unique, he did not totally reject the concept of the vital force. Assuming that the law of conservation of matter was also applicable to living things, Liebig attempted to determine what chemical events occurred in the living body by measuring and balancing all the components that were ingested and excreted by the body. This approach to metabolic phenomena was not unlike that of Santorio Santorio. Moreover, Liebig thought that muscle activity was directly dependent on protein degradation and that urea output could be used as an index of physiological activity. Although Liebig improved the analytical methods applicable to organic compounds, Claude Bernard said that his attempt to deduce the invisible metabolic phenomena that took place in the living body from quantitative analyses of input and output was like trying to deduce what happened inside a house by measuring what goes in the door and out the chimney.

Modern physiology owes much to Claude Bernard's mentor François Magendie (1783–1855). Although his work did not culminate in any broad or all-inclusive theory, Magendie virtually founded the field of experimental pharmacology. Magendie served as Professor of Medicine at the College of France, President of the French Academy of Sciences, and President of the Advisory Committee on Public Hygiene. Through his teaching, service, private lectures, and research, Magendie left his imprint on every branch of physiology.

Magendie was so closely associated with experimental physiology that reports of his ruthless experimentation invariably stimulated antivivisectionist activities. Generally, Magendie favored a purely mechanical explanation of vital phenomena, but he seems to have allowed for a vital force when explaining the physiology of the nervous system. His studies of the nervous system, which resulted in an acrimonious priority battle with Sir Charles Bell (1774–1842), included experiments that led to the discovery of the separate motor and sensory functions of spinal nerves. Perhaps Magendie's most important contribution to science was liberating his protégé Claude Bernard from his chronic apathy and inertia. Once aroused and engaged in experimental physiology, Bernard surpassed his mentor in many ways. Even Magendie had to admit that Bernard's skill in vivisection was superior to his own, and history has affirmed Bernard's ability to formulate significant generalizations.

Born into a family of poor peasants, Bernard was fortunate to have received instruction in classical subjects from his parish priest. After more advanced studies, Bernard taught language and mathematics at a Jesuit school while tutoring private pupils. Financial difficulties forced him to abandon teaching to take a position as assistant to an apothecary in Lyons. Sadly, then as now, performing menial tasks, such as sweeping the floors and washing glassware, often paid better than teaching. Delivering prescriptions was the only enjoyable part of his work, because it allowed him to stop and watch surgical operations performed at a nearby veterinary school. Finding many aspects of medicine quite ridiculous, Bernard wrote a short play revolving around theriac, a popular medicine so haphazardly compounded from over 60 ingredients that no two batches were ever the same. The success of this play made him aspire to greater things, and he began work on an ambitious five-act historical drama.

Soon after arriving in Paris with his play, Bernard was advised by a literary critic to learn another profession if he wanted to eat. Bernard chose medicine and did well enough at the Parisian School of Medicine to be awarded an internship in 1839. Nevertheless, Bernard, who ranked twenty-sixth out of the 29 taking the examinations, was not regarded as a brilliant student. Unable to obtain a professorship, and unwilling to practice medicine, Bernard served as Magendie's assistant for many years. Trying to rescue himself from financial

A portrait of Claude Bernard in 1866

difficulties, Bernard contracted an arranged marriage that became a source of intolerable tension, because his wife was adamantly opposed to vivisection. Despite persistent ill health and recurrent attacks of gastritis, Bernard's research career brought him success and the esteem of his colleagues. In 1854, he became a member of the Academy of Sciences and then assumed Magendie's professorship. At 47 years of age, exhaustion and illness forced him to retreat to his birthplace. During this period of enforced rest and reflection, he wrote about the broader implications of his work in *Introduction to the Study of Experimental Medicine*. This remarkably lucid and perhaps deceptively simple text has been more widely read and discussed than his numerous research papers and 14-volume *Lessons in Experimental Physiology Applied to Medicine*. When Bernard died, the French Chamber of Deputies voted to give him the first state funeral to honor a scientist. The novelist Gustave Flaubert (1821–1880) described the event as more impressive than the ceremonies held at the funeral of the Pope.

In the course of his researches Bernard illuminated physiological phenomena in new ways, demonstrating that many vital functions might better be seen in terms of chemistry than as aspects of animated anatomy. His most significant discoveries included the glycogenic function of the liver, the role of the pancreatic juices in digestion, the functions of the vasomoter nerves, and the nature of the action of curare, carbon monoxide, and other poisons. More importantly, Bernard placed his observations in a theoretical framework based on his concept of *determinism*, faith in the experimental method and its applicability to physiology, the science of life.

For Bernard, vitalism and mechanism were both essentially worthless distractions that led to endless disputes that obstructed scientific progress. When scientists characterized a phenomenon as vital, Bernard complained, they were essentially admitting that they could not explain its immediate cause or conditions. Bernard insisted that there was always a real physical-chemical basis for all vital phenomena, no matter how diverse and mystifying they might appear. Thus, serious physiological research could only proceed when scientists banished that capricious agent, the "vital force," that resisted the laws governing the inanimate world and put all acts performed by living organisms beyond the scope of science.

While praising doubt and an open mind, Bernard warned against excess skepticism. Scientists must believe in determinism, that is, in science itself, in order to reveal the complete and necessary relationships among phenomena in living beings and the inanimate world. The study of the science of life, Bernard wrote, was like coming into a superb and dazzlingly lighted hall, which could only be reached "through a long and ghastly kitchen." While it was impossible to define precisely what life is, the scientist must not allow speculations to interfere with true scientific work, which was to analyze and compare the manifestations of life. All too often scientists relied on the word "life" to mask their ignorance. If forced to define life in a single phrase, Bernard suggested the phrase "life is creation." An organism might then be described as a machine that works by means of the physicochemical properties of its constituent parts. The way to understand vital phenomena was through vivisection. Indeed, Bernard argued, the science of life could only be established by experimentation, and, since medicine was ultimately an aspect of experimental physiology, human beings could only be saved from death through the sacrifice of other beings. The ideal experimentalist envisioned by Bernard was so totally absorbed by the pursuit of scientific ideas that he was oblivious to the cries and blood of animals and could see only "organisms concealing problems which he intends to solve."

Perhaps then it is surprising that Bernard considered his demonstration of the glycogenic function of the liver his most important achievement. To do this he had to divorce himself from prevailing theories of plant and animal metabolism. At the time, scientists believed that sugar in the blood of carnivores must have been supplied entirely from ingested foods. Animals, whether carnivores or herbivores, supposedly used materials originally synthesized by plants in order to support a combustion that took place either in the blood or the lungs. Revolutionizing ideas about metabolism, Claude Bernard proved that animal blood contains sugar even when it was not supplied by foodstuffs. In tests of the theory that sugar absorbed from food was destroyed when it passed through the liver, lungs, or some other tissue, Bernard put dogs on a carbohydrate diet for several days and then killed the animals immediately after feeding. Large amounts of sugar appeared in the hepatic veins. To his surprise, animals in the control group, which had been fed only meat, had large amounts of sugar in their hepatic veins, but not in the intestines. Before concluding that current theories were wrong, Bernard did many additional experiments to ascertain the location of the tissue that supposedly served as the site of the destruction of carbohydrates. Through these experiments, Bernard discovered *gluconeogenesis*, the conversion of other substances into glucose in the liver. Further work led to the discovery of glycogen (the carbohydrate storage polymer of animals), as well as the synthesis and breakdown of glycogen. All animal tissues appeared to have ferments (enzymes) that acted on glucose. Thus, the investigation of glucose metabolism led to the concept of "internal secretions," which were products transmitted directly into the blood instead of being poured out to the exterior of the gland or organ secreting them.

French scientists generally ignored Theodor Schwann's metabolic theory of the cell, with its emphasis on the importance of the nutritive medium bathing the cells, but Bernard saw that this concept was applicable to the fundamental problem of physiology, the relationship between cells and their immediate environment. Bernard believed that he was the first scientist to insist that complex animals had two environments: an external environment in which the organism lived and an internal environment in which the cells functioned. Ultimately, vital phenomena occurred within the internal environment, bathing all the anatomical elements of the tissues. This was the basis of Bernard's well-known aphorism: "The constancy of the internal milieu is the condition for free and independent life." Higher animals were not totally independent of their external environment, but in close and intimate relation to it, so that their equilibrium was the result of continuous and exact compensatory adjustments.

The image of Claude Bernard that emerges from his *Introduction to the Study of Experimental Medicine* is that of the ideal scientist and sage, always lucid and rational, never at a loss for a reasonable hypothesis. A close examination of his research notebooks indicates that the path to each of his discoveries was much more arduous, confused, and tortuous than the published accounts admit. Although Bernard's skill in experimental surgery was superlative, deficiencies in his mastery of chemical techniques seem to have frustrated him throughout his career. Nevertheless, despite many years of struggle, obscurity, and failure, Bernard truly believed in himself as a reformer of physiology and a highly individualistic researcher.

Elucidating the complex pathways of metabolism and the enzymes, hormones, and neurotransmitters that regulate them is still an enormous and incomplete task, but Bernard's work helped physiologists wrestle with the persistent philosophical question: Can physiological functions be explained in physical-chemical terms? By applying appropriate methods to physiological problems, Bernard concluded that specific physiological processes were governed by the same physical-chemical laws that governed similar processes in inorganic systems. This approach provided at least a provisional resolution to ancient debates about whether all physiological functions could be reduced to physical properties and chemical reactions. Certain physiological functions, however, seemingly had no inorganic counterparts and could not be described in terms of the language of inorganic chemistry and physics.

Bernard's work marks the beginning of an era in which biochemistry nearly superseded animated anatomy as the key methodology of physiological research. Essentially, modern physiology, the parent science of cell physiology, endocrinology, and biochemistry, was liberated from its roots in anatomy and medicine. Following the approach established by Bernard, many physiologists explored vital phenomena in terms of self-regulatory processes. For example, through his studies of the mechanisms that regulate blood chemistry, Lawrence J. Henderson (1878–1942) helped establish the concept of the constancy of the internal environment as the key to physiological thought. Walter Bradford Cannon's (1871–1945) research on traumatic shock, the sympathetic branch of the autonomic nervous system, and neuroendocrinology illustrate the fruitfulness of the research program that began with the recognition of the internal secretions. Having coined the term *homeostasis* in 1926, Cannon popularized the

A portrait of Walker Bradford Cannon in 1918

concept well beyond the scientific community, especially in his well-known book *The Wisdom of the Body* (1932). According to Cannon, the term homeostasis did not mean something fixed and unchanging, but a relatively constant, complex, well-coordinated, and generally stable condition. Since Bernard and Cannon made homeostasis the guiding principle of physiological research, the concept has broadened considerably. New terms such as hormones, neuropeptides, neurotransmitters, feedback loops, servomechanisms, transfer functions, and cybernetics indicate that the concept is applicable to problems studied by endocrinologists, biochemists, engineers, sociologists, economists, ecologists, and mathematicians.

Endocrinology developed into a major area of research during the twentieth century, but it has remained closely linked to physiology and medicine. In contrast to Bernard's initial concept of internal secretions, hormones are generally thought of as products of the endocrine glands that are released into the bloodstream and regulate the activities of distant tissues and organs. Jacob Henle (1809–1885) first described the endocrine or ductless glands in the 1840s. Ernest Henry Starling (1866–1927) introduced the term hormone in 1905. In 1902 Starling and William Maddock Bayliss (1860–1924) isolated secretin, a hormone produced by the small intestine. Having proved that secretin caused the release of the pancreatic juices, Bayliss and Starling suggested that a combination of chemical regulation and nervous regulation controlled physiological processes.

Although the concepts formulated by Bayliss and Starling can be considered the first principles of the science of endocrinology, studies of the relationship between some hormones and the diseases associated with them began much earlier. Indeed, descriptions of some endocrine disorders, such as diabetes mellitus and goiter, and the use of glands or glandular extracts are very ancient. For example, some healers claimed that they could reverse male aging with extracts of animal testes. By removing specific glands and observing the outcome, scientists discovered the hormones produced by various glands. In some cases, extracts of the glands could be used to cure diseases caused by a deficiency in their production.

Comparative studies expanded the classical view of hormones as substances released into the bloodstream by endocrine glands to include hormonal regulation in plants and certain animals that lack a vascular system. In keeping with Aristotle's admonition that naturalists should be willing to study even the lowliest creatures with care and curiosity, studies of hormonal regulation in cockroaches led Berta Vogel Scharrer (1906–1995) to the discovery of neuropeptides. Through comparative studies of the regulation of the nervous system in vertebrates and invertebrates, Scharrer and her husband Ernest Scharrer (1905–1965) helped establish neuroendocrinology as a vital new field. Characterizing the complex pathways of metabolism and the enzymes, hormones, and neurotransmitters that regulate them remains a major goal of twenty-first century biology. Physiologists are particularly interested in understanding the mechanisms by which the nervous system and the endocrine system coordinate their regulatory mechanisms and interact with the immune system.

Based on a lifetime of research and reflection, Claude Bernard concluded, "there is only one way to live, only one physiology of all living things." By incorporating ideas and methods from many different fields, physiology has generated several related disciplines, such as biochemistry, biophysics, molecular biology, and physiological genomics, which are generally characterized by an analytical or reductionist focus. But despite the importance of physiological studies that focus on cells and biological molecules, many physiologists have retained their interest in an integrative or holistic approach to the activities of cells, tissues, organs, and the "physiology of all living things."

SUGGESTED READINGS

Anderson, W. C. (1985). *Between the Library and the Laboratory. The Language of Chemistry in 18th-Century France*. Baltimore, MD: Johns Hopkins University Press.

Benison, S., Barger, A. C., and Wolfe, E. L. (1987). *Walter B. Cannon*. Cambridge, MA: Harvard University Press.

Bensaude-Vincent, B., and Abbri, F., eds. (1995). *Lavoisier in European Context: Negotiating a New Language for Chemistry*. Canton, MA: Science History Publications.

Blasius, W., Boylan, J. W., and Kramer, K., eds. (1971). *Founders of Experimental Physiology*. Munich: Lehmanns.

Blum, D. (1994). *The Monkey Wars*. New York: Oxford University Press.

Brock, W. H. (1997). *Justus von Liebig: The Chemical Gatekeeper*. New York: Cambridge University Press.

Brooks, C. M., and Cranefield, P. F., eds. (1959). *The Historical Development of Physiological Thought*. New York: Hafner.

Cannon, W. B. (1932). *The Wisdom of the Body*. New York: Norton.

Coleman W., and Holmes F. L., eds. (1988). *The Investigative Enterprise: Experimental Physiology in Nineteenth-Century Medicine*. Berkeley, CA: University of California Press.

Cranefield, P. F. (1974). *The Way In and the Way Out: François Magendie, Charles Bell, and the Roots of the Spinal Nerves*. Mount Kisco, NY: Futura.

Descartes, R. (1972). *Treatise of Man*. Translated by T. S. Hall. Cambridge, MA: Harvard University Press.

Donovan, A. (1994). *Antoine Lavoisier: Science, Administration and Revolution*. Cambridge, MA: Blackwell Science Biographies.

Foster, M. (1970). *Lectures on the History of Physiology*. New York: Dover, 1970.

French, R. D. (1975). *Antivivisection and Medical Science in Victorian Society*. Princeton, NJ: Princeton University Press.

Fuchs, T., and Greene, M. (2001). *The Mechanization of the Heart: Harvey & Descartes.* Rochester, NY: University of Rochester Press.

Fulton, J. F., and Wilson, L. G., eds. (1966). *Selected Readings in the History of Physiology.* Chicago, IL: Charles C Thomas.

Fey, W. B. (1987). *The Development of American Physiology: Scientific Medicine in the Nineteenth Century.* Baltimore, MD: Johns Hopkins University Press.

Geison, G. L. (1978). *Michael Foster and the Cambridge School of Physiology: The Scientific Enterprise in late Victorian Society.* Princeton, NJ: Princeton University Press.

Grande, F., and Visscher, M. B., eds. (1967). *Claude Bernard and Experimental Medicine.* Cambridge, MA: Schenkman.

Graubard, M. A. (1953). *Circulation and Respiration: The Evolution of an Idea.* New York: Philosophical Library.

Hales, S. (1961). *Vegetable Staticks.* London: Oldbourne Science Library.

Hall, T. S. (1975). *History of General Physiology 600 B.C. to A.D. 1900.* 2 vols. Chicago, IL: University of Chicago Press.

Haller, A. von (1966). *First Lines of Physiology.* Translated by W. Cullen. New York: Johnson Reprint.

Holmes, F. L. (1974). *Claude Bernard and Animal Chemistry. The Emergence of a Scientist.* Cambridge, MA: Harvard University Press.

Holmes, F. L. (1985). *Lavoisier and the Chemistry of Life: An Exploration of Scientific Creativity.* Madison, WI: University of Wisconsin Press.

Korf, H.-W., ed. (1997). *Neuroendocrinology. Retrospect and Perspectives.* New York: Springer.

Langley, L. L., ed. (1973). *Homeostasis, Origins of the Concept.* Stroudsburg, PA: Dowden, Hutchinson & Ross.

Lawrence, C., and Weisz, G., eds. (1998). *Greater than the Parts: Holism in Biomedicine, 1920–1950.* New York: Oxford University Press.

Lenoir, T. (1982). *The Strategy of Life: Teleology and Mechanics in Nineteenth Century German Biology.* Boston MA: D. Reidel.

Medvei, V. C. (1993). *A History of Endocrinology.* Carnforth, UK: Parthenon Publishing Group.

Meites, J., and Donovan, B. T., eds. (1975). *Pioneers in Neuroendocrinology.* New York: Plenum Press.

Melhado, E. M. (1981). *Jacob Berzelius. The Emergence of His Chemical System.* Madison, WI: University of Wisconsin Press.

Mendelsohn, E. (1964). *Heat and Life: The Development of the Theory of Animal Heat.* Cambridge, MA: Harvard University Press.

Needham, D. (1971). *Machina Carnis. The Biochemistry of Muscular Contraction in its Historical Development.* Cambridge: Cambridge University Press.

Olesko, K., ed. (1989). *Science in Germany. The Interaction of Institutional and Intellectual Issues.* Osiris, Vol. 3. Philadelphia, PA: History of Science Society

Pagel, W. (1982). *John Baptista Van Helmont: Reformer of Science and Medicine.* Cambridge: Cambridge University Press.

Parascandola, J., and Whorton, J. , eds. (1983). *Chemistry and Modern Society.* Washington, DC: American Chemical Society.

Rossiter, M. W. (1975). *The Emergence of Agricultural Science: Justus Liebig and the Americans, 1840–1880.* New Haven, CT: Yale University Press.

Rothschuh, K. E. (1973). *History of Physiology.* New York: Krieger.

Schultheisz, E., ed. (1981). *History of Physiology.* Elmsford, NY: Pergamon Press.

Shea, W. R. (1991). *The Magic of Numbers and Motion. The Scientific Career of René Descartes.* Canton, MA: Science History Publications.

Todes, D. P. (2001). *Pavlov's Physiology Factory: Experiment, Interpretation, Laboratory Enterprise.* Baltimore, MD: Johns Hopkins University Press.

Wolfe, E. L., Barger, A. C., and Benison, S. (2000). *Walter B. Cannon, Science and Society.* Cambridge, MA: Harvard University Press.

7

MICROBIOLOGY, VIROLOGY, AND IMMUNOLOGY

It is virtually impossible to imagine the development of microbiology without the microscope, but the relationship between the instrument and the conceptual framework of modern microbiology is rather ambiguous. Seventeenth-century microscopists were certainly able to see a new world teeming with previously invisible entities, including protozoa, molds, yeasts, and bacteria. While Antoni van Leeuwenhoek was quite sure that he had discovered "little animals," that were produced by parents like themselves, others took exception to this conclusion. Indeed, questions concerning the nature, origin, and activities of the citizens of the microbial world were not clarified until the late nineteenth century. Several accounts of infusoria were, however, published in the eighteenth century. For example, in 1718, Louis Joblot (1645–1723) published an illustrated treatise on the construction of microscopes that described the animalcules found in various infusions. Carl von Linnaeus (1707–1778), the great arbiter of the names, classification, and (almost) the existence of living things, was quite skeptical of microscopic studies. Believing that an orderly arrangement of all species was the supreme achievement of a naturalist, Linnaeus found the creatures discovered by Leeuwenhoek, Joblot, and others rather a nuisance. Such creatures were tossed into the all-purpose category known as *Vermes*, in a class called *Chaos*. Sorting out these minute, but apparently infinitely diversified creatures provided a special challenge for nineteenth-century microscopists.

During the nineteenth century, microbes were studied by scientists investigating problems as abstract as the origin and evolution of life and as practical as fermentation and putrefaction. Theodor Schwann had predicted that fermentation could be successfully exploited as a model system for investigating the vital processes occurring in each cell of the higher animals and plants. Studies of microbes were also seized upon in the battle over evolutionary theory for insights into the great question called *biogenesis*, the origin of life. If microorganisms were the product of spontaneous generation, they might even fill the gap between living and nonliving things. Microorganisms could also be studied from the medical point of view. Infections and diseases could be analyzed as phenomena analogous to fermentation and putrefaction, processes that seemed to be associated with microorganisms.

The idea that corruption, impurity, or disease could be transmitted by means of contact is an ancient folk belief that was generally ignored in the Hippocratic texts. The establishment of the germ theory of disease, however, is often cast in terms of a conflict between *contagion theory* and *miasma theory*. Sharp distinctions between contagion and miasma theory, however, are actually misleading and anachronistic when applied to the period between the publication of *On Contagion* (1546) by the Renaissance physician and poet Girolamo Fracastoro (1478–1533) and the Golden Age of microbiology in the late nineteenth century. For medical writers of this period, the terms were often used interchangeably. When contagion encompassed harmful material that was indirectly, as well as directly transmitted, it was not incompatible with equally vague definitions of miasma as disease-inducing noxious air.

Fracastoro's *On Contagion* is regarded as a landmark in the evolution of the germ theory of disease, but it was Giovanni Cosimo Bonomo (d. 1697) who provided the first convincing demonstration that a contagious human disease was caused by a minute parasite close to the threshold of invisibility. Bonomo proved that scabies, commonly known as "the itch," was caused by a tortoise-like mite just barely visible to the naked eye. The mites could be transferred directly from person to person, or by means of bedding and clothing used by infested persons. The itch mite, however, was regarded as merely an interesting curiosity rather than a paradigm that might apply to other diseases.

Even after the microscope revealed a world of creatures invisible to the naked eye, most physicians and scientists regarded the notion of "disease-causing animalcules" as little better than ancient superstitions about disease-causing devils and demons. There was, moreover, considerable confusion about whether the minute entities observed under the microscope were the *product* of putrefaction, fermentation, and disease, or the *cause* of such phenomena. Agostino Bassi (1773–1857), however, argued that his studies of the silkworm disease known as muscardine indicated that a minute parasitic fungus caused the disease and that similar agents might cause other diseases. Intrigued by Bassi's work, Johann Lucas Schönlein (1793–1864) demonstrated in 1839 that a fungus could be found in the pustules of ringworm.

Living ferments have been used for thousands of years to produce beer, wine, and bread, but the nature of fermentation was obscure. Originally, the term *ferment* was used to refer to an active substance that could transform a passive (fermentable) substance into its own nature. For example, when yogurt (*ferment*) is added to milk (*fermentable substance*), the milk is transformed into yogurt. In the 1830s, Theodor Schwann (1810–1882) and Charles Cagniard-Latour (1777–1859) called attention to the association between the growth of yeast and the process of alcoholic fermentation and suggested that the development of yeast might be the cause of fermentation. Despite the lucidity of Schwann's reasoning and the ingenious nature of his experiments and observations, the great chemist Justus von Liebig (1803–1873) ridiculed the concept of cellular fermentation. To Liebig, Schwann and Latour were guilty of injecting occult vital phenomena into the purely chemical processes involved in decay, putrefaction, and fermentation. According to Liebig, certain chemicals had the ability to cause the decomposition of other substances. Taking up the attack on Liebig's position, the French chemist Louis Pasteur (1822–1895) defended the hypothesis that the activities of organized beings known as microorganisms cause specific fermentations.

LOUIS PASTEUR

Experiments on the nature of fermentation and the question of spontaneous generation eventually led Louis Pasteur to problems in medical microbiology. The work of Pasteur and his German rival Robert Koch (1843–1910) exemplify the establishment of the theoretical, methodological, and ideological principles of the new science of microbiology. Moreover, Pasteur's life and work reflect the interplay between scientific research, both basic science and applied, and the political and social milieu within which the scientist performs.

As a youth, Pasteur was a diligent student and a talented artist. His father, a tanner and former soldier in Napoleon's army, hoped that his bright and talented son would distinguish himself as a scholar or artist. Many of the portraits the young Pasteur made of family and friends are considered quite good, but at 19 years of age he gave up painting and devoted himself strictly to science. Not all of Pasteur's teachers were impressed with his aptitude; in chemistry he was rated as only mediocre. (Stories about such obvious errors in judgment are commonly encountered in the hagiographies of great scientists, perhaps to make teachers more humble and to offer hope to the truly mediocre.) In the competitive examination for the École Normale Supérieure of Paris, Pasteur ranked sixteenth. Admission to the school was considered an honor in itself, but Pasteur refused to enroll until he felt better prepared. In 1843 he competed again and advanced to fifth place. His first attempt at student life in Paris ended in homesickness so acute that he had to return to his family. Perhaps the most important lesson Pasteur learned as a student of chemistry and physics was the applicability of the experimental approach to a broad range of questions. This made it possible for him to explore problems in biology and medicine, areas in which he had no specific training.

While still a student, Pasteur became intrigued by new studies of crystal structure, stereoisomerism, and molecular asymmetry. These issues may seem quite remote from the biological and medical researches that made Pasteur's name a household word, but they were apparently the source of the impetus, inspiration, and insight that guided him through the labyrinth of his later research problems. The distinguished French scientist Jean-Baptiste Biot (1774–1862) had demonstrated that even in solution, some organic chemicals, such as tartaric acid, rotate the plane of polarized light. The German crystallographer Eilhard Mitscherlich (1794–1863) found that in addition to the common large crystals of tartaric acid there were smaller crystals that he called paratartaric acid or racemic acid. Although both types of crystals had the same chemical properties, racemic acid was inactive to polarized light. Pasteur seized upon this apparent inconsistency and devised experimental means of answering the question: How could chemicals be the same and yet different? After crystallizing different salts of the tartrates and paratartrates, Pasteur was able to separate out two kinds of crystals: left-handed and right-handed mirror images, which had equal and opposite polarizing properties. As he pursued this remarkable trait from the behavior of crystals to the process of fermentation, he came to see molecular asymmetry as a fundamental criterion that distinguished the chemical processes carried out by living organisms from those of the inanimate world. At the end of his life Pasteur expressed regret for abandoning the theoretical research on crystals that, he thought, might have led to the discovery of a fundamental cosmic asymmetrical force and perhaps even the secret of life itself.

Among the aphorisms attributed to Pasteur, the most quoted have to do with the importance of theory and the role of chance in discovery. "Chance favors only the prepared mind," has become the favorite cliché of many earnest teachers, but Pasteur also liked to say "luck comes to the bold." Although Pasteur accepted the idea that, for the good of France, scientific education should be made relevant to industrial and commercial needs, he insisted that the theoretical was as important as the practical. "Without theory," he argued, "practice is but routine born of habit." When asked the use of a purely scientific discovery, Pasteur liked to answer: "What is the use of a new-born child?" The path that led Pasteur from theoretical questions about molecular asymmetry to the behavior of microorganisms involved a progression that Pasteur saw as natural and almost inevitable. He claimed to be "enchained" by the "almost inflexible logic" of his studies. Yet Pasteur was involved in so many theoretical and practical problems that he certainly could have chosen many different pathways.

In 1849 Pasteur was appointed to the University of Strasbourg. While continuing his researches on crystals and preparing his lectures in chemistry, Pasteur met and married the daughter of the Rector of the Academy of Strasbourg. Five years later, having achieved considerable recognition for his chemical researches, Pasteur was appointed Professor of Chemistry and Dean of Sciences at the University of Lille in northern France. The mission with which he was charged included assistance to local industries. To make education more relevant to the needs of the district, Pasteur stressed laboratory experience and created a new diploma for those who wished to enter an industrial career at the level of foreman or overseer.

Through his studies of fermentation, Pasteur discovered that changes in the population of microorganisms were associated with healthy and spoiled fermentations. In addition, Pasteur noticed that fermentations involved the appearance of optically active compounds. From the clues obtained through his previous studies of organic crystals, Pasteur formed the hypothesis that fermentation was a process carried out by living ferments. This hypothesis was opposed by the most renowned chemists of the period, Justus von Liebig (1803–1873), Jöns Jacob Berzelius (1779–1848), and Friedrich Wöhler (1800–1882), who argued that fermentation was a purely chemical process and that microorganisms were the product rather than the cause of fermentation. After analyzing many kinds of fermentations, Pasteur suggested that microorganisms cause all fermentations and that each living ferment is specific for a particular kind of fermentation. Changes in environment, temperature, acidity, composition of the medium, and poisons affect different ferments in particular ways.

In 1857 Pasteur left Lille and joined the École Normale in Paris as Assistant Director in charge of administration and direction of scientific studies. When the Academy of Sciences awarded the Prize for Experimental Physiology to Pasteur in 1860, the great French physiologist Claude Bernard (1813–1878) wrote the report and emphasized the physiological significance of Pasteur's studies on alcoholic, lactic, and tartaric acid fermentation. By this time, Pasteur had begun his studies of spontaneous generation and was reporting his preliminary experiments to the Academy. This problem remained important to Pasteur for professional and political reasons, although he knew that attempting to prove a universal negative is logically absurd and thus neither a scientifically nor a philosophically rewarding proposition.

SPONTANEOUS GENERATION

The question of spontaneous generation was an ancient and controversial one, but Pasteur often argued that microbiology and medicine could only progress when the idea of spontaneous generation was totally vanquished. Belief in the spontaneous generation of life had been almost universal from the earliest times up to the seventeenth century. As suggested by the term "Mother Earth," many peoples have seen the whole world as a nurturing organism, capable of giving birth to living creatures. The lowest creatures, parasites and vermin of all sorts, which often appear suddenly from no known parents, seemed to be the result of some kind of transmutation of lifeless materials into organized beings.

Aristotle had supported the doctrine of spontaneous generation, even cataloging the particular species that would be generated from various substrates. According to Aristotle, heat was necessary for sexual, asexual, or spontaneous generation. While higher creatures reproduce by virtue of their animal heat, the lower forms arose from slime and mud in conjunction with rain, air, and the heat of the sun. A combination of morning dew with slime or manure produced fireflies, worms, bees, or wasp larvae, while moist soil gave rise to mice. Seventeenth-century physician and alchemist Jean Baptista van Helmont (1579–1644) believed that mice could be produced by incubating a flask stuffed with wheat and old rags in a dark closet. Questioning such assumptions, the Italian physician Francesco Redi (1626–1698), a member of the Academy of Experiments of Florence, initiated an experimental attack on the question of spontaneous generation. In keeping with the principles espoused by the Academy, Redi tested various kinds of substrates, raw or cooked, including flesh from animals as varied as lions and lambs, fishes and snakes. Noting the way that different flies behaved when attracted to these substances, Redi suggested that maggots might develop from the objects deposited on the meat by adult flies. While investigating the development of these larvae, he observed that different kinds of pupae gave rise to different species of flies. These studies were published in 1668 as *Experiments on the Generation of Insects.* These experiments did not altogether discredit the concept of spontaneous generation, but they did shrink the field of battle from the generation of macroscopic creatures to the small new world of infusoria and animalcules discovered by seventeenth-century microscopists.

Most seventeenth- and eighteenth-century preformationists rejected spontaneous generation as a contradiction of encapsulation theory and a dangerous materialistic theory. Some philosophers, most notably Gottfried Wilhelm Leibniz (1646–1716), circumvented this obstacle by positing the existence of living molecules, or monads. Following the precedent established by Redi, Louis Joblot (1645–1723) carried out a series of experiments that seemed to prove that infusoria are not spontaneously generated. After boiling nutrient broth, Joblot divided it into two parts. One half was sealed off and the other left uncovered. The open flask was soon teeming with little animals, but the sealed vessel was free of infusoria. To prove that the medium in the sealed flask was still susceptible to putrefaction, Joblot exposed it to air and followed the growth of infusoria. In support of spontaneous generation, the French naturalist Georges Buffon (1707–1788) and the English microscopist John Turbeville Needham (1713–1781) launched an attack on Joblot's work. When Needham repeated Joblot's

Francesco Redi

experiments, whether the flasks were open or closed, the water boiled or not boiled, the infusions placed in hot ashes or not, all vessels soon swarmed with microscopic life. Needham, therefore, concluded that there was a vegetative force in every microscopic bit of matter and every filament that had been part of living beings. When animals or plants died, they slowly decomposed to one common principle that Needham thought of as a universal semen from which new life arose. These experiments were published in the *Philosophical Transactions of the Royal Society* (1748).

The claims of Needham and Buffon did not, however, stand unchallenged for very long. Inspired by reading Francesco Redi's *Experiments on the Generation of Insects* (1668), Lazarro Spallanzani began a series of experiments that exposed Needham's fallacious assumptions and questionable techniques. By heating a series of flasks for different lengths of time, Spallanzani determined that various sorts of microbes differed in their susceptibility to heat. Whereas slight heating destroyed some of the larger animalcules, other very minute entities seemed to survive in liquids that had been boiled for almost an hour. Further experiments convinced Spallanzani that all animalcules entered the medium from the air. Convinced that a great variety of little "eggs" must be disseminated through the atmosphere, Spallanzani concluded that air could either convey the germs to the infusions or assist the multiplication of those already in them.

Although Spallanzani's experiments answered many of the questions raised by advocates of spontaneous generation and proved the importance of rigorous sterilization, his critics claimed that he had tortured the all important "vital force" out of the organic matter by his cruel treatment of his media. The vital force was, by definition, capricious and unstable, making it impossible to expect reproducibility in experiments involving organic matter. Another objection was posed by the French chemist Joseph Louis Gay-Lussac (1778–1850), who showed that the sterile vessels lacked oxygen. This suggested that oxygen was necessary for fermentation and putrefaction. Fortunately for the food industry, Nicolas Appert (1750–1841), a French chef, ignored the theoretical aspects of the dispute and applied Spallanzani's techniques to the preservation of food. He placed foods in clean bottles, corked them tightly, and raised them to the boiling point of water. His techniques were described in a book published in 1810, which heralds the beginnings of the canning industry.

During the nineteenth century, the design of experiments for and against spontaneous generation became increasingly sophisticated, as proponents of the doctrine challenged the universality of negative experiments. Theodor Schwann repeated many of Spallanzani's experiments, but added the refinement of heating the air as well as the medium. To prove that the vital principle had not been tortured out of the heated air, he demonstrated that a frog could live quite happily when supplied with such air. Still, Schwann's results were not wholly consistent or reproducible. In another attempt to purify the air used in such experiments, Franz Schulze (1815–1873) passed air through concentrated potassium hydroxide or sulfuric acid. Other attempts to perform these experiments gave equivocal results, while critics still claimed that Schulze's methods tortured the vital principle. Heinrich Schröder (1810–1885) and Theodor von Dusch (1824–1890) adopted a new approach; they filtered the air through a long tube of cotton wool before it entered a flask of putrefiable medium. While this treatment should have been gentle enough to satisfy the opposition, it was not uniformly successful. Because any single case of apparent spontaneous generation could allow proponents of the theory to maintain that it only occurred under special, favorable, perhaps even sympathetic conditions, opponents were always on the defensive.

One of the most vigorous defenders of spontaneous generation was Félix Archiméde Pouchet (1800–1872), Director of the Natural History Museum in Rouen, a member of many learned societies, and a respected botanist and zoologist. In 1858, he presented the first of a series of papers on *heterogenesis*, his new term for spontaneous generation, to the Academy of Sciences of Paris. Since Pouchet believed it was self-evident that nature employed spontaneous generation as one of the means for the reproduction of living things, his experiments were designed not to determine whether heterogenesis took place, but only to discover the circumstances under which it occurred. Adopting assumptions similar to those of Buffon and Needham, Pouchet rejected the idea that life arose de novo accidentally in solutions of inorganic chemicals. According to Pouchet, the factors that promoted heterogenesis were the vital force, organic matter, water, air, and the proper temperature. Any alterations in these prerequisites could affect the kind of organisms produced. Each factor, therefore, had to be examined systematically through a series of careful experiments. Pouchet's ingenious experiments on spontaneous generation generally gave positive results. Just as some gardeners seem to be blessed with a green thumb, Pouchet apparently had a heterogenetic thumb. Moreover, like his main opponent, Louis Pasteur, Pouchet was a vigorous and enthusiastic experimenter.

While many scientists criticized Pouchet's work, Pasteur forced his adversary to admit that the existence of germs in the air was the critical issue in establishing the experimental basis of the debate. Work on fermentation had shown Pasteur that the so-called ferments were microorganisms and that the air was the source of these entities. Rigorous experimentation, Pasteur argued, forced reasonable people to conclude that spontaneous generation is a chimera. All purported evidence in support of the doctrine was the result of flawed techniques. Demonstrating the germ-carrying capacity of air was, therefore, Pasteur's first priority. If the microbes associated with fermentation were brought to their substrate by the air, it should be possible to intercept them by sucking air through filters and trapping the dust particles. After washing such filters in a mixture of alcohol and ether, Pasteur collected the dust and examined it under the microscope. The number of microbes found varied with environmental factors. For example, the germ content of hospital air was quite high, while that of mountain air was low. Both Pasteur and Pouchet explored mountains and glaciers to test different kinds of air. Pasteur even considered a balloon ascent to literally rise above Pouchet, who had taken dust samples from the roof of the Cathedral of Rouen and the tomb of Rameses II.

Many of these contests were very theatrical, but Pasteur's most convincing demonstrations involved apparently simple experiments using flasks with peculiar long necks drawn out and bent into a curve resembling the neck of a swan. If liquids were heated in these flasks, the medium remained sterile even though ordinary air could enter via the swan neck. Germ-laden dust particles were, however, trapped in the curve of the neck. If the flask was tipped so that the medium sloshed through the bend in the neck, or if the neck was broken off and dust entered the flask, the medium soon seethed with microbial life. Experimenting with various media, Pasteur proved that even the most easily decomposable fluids, such as milk, blood, or urine, could remain sterile if proper precautions were taken. He also showed that microbes could be grown on a simple defined medium, that certain microbes grew only in the absence of oxygen, and that microbes remained true to their original type and were not transmuted into different species. These years of experimentation on the germ-carrying capacity of the air apparently caused some of Pasteur's peculiar mannerisms. Whether at home or dining out, he never used a plate or a glass without examining it for some speck of dust.

Clearly, Pasteur's experiments did not deal with the question of the ultimate origin of life, a process Thomas Henry Huxley named *biogenesis*. In practice, these experiments demonstrated that microbes do not arise de novo in properly sterilized medium under conditions prevailing today. In 1862 the Academy of Sciences awarded Pasteur a prize for this work, but Pouchet issued another challenge. A commission was appointed to settle the debate. Curiously, Pouchet withdrew from the showdown scheduled for 1864. When echoes of the old debate appeared again in the 1870s, Pasteur undertook a new series of experiments in his vineyards to prove that living yeasts were necessary for the fermentation that produced wine. In August, before yeasts become associated with the grapes, he had some hothouses built to seal off parts of the vineyard. When these grapes were harvested in October, they did not ferment unless yeasts were added.

The last major champion of spontaneous generation was Henry Charlton Bastian (1837–1915), professor of pathological anatomy at University College Hospital, London, and author of a 1000-page tome entitled *The Beginnings of Life*. Going beyond other heterogenesists, Bastian claimed he had evidence for *archebiosis*, the production of life de novo from inanimate matter even in the simplest solutions. In a famous letter to Bastian written in 1877, Pasteur explained that he felt it absolutely necessary to fight and conquer the advocates of spontaneous generation because their doctrine was an obstacle to progress in the art of healing. As long as doctors and surgeons continued to believe in the spontaneity of all diseases, their efforts at treatment and prevention would remain largely futile.

Further experimental refinements were introduced by John Tyndall (1820–1893), an Irish physicist who had studied the optical properties of particulate matter, a phenomenon now known as the light-scattering or Tyndall effect. Inspired by Pasteur's work, he carried out experiments on the germ-carrying capacity of the air and the efficacy of various sterilization methods. These studies were described in his *Essays on the Floating Matter of the Air in Relation to Putrefaction and Infection* (1881). Using a chamber in which light scattering by dust particles served as a measure of the optical activity, or biological purity of the air, Tyndall trapped dust particles with a layer of glycerine until the air was optically empty and, therefore, biologically pure. Test tubes filled with various infusions remained sterile when kept in this "Tyndall Box." In tests designed to probe the heat sensitivity of microbes, Tyndall discovered that some germs formed remarkably heat-resistant spores. These tests led to a method of sterilization involving cycles of heating and cooling known as fractional sterilization, or tyndallization. After destroying the active bacteria in the first step of sterilization, Tyndall allowed the medium to cool and then heated it again. Discontinuous boiling for one minute in five separate steps could sterilize medium containing microbial species that had resisted one hour of continuous boiling. Tyndall, who had been attacked by Bastian for intruding into the domain of biologists and physicians, was awarded an honorary Doctor of Medicine degree by the University of Tübingen. Further studies of the nature of microorganisms made it increasingly obvious that they were not simple blobs of organic material. Even the smallest bacteria appeared to be too complex to simply precipitate out of solution. The discovery of viruses stimulated a brief revival of the idea, but even viruses, while very small, are quite complex and endowed with definite mechanisms for replication.

Advocates of the doctrine of spontaneous generation have argued that some form of the doctrine is necessarily true in the sense that if life did not always exist on earth, it must have been spontaneously generated at some point. While avoiding the question of spontaneous generation on earth, some scientists and philosophers have argued that the seeds of life and epidemic disease come to earth from outer space in the form of spores, bacteria, or viruses. Other scientists contend that a process somewhat like spontaneous generation must have occurred here on earth when the planet was very young.

The idea that living beings were formed by means of the mixing of the primary elements in the nourishing environment of the ancient oceans appears in the Indian texts known as the *Rig Veda* and the *Atharva Veda*. Modern attempts to understand the origin of life, however, can be traced back to the 1920s, when A. I. Oparin (1894–1980), a Russian chemist, proposed an apparently plausible explanation for the chemical evolution of life. According to Oparin, organic molecules could have formed in the ancient seas, because the atmosphere of the ancient earth, a reducing atmosphere rich in methane, hydrogen, and ammonia, was very different from the oxygen-rich atmosphere of today. In a world devoid of living creatures and free oxygen, organic molecules could have accumulated until parts of the sea resembled a hot, thin soup. In this rich primordial soup, Oparin argued, the origin of life was not the result of some "happy chance," but "a necessary stage in the evolution of matter." To some extent, Oparin's speculations were vindicated by experiments performed in the 1950s by Stanley Miller (1930-), which proved that biological molecules can be produced when energy is added to a mixture of the gases that were presumably present in the atmosphere when the earth was young.

THE GERM THEORY OF DISEASE

Primitive ideas about *contagion* dealt with the general notion of transfer through contact and were not directly related to the modern germ theory of disease. Just as heat and cold were directly transferred to neighboring bodies, so too were putrefaction, uncleanliness, corruption, and disease. Epidemics were probably rare in small bands of primitive peoples, but they would have been terrifying events once population density increased sufficiently to produce and sustain them. Many peoples believed diseases were sent by the gods as punishment for their sins, but the Hippocratic physicians rejected supernatural explanations of disease and had little interest in the idea that disease was spread by contagion. Nevertheless, some Greek and Roman philosophers, poets, and architects thought that disease might be caused by tiny animals dwelling in swampy places or by contact with the sick and with contaminated articles. For example, in his discussion of hygienic regulations for selecting building sites, the Roman architect Marcus Terentius Varro (177–27 B.C.) warned against swampy locations. Tiny animals living in swampy places, so small as to be invisible, might enter the body through the mouth and nose and cause grave illnesses. More commonly, however, epidemic diseases were associated with comets, eclipses, floods, earthquakes, or major astrological disturbances that charged the air with poisonous vapors known as *miasmata*.

During the sixteenth century, Girolamo Frascastoro (1478–1553), poet, physician, and mathematician, attempted to analyze the kinds of evidence that seemed to support the miasma and the contagion theories of disease transmission. At the University of Padua, Fracastoro immersed himself in philosophy, literature, mathematics, and astronomy. After earning his M.D. in 1502, Fracas-

Girolamo Fracastoro

toro taught philosophy, established a private practice, and engaged in medical research. In 1530, he published the medical classic *Syphilis, or the French Disease*. The poem that gave the disease its modern name also described the natural history of syphilis, contemporary methods of treatment, and the controversy about the origin of the disease. Courting patronage in an age of uncertainty, Fracastoro dedicated his treatise to Cardinal Pietro Bembo, Secretary to Pope Leo X.

In his major medical work, *On Contagion and the Cure of Contagious Diseases* (1546), Fracastoro distinguished three forms of contagion and speculated about the existence of minute seeds of disease. Having observed the epidemics of syphilis, plague, and typhus that ravaged Italy in the sixteenth century, Fracastoro compared the miasmatic theory of disease with contagion theory. As part of his analysis, he described three different ways in which the seeds of diseases could be disseminated. Some diseases, such as syphilis and gonorrhea, were only transmitted by direct contact. Other diseases were transmitted by direct contact and by *fomites*, i.e., inanimate articles, such as clothing, that had been in contact with the sick. In the third category, he placed diseases such as tuberculosis and smallpox, which were apparently transmitted by direct contact, by fomites, and by contagions capable of infecting victims at a distance from the sick. Thus, the contagion of diseases could be compared to fermentation or putrefaction, which spread from grape to grape or apple to apple.

Although Fracastoro has been called the founder of the germ theory of disease, his writings were more ambiguous than this title would imply. In general, the idea of contagion existing in tiny germs or seeds of disease did not prove very useful in guiding medical practice. The contagion theory led to attempts to stop epidemics by quarantines, isolation, and disinfection, but these methods did not seem to be successful enough to justify their inconvenience. While such measures might have inhibited the spread of bubonic plague to some extent, they were quite ineffective against typhus fever, typhoid fever, and cholera. During epidemics, isolation of the sick had little or no effect. Public health reformers, therefore, seemed to have good reason to believe that filth and poisoned air spread such diseases.

In 1840 Jacob Henle (1809–1885), a prominent German pathologist, physiologist, and anatomist, revived the contagion theory and published his examination of the relationships among contagious, miasmatic, and miasmatic-contagious diseases. His contemporaries generally ignored his *On Miasmata and Contagia*, but after the establishment of the germ theory of disease it was retrospectively recognized as a landmark. Analyzing the patterns of transmission of various diseases seem to prove that malaria was a purely miasmatic disease, while smallpox, measles, scarlet fever, typhus, influenza, dysentery, cholera, plague, and puerperal fever were miasmatic-contagious. Diseases such as syphilis, foot-and-mouth disease, and rabies were acquired only through contagion.

Moreover, Bonomo, Bassi, and Schönlein had provided evidence that mites and mircoparasites transmitted scabies, muscardine, and ringworm. Clearly separating the concept of *disease* from the concept of *parasite*, Henle defined the material basis of contagion as an organic entity capable of living a separate existence, or a parasitic existence within the diseased body. That is, the contagion was not the disease itself but the inducer or cause of disease. Critically evaluating the experimental evidence, Henle discussed the nature of the proofs that would be required to establish a causal relationship between microbes and disease.

According to Henle, physicians blamed disease on miasma, which they defined as something that mixed with and poisoned the air, but, he argued, no one had ever demonstrated the existence of miasma. It was simply assumed to exist by exclusion, because no other cause could be demonstrated. A more likely hypothesis, Henle asserted, was that *contagia animata* (living organisms) caused contagious diseases because whatever the morbid matter of disease might be, it obviously had the power to increase in the afflicted individual. The natural history of epidemics could be explained by assuming that an agent excreted by sick individuals caused disease. If the lungs excreted this agent, it might pass to others through the air; if excreted by the intestines, it could enter sewers and wells. Given the fact that the pus from pox pustules could be used to infect a multitude of people, the contagion must be an animate entity that multiplies within the body of the sick person. Chemicals, organic or not, remain fixed in amount; only living things have the power to reproduce and multiply. While Henle's discussion of the question of contagion and miasma is superficially similar to that of Fracastoro, the context in which they worked and the centuries that separated them infused very different meanings into the terms miasma and contagion.

Using Schwann's metabolic theory of the cell as a point of departure, Henle compared the action of contagion to fermentation. Well aware of the difficulties confronting the contagion theory of disease, Henle warned that finding some microorganism in the sick did not prove that it had a causal role. The agent must be isolated and cultured so that it was free from any toxins or tissues of the diseased individual. Indeed, Henle came close to outlining what are usually called "Koch's postulates." Acknowledging the lack of rigorous evidence for the germ theory of disease, Henle argued that science should not wait for unequivocal proofs because scientists could only conduct research "in the light of a reasonable theory." A major problem in establishing the truth of germ theory was methodological. Obtaining pure cultures was a difficult and tedious procedure, almost impossible in the hands of any but the most meticulous experimentalists. Working out proper sterilization and culture procedures required a prior understanding of and commitment to the germ theory of disease. Given the state of the methods available until quite late in the nineteenth century, it was no wonder new animalcules seemed to appear spontaneously in the tempting broths set out for them, as well as in the raw wounds of surgical patients. Confused by the claims of the heterogenesists, physicians tended to reject the germ theory or dismiss it as a laboratory curiosity unrelated to clinical medicine. But as Pasteur's associate Émile Duclaux declared: "The great merit of a new theory is not to be true, because there is no such thing as a true theory, but to be fruitful."

Research on crystals led Pasteur to investigate the nature of fermentation and the diseases of wine, beer, and vinegar. His next practical challenge was a mysterious disease threatening the silk industry of France, causing great despair and hardship in many households and villages. Although Pasteur knew very little about silkworms and their diseases, he expected to resolve the problem by finding a germ. The problem was more complex than expected, because it involved two different disease organisms, as well as nutritional and environmental effects. The epidemic was actually the result of complex interactions among host, germ, and environment. From silkworm diseases, Pasteur progressed to the riddle of disease in higher animals and, finally, to rabies in humans. During this phase of his career, Pasteur was devastated by the deaths of two of his daughters, his son's war experiences, and a series of strokes that left him partially paralyzed.

Medical microbiology owes much to Pasteur's studies of chicken cholera, a disease unrelated to human cholera, except in the virulence of the infection. Chickens that picked at foods soiled with the excreta of the sick acquired the disease. Based on reports by veterinary surgeons that bacteria were present in the tissues of sick chickens, Pasteur isolated a microbe that could be cultured in a medium made of chicken gristle. A small drop of the fresh culture would quickly kill a chicken, but the microbe only caused a small abscess in guinea pigs. Eventually, Pasteur discovered that it was possible to produce weaker cultures of the microbe, which could be used as vaccines. Laboratory-created attenuated cultures did not cause disease when injected into chickens, but these inoculated animals resisted infection when they were later challenged by virulent cultures of the microbe.

Based on experiments with chicken cholera, Pasteur announced that it would soon be possible to create artificial vaccines that acted against virulent diseases just as Edward Jenner's (1749–1823) cowpox vaccine protected people from smallpox. That is, manipulating laboratory cultures could modify the virulence of pathogenic bacteria. Changing the laboratory conditions under which bacteria were grown would make it possible to decrease virulence and obtain a vaccine that caused only mild disease while inducing immunity to the natural disease. The microbe that caused chicken cholera could be attenuated by storing the cultures for variable intervals, but it was more difficult to attack the microbe that caused anthrax, because the bacillus was able to form protective spores.

Anthrax, a disease that primarily attacked sheep and cattle, also infected farmers and butchers. In many French provinces as many as 20 percent of the

Louis Pasteur dictating an article on the diseases of silkworms to Madame Pasteur

sheep succumbed to this disease. Some farms seemed especially cursed by the scourge and experienced even greater losses. Afflicted sheep often died within hours after the onset of symptoms. Autopsies revealed thick black blood and a black and liquid spleen. Several microscopists noted the presence of rod-shaped bacteria in the blood of anthrax victims, but Robert Koch was the first to provide unequivocal proof that the disease was caused by a specific pathogenic microbe. Although Koch later claimed that a veterinarian named Jean-Joseph Henri-Toussaint (1847–1890) invented anthrax vaccine, it was Pasteur who first publicly demonstrated the power of vaccination.

Pasteur hoped his work on animal vaccines would be extended to human diseases, but when contemplating the ethical dilemma of human experimentation he concluded that the experiments that could be carried out with animals would be criminal acts if applied to human beings. One way of escaping this impasse was to study a disease that attacked both humans and animals. If the disease was invariably fatal in humans, giving doomed patients a vaccine that had saved experimental animals could be justified. Human rabies is a disease that seemed to fit these criteria. The disease is very rare in humans, but once people exhibited symptoms of rabies, no matter what the treatment, the outcome was invariably a painful, agonizing death. Early symptoms of clinical rabies were vague: fatigue, malaise, irritability, pain at the site of the wound, severe headaches, and pain in the stomach or chest. The next stage revealed the involvement of the central nervous system: difficulty in breathing, hypersensitivity to light, thirst accompanied by intense aversion to liquids, gagging, choking, and convulsions. During the late stages of the disease, the victims appeared to be "mad," or ferocious, with convulsions, hallucinations, spitting, and excessive salivation. Cardiac arrest or respiratory failure ultimately caused death. Yet, it is very important to remember that not all humans bitten by rabid animals contract the disease. The incubation period is long and variable, and diagnosis is uncertain during the early phase. Some individuals, overcome by fear, exhibited what has been called "false rabies" long after they were bitten. Even with modern methods of testing, diagnosis is so difficult that sometimes it is made only through postmortem analyses.

The image of the rabid animal still exerts a powerful and terrifying grip on the popular imagination. In 1831, a wolf attack on people in Villers-Farlay and Arbois resulted in eight deaths from rabies. Those who had been bitten by the rabid wolf were subjected to the traditional treatment at the blacksmith's shop, including cauterization with a red-hot iron. Pasteur, who was only a child at the time, long remembered the howls of wolves in the countryside surrounding his village and the screams of the people being "branded." About 50 years later, the mayor of Villers-Farlay wrote to Pasteur that a rabid dog had attacked Jean-Baptiste Jupille, a 15-year-old shepherd. Jupille is generally considered one of the first patients to receive Pasteur's rabies vaccine. Critical studies of Pasteur's notebooks reveal that he had previously tested rabies vaccines on patients, but the results were inconclusive.

In his studies of other diseases, Pasteur had begun by identifying and then isolating the microbe that appeared to be the etiological agent. No visible agent could be found in preparations of materials capable of transmitting rabies, but Pasteur persisted by using live animals to "culture" the rabies virus. During the nineteenth century, the term *virus* was used in a nonspecific way in reference

Louis Pasteur studying rabies

to unknown agents of disease. In 1879, Pierre-Victor Galtier (1846–1908), a professor at the Veterinary School of Lyon, had reported that rabies could be transmitted from dogs to rabbits. In Pasteur's laboratory the incubation period between the bite of a rabid animal and the onset of disease was further shortened by inoculating material from rabid animals directly into the brain of dogs or rabbits. Although the "microbe of rabies" was invisible, a vaccine could be prepared by suspending the spinal cords of infected rabbits in a drying chamber. After about 2 weeks the material became virtually harmless. By proceeding from harmless to more virulent materials in a series of inoculations, Pasteur could protect dogs that had been bitten by rabid animals or inoculated with virulent preparations.

On July 6, 1885, 9-year-old Joseph Meister was brought to Pasteur after an attack by a rabid dog left him with deep wounds on his hands, legs, and thighs. Advised by physicians that the case was hopeless, Pasteur began a series of inoculations. Meister recovered completely, and within a year more than 2000 people received the Pasteur rabies vaccine. Even after statistics apparently confirmed the value of rabies vaccine, critics continued to argue the vaccine was neither safe nor effective. According to Pasteur's adversaries, his so-called vaccine was really an artificial virus that caused a new disease, i.e., laboratory or Pastorian rabies. Certainly rabies vaccines involve specific risks as well as unique ethical dilemmas. Because the sick person is not a threat to others, the vaccine it not a public health issue in the usual sense. The only function of the vaccine is to protect the specific patient. If the animal that has bitten the patient is not captured and tested, a person who was not really at risk for rabies might needlessly undergo a series of painful and dangerous inoculations. Estimates of the risk of rabies in people bitten by rabid dogs range from 50 to 60 percent. Bites by rabid wolves and cats are more dangerous than dog bites.

The rabies vaccine brought Pasteur international acclaim and led to the establishment of the Pasteur Institute. Unfortunately, when the Institute was dedicated in 1888, Pasteur was too weak to speak at the ceremonies. The Pasteur Institute began with microbiology laboratories and gradually added more buildings and laboratories, a hospital, and outpatient clinics. The first Institute served as the model for a network of Pasteur Institutes that were established in the former French colonies. For twentieth-century researchers like Nobel Laureate François Jacob (1920-), who began working at the Pasteur Institute in 1949, the Institute was still the heart of French biology, a place where innumerable discoveries had been made by almost legendary scientists. It remained a unique and highly flexible institution, unlike the universities and other organizations that invariably staggered under the burdens of bureaucracy. Workers at the Institute participated in an annual tribute to Pasteur by assembling at the Institute's oldest building, which housed Pasteur's crypt, and marching past his tomb. The ostentatious mausoleum contains mosaics depicting aspects of Pasteur's work on alcoholic fermentation, silkworm diseases, chicken cholera, anthrax, and rabies, and four angels depicting Faith, Hope, Charity, and Science. Marble panels on the walls of the tomb contain inscriptions recording Pasteur's victories: molecular dissymmetry, fermentation, spontaneous generation, studies on wine, silkworm disease, studies on beer, virulent diseases, vaccines, and prophylaxis against rabies. After the ceremonial visit to the tomb of Pasteur, the Pastorians march towards the simple tomb of Émile Roux (1853–1933), who succeeded Émile Duclaux (1840–1904) as director of the Institute.

ROBERT KOCH

Very different from Louis Pasteur in training, temperament, and philosophical approach, Robert Koch (1843–1920) was most successful at formulating the principles and techniques of modern bacteriology. While he lacked Pasteur's flair for staging dramatic presentations, his contemporaries considered the rather phlegmatic Koch "a man of genius both as technician and as bacteriologist." Unlike Pasteur, who began his career as a chemist, Koch approached bacteriology as a physician, and his research was primarily motivated by medical questions.

Koch was born in a small village in the Harz Mountains. He was the third of 13 children born to Herrmann Koch, a mining administrator, and his wife Mathilde. When Koch began his medical studies at the University of Göttingen, the faculty included many eminent scientists, but not even Jacob Henle seemed to have any interest in the relationship between bacteria and disease. Nevertheless, Koch's teachers did awaken in him a love of scientific research. After earning his degree and passing his state medical examinations in 1866, Koch spent several months in Berlin studying clinical medicine at the Charité hospital and attending lectures presented by Rudolf Virchow (1821–1902), the founder of cellular pathology. Dreaming of travel and adventure, Koch attempted to find work as a ship's doctor or military surgeon. But after becoming engaged to Emmy Fraatz, he seemed to be doomed to the dreary life of a country doctor and district medical officer. During the Franco-Prussian War of 1870, Koch volunteered for the medical service. Although military medicine disabused him of his romantic notions of adventure, like many other doctors Koch discovered that war was the ultimate medical education.

Despite the demands of his private practice and official duties, Koch explored new ideas in natural history, archaeology, photography, public health, hygiene, and bacteriology. He established a small laboratory next to his consultation room and purchased a Zeiss oil immersion microscope and the illuminating apparatus invented by the Ernst Abbe (1840–1905). Experimenting with the new aniline dyes, Koch was able to get clear images of bacteria and distinguish between what he called the "structure picture" and the "color picture." Thus, when an outbreak of anthrax occurred in his district, Koch was able to make

Robert Koch

his own tests of the relationship between the anthrax bacillus and the disease. Although anthrax is primarily a disease of sheep and cattle, it can cause severe, localized skin ulcers known as malignant pustules in humans, as well as a dangerous condition known as gastric anthrax, or a virulent pneumonia known as wool-sorter's disease. By 1860, Franz Pollender (1800–1879), Pierre Rayer (1793–1867), and Casimir Joseph Davaine (1812–1882) had observed bacteria in the blood of anthrax victims, but their evidence for an association between the bacteria and the disease remained largely circumstantial, despite attempts to transmit the disease to experimental animals by inoculating them with blood from diseased animals.

Bacillus anthracis, the causal agent of anthrax, is a rather large entity and appears in the bloodstream of infected animals in such large numbers that it can be detected by direct examination of blood. Thus, like his predecessors, Koch found bacteria in the blood of infected animals. After successfully transmitting the disease to rabbits and mice, he transferred the disease through a series of experimental animals. To provide further proof that the disease was caused by *Bacillus anthracis* itself, rather than some poison in the blood of sick animals, Koch cultivated tiny bits of spleen from infected animals in aqueous humor from the eye of a cow. This material served as the first link in a chain of cultures transferred through fresh aqueous humor. Even after 10 to 20 transfers, a bacterial culture grown in aqueous humor could kill a mouse just as quickly as blood taken directly from a diseased sheep. These experiments ruled out the possibility that the disease was caused by some poison from the original animal. Obviously, only an entity capable of multiplying in experimental animals and in laboratory cultures could create such a chain of infection. While studying anthrax bacilli trapped on microscope slides, Koch noticed that as the medium evaporated the thread-like chains of growing bacteria were transformed into bead-like spores. By adding fresh medium, he could reverse this transformation.

In 1876, when Koch was convinced that he had isolated the microbe that caused anthrax, established its life cycle, and explained the natural history of the disease, he contacted Ferdinand Cohn (1828–1898) for advice and criticism. Primarily a botanist, Cohn was the first prominent German scientist to take a special interest in bacteriological research. Of course, Koch was not the first amateur microbiologist to approach Cohn with "unequivocal evidence" that would solve the riddle of infectious disease. Despite his initial skepticism, Cohn invited Koch to come to the Institute of Plant Physiology at the University of Breslau and demonstrate his findings. Impressed by the rigorous nature of Koch's experiments, Cohn arranged for the publication of Koch's paper "The Etiology of Anthrax Based on the Developmental Cycle of *Bacillus anthracis*" in the journal *Contributions to Plant Biology*.

The complex life cycle of *Bacillus anthracis* immediately explained the mystery of the persistence of anthrax in pastures that farmers came to think of as permanently cursed by the disease. Because spores survived under harsh conditions, one contaminated carcass, dumped in a shallow pit, could serve as a reservoir of spores for many years. Spores were the most general means of disseminating the disease among animals, but people who slaughtered, butchered, or skinned infected animals were exposed to active bacilli. Having satisfactorily elucidated the role of *Bacillus anthracis*, Koch proposed various prophylactic measures to prevent the dissemination of the disease. The first step was the proper disposal of carcasses. Sheep buried in shallow wet graves served as excellent foci for the development and dissemination of spores because spore formation required air, moisture, and a temperature above 15°C. Since the average soil temperature at a depth of 1–10 meters in Germany was below the critical temperature, carcasses buried in deep dry trenches far from farms would become harmless. Surveys of disease patterns in sheep, cattle, and horses suggested that sheep were the normal reservoir for anthrax. Simply separating other animals from sheep interrupted the chain of infection.

Even after Koch's classic work on the anthrax bacillus, the controversy concerning the etiology of the disease and the role of *Bacillus anthracis* continued. In 1877 the French physiologist Paul Bert (1833–1886) announced that oxygen destroyed the anthrax bacillus, but tainted blood could still cause the disease. Therefore, Bert claimed, the bacillus was not the cause of anthrax. This argument drew Pasteur into the battle. Carrying out a series of 100 transfers from culture to culture, Pasteur purified the bacteria in media composed of broth or urine. Only a living organism capable of multiplying in the course of these transfers could be the virulent agent. That is, the dilution factor for this painstaking procedure was so great that no trace of any putative poison in the original sample could possibly remain. In explaining Bert's observations, Pasteur noted that much confusion had arisen because some experimenters used materials from long-dead carcasses. In such cases, their experiments involved the germs of putrefaction rather than anthrax. Another peculiarity of the anthrax bacillus was discovered in connection with a dispute as to whether chickens could be affected. Under normal conditions, chickens were not susceptible to the anthrax germ. To test the hypothesis that chickens did not acquire anthrax because of their high body temperature, Pasteur immersed an inoculated hen in a cold water bath. The chilled chicken took the disease and died. One control chicken had been inoculated but not placed in a similar bath; another had been placed in a bath but not inoculated. Later research suggested that Pasteur's hypothesis about resistance in chickens was incorrect, but the dispute drew attention to the complex phenomenon of differential susceptibility among various species.

Although it was Koch who demonstrated the bacterial etiology of anthrax and the critical problem of spore formation, it was Pasteur who developed a preventive vaccine. By growing bacteria under controlled laboratory conditions, Pasteur argued, a series of attenuated cultures could be established. In the case of the anthrax bacillus, cultivation in broth at 42–44°C was critical; spores could not develop at a temperature above 42°C, but the vegetative form died at 45°C. After several weeks of cultivation under these conditions, the anthrax bacillus gradually lost its ability to cause disease and could be used as a preventive vaccine. With the cooperation of the Agricultural Society of Melun, a public demonstration of the anthrax vaccine was held in May 1881 at Pouilly-le-Fort. The editor of the *Veterinary Press*, who called Pasteur the "prophet of microbiolatry," initiated a campaign for subscriptions to test Pasteur's anthrax vaccine. At the time of the well-publicized demonstration, Pasteur and the veterinarian Jean-Joseph Henri-Toussaint were competing to produce a successful vaccine. Opposed to the idea of chemically treated dead vaccines, Pasteur was furiously working on the development of a safe and effective live anthrax vaccine, but it was not ready at the time of the demonstration.

On May 5, 1881, before a crowd of supporters and skeptics, Pasteur and his associates inoculated 24 sheep, 6 cows, and 1 goat with anthrax vaccine; 24 sheep, 4 cows, and 1 goat were set aside as controls. In very public predictions, Pasteur insisted that all the vaccinated sheep would survive and the unvaccinated would die of anthrax. Late in May, all of the animals were inoculated with a highly virulent strain. In June, when the contest ended, Pasteur achieved a stunning success. The vaccinated animals remained free of anthrax, but the unvaccinated sheep and goat died and the unvaccinated cows became ill. Even among historians and philosophers of science 100 years later, there is little doubt that the demonstration at Pouilly-le-Fort was an event of high drama, deliberately planned and orchestrated by Pasteur as part of his strategy for effecting a revolution in biology, medicine, and hygiene. Understanding how Pasteur was able to mobilize support from the French public and much of the scientific establishment during what has been called the "high point of the scientific religion" is more problematic. Despite the widely held belief that Pasteur used a live attenuated vaccine at Pouilly-le-Fort, his research notebooks reveal that he actually used the chemically treated anthrax vaccine developed by his associates, Charles Chamberland (1851–1908) and Émile Roux (1853–1933).

By 1894, millions of sheep and cattle had been vaccinated against anthrax. Despite the apparent success of Pasteur's approach to the development of preventive vaccines, Koch used the pages of his journal *Record of the Works of the German Sanitary Office* to argue that Pasteur was incapable of cultivating pure strains of microbes. Constantly denigrating Pasteur's contributions to bacteriology, Koch ridiculed Pasteur's experiments on anthrax in chickens, his account of the role played by earthworms and spores in propagating anthrax, and even the preventive value of vaccination. In response to the honors awarded to Pasteur as a "second Jenner," Koch called attention to the obvious fact that his French rival was not a physician and noted that Jenner's vaccine saved human lives rather than sheep. Unfortunately, bitter and disruptive disputes between the French and German pioneers of microbiology were not uncommon and, at least in part, reflected the hostilities between their countries. Deeply troubled by the French defeat in the Franco-Prussian War, Pasteur blamed the neglect of science for the loss and saw his work as part of a much needed effort to raise the level of French science and industry. In retrospect, it is clear that Koch was correct about the superiority of the methods he had developed to identify and purify the bacterial agents of disease. Nevertheless, it was Pasteur who established practical methods of preventing anthrax, and, while quantity need not correlate with quality, it is interesting to note that Pasteur published 31 papers on anthrax, whereas Koch published only 2.

It should be noted that the anthrax bacillus, which can be disseminated in the form of spores, has been called the ideal biological weapon. Since the 1990s, the threat of biological and chemical weapons has primarily been associated with terrorist groups and nations that support such organizations. Bioterrorism, the intentional release of potentially lethal bacteria or viruses into the air, food, or water supply, is most likely to involve anthrax and smallpox. Germ weapons have been called the poor man's atom bomb. In the advanced, industrialized nations that might be the targets of bioterrorism, doctors are unlikely to recognize the symptoms of anthrax, plague, or smallpox. Proving that bioterrorists are preparing for germ warfare can be difficult, because plants that are ostensibly making vaccines can be using virulent strains of anthrax to develop biological weapons. More sophisticated programs might involve more deadly genetically modified anthrax in order to thwart the protection of existing vaccines.

Although research on important human diseases, such as typhoid fever and cholera, was inhibited by the lack of animal models, Koch confidently predicted that progress in bacteriology would soon lead to the control of the epidemic diseases. To overcome the skepticism of the conservative medical profession, Koch urged advocates of the germ theory of disease to abandon careless and speculative work, develop methods of cultivating pure strains of bacteria, and demonstrate the value of microbiology in the prevention and treatment of human disease. Thus, after his successful studies of anthrax, Koch turned to the general problem of wound infection and traumatic infective diseases. By this time, Joseph Lister (1827–1912) had introduced his antiseptic system as a means of applying Pasteur's germ theory to the prevention of postsurgical infection, and scientists had generally accepted the idea that traumatic fever, purulent infection, putrid infection, septicemia, and pyemia were conditions that shared the same essential nature. Nevertheless, disputes continued as to whether the microorganisms found in these infections were the cause or the result of disease processes or simply nonspecific entities. Part of the problem was the difficulty of determining whether microbes were present in the blood of healthy animals. With improved illumination and suitable stains, Koch was able to demonstrate that bacteria were not found in the blood or the tissues of healthy animals. Then when blood from infected animals was injected into healthy mice, septic diseases occurred as the bacteria multiplied. Having observed the subtleties of gangrene, spreading abscess, pyemia, septicemia, and erysipelas in mice and rabbits, Koch concluded that the criteria for associating these diseases with microbes had been adequately met. Many critics of the doctrine of specific etiology dismissed his work on traumatic infective diseases as a series of laboratory curiosities, but Joseph Lister championed Koch's work as proof of the applicability of germ theory to wound infection. In recognition of the importance of his work to German science, Koch was appointed Professor of Hygiene at the University of Berlin and Director of the University's Institute of Hygiene. The Institute for Infectious Diseases was created for him in 1891.

Each disease Koch studied was invariably associated with a definite form of microorganism. While many distinguished scientists believed that microorganisms were nonspecific and easily transformed into different types, Koch maintained that pathogenic bacteria existed as distinct species. His studies of the traumatic infective diseases were meant, at least in part, as a defense of the concept of bacteria as distinct, fixed species. The apparent generation of new forms in the laboratory, Koch contended, was not proof of transformation of type, but evidence of technical problems that allowed the introduction of contaminants. Clearly, it would be impossible to say that a specific bacillus caused a specific disease if bacteria did not exist as distinct species. Using pure cultures was essential, but the living body seemed to be the only reliable apparatus capable of growing many pathogenic microbes.

Growing bacteria outside the animal body and obtaining pure strains became an obsession for Koch. After many unsuccessful experiments, Koch concluded that it would be impossible to construct a universal medium in which all microorganisms could grow. Therefore, he attempted to find a means of converting the usual nutrient broths into a form that was firm and rigid so that individual bacteria would be fixed in place and would grow into colonies, like little islands on a sea of jelly. Medium solidified by the addition of gelatin was useful for cultivating many different organisms, but gelatin has several disadvantages; it melts at body temperature and can be digested by some bacteria. Far Eastern culinary techniques involving agar-agar provided a practical substitute. Jellied agar-agar could be added to hot soups without melting, and some Europeans had learned to use agar-agar to make jellies. Nutrient agar remains solid up to 45°C and resists bacterial digestion. A simple but very effective device for culturing microbes on solid media was introduced by Richard Julius Petri (1852–1921), the curator of Koch's Hygiene Institute. Although the use of solid media was initially called "Koch's plate technique," thanks to the universal adoption of the petri dish, Petri's name is probably more familiar to students of microbiology than Koch's.

While attending the Seventh International Medical Congress in London in 1881, Koch enjoyed the opportunity to demonstrate the superiority of his plate technique to an audience that included Louis Pasteur and Joseph Lister. Shortly after returning from this meeting, Koch committed all of his considerable energy and skill to the formidable task of proving that tuberculosis was an infectious disease, identifying the causal agent, and finding a means of prevention or cure. In contrast to Pasteur, who had taken rabies, one of the most dramatic but rare threats to human life, as his greatest challenge, Koch chose tuberculosis, a commonplace, indeed almost ubiquitous scourge, for his greatest work. Working in strict secrecy, Koch was ready to present his preliminary results concerning the discovery of the tubercle bacillus, *Mycobacterium tuberculosis*, in little more than a year. Accounts of his discovery and speculation about a possible cure quickly appeared in newspapers throughout the world.

To understand the excitement generated by Koch's research on tuberculosis, it is essential to appreciate the way in which this dreaded disease permeated the whole fabric of nineteenth-century life. When Koch began his work, epidemiologists estimated that tuberculosis was the cause of death for one of every seven human beings. Even in the 1940s, pathologists commonly found evidence of tuberculosis when performing postmortem examinations of the lungs and lymphatic glands of persons who had presented no overt evidence of the disease during their life. The impact of the disease on society was amplified by the fact that it was particularly likely to claim victims in what should have been their most productive adult years. Not unlike AIDS in the 1980s, tuberculosis seemed to claim the lives of a disproportionate number of young artists, writers, composers, and musicians. The romantic myth of a link between the fire of genius and the fever of tuberculosis was dispelled by the discovery of what Franz Kafka saw as the "germ of death itself" in the sputum of its many impoverished victims and the dust of their filthy, dark, crowded tenements. As in the case of AIDS, the connection between creativity and the disease was fortuitous, while the association between the dissemination of the pathogen and poverty was indicative of fundamental inequities in society.

Many other pathogens, Koch noted, had fallen into the hands of investigators "like ripe apples from a tree," while the tubercle bacillus remained out of reach. Special fixation and staining methods were needed before the fine rodlike forms could be distinguished from the nonspecific debris found in tuberculous tissues. The tubercle bacillus is only a tenth the size of the anthrax bacillus and is covered by a waxy layer that made staining very difficult. Great patience and a firm conviction that tuberculosis was caused by a specific bacterial species were needed in these studies, because the tubercle agent grew exceptionally slowly. Although tuberculosis could be found in many animals, including cattle, horses, monkeys, rabbits, and guinea pigs, not all strains of the bacillus are pathogenic for every species. Fortunately, the microbe that caused tuberculosis in humans could be transferred to the guinea pig. *Mycobacterium tuberculosis* could be isolated from the tissues of patients suffering from a bewildering array of disorders known as phthisis, consumption, scrofula, miliary tuberculosis, and so forth. Indeed, the tuberculosis bacillus can attack virtually every part of the body. Thus, after decades of controversy as to the nature of tuberculosis and whether it was a contagious disease, Koch's discovery vindicated the unitary theory of tuberculosis.

Koch presented his preliminary report on tuberculosis to the Physiological Society of Berlin on March 24, 1882. Contemporaries described the presentation as a masterpiece of scientific research and a deeply moving experience. In addition to identifying the microbe that caused tuberculosis, Koch had considered the problem of the dissemination of the disease. Although the bacillus could form spores, it did not do so as easily as the anthrax bacillus. The tubercle bacillus could only grow in specially constituted medium at temperatures greater than 30°C, but even under ideal conditions several weeks were needed for good growth on solid media. Tubercle bacilli were, therefore, true parasites, dependent on the animal body for their survival, rather than facultative parasites like the anthrax bacillus. Pulmonary tuberculosis, the most common form of the disease in humans, provided the most efficient means of transmission, because its victims coughed and spit up large quantities of germ-laden sputum. Tubercle germs disseminated through the air contaminated floors, furniture, clothing, and so forth. In poorly ventilated, dirty, dusty tenement rooms, rarely illuminated by the cleansing rays of the sun, the germs could remain viable and virulent for days, or even months. Although the milk of infected animals could be a source of infection, Koch underestimated this danger, largely because of his interest in the role of sputum in spreading the pulmonary form of the disease.

During his studies of tuberculosis, Koch formalized the criteria needed to prove unequivocally that a particular microbial agent causes a specific disease. These criteria are now known as *Koch's postulates*. First, it is necessary to find the microbe invariably associated with the pathological condition. A thorough study of the relationship of the microbe to the natural history of the disease provided good, albeit circumstantial evidence. Skeptics, however, were entitled to argue that even if the microorganism appeared in association with the disease, it might not be the cause of the disease. Rigorous proof of a causal relationship required complete separation of the microbe from the diseased animal, tissue fragments, body fluids, and all possible contaminants. After the putative pathogen had been grown as a pure laboratory culture, it should be introduced into healthy animals. If the disease was then induced with all its typical symptoms and properties, the investigator could assert that the microbe was truly the cause of the disease. Ideally, researchers would carry out all of these steps before making claims about the etiology of infectious diseases, but this might be impossible for human diseases lacking animal models. Although Koch's postulates were formulated to prove a causal relationship between bacteria and disease, his general approach has been extended to guide studies of other disease-causing agents, such as viruses, asbestos, lead, mercury, and other chemicals that are thought to induce disease, trigger allergic responses, or depress the immune system.

Despite the rigorousness of Koch's investigations of the tubercle bacillus, not all scientists were willing to accept the idea that a single pathogen could be responsible for an enigma as complex as tuberculosis. Although systematic postmortem examinations by René Laënnec (1781–1826) and other French physicians provided evidence that tuberculosis caused morbid effects throughout the body, Rudolf Virchow, the German "pope of pathology," insisted that pulmonary tuberculosis and miliary tuberculosis were different diseases. Questioning the value of Koch's work on the "so-called tubercle bacillus," Virchow argued that if the microbe were as widespread as Koch asserted while only certain individuals became consumptives, the true cause of tuberculosis could not be a contagion. Advocates of the germ theory of disease objected that one might as well say that bullets did not kill, because not every soldier on the battlefield was killed by a barrage of bullets. Some critics argued that Koch had not discovered anything new because others, such as the English epidemiologist William Budd (1811–1880) and the French physician Jean Antoine Villemin (1827–1892), had previously claimed that tuberculosis was contagious. Villemin had demonstrated that tuberculosis could be transmitted from humans to rabbits by means of sputum, blood, and bronchial secretions. Given the fact that tuberculosis seemed to run in families, many physicians were convinced that the disease was due to a noncontagious, hereditary "tubercular diathesis." In other words, people who are susceptible to tuberculosis are susceptible to tuberculosis. Whether a microbe or a hereditary weakness caused the disease, in the absence of a therapeutic agent, its victims faced a long, lingering, inexorable decline that inevitably terminated in death.

Nevertheless, identification of the tubercle bacillus helped dispel decades of perverted romantic sentimentalism. The pale, thin, "angel of phthisis," with eyes bright from fever, discreetly coughing up blood-stained sputum into her dainty lace handkerchief, before her inevitable, but redemptive death, was actually filling the air around her with clouds of bacteria. Worse yet, the tubercle bacillus was very similar in size, shape, and staining properties to the microbe associated with leprosy, a disease universally considered loathsome rather than romantic.

For those who accepted the relationship between the tubercle bacillus and the disease, Koch's discovery stimulated hope that appropriate medical guidance might break the chain of transmission and help victims in the early stages of disease. On the other hand, the ability of physicians to detect infections that would have gone undiagnosed previously might exacerbate public perceptions about the threat of the disease. Of course, a remedy for such a highly prevalent illness would be more welcome than sensible advice about personal hygiene and the fundamental social and medical reforms that would be required for a sustained and successful public health crusade. Koch had found the microbes that caused anthrax and tuberculosis, but it was Pasteur, his French rival, who had developed preventive inoculations for anthrax and rabies. Koch was, therefore, under considerable pressure to follow up his identification of the tubercle bacillus with a cure.

In 1890, the year that Germany served as host to the Tenth International Medical Congress, Koch announced that he had found a substance that could inhibit the growth of the tubercle bacillus in laboratory cultures and in guinea pigs. It should be noted that 1890 was also the year in which Emil von Behring (1854–1917) and Shibasaburo Kitasato (1852–1931) successfully initiated the new era of serum therapy based on their experiments with diphtheria and tetanus antitoxins. Given the preliminary nature of his findings, Koch may have been somewhat reluctant to discuss these studies in public. Although he apparently made his announcement under considerable pressure from government officials, some of his more cynical colleagues suggested that he had been bribed rather than coerced. News reports immediately referred to the mysterious remedy that Koch called "tuberculin" as "Koch's lymph." Three months later, after conducting some tests in humans, Koch published a "Progress Report on a Therapeutic Agent for Tuberculosis." Tuberculin was described as a stable, clear, brown liquid, but, despite German laws prohibiting secret medicines, Koch concealed information about its nature and preparation. Production of tuberculin was entrusted to Eduard Pfuhl, Koch's son-in-law. Hordes of consumptives and scientists flocked to Germany in hopes of gaining access to Koch's remedy. Joseph Lister, who brought his niece to Berlin for treatment, was impressed by German advances in medicine, especially the new methods for the prevention and treatment of diphtheria and tetanus. Before the safety and efficacy of tuberculin had been properly tested, the Grand Cross of the Red Eagle and other high honors were bestowed on Koch by a grateful nation. A wave of profound disappointment and condemnation soon followed. In 1905 Koch was awarded the fifth Nobel Prize for Physiology or Medicine in recognition of his pioneering work on tuberculosis. The first prize had gone to Emil von Behring for the development of serum therapy.

Tuberculin did not cure the sick, but it did discriminate between uninfected and infected people. In fact, human beings were much more sensitive to the agent than guinea pigs. While a healthy person had almost no reaction to the agent, those who harbored the tubercle bacillus experienced severe reactions, including vomiting, fever, and chills. When Koch tested tuberculin on himself, he had a strong reaction indicative of his own latent infection. Immunologists later identified the tuberculin response as part of the complex immunological phenomenon called delayed-type hypersensitivity. As Koch predicted, tuberculin proved to be an indispensable diagnostic tool for detecting early and asymptomatic cases of tuberculosis. But Koch's claim that tuberculin would be more valuable in therapy than diagnosis proved tragically misleading. Patients from all over the world flocked to Berlin for the new wonder drug, but miraculous cures rarely occurred, and some patients died from reactions to tuberculin. Public opinion rapidly turned against Koch when it became obvious that tuberculin was not a "magic bullet" for the cure of consumption. In 1891 Koch finally revealed the nature and means of preparation of tuberculin: it was essentially a glycerine extract of tubercle bacilli.

What came to be called the tuberculin fiasco could be regarded as a case study demonstrating the need for caution in evaluating potential remedies. One hundred years later, however, the AIDS crisis provided a powerful stimulus for calling into question the whole apparatus of controlled clinical trials. Like AIDS in the 1980s, tuberculosis in the 1880s was a mysterious, dreaded, and fatal disease. Clearly, withholding a drug that might cure a fatal illness would be a cruel and unethical act. However, the risks involved in offering futile remedies to victims of diseases as complex and uncertain in their course as tuberculosis or AIDS are more subtle. The threat of tuberculosis to society cannot, of course, be directly compared to AIDS, because everyone indulges in the major risk behavior for tuberculosis, which is breathing.

Up to the end of his life, Koch continued to hope that some improved form of tuberculin would provide a cure or a preventive vaccine, but this dream was never realized. While his work on the etiology of tuberculosis was regarded as his greatest accomplishment, Koch and his associates fought many other diseases, including cholera, malaria, rinderpest, and plague. Research on tropical diseases finally gave Koch the opportunity to journey to the exotic lands he had dreamed of as a youth. Eventually, Koch's methods were exploited successfully in the search for the agents of typhus, diphtheria, tetanus, pneumonia, dysentery, relapsing fever, and other infectious diseases. Similarly, studies of the marked variations in the virulence of different varieties of tubercle bacilli led to attempts to create preventive vaccines. The most widely used vaccine, BCG (bacille Calmette-Guérin), was derived from a live, attenuated strain produced by Albert Calmette (1863–1933) and his associates. One hundred years later, with the development of antibiotic-resistant strains of tubercle bacilli, the World Health Organization was still recommending use of BCG, despite continuing questions about the efficacy of the vaccine.

While colleagues generally considered Koch a patient, conscientious researcher and a master of bacteriological technique, after achieving his initial successes he seems to have become increasingly arrogant, dogmatic, and opinionated. Despite his professional success, his private life was apparently unfulfilling. Koch and his first wife had been estranged for some time when he met 17-year-old Hedwig Freiberg. Colleagues at the 1892 Congress of German

Robert Koch and his second wife Hedwig in Japanese costume, photographed in 1903

Physicians later admitted that there was more excitement about Koch's love affair with the young artist's model than the scientific papers. When he remarried in 1893, Koch was almost 50 and Hedwig was 20. As Koch grew increasingly jealous of the success of his young colleagues, he lashed out at bacteriologists who disagreed with positions he had taken.

When Emil von Behring and the American bacteriologist Theobald Smith (1859–1934) warned that milk from cows with bovine tuberculosis was a danger to children, Koch insisted, quite incorrectly, that humans could not be infected with the bovine tubercle germ. In 1898, Smith published "A Comparative Study of Bovine Tubercle Bacilli and of Human Bacilli from Sputum." Controversy about the existence of two distinct species of tubercle bacilli in mammals continued for several years. Although Koch accepted Smith's conclusion about the existence of distinct bovine and human forms of the tubercle bacillus, even at the 1908 International Tuberculosis Congress Koch continued to argue that the question of contaminated milk was irrelevant to human tuberculosis.

Theobald Smith, who worked in the Veterinary Division of the U.S. Department of Agriculture, was also a pioneer in demonstrating the role of ticks as vectors of disease. The success of his work on Southern cattle fever demonstrates the problem of the unexpected consequences of the battle against disease. In 2000, Florida Department of Environmental Protection Agency officials estimated that more than 3000 sites throughout the state had been polluted by the cattle-dipping procedures that had been used in tick-control programs. Contaminants left by the cattle industry included arsenic, DDT, and other pesticides. Even at sites that had not been used for 80 years, contaminants remained in the soil and water. After Smith established the role of ticks in Southern cattle fever in 1906, voluntary dipping programs began. But skeptical cattlemen resisted such programs until mandatory dipping programs were established in 1932. Through the action of federal, state, and local governments, tick fever was eradicated in Florida by 1961.

During the last years of his life, Koch resumed his research on tuberculin, but his health rapidly deteriorated and he died of heart disease in 1910. Two years after his death, the Institute for Infectious Diseases was renamed the Robert Koch Institute. Eighty years later, according to the World Health Organization, tuberculosis, a disease that public health workers in the 1960s thought could be eradicated, was still claiming 3 million victims each year while causing an additional 8 million cases on a global basis. In the 1990s United States public health authorities found themselves dealing with significant increases in the incidence of tuberculosis, often in association with AIDS. The resurgence of tuberculosis in the 1990s and the global emergence of virulent, multidrug-resistant bacteria reflect the misplaced optimism that led to the neglect of the systematic public health measures that are needed to control infectious epidemic diseases.

IMMUNOLOGY: THE ART OF SELF-DEFENSE

On the foundations built by Pasteur, Koch, and others, microbiology exploded and splintered into numerous subdisciplines. Physicians and scientists were fascinated with the search for the microbial agents of disease and the toxins that some of them produced as well as the implications of these discoveries for preventing and curing epidemic diseases. As the nature of infectious diseases became more clearly understood, ancient doctrines about the body's mysterious healing powers could be investigated in the light of new theories. Given the ubiquity of the microbial hordes, it was obvious that the human body must be equipped with powerful weapons to fight off microscopic invaders. The successful use of vaccinations for smallpox, anthrax, rabies, and other diseases helped establish the new science of immunology.

The Latin term "immunity" originally referred to an "exemption" in the legal sense, but the concept was obviously applicable in medicine as well as law. Even during the most deadly of epidemics, not all of the afflicted succumbed to the disease. Learned physicians and laymen knew that, in many cases, people who had survived one attack of a disease were unaffected during subsequent epidemics. The germ theory of disease allowed scientists to demonstrate that the virulence of infectious diseases varied with many factors, including the means and duration of exposure, the way in which the germ entered the body, and the physiological status of the host. Moreover, it was clear that protective vaccines took advantage of the body's own defense mechanisms.

By the end of the nineteenth century, the fundamental question concerning scientists investigating the immune response was: Is immunity based on humoral or cellular agents? According to the humoral theory, immunity was dependent on the induction of factors, now known as antibodies, which circulate in the blood and fluids of the body. In the case of diphtheria and tetanus, proponents of the humoral theory proved that immunity could be induced by serum therapy. In contrast, the Russian zoologist Élie Metchnikoff (1845–1916) argued that serum factors were insignificant compared to phagocytes, special blood cells that served as soldiers in the body's war against invaders.

Emil Adolf von Behring and Shibasaburo Kitasato proved that active and passive immunity to diphtheria and tetanus could be induced by an approach they called serum therapy. Their work raised great hope that physicians would have access to a powerful and perhaps universal therapeutic approach to all infectious diseases.

Like many ambitious but impoverished students, Behring obtained his medical degree from the Army Medical College in Berlin in exchange for a 10-year commitment to the Prussian military service. After working with Robert Koch at the Institute for Infectious Diseases, Behring held professorships at Halle and Marburg. In 1914 Behring, who took a very entrepreneurial approach to serum therapy, founded the Behringwerke for the production of sera and vaccines. In many ways his career provides a paradigm for an era in which the search for profits, glory, and patents may well be pursued more energetically than the search for scientific knowledge. Having observed the deficiencies of clinical medicine while working as a military doctor, Behring became interested in the possibility of using "internal disinfectants" to kill the microbial agents of disease.

By the 1890s Émile Roux and Alexandre Yersin (1863–1943) discovered that although diphtheria bacteria generally remained localized in the throat, toxins that were released into the bloodstream damaged distant tissues. Shibasaburo Kitasato, a Japanese physician working with Robert Koch, isolated the tetanus bacillus and proved that it also produced a specific toxin. Further work proved that diphtheria and tetanus toxins provoked the symptoms of the disease when they were injected into the body.

Realizing the importance of these observations, Behring and Kitasato showed that the blood of animals that had been challenged by a series of injections of the toxins produced *antitoxins*, that is, chemicals in the blood that neutralized the bacterial toxins. Antitoxins produced by experimental animals could be used to immunize other animals and to treat patients suffering from diphtheria. Encouraged by his early results, Behring predicted that his technique would soon lead to the eradication of diphtheria. Although horses and sheep could be used as antitoxin factories, Behring's antitoxin preparations were too variable and weak for commercial production or routine use. Afraid that French scientists would make further advances in serum therapy, Behring sought help from Paul Ehrlich (1854–1915), who had used specific plant toxins to work out systematic methods of immunization. Ehrlich's quantitative methods for producing highly active, standardized sera made serum therapy practical.

When Behring won the first Nobel Prize in Physiology or Medicine in 1901 for his work on serum therapy, the Nobel committee honored him for creating a "victorious weapon against illness and deaths." In his Nobel lecture, "Serum Therapy in Therapeutics and Medical Science," Behring made a special point of reviewing the history of the dispute between cellular and humoral pathology. Characterizing antitoxin therapy as "humoral therapy in the strictest sense of the word," he predicted that serum therapy would lead to complete triumph over the infectious diseases. Indeed, by the end of the twentieth century, at least in the wealthy industrialized nations, massive immunization campaigns had almost eliminated the threat of diphtheria. No other major epidemic disease of bacterial origin has been so successfully managed by preventive immunizations. Nevertheless, serum therapy did not provide a universally effective weapon in the battle against infectious diseases. Even the originality of scientific studies of toxins was questioned when traditional healers in Africa told European explorers that they routinely protected people against snake venom with potions containing snake heads and ants. By exploiting the fact that ants contain formol, traditional healers apparently anticipated Léon Ramon's (1886–1963) 1928 discovery that formaldehyde can modify various toxins and venoms.

Not long after Behring spoke so confidently about the promise of serum therapy, euphoria was replaced by profound frustration and the dawn of a period that has been called the Dark Ages of Immunology. However, by the end of the twentieth century genetic engineers were exploiting the "naturally engineered" properties of various bacterial toxins in order to create hybrid molecules linking toxins to specific antibodies. Such novel immunotoxins have been referred to as "poisoned arrows" or "smart bombs," which, at least in theory, could deliver more fire power than the "charmed bullets" first synthesized in the laboratory of Paul Ehrlich, the founder of chemotherapy.

Some historians have argued that the Russian-born biologist Élie (Ilya Ilyich) Metchnikoff should be considered the founder of modern immunology because he asked new questions about the nature of the active responses that the organism mobilized to protect itself against infection. Previous speculations about immunity dealt with the nature of the infective agent and its development but ignored the question of an active defensive mechanism in the host. Honored for his discovery of phagocytes (eating cells, from the Greek *phagos*, to eat, and *cyte*, cell) and the process of phagocytosis, Metchnikoff remained for many years the advocate of what seemed to be an eccentric and unproductive approach to immunology. Indebted almost as much to the evolutionary theories of Charles Darwin as to the germ theories of Pasteur and Koch, Metchnikoff came to the concept of "active host immunity" through his interest in digestion, comparative embryology, and evolution. Phagocytic cells in primitive organisms played a nutritive, digestive role. In higher animals, which had specialized digestive organs, phagocytes could undertake active new roles throughout the life cycle. Phagocytes were responsible for digestion in the embryo and, subsequently, for protecting the organism from infection.

While Metchnikoff was not the first to observe bacteria inside white blood cells, he was the first to see the link between the intracellular digestion of microbes and the body's defense mechanisms. Other observers thought that white blood cells might accidentally carry and transport pathogens. When Koch saw anthrax bacilli inside blood cells, he assumed that the bacteria must have penetrated white blood cells in order to reproduce within them. In contrast, Metchnikoff thought that the white blood cells must have deliberately devoured the parasites. Furthermore, Metchnikoff argued that white blood cells were deliberately consuming foreign invaders and damaged, dying, or malignant cells in order to defend the body. In opposition to prevailing beliefs, Metchnikoff argued that inflammation was a defensive phenomenon, rather than a passive response to noxious stimuli. Heat, swelling, redness, and pain, which are characteristics of the inflammatory response, were all aspects of a beneficial and protective, but imperfect physiological process.

Élie Metchnikoff

Metchnikoff did not react well to criticism or academic restrictions, and his relationships with mentors and colleagues were often stormy and acrimonious. Oscillating between extremes of enthusiasm and depression, his volatile nature led to several suicide attempts and to bitter and protracted battles with critics. During one bout of depression, Metchnikoff decided to make his death useful to science by injecting himself with the spirochete that causes relapsing fever. Ironically, instead of killing him, the inoculated illness restored his energy and optimism. During this troubled period, Metchnikoff made his major discovery concerning the defensive role of the white blood cells. He concluded that phagocytosis might be a general physiological defense mechanism in which phagocytes served as the "bearers of nature's healing power." Although Rudolf Virchow supported Metchnikoff's theory, many physicians remained very skeptical. Critics argued that the all-too-common pus cells seen in wounds and post-surgical patients were dangerous traitor cells that spread the seeds of infection throughout the body, rather than valiant soldiers trying to fight off the enemy. Given the optimism associated with serum theory, it was not surprising that Metchnikoff's revolutionary cell-based theory of immunity seemed to be totally eclipsed by Behring's humoral theory.

Despite years of illness and depression, Metchnikoff remained confident that science would ultimately better the human condition, save the young from infectious disease, and prevent the degenerative diseases of middle and old age. In his books and lectures, Metchnikoff explored an approach to health and longevity that he called *orthobiosis*. He also proposed the term *gerontology* for a new scientific discipline dedicated to the study and control of senescence. After comparing the life spans of various animals, Metchnikoff concluded that longevity was determined by the organs of digestion. Specifically, microbial mischief in the large intestine produced harmful fermentations and putrefaction. According to Metchnikoff, every sign of aging, from gray hair and wrinkles to weakness of muscle and memory, was caused by the transformation of friendly phagocytes into the agents of chronic degeneration. Phagocytes became agents of senility when they were damaged by toxin-producing bacteria lodged in the large intestine. Intrigued by the possibility that it might be possible to reverse the aging process, Metchnikoff suggested disinfecting this "useless organ," with a hygienic diet to neutralize the deleterious effects of bacteria harbored by this perfidious and useless organ. Frustrated by evidence that traditional purges and enemas did more damage to the intestines than to the noxious intestinal flora, Metchnikoff recommended the consumption of large quantities of fresh yogurt as a means of introducing beneficial ferments into the digestive system. Other lines of evidence supported Metchnikoff's warning that the immune system might not be wholly beneficent. In addition to its inability to defend the body against some pathogens, the immune system was also capable of reacting to certain antigens with life-threatening hypersensitivity. The phenomenon known as *anaphylaxis* ("against protection") was discovered in 1902 by Charles Robert Richet (1850–1935) and Paul Jules Portier (1866–1962). Richet, who won the Nobel Prize in 1913, proved that in its most severe form anaphylactic shock killed within a few minutes after exposure to the offending antigen.

In 1908, Élie Metchnikoff and Paul Ehrlich shared the 1908 Nobel Prize for Physiology or Medicine for their work on immunity. The award represents

PROFESSEUR ROUX
Directeur de l'Institut Pasteur, Paris.

PROFESSEUR METCHNIKOFF   PROFESSEUR CALMETTE
Sous-Directeur de l'Institut Pasteur, Paris.   Directeur de l'Institut Pasteur, Lille.

Directors of the Pasteur Institute: Émile Roux, Élie Metchnikoff, and Albert Calmette

a landmark in the history of immunology, but the two laureates had very different ideas about the nature of the body's immunological defenses and the ways in which scientists could help the body fight disease. Rather than exploring the possibility that their work might represent complementary aspects of a new integrated theory of immunity, Metchnikoff rejected the suggestion that immunity involved both humoral factors and phagocytes. While Paul Ehrlich's Nobel lecture acknowledged the key role that cell theory had played in nineteenth-century biology, he argued that further progress in therapeutics and the life sciences would required a new focus on the chemical nature of the processes that occur within the cell.

The germ of Ehrlich's guiding principle, the realization that specific chemicals can interact with specific tissues, cells, subcellular components, or microbial agents, can be found in his doctoral thesis "A Contribution Towards the Theory and Practice of Histological Staining." As a medical student, Ehrlich had conducted systematic studies of the aniline dyes that led to the discovery of different types of white blood cells and a means of classifying the leukemias in terms of the prevailing cell type. Like Pasteur, Ehrlich did not separate his theoretical interests from the practical problems to be found in immunology, toxicology, pharmacology, and therapeutics. Indeed, Ehrlich suggested that there was a direct link between the growth of the chemical industry and progress in histology. On this foundation it would be possible to establish a better understanding of the body's natural immunological defenses, develop means of augmenting them, and create a new experimental pharmacology dedicated to the design of "magic bullets," that is, compounds which could specifically interact with pathogens without damage to normal cells.

In addition to developing the antisyphilis drug known as Salvarsan, Ehrlich's achievements include clarification of the distinction between active and passive immunity, recognition of the latent period in the development of active immunity, and an ingenious conceptual model for antibody production and antigen-antibody recognition. For his role in assuring the success of Behring's serum therapy approach to diphtheria and tetanus, Ehrlich was appointed director of the Institute for Serology and Serum Testing (later the Royal Institute for Experimental Therapy) and the Georg Speyer Institute for Chemotherapy. With only a small laboratory and a very small budget, Ehrlich and his associates carried out a prodigious amount of work. Ehrlich often said that he could work in a barn as long as he had a water tap, a flame, and some blotting paper. Cigars and mineral water should have been on the list, because he practically lived on these two items. His thought processes were so dependent on smoking that he always carried an extra box of cigars under his arm. Every day Ehrlich gave each of his laboratory workers a series of index cards in various colors containing instructions for the day's experiments. Any assistant who dared to ignore the instructions was severely and publicly reprimanded.

Paul Ehrlich

To explain the generation and specificity of antibodies, Ehrlich proposed the ingenious "side chain" theory of immunity, which is generally considered the first theory of antibody formation to offer a chemical explanation for the specificity of antibody-antigen interactions. Antibody-producing cells, according to this theory, were studded with side chains, that is, groups that could combine with antigens, such as diphtheria or tetanus toxins. When a specific antigen entered the body, it reacted with a specific side chain. In response, antibody-producing cells committed themselves to maximal production of the appropriate side chain. Excess side chains were released from the surface of the cell into the body fluids, where they bound to and neutralized the antigen. The fit between antigen and antibody was as specific as that between a lock and a key, although it was presumably the result of accident rather than design. Having demonstrated that the body produced antibodies to poisons such as ricin, as well as to pathogens and bacterial toxins, Ehrlich predicted that organisms must possess some means of preventing the immune response from acting against its own constituents.

In 1906, while expectations for serum therapy were still high, Ehrlich introduced the term *immunotherapy*, but as the limitations of this approach became clearer he shifted his emphasis to *chemotherapy*. Whereas immunotherapy depended on the antibodies that serve as nature's own magic bullets, chemotherapy was an attempt to imitate nature by creating synthetic magic bullets. If the body could not produce effective antibodies for every pathogen, then the goal of biomedical science must be to design and synthesize chemotherapeutic agents that would act to supplement and sustain the natural defense mechanisms. The search for chemotherapeutic agents began with a drug called atoxyl, an arsenical compound that killed trypanosomes, the causative agents of African sleeping sickness, Gambia fever, and nagana, and spirochetes, the microbes that cause chicken spirillosis, relapsing fever, and syphilis. Unfortunately, although the drug was effective in the test tube, it caused neurological damage and blindness when administered to patients. After synthesizing and testing hundreds of atoxyl derivatives, Ehrlich and Sahachiro Hata (1873–1938) discovered that Preparation 606, popularly known as Salvarsan, was effective in the treatment of syphilis. Despite charges that the drug was itself a dangerous poison, until penicillin became available after World War II, Salvarsan, in conjunction with mercury and bismuth, remained the only effective remedy for syphilis.

Further attempts to create magic bullets were largely unsuccessful until the 1930s, when Gerhard Domagk (1895–1964) discovered that a sulfur-containing red dye called Prontosil protected mice from streptococcal infections. This led to the synthesis of a series of related drugs called the sulfonamides. Domagk, who was the director of research in experimental pathology and bacteriology at the German chemical firm I. G. Farben, did not publish his results until 1935. Researchers at the Pasteur Institute soon discovered that the antibacterial activity

Paul Ehrlich

Gerhard Domagk

resided in the sulfonamide portion of the red dye. Eventually, thousands of sulfonamide derivatives, or "sulfa drugs," were synthesized. So few of them proved to be safe and effective enough for use in patients that chemists began to compare the odds of finding a new magic bullet with winning the lottery. Although some of the sulfa drugs were effective in the treatment of pneumonia, scarlet fever, and gonococcal infections, drug-resistant strains of bacteria began to appear almost as rapidly as new drugs. By the end of World War II, these "miracle drugs" were considered largely obsolete. In the Nobel lecture that Domagk gave in 1947, he suggested that the rapid spread of resistant strains of bacteria might have been caused by wartime stress and upheaval, and he warned that the same disappointments would soon follow the careless use of penicillin. Domagk was awarded the Nobel Prize in 1939, but Nazi officials refused to allow him to attend the ceremonies in Stockholm.

Many years before the discovery of Salvarsan and the sulfa drugs, scientists had observed examples of "antibiosis," the struggle for existence between different microorganisms. Serious attempts to exploit this phenomenon began in 1928 when Alexander Fleming (1881–1955) discovered that the mold *Penicillium notatum* inhibited the growth of certain bacteria. Although Fleming prepared crude extracts of penicillin, he was unable to purify the active principle and perform clinical tests. Therefore, penicillin remained a laboratory curiosity until World War II, when Walter Florey (1898–1968) and Ernst Boris Chain (1906–1979) isolated penicillin and proved that it was a chemotherapeutic agent with unprecedented activity against pathogenic bacteria. The path from laboratory curiosity to wonder drug was full of obstacles, not all of them technical and scientific, primarily because large-scale production of penicillin was closely associated with the Allied war effort. In 1945, the Nobel Prize for Medicine or Physiology was awarded to Alexander Fleming, Howard Walter Florey, and Ernst Boris Chain "for the discovery of penicillin and its curative effect in various infectious diseases."

Inspired by the success of penicillin, researchers sifted through samples of dirt from every corner of the world in search of new "miracle molds." Selman A. Waksman (1888–1973), who discovered streptomycin, neomycin, and several other antibiotics that were too weak or too toxic for clinical use, coined the term antibiotics to refer to compounds produced by microorganisms that inhibit the growth of other microorganisms. Streptomycin was especially valuable, because it was effective against tuberculosis. Despite optimistic predictions that more active and less toxic agents would continue to appear, the most prolific phase of the antibiotic gold rush seemed to end during the 1950s. Nevertheless, by the 1960s many physicians were convinced that antibiotics had eliminated the threat of bacterial diseases and that viral diseases could be eradicated by vaccines. This optimistic belief was discredited by the rise of multidrug-resistant microbes and the emergence of new diseases, including AIDS.

Alexander Fleming

INVISIBLE MICROBES AND THE
SCIENCE OF VIROLOGY

As bacteriologists successfully exploited the methods of Pasteur and Koch to identify the specific causative agents of epidemic infectious diseases, it seemed reasonable to accept Pasteur's assertion that "every virus is a microbe." But the meaning of the word *virus* as used by Pasteur was quite different from the modern definition of a virus as a minute entity composed of a core of nucleic acid and an envelope of protein that can only reproduce within living cells. Originally, the Latin word virus referred to something that was slimy and repulsive, but not necessarily harmful. Medical writers, however, tended to use the term in reference to obscure infectious agents or things that were dangerous to health, such as poison, venom, or contagion. In this sense, both the first-century Roman encyclopedist Celsus and the nineteenth-century chemist Louis Pasteur could speak of the virus of rabies.

Long before the nature of viruses was understood, traditional practitioners, at least in Africa, Asia, India, and the Mediterranean region, had discovered means of combating certain viral diseases. In the 1720s, Lady Mary Wortley Montagu (1689–1762) and the Reverend Cotton Mather (1663–1728) attempted to draw attention to the ancient use of smallpox inoculation. Lady Mary, wife of the British Ambassador to the Turkish court, had been introduced to the Turkish method of inoculation against smallpox while living in Constantinople. Cotton Mather learned about smallpox inoculation from a young African slave given to him by members of his congregation. Inoculation, or ingrafting, involved taking fresh material from smallpox pustules and inserting it into cuts in the skin of a healthy individual. Generally, inoculated smallpox was less dangerous than the naturally acquired infection, and those who survived were rewarded with lifelong immunity. In 1798 Edward Jenner (1749–1823) investigated the folk belief that the mild disease known as cowpox, or *variolae vaccinae*, provided protection against smallpox. Experiments proved that inoculations of cowpox did indeed confer immunity to smallpox. Jenner named the new procedure *vaccination* in honor of the cow. Thus, when Pasteur discovered that attenuated cultures of the microbe that causes chicken cholera could immunize experimental animals against the disease, he assumed that the phenomenon was analogous to Jenner's use of cowpox vaccine.

By the beginning of the twentieth century, the techniques of microbiology were sufficiently advanced for scientists to state with some degree of confidence that certain diseases were caused by specific pathogenic microbes. Still, the pathogens for some diseases refused to be isolated by the most sophisticated bacteriological techniques available. When such diseases seemed to be associated with submicroscopic infectious agents, the term virus was enlisted to stand for these unknown entities. Working on many exotic tropical diseases, Koch himself came to the conclusion that not all infectious diseases could be attributed to ordinary bacteria. Some infectious agents had complex life cycles and growth requirements that could not be met by conventional laboratory techniques. Pathogenic protozoa were larger than bacteria, but it was also possible that some pathogens might be smaller than bacteria.

Providing unequivocal proof that unconventional agents caused disease was difficult because it was virtually impossible to grow these entities in laboratory cultures. Nevertheless, the rickettsias, chlamydias, mycoplasmas, and brucellas were eventually added to the ranks of the pathogenic bacteria, fungi, and protozoa. The infectious agents that caused many important diseases of humans and animals, however, remained in the unsatisfactory category of "filterable viruses." The criterion of filterability was the outcome of work conducted by Pasteur's associate Charles Chamberland (1851–1908), who discovered that a porous porcelain vase could be used to separate visible microorganisms from their culture medium. This technique could be used in the laboratory to prepare bacteria-free liquids and in the home to prepare pure drinking water. Chamberland was also instrumental in the development of the autoclave, a device for sterilizing materials by means of steam heat under pressure. The availability of these techniques made it possible to establish, by means of operational criteria, a new category of infectious agents, the invisible-filterable viruses. Although many important diseases of humans and animals were thought to be caused by these obscure agents, studies of a disease of tobacco plants established the foundations of modern virology. As a result of these studies, scientists began to consider the possibility that filterable viruses might differ from other microbes in properties more fundamental than size. That is, they might be obligate parasites of living organisms that could not be cultured in vitro on any cell-free culture medium.

In 1892 Dimitri Ivanovski (1864–1920) read a paper to the Academy of Sciences of St. Petersburg on tobacco mosaic disease (TMD). Using the filter method of Chamberland, Ivanovski found that a filtered extract of infected tobacco leaves produced the same disease in healthy plants as an unfiltered extract. He thought that a toxin in the filtered sap might explain his observations, but he did not exclude the possibility that unusual bacteria might have passed through the filter. According to his Russian biographers, Ivanovski was the true founder of virology, but because he was excessively shy and self-effacing, the rest of the scientific world tended to think of him as "Ivanovski the obscure." Unable to isolate or culture the invisible tobacco microbe, Ivanovski abandoned this line of research and turned to studies of alcoholic fermentation. Apparently unaware of Ivanovski's work, Martinus Willem Beijerinck (1851–1931) also reported that a filterable agent caused TMD. Thinking about the way in which a small quantity of filtered plant sap transmitted the disease to a large series of plants, Beijerinck concluded that TMD must be caused by a *contagium vivum fluidum* that could pass through filters and reproduce within living plant tissues. Based on reports in the botanical literature, Beijerinck thought soluble germs could cause many other plant diseases.

Similar observations were made by Friedrich Loeffler (1862–1915) and Paul Frosch (1860–1928) in their studies of foot-and-mouth disease (FMD), the first example of a filterable virus disease of animals. Attempts to culture bacteria from lesions in the mouths and udders of sick animals were unsuccessful. Even after passage through a Chamberland filter, however, the apparently bacteria-free fluid from FMD lesions could transmit the disease to experimental animals. These animals could transmit the disease to other animals. Thus a living agent, rather than a toxin, must have been in the filtrate. Loeffler and Frosch suggested that other infectious diseases, such as smallpox, cowpox, and cattle plague, were caused by similar filterable microbes. Nevertheless, they continued to think of the infectious agent as a very small and unusual microbe rather than a novel entity that could only multiply within the cells of its host. Scientists later demonstrated that FMD is a highly infectious, airborne viral disease that attacks cloven-hoofed livestock animals like cows, sheep, goats, and pigs. The disease is generally not regarded as a threat to humans who consume meat or pasteurized milk from affected animals, but people in close contact with infected animals can acquire the disease. Although many authorities assumed that FMD had been essentially eliminated by the end of the twentieth century, a major epidemic of FMD in 2001 led to the destruction of hundreds of thousands of livestock animals throughout affected areas in Europe in an attempt to prevent the spread of the virus. Despite the epidemic, governments resisted vaccination because vaccinated animals test positive for FMD antibodies and nations that vaccinate are unable to claim disease-free status.

In 1915 Frederick William Twort (1877–1950) reported that even bacteria could fall victim to diseases caused by invisible viruses. As Jonathan Swift (1667–1745) had suggested in a satirical poem on the use of the microscope, or flea-glass, naturalists found that fleas were preyed on by smaller "fleas" that were, in turn, attacked by still smaller "fleas." While trying to grow viruses in artificial medium, Twort noted that colonies of certain bacteria growing on agar sometimes became glassy and transparent. If pure colonies of this micrococcus were touched by a tiny portion of material from the glassy colonies, they too became transparent. Twort later became obsessed with speculative work on the possibility that bacteria evolved from viruses that had developed from even more primitive forms.

Working independently, Félix d'Hérelle (1873–1949) also discovered the existence of bacterial viruses. In 1917, he published his observations on "An Invisible Microbe That Is Antagonistic to the Dysentery Bacillus." Because the invisible microbe could not grow on laboratory media or heat-killed bacilli, but grew well in a suspension of washed bacteria in a simple salt solution, d'Hérelle concluded that the antidysentery microbe was an *obligate bacteriophage*, that is, an eater of bacteria. Bacterial viruses were sometimes called Twort-d'Hérelle particles. The invisible microbe was found in stools samples of patients recovering from bacillary dysentery. When an active filtrate was added to a culture of Shiga bacilli, bacterial growth soon ceased and bacterial death and lysis (dissolution) followed. A trace of the lysate produced the same effect on a fresh Shiga culture. More than 50 such transfers gave the same results, indicating that a living agent was responsible for bacterial lysis.

Speculating on the general implications of the phenomenon he had discovered, d'Hérelle predicted that bacteriophages would be found for other pathogenic bacteria. Although the natural parasitism of the invisible microbe seemed species specific, d'Hérelle believed that laboratory manipulations could transform bacteriophages into microbes of immunity with activity against human pathogens. The hope that bacteriophage could be trained as weapons in the war on bacteria was not realized in the twentieth century, but researchers continue to explore the possibility that viruses might be recruited to attack drug-resistant bacteria. Until pressed into service as the experimental animal of molecular genetics, bacterial viruses remained a laboratory curiosity. In the hands of molecular biologists, viruses became the "living molecules" that served as a kind of scientific Rosetta stone and made it possible to decipher the instructions for cell structure and function.

Bacterial viruses helped reveal the nature of the gene, but it was the tobacco mosaic virus (TMV) that provided the key to understanding the chemical nature of viruses and, eventually, the gene. In 1935, Wendell M. Stanley (1904–1971) reported that he had isolated and crystallized a protein having the infectious properties of TMV. Starting with a ton of infected tobacco leaves and utilizing the tedious conventional purification techniques then available, Stanley obtained a spoonful of crystalline material that still possessed the attributes of the TMV found in cell sap. Since the crystalline material seemed to be mainly protein, Stanley concluded that TMV was an autocatalytic protein that could only multiply in the presence of living cells. Further work showed that the virus was actually a combination of protein and nucleic acid. Stanley won the Nobel Prize in Chemistry in 1946 for this provocative demonstration that a chemical substance could behave as if alive.

Generally ignoring metaphysical debates about the nature of viruses and life, biochemists and geneticists described viruses as particles composed of an inner core of nucleic acid, which functioned as the genetic material, and a protein overcoat. Perhaps the most appropriate commentary on the great philosophical debate was Nobel Laureate André Lwoff's paraphrase of a famous line by Gertrude Stein: "Viruses should be considered viruses because viruses are viruses." While viruses may still be considered "living molecules" on the borderline between macromolecules, genes, and cells, research on diseases that have been attributed to slow viruses, viroids, and prions suggests that other obscure, intriguing, and perhaps dangerous creatures may well populate the submicroscopic world.

Prions, the most bizarre of all the infectious agents discovered in the twentieth century, even challenge the idea that "viruses are viruses," at least in the case of disorders that were originally attributed to "slow viruses." In 1982, Stanley B. Prusiner (1942-) coined the term prion, which stood for "proteinaceous infectious particle." All of the diseases attributed to prions are known as transmissible spongiform encephalopathies (TSE), that is, degenerative diseases of the central nervous system. Prion diseases of animals include scrapie in sheep and goats, transmissible mink encephalopathy, chronic wasting disease of mule deer and elk, feline spongiform encephalopathy, and bovine spongiform encephalopathy (BSE, commonly known as mad cow disease). Human diseases attributed to prions include Creutzfeldt-Jakob disease (CJD) and a new variant (vCJD) that apears to be related to BSE, fatal familial insomnia (FFI), Gerstmann-Sräussler-Scheinker syndrome, and kuru. In the 1960s, Carleton Gajdusek (1923-) suggested that kuru, a disease found only among the Fore people of New Guinea, was transmitted by cannibalistic mourning rituals. Laboratory experiments by Gajdusek and others suggested that kuru, scrapie, and CJD might be caused by similar infectious agents, which were called slow viruses. Gajdusek, who was awarded a Nobel Prize in 1976, thought that the infectious agent must be an unconventional virus.

Many aspects of the history of kuru seem relevant to the still unfolding story of BSE and vCJD. Scrapie, an old Scottish name for a disease of sheep and goats, has been known since the eighteenth century, but until the 1980s when BSE first appeared in England, there was no evidence of transmission to cows or humans. Scientists believe the BSE epidemic began when a nutritional supplement containing the rendered remains of sheep and cows was added to cattle feed. By essentially transforming domesticated herbivores into carnivores, or even cannibals, the new dietary regimen presumably created an unprecedented opportunity for scrapie agents to infect cattle. As the mad cow epidemic reached its peak in 1992, millions of cattle were destroyed, but by then contaminated meat products had probably entered the food chain. Not all prion infections can be transmitted from one species to another, but a new variant of CJD (vCJD) has been attributed to the consumption of beef from animals with BSE. The panic caused by mad cow disease has raised awareness of all the prion diseases. Perhaps the emergence of new diseases such as BSE and vCJD might be related to the ways in which human beings have affected the environment, especially through global exchanges of previously isolated plants, animals, and infectious agents.

In 1972 after one of his patients died of CJD, Stanley Prusiner began studying the literature that linked CJD to kuru and scrapie. CJD appears to strike sporadically, affecting about one in a million people over the age of 60 throughout the world. When Prusiner isolated the scrapie agent from the brains of diseased hamsters, he was surprised to find that it apparently consisted of a specific protein. All previously known infectious agents, even the smallest viruses, contained genetic material in the form of nucleic acids, either DNA or RNA. Prusiner's "protein-only hypothesis" was initially regarded as heresy, but within a few years genes that encoded prion proteins were found in all animals tested, including humans.

Despite continuing skepticism and controversy, by the early 1990s many scientists had accepted Prusiner's prion hypothesis. In 1997, Prusiner was awarded the Nobel Prize in Physiology or Medicine for discovering prions and establishing a new genre of disease-causing agents. According to Prusiner's theory, prion proteins can exist in two distinct conformations, one of which is quite harmless. Prion proteins can also exist in altered conformations that act as rogue proteins, or "evil twins." Moreover, in the altered conformation prion proteins are capable of inducing their benign counterparts to undergo the same Dr. Jekyll-to-Mr. Hyde transformation. As the transformed proteins accumulate and aggregate, they form thread-like structures that ultimately destroy nerve cells and result in fatal brain diseases. Despite many uncertainties about the way in which prions cause brain disease, Prusiner suggests that understanding the three-dimensional structure of prion proteins might lead to useful therapeutic interventions. Moreover, the success of the prion hypothesis in explaining infectious, hereditary, and sporadic forms of scrapie-like diseases suggests that similar mechanisms might play a role in other disorders, including Alzheimer's disease, Parkinson's disease, and amyotrophic lateral sclerosis.

SUGGESTED READINGS

Bäumler, E. (1983). *Paul Ehrlich: Scientist for Life*. Translated by G. Edwards. New York: Holmes & Meier.

Brock, T. D. (1988). *Robert Koch. A Life in Medicine and Bacteriology*. Madison, WI: Science Tech.

Conrad, L. I., and Wujastyk, D., eds. (2000). *Contagion: Perspectives from Pre-modern Societies*. Brookfield, VT: Ashgate.

Daniel, T. M. (2000). *Pioneers in Medicine and Their Impact on Tuberculosis*. Rochester, NY: University of Rochester Press.

Dubos, R. (1988). *Pasteur and Modern Science*. Madison, WI: Science Tech Publishers.

Ehrlich, P. (1956–60). *The Collected Papers of Paul Ehrlich*. Edited by F. Himmelweit. 3 volumes. London: Pergamon.

Farley, J. (1977). *The Spontaneous Generation Controversy from Descartes to Oparin.* Baltimore, MD: Johns Hopkins Press.

Fenner, F., and Gibbs, A., eds. (1988). *A History of Virology.* Basel: Karger.

Fox, S. W., ed. (1965). *The Origins of Prebiological Systems.* New York: Academic Press.

Garrett, L. (2000). *Betrayal of Trust: The Collapse of Global Public Health.* New York: Hyperion.

Geison, G., L. (1995). *The Private Science of Louis Pasteur.* Princeton, NJ: Princeton University Press.

Grafe, A. (1991). *A History of Experimental Virology.* Trans. by E. Reckendof. Berlin: Springer-Verlag.

Hare, R. (1970). *The Birth of Penicillin.* London: George Allen & Unwin.

Henle, J. (1938). *Jacob Henle: On Miasmata and Contagia.* Trans. by George Rosen. Baltimore, MD: Johns Hopkins University Press.

Hughs, S. S. (1977). *The Virus: A History of the Concept.* New York: Science History Publications.

Iterson, C. Van, Den Dooren de Jong, L. E., and Kluyver, A. J. (1983). *Martinus Willem Biejerinck: His Life and His Work.* Madison, WI: Science Tech Publishers.

Kay, L. E. (1986). W. M. Stanley's Crystallization of the Tobacco Mosaic Virus, 1930–1940. *ISIS* 77:450–472.

Koch, R. (1932). *The Aetiology of Tuberculosis.* Trans. of the 1882 paper by Dr. and Mrs. Max Pinner. Intro. by A. K. Krause. New York: National Tuberculosis Association.

Langton, C. F., ed. (1989). *Artificial Life: Proceedings of an Interdisciplinary Workshop on the Synthesis and Simulation of Living Systems.* Redwood City, CA: Addison-Wesley Publishing Co.

Latour, B. (1988). *The Pasteurization of France.* Trans. by A. Sheridan and J. Law. Cambridge, MA: Harvard University Press.

Lechevalier, H. A., and Solotorovsky, M. (1974). *Three Centuries of Microbiology.* New York: Dover.

Metchnikoff, E. (1968). *Lectures on the Comparative Pathology of Inflammation.* Delievered at the Pasteur Institute in 1891. New York: Dover.

Metchnikoff, O. (1921). *Life of Elie Metchnikoff, 1845–1916.* Trans. by R. Lankester. Boston: Houghton, Mifflin.

Miller, J., Engelberg, S., and Broad, W. (2001). *Germs: America's Secret War Against Biological Weapons.* New York: Simon & Schuster.

Oparin, A. I. (1964). *The Chemical Origin of Life.* Springfield, IL: Charles C Thomas.

Parascandola, J., ed. (1980). *The History of Antibiotics: A Symposium.* Madison, WI: American Inst. Hist. Pharmacy.

Prusiner, S. B. (1999). *Prion Biology and Diseases.* Cold Spring Harbor, NY: Cold Spring Harbor Laboratory Press.

Rabenau, H. F., Cinatl, J., and Doerr, H. W., eds. (2001). *Prions: A Challenge for Science, Medicine and the Public Health System.* New York: Karger.

Sagan, D., and Margulis, L. (1988). *The Garden of Microbial Delights. A Practical Guide to the Subvisible World.* New York: Harcourt.

Sheehan, J. C. (1982). *The Enchanted Ring: The Untold Story of Penicillin.* Cambridge, MA: MIT Press.

Silverstein, A. M. (1989). *A History of Immunology.* New York: Academic Press.

Tauber, A. I., and Chernyak, L. (1991). *Metchnikoff and the Origins of Immunology. From Metaphor to Theory.* New York: Oxford University Press.

Vallery-Radot, R. (1923). *The Life of Pasteur.* Trans. by R. L. Devonshire. New York: Doubleday, Page.

8

EVOLUTION

The publication of Charles Darwin's *Origin of Species* in 1859 and the controversy that it ignited have overshadowed other aspects of the development of the life sciences during the nineteenth century, a period in which the new science of biology replaced traditional natural history and the study of nature became the domain of the professional scientist. The term *biology*, which was first introduced at the beginning of the nineteenth century, was popularized in the writings of the French zoologist Jean Baptiste Lamarck (1744–1829) and the German naturalist Gottfried Reinhold Treviranus (1776–1837). According to Lamarck, biology encompassed the study of all that pertained to living bodies, their organization, development, special organs, and vital movements. In describing the goals of a new science of life, Treviranus predicted that, in contrast to the old natural history, biology would organize all observations of vital phenomena into a unified and harmonious whole. Auguste Comte (1798–1857), the French social philosopher, considered biology one of the principal sciences of positive philosophy. The idea of progress was central to Comte's system, and his law of the three stages called for the intellectual evolution of humankind from theological to metaphysical and finally to scientific knowledge. His classification of the sciences was based on their increasing complexity from mathematics, astronomy, physics, and chemistry, to biology and sociology. Biologists were urged to turn away from work that was merely descriptive toward the study of the vital functions of plants and animals. Eighteenth-century naturalists had exemplified that era's passion for the accumulation and classification of masses of facts about plants, animals, and peoples from all over the world, but the new task of biologists was to apply physical theories to the rational solution of physiological problems.

All human societies have wrestled with the problem of how to group individual objects into useful general categories. Simple classification systems are usually based on easily comprehensible, practical properties, such as moist and dry, edible and inedible. Plato and Aristotle were among the first to attempt to create systematic classifications based on *form* and *idea*, *genus* and *species*. The search for systematic schemes of classification dominated the life sciences during the seventeenth and eighteenth centuries. Several factors contributed to this new preoccupation with organization, but the sheer bulk of new materials certainly provided a powerful stimulus. Millions of different species may exist, but until 1600 only about 6000 species of plants were known. By 1700, botanists had 12,000 new ones to assimilate. Zoologists also faced the problem of information overload. It was clear, moreover, that eighteenth-century natural history lagged far behind the physical sciences. Physicists had created an orderly universe, bound by natural laws. Certainly, naturalists argued, there must be laws that would create order from the chaotic mass of information about plant and animal species, as well as some means of purging traditional bestiaries and herbals of exotic and dubious entities. Even Conrad Gesner (1516–1565), the greatest of the sixteenth-century encyclopedic naturalists, described creatures like the tree goose and the basilisk in his massive *Historia Animalium*. Many herbals merely listed plants in alphabetical order, or by relative importance, but some botanists attempted to organize plants according to overall affinities or by the properties of particular parts.

The first naturalist to develop a taxonomic system applicable to plants and animals, using species as a unit, was John Ray (1627–1705). His mother's local success as an herbalist and healer probably stimulated Ray's early interest in the medicinal virtues of plants. A man of strong convictions, Ray was willing to lose his position and livelihood rather than sign the Act of Uniformity during the reign of Charles II. Reduced to barely genteel poverty, Ray spent the rest of his life studying natural history and composing sermons, religious essays, handbooks on the classics, and treatises on folklore and natural science. Whatever the dangers and uncertainties of the political and religious milieu of the seventeenth century, in the study of plants and animals Ray was able to find his bedrock of order and purpose in accord with God's own design. Dedicating his *General History of Plants* to completing Adam's task of naming all of God's creatures, Ray also undertook the more complex labor of describing, classifying, and interpreting the meaning of form and function in living beings. In 1691 he published his most popular and influential work, *The Wisdom of God manifested in the Works of the Creation*.

The period between the publication of *The Wisdom of God* and Charles Darwin's *Origin of Species* (1859) might well be considered the golden age of natural theology, especially in England. Scientists and clergymen shared the comfortable conviction that, when properly interpreted, studies of the works of nature, God's universal and public manuscript, would always be in harmony with the Bible, God's written word, because both revealed the attributes of God to human reason. In nature Ray found a stable and perfect world created by God, but his religious convictions rarely interfered with his studies of form and function.

Given the conviction that God in his plentitude had filled the entire Scale of Nature, Ray argued there could be no great gaps between living forms. Eventually naturalists would discover all the species that had been established by God at the creation. The total number of species must be fixed, because God had rested on the seventh day after the work of the creation. Nevertheless, Ray admitted that new varieties of plants seemed to arise by means of degeneration. Thus, although all species were descendants of pairs originally created by God, there was still some room for flexibility in Ray's system. Having concluded that the species was the fundamental unit of classification, Ray suggested that breeding tests could be used to determine whether plant specimens were members of the same species. Natural theology incorporated the belief that God always preserved the species, despite the death of individuals. Fossilized specimens with no known living counterparts, therefore, presented a problem for this doctrine. Despite the doubts raised by fossilized plant and animal remains, Ray comforted himself with the belief that man was God's most recent creation. It is possible that Ray's ideas, if further developed, could have led to natural systems of classification that might have yielded a form of developmental taxonomy more compatible with evolutionary ideas than the well-known Linnaean system.

For Carolus Linnaeus (Carl von Linné, 1707–1778), taxonomist, botanist, entomologist, zoologist, and physician, the task of the naturalist was to complete the assignment given to Adam: to name the plants and animals, contemplate the handiwork of God, and marvel at His creations. While Ray's work might fall short of Linnaeus's in some aspects, Ray seems to have had broader goals than Linnaeus, who more rigidly defined the fundamentals of biology in term of nomenclature and classification. As a youth Linnaeus had little affinity for classical studies, schools, schoolmasters, or a lifetime spent as a country parson, like his father. Primarily interested in botany and natural science, Linnaeus studied medicine at Lund and Uppsala. The most important event in this period of his life, however, was his extensive field trip through Lapland to collect specimens of plants and animals. During the trip he also collected anthropological data and met his future wife, the patient Sara-Lisa Moraeus, to whom he was engaged for 8 years because her father would not allow the marriage to take place until Linnaeus earned his doctor's degree.

The publication of his *Systema Naturae, or The Three Kingdoms of Nature Systematically Proposed in Classes, Orders, Genera, and Species* (1735) led to recognition from the world of science and freed him from the need to ever practice medicine. The first edition of the *System of Nature* contained the basic principles of the Linnaean binomial system, although the text was often revised and expanded in the light of new information and ideas. Well aware that his classification system was not natural, in the sense called for by Aristotle, Linnaeus emphasized the virtues it undeniably had. Above all, the system was practical and easy to use; plants and animals were grouped into a hierarchy of varieties, species, genus, order, class, and kingdom. While Linnaeus developed a simple "sexual system" for classifying plants, he was forced to use an assortment of rather arbitrary morphological criteria to organize the animal kingdom. Unsure about the classification of monkeys, apes, and humans, Linnaeus noted that humans and apes seemed to form a continuum. Although he saw humans as a part of nature, he believed that they had a divine right to dominate, improve, and decorate the earth. Unlike Descartes, Linnaeus thought that animals had souls and were not mere machines. While Linnaeus praised animals as the "highest and most perfect works of the Creator," he thought that in his studies of procreation in plants he had discovered "the very footsteps of the Creator." Censors today would find this subject somewhat less than inflammatory, but Linnaeus must have shocked many readers with his poetic descriptions of the "celebration of love" in the plant world, including scandalous episodes of adultery as well as pure marriage. Nevertheless, classification required careful attention to the reproductive organs and sex life of the plants because plants were divided into *classes* according to the number and character of the stamens, and classes were divided into *orders* by the number and character of the pistils. Physician, scientist, and poet Erasmus Darwin (1731–1802), grandfather of Charles Darwin, made the Linnaean sexual system seem almost pornographic in his satirical treatise "The Loves of the Plants" (1789), but his *System of Vegetables* (1783) treated the Linnaean system with all due respect.

Many contemporary naturalists, overwhelmed by the great diversity of life on earth, criticized the Linnaean system and warned that it was impossible to create a universal system of plants and animals. Having studied and named thousands of plants and animals, Linnaeus sadly concluded that he had not yet discovered the true natural order. Nevertheless, even if his critics thought his sexual system as artificial as an alphabetic list, he was sure that it was both useful and necessary. Despite its limitations, the binary nomenclature standardized by Linnaeus became the most widely accepted method of naming plants and animals. When Linnaeus applied for a professorship at Uppsala University, he pointed out that his fame was already equal to that of Galileo, Leibniz, and Newton. His admirers included Johann Wolfgang von Goethe and Jean-Jacques Rousseau. The English literary magazine *Monthly Review* called him "the greatest Botanist that the world ever did or probably ever will know." Although

Sweden rewarded his efforts with many honors, Linnaeus often complained that his fame prevented him from undertaking new expeditions. Instead, he sent his students and disciples on dangerous voyages to exotic locations. Half of the young soldiers in his "botanic army" died, and many of those who survived were permanently damaged in mind and body.

Throughout the eighteenth century, famine, malignant fevers, and dysentery caused high mortality rates in Scandinavia, especially among poor peasants. Trying to avert suffering and death when crops failed, Linnaeus advised the poor to learn how to prepare native plants, such as acorns, Iceland moss, tulip bulbs, seaweed, burdock, bog myrtle, nettles, thistles, and the bark, sap, and resin of appropriate trees. For melancholy, he recommended marijuana as a substitute for expensive aquavit. Using his "new science," Linnaeus hoped to establish profitable silk, cotton, dye, and pearl industries in Sweden. In diaries, notebooks, and letters, Linnaeus revealed his most cherished goals, including remarkable plans to "teach" plants like tea, rice, quinine, olives, pepper, and saffron to grow in the harsh climate of Scandinavia. Teaching tropical plants and animals to grow in Lapland and Sweden, through "botanical alchemy" or "transmutational zoology," was probably about as plausible as turning lead into gold, but some exotic crops, such as tobacco and rhubarb, had been grown in Europe. Through a commitment to scientific principles, Linnaeus predicted, the Swedish people and the state economy would achieve self-sufficiency and wealth. Unfortunately, his plans for the economic advancement of Sweden produced little or no practical success.

Critics called Linnaeus an arrogant man who wrote as if he had personally witnessed the creation, but in private letters he expressed doubts about the fixity of species and complained about the conservative religious authorities who could punish those who expressed unorthodox opinions. Probably Linnaeus was not entirely rigid, unimaginative, and sure of himself; the difficulty of distinguishing between true species made at the creation and the varieties and "sports" known to gardeners and animal breeders apparently troubled him. Perhaps life in a harsh environment, or his own periods of dark depression, led Linnaeus to question certain theological assumptions about design in nature. In his public lectures he expressed conventional views about the economy of nature, but in secret he wondered whether crop failures, famine, disease, and the death of young children proved that God had designed nature for himself, rather than for the benefit of man. In his published work, however, Linnaeus generally accepted the doctrine that all species had been created by God and that no new species had made its appearance since the creation. Practically a national monument by the time of his death, he was buried with the honors proper to royalty. Quarrels over the disposition of his collections soon followed, and, unfortunately for Sweden, his widow sold his collections and books to an Englishman. Legend has it that Sweden sent a warship out to recapture this national treasure, but the British ship escaped. In 1829 the Linnaean Society of London acquired these materials.

EVOLUTIONARY SPECULATIONS

Searching for theories that were subsumed by the Darwinian revolution, historians of science have traced the tenuous threads of evolutionary thought through the writings of numerous naturalists and philosophers. Lists of Darwin's "precursors" include Anaximander and Empedocles, among the ancient Greeks; Benoit de Maillet, Étienne Geoffroy Saint-Hilaire, Georges Buffon, Jean de Lamarck, and Pierre Maupertuis, among French naturalists; and Erasmus Darwin, William Charles Wells, Patrick Matthew, Robert Chambers, and Alfred Russel Wallace, among the English. Some of these individuals are fairly well known, while others have been largely forgotten in spite of their previous notoriety. For example, a once scandalous book entitled *Telliamed: Conversation Between an Indian Philosopher and a French Missionary on the Diminution of the Sea, the Formation of the Earth, the Origin of Men and Animals, etc.*, first appeared in 1748, 10 years after the death of the author, a diplomat and traveler named Benoit de Maillet (1656–1738). Even though he had suppressed the book during his lifetime, de Maillet disguised his heretical ideas by ascribing them to a mythical foreign sage named Telliamed (the author's name spelled backwards). This exotic spokesman could safely express unorthodox ideas that his interrogator, a respectable, but curious French missionary, knew to be contrary to religious dogma. Popular in France and England, *Telliamed* brought theories of cosmic and organic evolution to the attention of readers who might have little interest in purely scientific texts.

De Maillet attempted to link cosmic and biological evolution in an all-embracing cosmology reminiscent of the pre-Socratic Greeks. As in those ancient cosmologies, Telliamed's world was one of eternal cycles of creation and destruction. Incorporated into the text were discourses on the inheritance of acquired characteristics, the successive deposition of geological strata, the nature of fossils, and the great age of the earth. The all-important seas were the agents responsible for changes in the earth's crust as well as the site of the origin of life. Living beings developed only in response to favorable conditions and adapted to their changing circumstances. Thus, when the seas began to recede, living beings were able to change and make their way onto the land. Finding it difficult to reconcile the apparent fixity of species with his conviction that species were always changing, de Maillet suggested that changes in species are never directly observed because they take place over long periods of time. Since the time of Galileo it had become irrefutable that the allegedly immutable heavens were indeed subject to change. Profound changes in the earth and its living beings were not, therefore, totally implausible. De Maillet suggested that microscopic living entities might be the primeval seeds of life that evolved into ever more complex creatures. Improvement and progress were not ultimately inevitable, because the continued diminution of the seas would eventually result in a period of universal volcanic action that would cause the earth to catch fire and become another sun.

In a remarkable little book entitled *The Earthly Venus*, Pierre Louis Moreau de Maupertuis (1698–1759) promised to explore the great riddle of life as an anatomist, rather than as a metaphysician. Best known as a mathematician, the ingenious Maupertuis was also interested in the problem of heredity and the origin of species. After evaluating ovist and spermist theories of generation, Maupertuis proclaimed that both doctrines were inadequate and incorrect. According to preformationist theories, only one of the parents could contribute heritable traits to the offspring, but there was considerable evidence to the contrary, especially the appearance of mules and mulattos. Maupertuis suggested that mating novel varieties for several generations might result in the production of new species. New varieties presumably arose by chance, but climate and food might have some influence as well. For example, the heat of the tropics might be more favorable to the production of varieties with pigmented skin. Maupertuis used this argument to explain the development of the various "races" of mankind from what he thought of as the original human type. Not surprisingly, Maupertuis believed that the "European race" was the original human type and that other "races" had formed by degeneration, under the influence of adverse conditions found in other parts of the world. These and other speculative ideas were further developed in his *System of Nature*, which was published under a pseudonym in 1751. Many of Maupertuis's ideas were explored and extended by Georges Louis Leclerc, Comte de Buffon (1707–1788).

Born into a wealthy and power family, Buffon enjoyed a reputation for charm and elegance, but after a youthful love affair led to a duel, he was forced to leave France. Escaping from the rigidities of the ancien régime, Buffon encountered the excitement of the new experimental sciences in England. He demonstrated his interest in Newtonian physics and English science by translating Isaac Newton's *Fluxions* and Stephen Hales's *Vegetable Staticks* into French. Although his first scientific interest was physics, Buffon was drawn to biology. A large inheritance freed him from the nuisance of having to earn a living and allowed him to devote his life to science. He died in 1788, 10 years after the death of Linnaeus and one year before the French Revolution claimed the life of his only son.

In 1739 Buffon became an associate of the French Academy of Science and was appointed Keeper of the Jardin du Roi, which he turned into a valuable research center. His work resulted in the publication of a multivolume compendium of natural history. The 44 volumes of Buffon's popular *Natural History*

(1749–1804) presented a complete picture of the cosmos, from the stars and solar system to the earth and its organic and inorganic constituents. Known as an elegant writer, Buffon managed to touch on many dangerous and forbidden topics so delicately that he nearly concealed his unorthodox views by pointing out how the facts might lead one to support an evolutionary interpretation, *if* one did not know what Genesis and the True Church taught. Despite his cryptic style, Buffon aroused the suspicion of orthodox scientists and theologians. He was called before the Syndic of the Sorbonne, the Faculty of Theology, in 1751 and warned that certain parts of the *Natural History* were in conflict with religious dogma. Acknowledging the infallible authority of the church, Buffon promised to subdue his heterodox speculations and express himself in a more circumspect manner.

Completely won over by Newtonian science, Buffon struggled to apply natural law to the living world. Seeing the great complexity and diversity of nature, he ridiculed the oversimplifications and arbitrary arrangements found in Linnaean taxonomy. At one stage of his thinking, Buffon argued that the conceptual basis of Linnaean taxonomy was artificial, arbitrary, and mischievous, because genus and species were no more real than the lines of longitude on maps. In nature, he asserted, only individuals exist. Further reflections concerning animal breeding forced Buffon to revise his ideas and define species operationally as groups of animals that could interbreed to produce fertile offspring rather than sterile mules. Species appeared to be nature's only "objective and fundamental realities," but the fixity of species was called into question by evidence of imperfect, useless, and rudimentary organs. Why should such structures exist if all species had been perfectly designed at the Creation?

Exploring the ways in which the earth had developed into a place suitable for living creatures, Buffon offered an alternative to conventional biblically based calculations that had fixed the creation at 4004 B.C. Scriptural authority could be maintained by treating the 7 days of creation as 7 epochs of indeterminate length. During the first epoch, the earth and the other planets were formed as the result of a collision between the sun and a comet. Rotating and cooling for some 3000 years, the incandescent gases torn from the sun turned into spheroids. During the second epoch, the earth hardened and congealed into a solid body. In the third epoch, the earth was covered by a universal ocean. As the waters began to subside during the fourth epoch, volcanic activity transformed the planet. Once the land surface was exposed, plants and land animals began to appear. During the sixth epoch, the original landmass began to break up and the continents separated from each other. Finally, in the seventh epoch human beings appeared. Simple experiments provided a tenuous basis for this speculative time frame. Spheres made of a variety of materials, such as iron, copper, granite, or limestone, were heated in a blacksmith's furnace and placed on wire racks so that Buffon could measure the time they took to cool. He then estimated the length of time it would have taken before the red-hot earth became cool enough to support life.

With his provocative speculations and elegant writing, Buffon helped to spread new ideas about comparative anatomy, geology, physics, and natural history. For example, Buffon's denigration of the native animals and peoples of the Americas provoked Thomas Jefferson to include a rebuttal in his *Notes from Virginia*. Special comparative tables demonstrated American superiority in the size of various mammals. Jefferson went to great trouble and expense to have the skeleton of a 7-foot moose and the bones of American caribou, elk, and deer sent to Paris to prove the "immensity of many things in America."

By the turn of the century, scientists were ready to explore the idea of species transformation more consistently and openly than Buffon. Jean Baptiste Pierre Antoine de Monet, Chevalier de Lamarck (1744–1829), was certainly the best known and most misunderstood of all the eighteenth-century advocates of evolutionary theory. Although he was the youngest of 11 children, the family title eventually devolved on him because of a series of deaths among his brothers. Unfortunately for Lamarck, there was no fortune to go along with the title. As a young man with no prospect of an inheritance, Lamarck was sent to a Jesuit school to become a priest. When he was 16 and the Seven Years' War was drawing to an end, his father's death brought him just enough money to buy a horse and join the army. According to Lamarckian family tradition, all the officers of his company were killed in battle and Lamarck assumed command. His bravery was so outstanding that he was immediately promoted, but at age 22 he was forced to leave the army because of ill health. While convalescing in Paris, Lamarck studied medicine, botany, and music. During this period, he met the philosopher Jean-Jacques Rousseau (1712–1778), who may have had some influence on Lamarck's ideas about nature.

In 1778, Lamarck published a popular guide to the flowers of France. Impressed by this study, Buffon helped to establish Lamarck's scientific career. The French Revolution destroyed his patron's family, but in 1793, when the King's Garden became the National Museum of Natural History, Lamarck and Étienne Geoffroy Saint-Hilaire (1772–1844) were appointed to new professorships in zoology created by the National Convention. Dividing the animal kingdom between them, Geoffroy Saint-Hilaire took the vertebrates and Lamarck became professor of insects, worms, and microscopic animals. Despite this advance in his career, Lamarck remained modest, retiring, and poor. Married and widowed four times, Lamarck lived long enough to see most of his children die. After becoming blind in 1817, he had to give up his professorship. He died in Paris in 1829 and was buried in a pauper's grave; his bones were thrown in a trench to mingle with the skeletons of other nameless unfortunates. Worse yet

M. LE CHEVALIER DE LAMARCK,

Professor of Botany of the National Institute

Jean Baptiste Pierre Antoine de Monet, Chevalier de Lamarck

Lamarck, old and blind

for Lamarck's reputation, the Biographical Memoir prepared by Georges Cuvier (1769–1832), and presented to the French Academy of Science in 1832, was a vicious attack on Lamarck's work.

As a self-taught zoologist, Lamarck indulged in interests so broad his researches in any particular area were likely to be lacking in depth. Little is known about the sources of Lamarck's ideas, because few of his papers survived and his published works rarely contained references to other authors. His ideas can be interpreted, however, as refutations of the natural theological traditions of eighteenth century-France. His *Memoirs on Physics and Natural History, Based on Reason Independent of Any Theory* was an attempt to formulate a general theory of existence based on physics, chemistry, and biology, a term he had coined. Considered primarily a taxonomist by his contemporaries, Lamarck was more excited by considerations of comparative anatomy and evolution. His views on the transformation of species were generally ignored or ridiculed until the British geologist Charles Lyell (1797–1875) introduced Lamarck's ideas to the English speaking world in the 1830s. Taking his clue from his friend Lyell, Charles Darwin gave Lamarck credit for being the first writer to create significant interest in the study of the origin of species. But Darwin noted that his own grandfather, Dr. Erasmus Darwin, had largely anticipated Lamarck's ideas.

The concept of the inheritance of acquired characteristics is so firmly associated with Lamarck that it is often assumed that he invented it. Actually what Lamarck did was to apply a very old assumption about heredity in an original way to the theory of evolution. According to Lamarck, as conditions on earth changed, geographic and climatic alterations affected plant and animal life. Given time without limit, transmutation of species occurred in accordance with Lamarck's laws of evolution. In *Zoological Philosophy* (1809), evolution was attributed to a fundamental innate tendency of living beings to evolve towards increasing complexity, in conjunction with the transmission of these increasingly complex structures to each generation. Simple creatures like worms and infusorians, which were thought to be the products of spontaneous generation, were imbued with the innate tendency to evolve into higher life forms. If the environment were stable, this innate drive would itself lead to more complex life forms. When the environment changed, living beings were forced into new kinds of exertions, which eventually produced modifications. Increased usage of particular parts would, for example, lead to an increase in the size of the organs involved. Changes in size, strength, or structure brought on by increased usage would then be passed on to future generations.

In *A Natural History of Invertebrate Animals* (1815), Lamarck explicated his thoughts on the causes of organic diversity and increasing complexity. The most important and misunderstood aspect of his theory was the assertion that under new conditions, animals experienced a strong physiological need for new organs. The French word *besoin* was often mistranslated into English as *desire* rather than *need*, although Lamarck did not believe that living beings evolved in response to "wishes" or "will." For Lamarck, habits and actions were critical to changes of form, but some Romantics and Transcendentalists found the idea that *desires* drove evolution quite inspiring, as demonstrated by Ralph Waldo Emerson's (1803–1882) poetic explanation of this putative tendency towards greater perfection:

> And striving to be man, the worm
> Mounts through all the spires of form.

When all living animals were known, Lamarck predicted, they could be arranged in a linear series with some small collateral branches. Such an arrangement would reflect the natural order of animals in nature, encompassing all forms from the lowest monad to the highest creature on the scale of nature, the human being. If all species were included on the ladder of creation, the gradations between neighboring species would be nearly imperceptible; apparent gaps would represent undiscovered species. The relationship between humans and other animals was somewhat obscure, although Lamarck seems to have suspected a common origin for humans and the great apes. For example, he speculated about how an ape might have learned to stand more erect in order to see further. The new posture, along with other changes in needs and behavior, could have caused heritable structural modifications. Thus, it was possible that, in the future, through training and practice the apes could be brought up to a higher level of intelligence. After all, Lamarck pointed out, few individuals in all of history had achieved real intelligence and the rest remained in a state approximating bestial ignorance.

Lamarck's colleague Geoffroy Saint-Hilaire was the author of equally speculative and imaginative theories. Before his appointment as a professor of zoology in 1792, Geoffroy had been a mineralogist, but his interests included chemistry and natural history, as well as crystallography. Nature philosophy shaped much of Geoffroy's philosophy of biology. Basically, he was convinced that all animal species could be thought of as variant forms of a fundamental archetype. Although he tended to exaggerate possible affinities, his theory of the archetype served many comparative anatomists as a workable alternative to either a strict creationist or evolutionary framework. According to his son Isidore, who adopted and extended many of his ideas, Geoffroy began to suspect as early as 1795 that modern species were degenerations of a primeval organism. It was not until 1828, however, that he publicly discussed this idea.

Erasmus Darwin, grandfather of both Charles Darwin and Francis Galton, was a physician, author, poet, inventor, and a member of the remarkable band of prodigies comprising the Lunar Society. Known as a man with a great appetite for food and ideas, Erasmus Darwin explored a broad range of botanical, zoological, physiological, and medical questions. More radical, if less system-

Erasmus Darwin

atic a thinker than Charles Darwin, Erasmus Darwin proposed a theory of the transmutation of species in which all organisms were descendants of some primordial filament. His speculations about the nature of life were presented in *Zoonomia; or the Laws of Organic Life* (1794), *The Botanic Garden; a Poem in Two Parts* (1789, 1791), and *The Temple of Nature; or Origin of Society* (1803). Although his writings might not excite much attention today if not for his grandson, at the time of their publication these books were considered threatening enough to earn a place on the *Index of Prohibited Books*. Translated into several languages, *Zoonomia* was especially popular among German natural philosophers. Charles Darwin admitted to reading it twice, and it probably had more influence on his own thoughts about the transmutation of species than he cared to admit. Nevertheless, while condescendingly expressing his disappointment at the disproportion between the amount of speculation and the number of facts in his grandfather's writings, Charles Darwin came close to accusing Lamarck of stealing ideas from Erasmus Darwin. It is more likely that both Erasmus Darwin and Lamarck had been inspired by the work of Georges Buffon and others, as well as their own observations and inferences drawn from common knowledge about plant and animal breeding. Always rather self-conscious and even defensive about his own originality, Charles Darwin seems to have underestimated his grandfather's curiosity, intelligence, and achievements.

Unlike Charles Darwin, who focused his attention on the origin of species from preexisting, but still complex antecedents, Erasmus Darwin boldly envisioned the history of the earth in terms of millions of ages during which "one living filament" was transformed into all the myriad forms of animals that have ever existed, all such transformations being the result of "the great first cause," which endowed a primordial living filament with the ability to acquire new parts and properties. Acted upon by various stimuli, these new entities acquired further improvements and passed these changes on to their posterity "world without end." Although the "whole warring world" might now appear to be "one great slaughter-house," Erasmus Darwin suggested that sympathy should exist among all life forms because of their relatedness through descent. Rudimentary structures and vestigial organs revealed the "wounds of evolution." Ultimately, through the study of analogies among contemporary species, we would realize that all living beings were one family descended from a common ancestor. Even the most arrogant man, who saw himself as the image of God, should realize that human beings too had arisen from "rudiments of form and sense, an embryon point, or microscopic ens!" (Philosophers used the term "ens" for that which has being in its most general sense.)

Naturalists like Lamarck and Erasmus Darwin were intrigued by the eighteenth-century idea that unlimited progress and organic change were possible, but the fears generated by the excesses and terrors of the French Revolution did much to eclipse the hope of progress. Stability in society and nature seemed more desirable than limitless, unpredictable change. Indeed, evolutionary theories and their advocates were rejected and ridiculed by one of France's most eminent scientists, Georges Leopold Chrétien Frédéric Dagobert, Baron Cuvier (1769–1832). Georges Cuvier, preeminent comparative anatomist and founder of paleontology, was an implacable opponent of Lamarckian ideas in general and evolutionary theory in particular. A child genius who had been reading since the age of 4, Cuvier entered the Academy of Stuttgart at 14 and distinguished himself by his prodigious memory, hard work, and self-discipline. His biology teacher, Karl Friedrich Kielmayer, taught that taxonomic order would be achieved when comparative anatomists discovered the archetype of all living creatures. Cuvier's attempts to bring order to the study of marine animals came to the attention of Geoffroy Saint-Hilaire and led to Cuvier's appointment as assistant in comparative anatomy at the Museum of Natural History in Paris in 1795.

Despite Cuvier's lack of formal preparation for his duties, which included training medical students, he soon surpassed Geoffroy Saint-Hilaire and Lamarck in prestige, wealth, and fame. To his contemporaries he was no less than a second Aristotle. In addition to his research and teaching, he took on the reform of the French educational system as Inspector General in the Department of Education, and also served as Chancellor of the Imperial University and Councilor of State. While Cuvier disentangled the taxonomic relationships of many different groups of animals, his most exciting studies were of fossil mammals and reptiles. Georg Bauer Agricola (1494–1555) had used the term *fossil* in *De re metallica* (1555) to refer to anything dug out of the earth. Things that resembled animal or plant forms were generally regarded as sports of nature, a kind of art form that nature practiced rather capriciously to amuse and mystify human beings.

Understanding fossil forms was quite difficult because these entities were rare, often incomplete, and usually in poor condition. By the end of the eighteenth century, however, large quantities of fossil teeth and bones had been collected, especially from caves in the Harz Mountains and French Alps. Caves appeared to be primordial features of the planet that gave access to the remote bowels of the earth, the archival vault of its antediluvian history. Some naturalists thought that fossilized bones had been washed into caves during great floods; others argued that humans had deposited peculiar bones in caves. Goethe suggested that the close study of caves and fossils could provide profound insights into distant chapters of the still unfolding history of the earth. Johann Friedrich Blumenbach (1752–1840), whose work in paleontology was eventually eclipsed by that of Cuvier, attempted to use fossils as a means of classifying the geological epochs of earth history. In his popular *Textbook of Natural History* (1779) and his *Contributions to Natural History* (1790, 1806), Blumenbach called attention to the uses of comparative anatomy for paleontology and geology. He noted that certain species such as the dodo had become extinct in historical times and suggested that extinctions had probably occurred during the most ancient epochs.

Most collectors of fossils knew too little about the structure of living beings and their geographic distribution to say whether or not particular bones might be identical to those of known species. Many of those who had the opportunity to observe fossils within rock formations and caves were primarily interested in practical aspects of mineralogy and mining. In contrast to the antiquity of mineralogy and mining technology, modern paleontology essentially began in the late eighteenth century. Paleontology eventually became part of the domain of historical geography, but it was originally considered a part of anatomy and was, therefore, associated with the medical sciences. To understand fossil animals, Cuvier analyzed the bones of living creatures for comparative purposes. Most importantly, Cuvier used his knowledge of the functional relationships between form and function in living animals to analyze fossils. He called this approach his *correlation theory*. Comparative anatomy provided the clues that allowed Cuvier to reconstruct the form and function of unknown animals from surviving bits and pieces. For example, Cuvier proved that fossil bones that had been identified as the remains of a man who had died before the great flood of Noah were really the remains of a gigantic extinct salamander.

Recognizing discontinuities between extinct and living species, Cuvier explained changes in the animal world in terms of catastrophes or revolutions that had destroyed whole populations of living things in prehistoric times. Many cultures have legends about global catastrophes, particularly floods, which supposedly took place in the distant past. Both the physical evidence of local geological strata and recent political and social events in France seemed to demonstrate the power of sudden, violent transitions. The sharp lines of demarcation between various geological strata, the inversion and disorder of strata at some sites, and the great numbers of fossils in some layers could readily be interpreted as evidence of mass extinctions and rapid changes in structural features of the earth. The discovery of frozen mammoths in Siberia, for example, seemed to provide evidence of such sudden and violent changes. To account for such observations, Cuvier proposed a series of catastrophes in the history of the earth, culminating with one about 6000 years ago that had established the geology of the present epoch.

Despite much authoritative condemnation, evolutionary theories continued to attract the attention of imaginative amateurs such as Robert Chambers (1802–1871), the author of a much maligned book called *Vestiges of the Natural History of Creation*. Chambers admitted that his goal as a writer was to be "the essayist of the middle class," but he was fascinated with the idea of anonymity, which was a prominent theme in nineteenth-century fiction, journalism, and political writing. Special precautions were taken to maintain the anonymity of "Mr. Vestiges" until 1884, when the posthumous twelfth edition of *Vestiges* was published. Very little is known about how and why Chambers came to write *Vestiges*, because Robert destroyed many relevant documents and the biography written by his brother William provided hardly any clues.

Robert and William Chambers published a very successful series of journals and textbooks. *Chambers's Educational Course*, including Robert's (anonymous) *Introduction to the Sciences* (1836), which was widely used in schools. Publicly, Chambers did not admit support for evolution because advocating radical ideas might have been detrimental to his family's commercial ventures. The Chambers brothers had a very personal interest in the problem of variety in nature; both were complete hexadactyls, that is, they were born with six digits on their hands and feet. Both were operated on to correct this anomaly, but Robert's convalescence was long and painful. After suffering a mental breakdown allegedly caused by excessive mental exertion, Chambers left Edinburgh in the 1840s to settle in the peaceful village of St. Andrews with his wife Anne and their 11 children. Enforced convalescence and isolation gave Robert time to work on *Vestiges*. The sources of his ideas about evolution apparently included studies of phrenology, especially as explicated by George Combe, author of *Constitution of Man* (1828), and the nebular hypothesis as described by John Pringle Nichol in his *View of the Architecture of the Heavens* (1837). Beginning with the nebular hypothesis concerning the formation of the universe, Chambers might have envisioned the law of progress acting on the whole realm of nature, including zoology, botany, physiology, archeology, anthropology, and prehistory.

Vestiges, a layman's guide to scientific theories concerning the development of nature as a whole, from the evolution of the solar system to that of plants, animals, and human societies, created a sensation when it appeared in 1844. Clergymen attacked *Vestiges* as a work of "black materialism" that threatened the very foundations of morality and religion, intellectuals thought the book would cause "sensation and confusion," and scientists dismissed it as "bad science." Nevertheless, *Vestiges* was read by hordes of middle- and working-class readers in Europe and America, as well as distinguished individuals such as Abraham Lincoln, Frances Cobbe, Benjamin Disraeli, Charles Darwin, Alfred Russel Wallace, and Thomas Henry Huxley. Disraeli, who found the book quite enchanting, satirized it in his novel *Tancred* (1846), where it appears as *The Revelations of Chaos*. The English geologist Adam Sedgwick (1785–1873) said that he loathed and detested *Vestiges* and that the book was so stupid he knew the author must be a woman. Others suspected that Prince Albert was Mr. Vestiges. Charles Lyell credited Mr. Vestiges with familiarizing the English reading public with Lamarckian views of transmutation. On the other hand, Lyell noted that despite the author's clear and attractive prose, he presented no new facts or original arguments, and he failed to refute the principal objections previously presented by distinguished scientists. Huxley was very much on target when he said that the author of *Vestiges* obviously knew no more about science than what could be picked up in *Chambers' Journals*. Clergymen and naturalists denounced the book so loudly that they increased the appetite of the public for more discussions of the presumably dangerous topics discussed in *Vestiges*. While scientists and clergymen competed in vilifying Mr. Vestiges, his book went through 10 editions in as many years and was translated into several languages.

Conventional references to divine providence were incorporated into Chambers's general writings, but religion was apparently a minor influence in his own life. Presumably, the writing of *Vestiges* was not a response to a personal religious crisis, but a product of Chambers's interest in physiology, phrenology, and physics. By extrapolating from the nebular hypothesis concerning the formation of the universe, Chambers attempted to apply the "law of progress" to the whole realm of nature, from geology and biology, to anthropology and archaeology. Seeing a fundamental unity in the development of the universe, the earth, and living beings, Chambers contended that his theories did more to glorify the Creator than the idea that God was directly involved in fashioning every kind of creature from the elephant to the dung beetle. Special creation entailed the embarrassing juxtaposition of a perfect and efficient Creator with incontrovertible evidence of created species burdened by the "blemishes or blunders" manifested in vestigial and rudimentary organs.

Even if readers believed that the theory of evolution was not in complete harmony with a literal interpretation of the Bible, Chambers suggested that the theories discussed in *Vestiges* were completely compatible with the spirit of the scriptures. That is, God was the First Cause and the laws of nature were His mandates for the working out of the divine will. The processes of cosmogony were not, however, necessarily fixed at one moment in creation. The universe was, therefore, always developing according to laws for which physicists could establish a sound mathematical basis. There was no reason, Chambers wrote, to deny the fact that God had created living beings, but this did not mean that God was constantly involved in the creation of the myriad forms of life. This could be done more economically through the operation of natural laws that were expressions of the will of God.

While Chambers seemed quite confident that his evidence proved that evolution had occurred, he was unable to invent a new mechanism or driving force. Generally, Chambers ascribed improvement or degeneration of individuals and species to environmental influences. Citing the well-known fact that bees could produce a queen or a worker from the same starting material by changing the conditions of development, he argued that similar influences must be at work in other cases. Such influences might even lead to the evolution of human beings. Assuming that the laws of evolution explained the development of human races and civilizations, Chambers proposed a hierarchical arrangement of the "races." Using concepts gleaned from the work of Karl von Baer, Chambers argued that just as the embryo passed through stages corresponding to the lower animals during development, so too did the human brain develop through the stages represented by human "racial types." He had, however, argued that speech was the only real novelty attending the creation of human beings. Because studies of language had shown that some so-called primitive people had more complex language systems than Europeans, Chambers placed himself in the awkward position of denigrating the difficulty of acquiring the one truly unique human trait. Complexity of language, he rationalized, is not directly correlated with intelligence, because little children learn whatever language their parents speak. Man was fundamentally, Chambers concluded, a "piece of mechanism, which can never act so as to satisfy his own ideas of what he might be." Well aware of the fact that evolutionary change is not necessarily what humans would think of as progress, Darwin wrote "Never use the words higher or lower" in his copy of *Vestiges*.

THE HISTORY OF THE EARTH

Geology, the science that led many biologists to consider unorthodox thoughts about the history of the earth and its inhabitants, was also a source of inspiration for advocates of Special Creation. Indeed, many geologists considered their science the true handmaiden of theology. Although applied knowledge of the earth is very old, involving mining, metallurgy, and surveying, a new scientific discipline grew out of eighteenth-century controversies about the nature of the forces that had shaped the earth. The emerging science of geology gave rise to two major schools of thought, which have been called *Neptunist* and *Vulcanist*. The disciples of Abraham Gottlob Werner (1750–1817), the founder of geognosy and Neptunist geology, believed that the formation of the crust of the earth could best be explained by the action of water, such as the great flood of Noah described in the Bible. Despite the limited extent of his own fieldwork, Werner exerted a tremendous influence on the study of geology through his success as a lecturer at the new Mining Academy of Freiberg, where he taught for 40 years. Satisfied that his contributions to geognosy, as explicated in his treatise *On the External Characteristics of Fossils* (1774), had established an orderly system for the classification of rocks and minerals, Werner turned his attention to more complex problems. Reflecting on the history of the earth, Werner suggested that about one million years ago a universal ocean had covered the earth. The oldest crystalline rocks had, therefore, originated by chemical precipitation out of the water. Eventually the ocean subsided and more new rock originated mechanically as older formations were exposed to the air and the process of erosion. Exactly how the ocean had held the great mass of minerals in solution and where the great waters had gone were problems the Neptunists had difficulty explaining.

When Charles Darwin was a student, the most popular English geologists, William Buckland, Adam Sedgwick, and William Daniel Conybeare, often lectured about the marvelous harmony between Neptunist geology and biblical accounts of the creation and the great flood of Noah. But such ideas had already been challenged by James Hutton (1726–1797), the guiding spirit of the Vulcanists, or Plutonists, who emphasized the action of heat in shaping the earth. Although Hutton had studied medicine at Edinburgh, Paris, and Leiden and received his Doctor of Medicine in 1749, he never practiced the healing art. Hutton preferred to supervise his family farm and study agriculture, mineralogy, and natural history. In 1785 Hutton presented his novel ideas about the history of the earth to the Royal Society of Edinburgh. Ten years later Hutton published his landmark treatise *Theory of the Earth, with Proofs and Illustrations.* Except for attacks by Werner's followers, the book attracted little attention; fewer than 500 copies were printed.

Hutton formulated the two major tenets of geology: the geological cycle and the principle of uniformitarianism, which states that scientists must explain the history of the earth in terms of the natural processes acting in the present. Observing that heat, pressure, and water were continuously but gradually shaping and reshaping the earth, Hutton assumed that its present form must be the outcome of similar changes that had taken place over limitless stretches of time. Most rocks, Hutton argued, had been formed by the erosion of land. Eroded materials were deposited on the ocean floor, where they were turned into rocks by heat from deep within the earth. Eventually these rocks were uplifted and become dry land once again. In other words, the earth as we know it today is the product of an endless cycle of erosion, deposition, lithification, and uplift. According to Hutton, the earth was a self-renewing machine operating in a lawbound manner as a part of the great cosmic engine. When Hutton investigated the history of the earth, he found "no vestige of a beginning, no prospect of an end."

Uniformitarian geologists, who modeled their work on the concepts of Lyell and Darwin, tended to portray Hutton as the hero and Werner as the villain in a battle between scientific, secular geology and an archaic, biblically based Neptunist cosmogony. But Hutton's work was also deeply teleological in outlook. While he ridiculed those who used the eruption of volcanoes as a way to frighten superstitious people into orthodox piety, he too saw the structure of the earth as evidence of a grand design reflecting the wisdom that had produced "a habitable globe." Few people became acquainted with Hutton's ideas until his friend John Playfair (1748–1819), professor of natural philosophy at the University of Edinburgh, published his popular *Illustrations of the Huttonian Theory of the Earth* in 1802. It was Playfair who enthused that the mind "seemed to grow giddy by looking into the abyss of time." While defending Hutton's theory, Playfair attacked the champions of the Neptunian system for misusing the weapons of both theology and science and thus doing as much damage to the dignity of religion as to the freedom of scientific inquiry. By pretending to carry their own theory of the earth to a period prior to the present world of causes and effects, such critics had committed crimes against true empirical science. Claiming that they had been guided by a superior light, Neptunists cast their geological speculations in the form of a commentary on the books of Moses. Creation itself, Playfair insisted, would always be hidden from scientific, philosophical, empirical enquiry.

Playfair's explication of Huttonian theory was considered a major landmark in the history of British geology but it was William Smith (1769–1839), a largely self-taught drainage engineer and surveyor, who was awarded the title "father of English geology" and the first Wollaston Medal of the Geological Society (1831). Smith's great insight into the arrangement of rock strata was that beds of the same kind of stone, at different levels of succession, could be recognized by means of the fossils they contained. By 1799 Smith had recognized the relationship between fossils and geological strata and had explained his ideas to some of his friends. Seeing himself as a practical field geologist, with a profound aversion to writing, rather than a scientist, Smith failed to publish his famous map of the strata of England and Wales until 1815. His friend, the Reverend Joseph Townsend, however, used Smith's ideas as the foundation for his own treatise *The Character of Moses Established for Veracity as an Historian, Recording Events from the Creation to the Deluge* (1813). Avoiding speculation, Smith contended that science, especially geology, should be useful as a means of promoting industry and manufacturing. His geology was essentially descriptive and claimed no higher, broader vision or challenge to biblical creation.

Many early structural geologists, such as the Reverend Adam Sedgwick (1785–1873), professor of geology at Trinity College, Cambridge, were clergymen eager to study the great book of nature in order to illuminate the attributes of the Creator. When Sedgwick was appointed Woodwardian Professor of Science in 1818, he admitted he knew nothing about geology. Another candidate had been nominated for the position, but Sedgwick knew his rival had no chance because he was burdened by a great deal of knowledge, all of which was wrong. Energetically immersing himself in research and fieldwork, Sedgwick found evidence to support the biblical account of creation and the great flood everywhere he looked. This was to be expected, he explained, because truth must always be consistent with itself. Thus, when science was properly interpreted it must be consistent with the scriptures. Ironically, it was a summer spent exploring the geology of North Wales under Sedgwick's guidance that stimulated Charles Darwin's interest in the history of the earth.

Another important influence on the young Darwin was the work of Charles Lyell (1797–1875), the geologist who became known as the great high priest of uniformitarianism, an ironic title considering it was Lyell's work that eventually severed the chains that bound geology to theology. As a young man Lyell was not a particularly good student, but he was zealous in pursuit of his hobbies, such as lepidoptery and geology. While enrolled at Oxford he undertook several geological tours of England, Scotland, and the continent. His plans for a career in law were abandoned because of an eye disease, but this apparent misfortune left him free to pursue his real interests. In his elegant writings, Lyell reasserted Huttonian principles so persuasively that the concept of the historical and physical continuity of nature became inextricably linked with his name. Generous to his predecessors, Lyell suggested that the revolutionary events of the eighteenth century had promoted a religious and political climate in which the work of Hutton and Playfair was viewed with suspicion. In contrast, Neptunianism and Wernerianism appealed to orthodox religious principles and assumptions. During his own era, Lyell noted, the intellectual climate had become more propitious for uniformitarian geology. Lyell's three-volume *Principles of Geology* was kept up to date with constant revisions; the twelfth edition was published posthumously in 1875.

Like Hutton, Lyell adopted a strictly uniformitarian approach towards geology, insisting that geological data must be interpreted in accordance with forces presently known to be acting on the earth. He did not, however, extend this concept to the natural and historical succession of plant and animal life forms. Although Charles Darwin considered Lyell one of his most important friends and guides, for most of his life Lyell resisted a uniformitarian system of organic evolution. Lyell believed that each species had somehow been created for its proper niche and could neither change nor adapt to altered conditions. If a species were not already suitable for an altered environment, it would be replaced with other, more suitable species. The creative force that originally produced species was, Lyell believed, outside the proper scope of scientific inquiry.

Convincing Lyell of the validity of the transmutation of species by means of natural selection was one of Darwin's proudest achievements. Lyell disagreed with Darwin's theory of human evolution, but he had no great difficulty accepting evidence for the "antiquity of man" and the vast distance of time separating Stone Age humans from civilized Europeans. In the first stage of existence, Lyell argued, man would have been "just removed from the brutes" in terms of intellectual powers. It was impossible to set dates for the beginning and end of the first Stone Age, but Lyell was sure that the original stock of mankind lacked the intellectual powers, inspired knowledge, and "improvable nature" enjoyed by their modern counterparts. Once civilization had been achieved, however, the rate of progress in the arts and sciences would have increased geometrically as knowledge increased. In analyzing various theories of human progression and transmutation, including those of Lamarck, Lyell concluded that all the major human races were the descendants of a single pair of ancestors. Rather than using this concept of a common human heritage to argue for a strict biblical chronology and a young earth, Lyell saw it as evidence that a great amount of time had been necessary to form the modern human races.

Tracing the lines of progress in biology from Lamarck to Darwin forced Lyell to face the question of whether the general hypothesis of transmutation also applied to the human race. Were human beings the product of a continuous line of descent from the lower animals? This conclusion seemed inescapable, at least for physical characteristics, but might not apply to specifically human moral and intellectual attributes. While not abandoning the hypothesis of variation and natural selection, Lyell objected that it was not necessary to assume a gradual and unbroken line of improvement in intelligence from the inferior animals up to human reason. Lyell called attention to the occasional birth of an individual of "transcendent genius" to parents of average intellectual capacity. Perhaps, Lyell reasoned, similar rare and anomalous events had established the great chasm between the unprogressive intelligence of the lower animals and the first stage of the improvable reason found in humans. Citing the work of Asa Gray (1810–1888), Darwin's chief advocate in America, Lyell concluded that the Darwinian doctrine of variation and natural selection did not weaken the foundations of Natural Theology. The Darwinian hypothesis of the origin of species could be held along with the belief that the deity might have initiated a chain of events that could proceed without any subsequent interference, or with some occasional direct intervention. Thus, Lyell was sure that the idea of change and progress in the history of the earth need not foster a materialistic tendency, but could be seen as "the ever-increasing dominion of mind over matter."

THE DARWINIAN REVOLUTION

In formidable waves and faint ripples, the Darwinian revolution inundated virtually every domain of human thought and belief: scientific, religious, political, philosophical, historical, social, and literary. Like Hutton's vision of the history of the earth, there seems to be "no prospect of an end" of books, articles, and new interpretations where Charles Darwin and the Darwinian revolution are concerned. As Ralph Waldo Emerson warned his contemporaries: "Beware when the great God lets loose a thinker on this planet." Modern historians of science tend to believe that all ideas and theories are largely socially determined, but this hardly explains how and why Darwin formulated a theory of such power and notoriety. Although his family tree was a veritable Victorian *Who's Who*, Charles Robert Darwin (1809–1882), the sixth of eight children born to Susannah Wedgwood Darwin and Dr. Robert Waring Darwin, was not a very promising young man. Yet Darwin's life and work remain the subjects of intense interest, controversy, and scrutiny more than 100 years after his entombment in Westminster Abbey, not far from the monument to Sir Isaac Newton.

Darwin's son Francis speculated that Charles Darwin inherited his sweetness of disposition from the Wedgwood side of the family and his genius from grandfather Erasmus Darwin. When Darwin reflected on his education, he complained that his early schooling had been detrimental to the development of his mind. Moreover, he was well aware of the fact that his teachers and his formidable father considered him a very ordinary boy, below commonly accepted academic standards, who would probably be a disgrace to the family. Nevertheless, Charles was dispatched to Edinburgh to study medicine in the tradition of his father and grandfather. It was soon obvious that he was temperamentally unsuited for a career in medicine; anatomy and Latin merely bored him, but the sight of blood made him ill. Seeing operations performed at a time when anesthesia and antisepsis were not yet part of the art of surgery convinced him that medicine was a "beastly profession." In any case, Darwin did not feel the need to exert himself in preparation for any career, because he expected to inherit enough money to live well without working. He did, however, actively pursue his interest in natural science. While Darwin was at Edinburgh, an anonymous paper supporting the theory of the transmutation of species appeared in the *New Philosophical Journal*. The author was Darwin's friend Dr. Robert Grant, who had astonished Darwin by his enthusiastic support for Lamarck's evolutionary theory. Despite his association with Grant and his reading of *Zoonomia*, Darwin later claimed that in his youth he had not entertained any doubts about the strict and literal truth of every word in the Bible, augmented by close readings of William Paley's *Natural Theology: or, Evidences of the Existence and Attributes of the Deity, Collected from the Appearances of Nature* (1802).

Accepting the fact that Charles could not be turned into a respectable physician, his father decided that the next best profession was that of clergyman, although Robert Darwin was himself a skeptic where religion was concerned. Dutifully, Charles began to prepare himself to take holy orders, but his attitude towards education and the need for career planning was unchanged. The event that transformed Darwin's life was an offer to serve as unpaid naturalist on the *H.M.S. Beagle*. The voyage of the *Beagle*, which lasted from December 1831 to October 1836, stands out as the great adventure and determining influence in Darwin's life. Indeed, the first sentence of the *Origin of Species* reflects back on this experience and how the distribution of species in South America and the geological relationship between the region's living beings and its fossil forms seemed to throw light on the great "mystery of mysteries," the origin of species.

It was only through a peculiar chain of accidents and compromises that Darwin managed to secure a place on the ship. The offer was extended to this untested young man because more suitable candidates had turned down the posi-

Charles Darwin

tion. The geologist John Henslow recommended Darwin despite his youth and inexperience, but Darwin's father objected to this wasteful enterprise, and his obedient 24-year-old son was ready to comply with his father's wishes. Dr. Darwin said he would change his mind only if a man of common sense would support his son's strange idea. Fortunately, Uncle Josiah Wedgwood, unquestionably a man of common sense—and uncommon means—gave Charles his invaluable support. The ship and its captain posed another set of problems. Captain Robert Fitz Roy, a follower of phrenology, thought that the shape of Darwin's nose suggested a certain lack of energy. The *Beagle*, a 10-gun brig with rotting timbers and decks, refitted for the surveying mission as a three-masted bark, needed thorough reconstruction. The mission of the *Beagle* was to complete the survey of Patagonia and Tierra del Fuego. While circumnavigating the earth, the ship was to make stops at the Cape Verde Islands, South America, the Galapagos Islands, Tahiti, New Zealand, Australia, Mauritius, and South Africa. After two unsuccessful efforts to set sail, on December 27, 1831, the *Beagle* finally departed with 74 persons crammed into every available nook and cranny. Accommodations, even for the captain and the naturalist who was to serve as his companion, were far from luxurious. Darwin, who slept in a hammock slung over the chart table in the poop cabin, soon discovered that he was very susceptible to seasickness and that his quarters were in the roughest-riding section of the ship.

As the *Beagle* proceeded southwards along the coast of South America, Darwin noticed how plant and animal forms differed with geographical location. Like Chambers, he must have worried about the uneconomical way the direct creation of all these slightly different forms would have taken up God's time and energy. These questions became most striking when he investigated the Galapagos Islands, a group of about 20 barren islands of volcanic origin on the equator, about 600 miles off the west coast of South America. In general, the animals on these desolate, geologically young islands were similar to those on the mainland of South American, but many species were unique. The Galapagos Islands served as a living laboratory of evolution where reproductive isolation produced peculiar species living under essentially identical environmental conditions; different species of birds, tortoises, and so forth were associated with specific islands. Studying the creatures living on these bleak islands was probably the single most decisive experience in Darwin's conversion from a naively orthodox future clergyman into a scientist of infinite curiosity and endless patience.

Darwin always regarded the voyage as his first real education; he discovered that the pleasure he found in "observing and reasoning" could be attained only by hard work and concentrated attention. All former interests gave way to a love of science. By the time he returned to England, his mind and habits had been so thoroughly transformed that when Robert Darwin first saw his son again he resorted to a diagnosis rooted in phrenology and exclaimed: "Why, the shape of his head is quite altered." The change in Darwin's formerly rugged and indefatigable constitution was equally profound; the bold adventurer of the *Beagle* was transformed into a frail, nervous, reclusive invalid. When Darwin visited the *Beagle* for the last time in 1837, he said that he would gladly sail around the world again, if not for seasickness, but he never left England again.

The nature of Darwin's mysterious diseases and chronic ill health has long fascinated scholars, especially for the insights such studies might offer into the nature of the creative process. Many diagnoses and theories have been advanced, including the suggestion that Darwin's recurrent bouts of debilitating illness were psychological in origin. Some of the evidence suggests that he enjoyed the iron constitution of the true hypochondriac, but hereditary weakness, anxiety attacks, depression, arsenic poisoning due to his medications, and a debilitating malady peculiar to South America known as Chagas' disease have been blamed for the suffering he recorded in his meticulously detailed diary of ill health. The chief vector for the parasite that causes Chagas' disease might well have been the "great black bug of the Pampas" that attacked Darwin as he explored Mendoza, a province of Argentina. Exceptionally severe episodes of illness were likely to occur when Darwin finally completed a special phase of his work, when he was overwhelmed by concern about the health of his wife and children, or when there was a death in the family. Attempts to find relief included rest cures, visits to spas, hydrotherapy, electrotherapy, and the usual panoply of noxious Victorian drugs.

Like his father, Darwin had the good sense to marry a Wedgwood. Cousin Emma was the daughter of Uncle Josiah, who had convinced Robert Darwin to allow Charles to serve as naturalist on the *Beagle*. The marriage that took place in 1839 provided a handsome dowry and an annual allowance. With a comfortable guaranteed income, Darwin was able to devote himself to research while Emma devoted herself to worrying about his health and her endless series of pregnancies. To avoid the stresses of city life, the Darwins left London and moved to the village of Down, where they lived in almost total seclusion. Despite chronic ill health, punctuated by devastating episodes of mysterious debilitating crises, Darwin produced a prodigious amount of scientific work, a revolutionary theory, and 10 children, 7 of whom survived to adulthood.

Using the diaries and notebooks accumulated during the years spent as Captain Fitz Roy's naturalist, Darwin wrote several rather turgid volumes on geology and *The Voyage of the Beagle*, which became a popular success. His scientific reputation was first built on his geological work, especially his explanation of the formation of coral atolls. During his years in London, Darwin had served as secretary of the Geological Society and met many important leaders of the scientific and intellectual community. Charles Lyell, the geologist, and

Joseph Dalton Hooker (1817–1911), the botanist, became his closest friends. Lyell and Hooker served as confidants, as critics, and as Darwin's special agents in a narrowly averted priority dispute with Alfred Russel Wallace (1823–1913).

Although Darwin later said that his conversion from creationism to evolution occurred during the voyage of the *Beagle*, 20 years elapsed before he published his theory of the origin of species. While Darwin may have convinced himself of the fact of evolution as early as 1837, he did not write out his preliminary sketch of the idea until 1842. Alluding to the persecution of Galileo, Darwin later recalled his reluctance to make public views that his contemporaries, especially his wife Emma, would find unorthodox and offensive. But surely Darwin knew that the last burning for scientific heresy was that of Giordano Bruno (1548–1600) in 1600. Moreover, being independently wealthy, Darwin did not have to worry about losing a job. Probably, Darwin was less concerned with being labeled a heretic than he was afraid that he might not be able to convince his peers of the scientific validity of his ideas.

Of course Darwin did not claim that the theory of transmutation of species was in itself new and original. The theory remained incomplete, because none of the advocates of the theory had proposed a compelling mechanism. However, the question of *whether* evolution had occurred could be separated from the question of the *mechanism* by which it might occur. Just how Darwin worked out the details of his theory is unclear. Perhaps the most important moments in the creative process are never recorded, but there are surprising gaps in the written record, despite the fact that Darwin was a compulsive hoarder. Almost all of Darwin's voluminous correspondence was preserved, along with journals of his research, annotated books and articles sent to him by fellow scientists, minutely detailed financial and family records, and bits of paper containing scribbled notes.

By the time Darwin published *Origin of Species*, he was finally beginning to believe that he had enough evidence to build a persuasive case for his theory, if not a compelling one. His contemporaries acknowledged that for every fact adduced by Lamarck, Darwin presented 100. He had, moreover, provided a mechanism for evolution quite different from that of Lamarck. In reflecting on the major literary sources of his inspiration, Darwin called attention to Charles Lyell's *Principles of Geology* and *The Essay on Population* by that cheerful prophet of doom Robert Malthus (1766–1834). In his *Autobiography*, Darwin states that while reading the *Essay on Population* for diversion in September of 1838, he was struck by the way the Malthusian struggle for existence could lead to positive as well as negative results. It is rather difficult today to imaging anyone reading a work so ponderous and gloomy for amusement, but the *Essay on Population* was apparently a great stimulus for evolutionary thought. The *Essay on Population* was also the inspirational spark that led Wallace to the theory of natural selection. While Darwin's *Autobiography* gives the impression that the idea of natural selection came to him in a flash of insight, his letters and notebooks indicate that he was already thinking in terms of "selection" months before he turned to Malthus.

When critics wanted to denigrate Charles Darwin's achievements, they contemptuously charged that only a Victorian gentleman could have produced a book an unoriginal as *Origin of Species*, a mere application of the doctrine of Malthus to the world of plants and animals. Robert Malthus had written the *Essay on Population* after a dispute with his father concerning the nature of man and the prospects for human society. Daniel Malthus had been a friend of the philosopher Jean Jacques Rousseau (1712–1778) and had attempted to instill in his son a belief in the perfectibility of human beings. But Robert rejected such hopeful visions of the future and developed an awesome and apparently inexorable calculus of human misery.

The thesis propounded in the *Essay on Population* was that all utopian visions of future human societies were impossible, because in any society innate and immutable defects in human nature inevitably led to misery and vice, sickness, war, perversion, and promiscuity. All of this misery was the inevitable outcome of two basic needs: food and sex. Misery was the natural state of a major portion of humankind because population increases invariably tended to exceed the available resources, leading to violent competition for food and other necessities. As a consequence of the fixed relationship between land and production, food supplies increased arithmetically in the series 1, 2, 3, 4, 5, and so forth. But, Malthus explained, populations increased geometrically so that the human population, if left unchecked, would tend to increase every 25 years in the series 1, 2, 4, 8, 16, 32, 64, and so forth. As precise and unarguable as a multiplication table came the proof that in 200 years the ratio of population to food would be 256/9. In 300 years this would reach 4096/13. Obviously, the poor would be more severely affected by the rigors of competition than the rich. The misery of the poor classes, therefore, would relentlessly tend to increase. An unsettling paradox is embedded in Malthus's seemingly definitive exposition of the future: the ratios he calculated were those that would obtain *if population growth continued unchecked*. Yet Malthus also argued that misery and vice provided constant and unavoidable checks on population growth. In later editions of the *Essay*, Malthus compounded this inconsistency by conjuring up another check on population growth that he called *moral restraint*, a very fallible form of birth control.

The Malthusian vision of the unceasing competition for resources and the ensuing struggle for survival could be translated into the theory of evolution by natural selection. Since variation and fertility clearly exist in abundance, and different varieties have different degrees of fitness in the struggle for existence, it must follow that selection will operation on all varieties and allow only the most fit to survive. The varieties that survive will leave behind the most offspring; if their descendants inherit favorable variations, they and their descendants will in turn be favored in the battle for survival. Over a fairly short period of time, Darwin reasoned, human beings had domesticated plants and animals with useful traits by acting as agents of *artificial selection*. When selection was carried out by nature over unlimited periods of time, the effects must be superior to those produced by humans. *Natural selection* had two effects on populations: it led to divergence of characters, or speciation, and to the extinction of less fit varieties and species.

As early as 1844, Darwin was hinting in rather melodramatic tones to his friend Joseph Hooker that he was working on the species problem from a very unorthodox viewpoint. Comparing his heresy to the act of confessing to a murder, Darwin admitted to Hooker that he was almost convinced that species were not immutable. For Linnaeus and other taxonomists, defining the nature of *species* was central to biology; but once organic life was seen as the product of long eons of change and diversification, species became a rather arbitrary term used primarily for the sake of convenience. While it might be necessary for cataloging and descriptive purposes, from an evolutionary viewpoint the species was merely a group of organisms that resembled each other at a given point in time. *Varieties*, which Linnaeus had dismissed as fluctuations of no real importance, now took on an important role as incipient species. Viewing living entities in this dynamic fashion rather than as static creations suggested a truly natural system of classification based on descent from common ancestors. Nevertheless, the practical problems of taxonomy were not easily resolved, because in most cases all the information available to taxonomists was morphological. Descent, therefore, had to be traced by inference from morphology. Many enthusiastic evolutionists were, however, eager to turn morphological and embryological relationships into lines of descent. Before the *Origin*, biological classification systems were generally based on morphology, but Darwin's goal was to establish a classification system based on genealogy.

The "origin" in Darwin's *Origin of Species* clearly referred to the transmutation of one species into another and not to a process in which all living creatures were derived from some primordial fluid, fiber, or living molecule. Speculation about the ultimate origin of life was outside of the scientific problem that Darwin chose to explicate. The starting point for his argument was the great abundance of variation in nature, a finding that should be obvious, he wrote, to even the most casual observer. The existence of evolution as change was, Darwin suggested, essentially self-evident. The more difficult problem was to explain the mechanism of change. This was where Darwin believed he had made his greatest contribution: natural selection, he contended, was the engine of evolution.

Having convinced himself that modern species were the products of evolution, that is, descent with modification, Darwin wrote out two trial essays for his own use in 1842 and 1844. It was not, however, until about 1854 that he finally attempted to bring order to his great collection of evidence. His friend Lyell, although still skeptical about evolution, urged Darwin to publish his theory before someone else anticipated him. Slowly and methodically, Darwin began to write out his ideas in great detail. About halfway into this project, Darwin received a letter from Alfred Russel Wallace (1823–1913), a naturalist working in Malay, along with an essay entitled "On the Tendency of Varieties to Depart Indefinitely from the Original Type." Wallace wanted Darwin's opinion of his theory. Should it have any merit, Wallace requested that Darwin send the essay to Lyell for review and publication.

Reduced to a state of shock and profound disappointment, Darwin realized that Lyell's prediction had been confirmed. Not only had Wallace developed a theory of transmutation by natural selection, but his clear and concise essay could be seen as an excellent abstract of Darwin's 1842 manuscript on the theory of evolution. Left to himself in this crisis, Darwin might have withdrawn from the potential priority battle. Luckily for Darwin's reputation, his loyal friends Charles Lyell and Joseph Hooker took over as mediators and champions of fair play. A joint presentation of the papers written by Wallace and Darwin was quickly arranged, as was simultaneous publication in the August 1858 *Proceedings of the Linnaean Society*. With this unsettling experience to focus his attention, Darwin polished off the *Origin of Species* in little more than 13 months. In his autobiography, Darwin declared that he did not really care whether others attributed originality to him or to Wallace, but in his letters to Lyell and Hooker he admitted how hard it was to think of seeing 20 years of priority lost and his claims to originality smashed. Later, when critics attacked the *Origin*, Darwin complained that he had been rushed into premature publication.

With the help of his friends, Darwin was awarded the lion's share of the honors for revolutionizing ideas about evolutionary theory. Wallace came close to becoming the forgotten man, but his role in the discovery of natural selection was often used to belittle Darwin's achievements or to prove that the basic idea of survival of the fittest was already commonplace in the intellectual milieu shared by men like Darwin and Wallace. Whereas Thomas Henry Huxley has been called "Darwin's bulldog," Wallace has been called "Darwin's moon." Reacting with stoic grace, Wallace accepted his secondary role as just another disappointment in a life full of bad luck. Even though Wallace apparently believed that his essay alone would not have created the revolution associated with the *Origin*, others charged Darwin and his friends with the greatest sins that scientists can commit—the misappropriation of another scientist's ideas and a conspiracy to cover up the crime. Suggestive evidence of opportune gaps in the surviving collections of pertinent letters and manuscripts, as well as odd discrepancies in Darwin's records of important dates, need not indicate a full-

Alfred Russel Wallace

fledged conspiracy, but it does demonstrate the need for a fuller appreciation of Alfred Russel Wallace as a scientist and a highly original thinker in his own right.

One of the nineteenth century's most colorful and unlucky scientists, Wallace endured misadventures in the jungles of the Amazon and the Malay Archipelago that made the adventures of Charles Darwin on the *Beagle* look like a picnic in the park. Unlike Darwin, who came from a background of wealth, privilege, and high expectations, Wallace had a rather meager formal education, and he spent his entire adult life in pursuit of a variety of modest employments. By the time Darwin had written his preliminary essays on evolution, Wallace had become a teacher at the Collegiate School at Leicester. In 1848 Wallace was able to persuade the entomologist Henry Walter Bates (1825–1892), remembered for his discovery of mimicry in animal coloration, to accompany him on an expedition to the Amazon. The two planned to study natural history in the tropics and then pay for their trip by selling their collections. While planning this expedition, Wallace was already thinking about the question of the origin of species, having been inspired by the writings of Alexander von Humboldt (1769–1850), who has been called the second discoverer of the Americas, and those of Charles Darwin. Wallace was sure that he would find the answer to the "question of questions" in the lush and exotic world of the tropics. While the venture added immeasurably to his knowledge of natural history, his collections, obtained with such difficulty, were lost when his ship caught fire and sank in the middle of the Atlantic on the return voyage in 1852. After spending 10 days in a leaky lifeboat, Wallace and his shipmates were rescued by another decrepit and unseaworthy ship. Recovering from this disaster, Wallace began a new phase of his research in the Malay Archipelago (now Indonesia), observing and collecting. During an attack of a tropical fever, Wallace experienced a flash of insight in which he realized that natural selection could serve as the mechanism of evolution. Within a few days he had completed his essay on natural selection and sent it to Darwin.

Having ceded priority to Darwin, Wallace continued to publish works on natural history and travel, such as *The Malay Archipelago* (1869), *Contributions to the Theory of Natural Selection* (1870), *Geographical Distribution of Animals* (1876), and *Island Life* (1880), while spreading word of the new gospel of evolutionary theory as a popular scientific lecturer. Wallace bombarded the journals with articles about geology, geography, taxonomy, astronomy, anthropology, education, social and economic problems, land ownership, the occult, and the evils of vaccination. Despite major differences in their approach to scientific questions, Darwin and Wallace remained good friends and collaborators. It was Wallace who called evolution by means of natural selection "Darwinism" in order to distinguish this theory from its predecessors. With excessive modesty, Wallace said that his own discovery had a ratio to that of Darwin's work of "one week to 20 years." In many ways, Wallace remained more "Darwinian" than Darwin. Their differences in emphasis and interpretation involved questions related to altruism, or cooperative behavior, and sexual selection, as well as human evolution, which are still unresolved.

In his later years Wallace supported a variety of unorthodox and radical causes, such as women's suffrage and socialism. These activities, along with his studies of spiritualism, led critics to describe him as half genius and half crank. Gilbert Keith Chesterton (1774–1936), poet, polemicist, and journalist, called Wallace one of the two most important figures of the nineteenth century, having the unique distinction of being a leader of a major revolution in thought for his advocacy of materialistic evolutionary theory and its own counterrevolution as a champion of antimaterialistic social theories and psychic research. Invariably described as a modest, self-effacing man, the aging and financially distressed Wallace was enormously pleased when he was notified in 1881 that, thanks to the efforts of Charles Darwin and other eminent scientists, he would receive a Civil List Pension of £200 a year.

At one point an admirer of Herbert Spencer's (1820–1903) doctrine of individualism, Wallace eventually rejected Spencerian Social Darwinism. Wallace had been impressed by Spencer's assertion in *Social Statics* (1851) that improved adaptation of human beings to the social state would eventually produce a society in which no governmental restraints on individual liberties would be needed because of the innate moral sense of each citizen. The essence of progress, according to Spencer, was individuation with differentiation and mutual adaptation; the process was the same ultimately for cells in the animal body and human beings in society. But Wallace later became an enthusiastic follower of Robert Owen's (1771–1858) utopian philosophy of communitarian socialism and his cooperative commercial ventures. Natural selection, according to Wallace, could explain the existing diversity among human races. Understanding the operation of natural selection made it possible to predict that in the distant future all human beings would reach a new and exalted intellectual and moral level. In speaking to the Anthropological Society in 1864, Wallace merged Darwinian science and Spencerian philosophy and grandly predicted that natural selection would eventually eliminate all but the most "intellectual and moral" race; the remaining men (sic) would then be so well adapted to the social state that compulsory government would be unnecessary. Later, in reviewing Darwin's *Descent of Man*, Wallace came to the unhappy realization that there might not be an evolutionary guarantee of future moral and social fitness and happiness. To the contrary, Darwin's work suggested that selection might increase a tribe's competitive fitness and survival without increasing its moral sense.

As the Darwin/Wallace priority question indicates, the core of modern evolutionary theory appears to be simple enough to be contained within the pages of Wallace's 1858 essay. While the *Origin* is a challenging and complex book, Darwin always thought of it as merely an abstract of the comprehensive treatise he intended to write. It can also be seen as simply the first installment of the long argument with nature that consumed the rest of Darwin's life. After publishing the first edition of the *Origin*, Darwin returned to the book many times, struggling in each new edition to answer his critics. The sixth and last edition was published in 1872; Darwin had reached the point of exhaustion and could not face further revisions. His audience, however, continued to grow, and reprints of both the early and later editions, with introductions and commentaries by distinguished scientists and scholars, continue to appear.

Although Darwin's work was criticized by many scientists, philosophers, and theologians, perhaps the attacks that troubled him most were those of the eminent physicist William Thomson (1824–1907), better known as Lord Kelvin, and Fleeming Jenkin (1833–1885), professor of engineering at Glasgow University. Although largely forgotten today, Jenkin was memorialized by his friend Robert Louis Stevenson for his achievements as a scientist, engineer, economist, educator, and pioneer of electrical engineering. Lord Kelvin, one of the founders of thermodynamics, challenged Darwin's core assumption that evolution proceeded through continuous small steps by declaring that the earth was not old enough to allow for such gradual transmutations. Biology and geology, according to Lord Kelvin, must defer to the superior authority of physics. Jenkin used the prevailing theory of inheritance to challenge Darwin's assumption that individuals with favorable traits could pass them on to future generations. Given a "blending" theory of inheritance, it was reasonable to assume that a new trait would be swamped out of existence because of matings between rare mutants and normal individuals. If the laws of inheritance precluded the proliferation of favorable traits, natural selection could not act on novel traits. Although Darwin did not see blending inheritance as a fatal flaw, he was quite distressed by Fleeming Jenkin's review of the *Origin*. But as Wallace noted, variations were probably more common than Jenkin or even Darwin supposed. Eventually, Darwin acknowledged Jenkin's argument and confessed that he saw no simple way around it, except to argue that given enough time and slight variations, modifications would accumulate and new species would appear. In some instances, Darwin adopted a more Lamarckian position in which the effect of the environment, along with the effects of use and disuse, also contributed to evolutionary change. Still, Darwin, Lyell, and their supporters refused to succumb to "physics envy." Geologists and Darwinians believed that their vision of the great, virtually unlimited abyss of time was true and hoped that some way would be found to refute the apparently devastating logic of the physicists.

Lord Kelvin, a prodigy who had been appointed to the Chair of Natural Philosophy at the University of Glasgow at the age of 22, was regarded as the greatest Victorian physicist. Three years after the publication of *Origin of Species*, Kelvin wrote an article entitled "On the Age of the Sun's Heat" and published it in a popular magazine. Citing established physical laws and apparently impeccable calculations, Kelvin argued that the sun could not have illuminated the earth for more than 100–500 million years. Worse yet, solar light and heat would not illuminate and warm the earth for many more millions of years "unless sources now unknown to us are prepared in the great storehouse of creation." Convinced that physics was the superior science, Kelvin contended that given a choice between the estimates of time made by a physicist and those based on geology, any rational person must accept the work of the physicist. Rather contemptuously, Kelvin compared Lyell's steady-state model of the earth to a perpetual motion machine operating in violation of the laws of thermodynamics, especially Kelvin's own second law. In later work dealing with the relationship between the distribution of heat within the earth and the age of the earth, Kelvin challenged biologists and geologists to accommodate themselves to his best estimate of the age of the earth, i.e., about 100 million years. Of course Darwin never entered into a direct battle with Lord Kelvin, but Thomas Henry Huxley, who saw himself as the "gladiator-general" of Darwinism, insisted that physics need not be privileged over geology in estimating the age of the earth. Both biology and geology, Huxley argued, provided independent evidence of processes operating over immense time periods. Essentially what Huxley said in his elegant nineteenth-century prose was "garbage in, garbage out." Lord Kelvin was obviously missing crucial evidence. Ultimately, the discovery of radioactivity by Henri Becquerel (1852–1908) in 1896 established a powerful new rationale for rejecting Kelvin's pronouncements on the age of the earth.

Radioactive decay not only provided a source of huge quantities of heat energy unknown to nineteenth-century physicists, it also revealed the existence of radioactive "clocks" that have been extremely important to biologists, paleontologists, and archaeologists. By 1913, Arthur Holmes (1890–1965), a geologist who pioneered the use of radioactive decay methods for rock dating, estimated the age of the earth at 1.6 billion years. Radiometric dating based on estimates of the relative percentages of "parent" and "daughter" isotopes provided the first quantitative geological time scales. Early methods based on uranium and thorium were later supplemented by techniques that used potassium-argon, rubidium-strontium, and carbon-14. In 1959, after further refinement of radiometric dating techniques, Holmes suggested that more accurate methods suggested that the age of the earth was actually three times as much as previous estimates. Using the uranium-lead technique, Holmes was able to establish absolute ages for fossils whose relative (i.e., stratigraphical) ages had been previously established. Holmes also developed the concept of convection currents in the mantle of the earth caused by heat from radioactive decay. Through these studies, Holmes became one of the early supporters of the theory of continental drift. Shortly before his death, Holmes published a revised edition of his classic textbook, *Principles of Physical Geology*. The discovery of radiocarbon dating by

Willard Frank Libby (1905–1980), 1960 Nobel Laureate in Chemistry, created a revolution in archeology. Scientists continue to wrestle with uncertainties about the age of the earth, but there is general agreement that the earth was formed as a terrestrial planet more than 4 billion years ago by the aggregation of swirling space debris and gas. Oceans might have existed 200–300 million years after the earth formed, and life might have appeared 3.9 billion years ago.

If Darwin was reluctant to publish *Origin of Species* because of the attacks that were sure to follow, he must have dreaded the reaction to *Descent of Man and Selection in Relation to Sex* (1871). In *Origin* Darwin merely hinted that his ideas might throw some light on the origins of human beings. Probably he hoped that someone else would assume the burden of writing the book that would explore this theme. But he discovered that while other evolutionists were willing to write about man's place in nature and the survival of the fittest in human society, they were not willing to analyze human evolution as a purely biological process. Many supporters of evolutionary theory, Darwin realized, reserved some role for divine intervention where human intelligence and the soul were concerned. Allowing no such exceptions, Darwin's analysis of the evidence led to the conclusion that "man is descended from a hairy, tailed, quadruped, probably arboreal in its habits."

Arguments presented in the *Origin* that related to the common descent of species were extended in the *Descent of Man* to examine the question of whether human beings, like every other species, were the descendants of some preexisting form, as well as the manner of human development and the nature of the differences between the "so-called races of man." Through the analysis of behaviors shown by animals, such as curiosity, memory, imagination, reflection, loyalty, and the tendency to imitate, Darwin elucidated the precursors of human qualities. Because he was very cautious and vague about the time and place, as well as the racial identity of the first humans, it is difficult to be sure where Darwin stood on many issues. He suggested so many possibilities and ranged so widely that adherents of all kinds of ideologies were able to cite Darwin in support of their particular cause. As indicated by the complete title, a major portion of the book was devoted to the subject of sexual selection rather than human evolution.

The original theory of natural selection, Darwin reasoned, had failed to explain the evolution of secondary sexual characteristics because such variations were not directly involved in the struggle for existence. According to the concept of sexual selection, traits involved in the competition between males for attracting the females of their own species were not, strictly speaking, useful in the struggle for existence carried out between species. Some secondary sexual traits, such as the colorful plumage of male birds or the large decorative antlers of certain deer, might even be dangerous or cumbersome.

Like Aristotle, Darwin assumed that strength, bravery, and intelligence were "male traits." Females, in contrast, were described as passive, weak in body, and deficient in brains. Yet even here Darwin was sufficiently ambiguous to leave room for various interpretations. For example, Darwin noted that the male brain was larger, but he admitted that he did not know whether or not the alleged advantages of an absolute difference in brain size were lost when the difference in body size between males and females was taken into account. Since the theory of sexual selection depended on intrasexual competition for mates, Darwin overemphasized male traits and generally neglected the very demanding female functions involving the care, feeding, and training of the young. Needless to say, Darwin's assertions of scientific evidence for male superiority won him many supporters. Even the centennial tribute to Darwin entitled *Sexual Selection and the Descent of Man 1871–1971* presented essentially the same perspective and emphasis. Animals in which the average female is larger than the average male tend to be ignored, although that is the case in about 30 of the 122 families of living mammals, including some rabbits, hares, bats, whales, seals, antelopes, and the hairy-nosed wombat. No single theory has so far been able to account for sexual dimorphism with respect to size, the diversity of mating systems, and intrasexual competition. These questions are obviously quite complicated since related species may differ as to the relative size of the two sexes despite similarities in habitat, food preferences, and behavior.

Darwin considered the idea of sexual selection essential to his theory of evolution as a means of explaining elaborate and apparently nonadaptive sexual characteristics, but Wallace did not find the idea very persuasive. Indeed, the theory was largely ignored until a resurgence of interest occurred in the 1960s. Generally, scientists found the idea of "choice" by animals, especially female animals, incompatible with their view of behavior. However, sexual selection has been invoked as a means of providing Darwinian explanations for the evolution of traits, such as altruism, cooperation, and sexual ornamentation, that might be seen as counterproductive in the struggle for existence.

Given the fact that human paleontology and cultural anthropology were still in their infancy at the time Darwin was working on *Descent of Man*, his views of human evolution were remarkably prescient. From the fragmentary evidence of comparative anatomy and psychology, Darwin surmised that the ancestor of modern human beings was related to the gorilla and the chimpanzee, that the first humans originated in a hot climate, probably in Africa, and that the early human ancestor appeared between the Eocene and Miocene periods. Before the publication of *Origin*, naturalists were divided as to whether all human "races" had descended from the same ancestors or different races were the products of separate creations. In spite of these uncertainties, European scientists were sure that there was a hierarchy of race, with the Hottentot at the bottom and the white man at the top. After the triumph of Darwinian evolution, most scientists accepted a common origin for all human races but assumed that fixed and ineradicable differences had accumulated during evolution. Despite the lack of any significant scientific evidence of the existence of a biological hierarchy based on race, such specious assumptions were widely used to provide an allegedly natural basis for a hierarchy of power and privilege. Given the fragmentary nature of the fossil evidence of human origins, efforts to reconstruct the family tree of modern humans remain speculative and controversial. New biochemical techniques, in conjunction with more extensive fossil evidence, however, tend to support the "out of Africa" hypothesis and a rather recent origin for all human races.

Discouraged by the unceasing attacks on his work and reputation, Darwin began to doubt "whether humanity is a natural or innate quality." When scientists, philosophers, and clergymen asserted that human origins could never be discovered through science, Darwin admonished them to remember that such confidence was more often the product of ignorance than knowledge. "It is those who know little, and not those who know much," he wrote, "who so positively assert that this or that problem will never be solved by science." Always seeking new evidence and arguments to support the theory of evolution, Darwin was obsessed by his "One Great Subject," even though he also worked on more conventional topics before publishing *Origin*. For 8 long years he studied the taxonomy and physiology of living and fossil barnacles. Later, when he wondered whether this respectable but extremely tedious study had been worthwhile, his friend Huxley called it an excellent form of self-education and discipline. After the publication of *Origin*, Darwin produced a series of volumes on a wide variety of subjects sharing the evolutionary theme. The *Expression of the Emotions in Man and Animals* established the foundations of modern research in ethology (animal behavior) and ethnology (comparative anthropology): studies of infants, the insane, paintings and sculptures, photographs of expression evaluated by different judges, and comparative studies of expression among different peoples. Darwin argued that the evolution of behavior is like the evolution of organs. Thus, because behavioral patterns were subject to the laws of inheritance and selection, some behavior patterns might outlive their original function and become vestigial, just like certain physical characteristics inherited from distant ancestors.

Observations of animal behavior had long been part of the domain of natural history and amateur nature studies; Darwin's work made it a key aspect of evolutionary theory. That is, Darwinian evolution suggested that an animal's behavioral adaptation to its environment was as important as its anatomy and physiology. The establishment of ethology—the study of behavioral patterns—as a new science occurred in Germany, but it soon found affinities with British traditions of natural history. Konrad Lorenz (1903–1989), best known for his discovery of imprinting in Greylag geese, was one of the key figures in the development of ethology. Like many other species of birds, young Greylag geese do not recognize birds of their own kind (conspecifics) unless they are introduced to them at a specific point in their development. If such geese see human beings rather than geese when they first hatch, they regard the human beings as their parents. Unlike learned routines, which characteristically can be unlearned, imprinting becomes permanent. In 1973 the Nobel Prize in Physiology or Medicine was awarded to Lorenz, along with Nikolaas Tinbergen (1907–1988) and Karl von Frisch (1886–1982), "for their discoveries concerning organization and elicitation of individual and social behavior patterns." Unfortunately, the study of ethology has often generated facile comparisons between animal and human behaviors and questionable conclusions about human evolution and human nature. Ethology has, essentially from its inception as a scientific discipline, been burdened with dubious political and social assumptions about the nature of learning and inheritance. By the end of the twentieth century, some cognitive scientists were questioning earlier concepts of animal behaviors and intelligence. New research approaches in cognitive psychology, infant development, and neuroscience suggested that some animals demonstrate self-awareness and mastery of certain abstract concepts. Indeed, some animals can apparently master skills previously regarded as unique to humans, including understanding language, performing simple arithmetic tasks, forming mental maps of their environment, exchanging complex messages with other members of their species, creating tools and teaching others how to use them, and understanding complex social relationships. Positron emission tomography (PET) scans and magnetic resonance imaging (MRI) reveal similar patterns of electrical activity in human and animal brains, which should not be too surprising, considering that chimpanzees share 99% of their DNA with humans.

Perhaps because of the intensity with which Darwin approached complex, controversial scientific ideas, he has often been portrayed as an obsessive, lonely, and even neurotic man. But more detailed and careful explorations of the way in which Darwin wrestled with "the question of questions" suggest a strong, tenacious mind and a generous, warm personality. Focusing on evolutionary theory to the exclusion of almost everything else, Darwin complained to his friend Hooker that "it is an accursed evil to a man to become so absorbed in any subject as I am in mine." Ostensibly a modest, reticent, and private man, Darwin provided many surprisingly candid insights into his life and thought in his letters, journals, and autobiography. According to Thomas Henry Huxley, all autobiographies are essentially works of fiction, but despite some apparent discrepancies between his memories and the written records, Darwin's autobiography has been called one of the "most sincere" ever written. In 1876, at the age of 67, Darwin began to compose an autobiography for his children and their descendants. Salient passages on religion, as well as various overly candid references to friends and critics, were deleted when the *Autobiography* was published in 1887 as part of the *Life and Letters of Charles Darwin*. The normally close, warm, and loving Darwin clan was brought to the brink of litigation when

Francis Darwin, who had been entrusted with the manuscript of his father's autobiography, attempted to publish the unexpurgated text. Within the Darwin family, religion was strictly relegated to the woman's sphere. Thus, Francis was quite surprised by the violent objections his mother and sisters raised to the section headed "Religious Belief." A compromise was achieved by deleting this section and placing parts of it in a separate chapter of the *Life and Letters*. As part of the celebration of the centenary of the publication of the *Origin*, Darwin's granddaughter Nora Barlow published an unexpurgated edition of the *Autobiography*, in which some 6000 words were restored.

In a discussion of his religious beliefs, Darwin explained that he had lost his faith in Christianity, and, furthermore, he could not see how anyone would want to believe in a "damnable doctrine" that purported to consign eminently good, thoughtful, and moral people, such as his grandfather, father, and many of his best and brightest friends, to everlasting punishment. Perhaps part of Darwin's belief in the inferiority of the female mind was the result of the conviction, handed down from Erasmus and Robert Darwin, that the female part of the family maintained religious beliefs that the more enlightened males had discarded. For Emma, the central belief of Christianity was not eternal damnation but eternal salvation and the hope of being reunited with her loved ones beyond the grave. This conviction helped her deal with the death of her oldest daughter, 10-year-old Annie, in 1851, a loss that was totally devastating to her husband.

Given the profound psychological and intellectual difficulties involved in constructing a complex scientific theory, the scope and quantity of the work carried out by Charles Darwin, a gentle, modest, and chronically sick man, was truly extraordinary. He answered his own intriguing questions in a manner so direct and apparently simple that both his critics and allies tended to underestimate the quality of his work. For example, Darwin's studies of earthworms, including observations of a pot of worms kept on Emma's beloved piano, called attention to the great effects that resulted from "small causes often repeated." Indeed, when Thomas H. Huxley first read *Origin*, he expressed the reaction of many eminent naturalists: "How extremely stupid not to have thought of that oneself." Disciples and adversaries understood that Darwin had provoked a revolution in thought by challenging the long-cherished picture of nature as the result of God's design, intention, and direct intervention and the belief that earth was a place created especially for man. The *Origin* was cited by many who had never read the text, but an amorphous concept called Darwinism was invoked in support of many causes and quite contradictory ideas, from atheism to new forms of Christian faith, from socialism to fascism, from robber-baron capitalism to Marxist economic theory. Evolution by means of natural selection, the original insight at the heart of Darwin's work, was largely eclipsed by debates about the implications of Darwinian ideas for theology, philosophy, literature, anthropology, and so forth.

It has been said that so much has been written for and against Darwin's theory of evolution because it appears to be so simple that almost anyone can misunderstand it. Familiar arguments from natural theology were turned inside out to create a new worldview in which the fit between creatures and their environment was not proof of the wise design of the Creator, but the result of natural selection acting blindly and impersonally on accidental variations over the course of vast periods of time. Nevertheless, some theologians accepted the idea of evolution, or at least abandoned the idea that all plants and animals were the unchanged and unchangeable descendants of those species fabricated during the first 6 days of Creation. As Robert Chambers and others had suggested, it was not impossible to see evolution as consistent with a more glorious and efficient, if more distant and abstract, concept of God. Many others were, however, so deeply committed to arguments from design and the literal truth of the scriptures that they could not accept any compromise or alternative.

Because excitement, discord, or arguments exacerbated his chronic ill health, Darwin could not enter into direct confrontations with opponents of evolutionary science. Fortunately for Darwin, the battle was taken up by some extremely pugnacious, determined, and ingenious scientists. Chief among them were Thomas Henry Huxley and Ernst Haeckel. The amazing impact of Haeckel's writings on a young mind was well illustrated by geneticist Richard Goldschmidt's (1878–1958) reminiscences in *Portraits from Memory*. On reading Haeckel's history of creation as a young man, Goldschmidt immediately felt that all the problems of heaven and earth had been solved simply and convincingly. Every question that had troubled him had been answered. All previous beliefs and creeds could be discarded, thought Goldschmidt, because evolution was "the key to everything."

For his role as leader of the evolutionists' army and debating team, the irrepressible Thomas Henry Huxley (1825–1895) was awarded the title "Darwin's bulldog." Huxley was, however, an eminent scientist in his own right, who was definitely not a slavish follower of Darwinian hypotheses. Indeed, Darwin considered converting Huxley to evolutionary theory one of his greatest accomplishments. Long after Huxley accepted descent with modification, he continued to think of natural selection as a working hypothesis. Unlike Darwin, whose wealth gave him access to the best available education (even if he considered it dull and worthless), Huxley had only a few years of formal schooling. Huxley's father, an impoverished schoolmaster with eight children, could offer his son little more than lectures on the importance of education and hard work. When only 15 years of age, Huxley designed his own rigorous and highly successful program of self-instruction. Later, Huxley had to struggle to find a position with a salary sufficient to bring his fiancée Henrietta Anne Heathorn to England from her home in Australia. Physicians told Huxley that his frail young bride was unlikely to live another 6 months, but their estimates proved to be wrong by about 50 years. Even as Huxley anxiously awaited the birth of his

Thomas Henry Huxley

first child on the last night of 1856, he was outlining a plan of study to provide a "new and healthier direction to all Biological Science." Like Darwin, Huxley experienced a period of profound grief at the death of a much-loved child, which apparently intensified his skeptical approach to orthodox theology. Known for his love of controversy, Huxley dedicated himself to the campaign to "smite all humbugs," large or small.

Like Darwin and Wallace, Huxley developed his talents as a naturalist through a long voyage to exotic places. A prodigious worker, Huxley turned out over 150 research papers covering fields as diverse as zoology, paleontology, geology, anthropology, and botany, as well as numerous books and essays. While Darwin was wholly consumed by his work on evolutionary theory, Huxley became a key figure in debates about virtually every aspect of English civilization and culture, including education, metaphysics, politics, and theology. His research, writings, and position as president of the Royal Society made him one of the most powerful scientists in Britain. He was called the fiery prophet of Victorian science, as well as the gladiator-general with the razor-sharp claws and beak who was always ready to do battle with anyone who attempted to obstruct the sacred cause of science. A master stylist and one of the most influential men of his generation, Huxley coined words such as *biogenesis, agnostic,* and *bishopophagous* (to describe his ability to eat up conservative clergymen) when the English language failed to provide a term precise enough to suit his needs. Part of Huxley's success derived from his gift for language, spoken or written. The gift was apparently handed down through three generations of Huxleys, most notably in the case of his grandsons Aldous Huxley, Sir Julian Huxley, and Sir Andrew Fielding Huxley, who independently earned their places in what has been called the world's "reigning dynasty of the mind." By the time Huxley began what might be called his second career, that of cultural critic of English society, his professional reputation as anatomist, physiologist, and defender of science was already well established. Perhaps the greatest warrior in the battle to convert the world to evolutionary thought, "Darwin's bulldog" was also given the ironic title "Pope Huxley" in recognition of the influence he wielded as a scientific humanist, educator, and social critic.

One of Huxley's most famous debates involved Samuel Wilberforce (1805–1873), the Bishop of Oxford, a dignitary Lyell called that "Jesuit in disguise." Others preferred to use the nickname "Soapy Sam" in recognition of his gift for "overfacile eloquence." The occasion for the "Battle of Oxford" was the June 1860 meeting of the British Association. Huxley did not like the medieval atmosphere of Oxford, but Robert Chambers, the still anonymous author of the infamous *Vestiges,* talked him into attending the meeting. Darwin, of course, was conspicuously absent. After reexamining accounts of the verbal duel between Huxley and Wilberforce, some scholars have consigned the best parts of the story to the realm of myths and legends. There is, however, general agreement that hundreds of people packed into the lecture room to hear the Bishop denounce the dreadful "monkey theory" presumed to be at the heart of *Origin of Species.* Wilberforce, who attempted to appeal to scientific authority, had been coached by the distinguished anatomist Richard Owen (1804–1892). In conclusion, the Bishop turned to Huxley and politely asked whether he claimed his descent from a monkey through his paternal or maternal lineage. Relishing the chance to respond to this gratuitous insult, Huxley explained Darwin's ideas and, according to the venerable myths surrounding the debate, closed the discussion by saying that he would "rather have a miserable ape for a grandfather" than a man who used his great gifts and influence "to introduce ridicule into a grave scientific discussion."

In this episode, as well as others less distorted by the shadows and shards of myth and memory, Huxley gave an impressive performance in his role as the preeminent champion of the Darwinian worldview. Devoted to the task of establishing a new morality founded on natural knowledge, Huxley the agnostic preached the gospel of a moralizing naturalism. Instead of a pulpit, Huxley saw himself standing at the lectern of "Nature's university," the world institution in which all mankind was inescapably enrolled. Of course, Huxley and his contemporaries assumed that the curriculum and opportunities available to womankind were very different from those appropriate to mankind. Despite Huxley's reputation as a champion of enlightened thought and education for women, his actions in late nineteenth-century disputes concerning the professionalization of science, the control of Victorian anthropology, and the role of women reveal the depth of his very conventional belief in female inferiority.

While Huxley and his beloved wife Henrietta believed in educating their daughters in science so that they would be "fit companions of men," Huxley was sure that the vast majority of women would remain in the "doll stage" of evolution. Huxley and his contemporaries took it for granted that anthropology, the science of man, along with any discussion of sex and reproduction, anatomy and physiology, heterodox religious ideas, and agnosticism were not fit subjects for women. In 1871, the year in which Darwin published the *Descent of Man*, the Ethnological and Anthropological Societies merged, thanks to Huxley's efforts, and the Darwinians, safe from the intrusions of frivolous females, gained control of the "science of man."

Like the physical anthropologist Carl Vogt (1817–1895), Huxley was convinced that scientific studies of the structure of the brain had provided proof of the inferiority of women and the "lower races." Vogt claimed that the crania of men and women were so different that it was scientifically appropriate to classify them as different species. According to Vogt, the crania of adult women were invariably more childlike than those of men of the same race; this difference increased with racial evolution and the progress of civilization. Thus, the difference was most marked in the advanced European white race and less well established in the black race. The arguments made by Darwin in *Descent of Man* and Huxley in "Emancipation—Black and White" (1865) concerning female inferiority may have been couched in more chivalrous language than that of the openly racist Vogt, but they were not fundamentally different.

In general, Huxley asserted, the cerebral convolutions of women and those of males and females of the "lower races" were less complex than those of white European males. Given the central role of the brain in all of that was truly human, the average woman was, therefore, invariably inferior to the average man in every important mental and physical characteristic. Nature's irrevocable laws had, Huxley assured his contemporaries, given the human male all of the advantages of physical size and strength, as well as a more massive brain. With such a guarantee of their inevitable superiority, men should not deliberately add to the biological burden already placed on women. After all, Huxley insisted, no amount of education or opportunity would allow women to compete successfully with men.

THE IMPACT OF DARWINISM

Although it could be argued that other branches of nineteenth-century biology, such as bacteriology and physiology, have had a greater effect on life and health than evolution, no other science has had the same impact on human thought. Evolutionary theory was extended into many fields far removed from the biological limits to which Darwin generally confined himself. Darwin's writings were cited as evidence for movements and philosophies as dissimilar as socialism, fascism, communism, capitalism, atheism, and evolutionary theology. Scholarship in linguistics, biblical studies, comparative mythology, jurisprudence, and other humanistic disciplines also reacted to evolutionary theory. Intrigued by the implications of Darwin's picture of the branching evolutionary tree, linguists built genealogies of languages that featured branching descent from a common ancestor. British empirical philosophy and associationist psychology, which argued that ideas arise through the interaction of the senses and the environment, seemed to resonate with Darwin's discussion of mental evolution. Darwinian myths and metaphors apparently insinuated themselves into the novels of George Eliot (born Marian Evans, 1819–1880), especially *Middlemarch* (1872) and *Daniel Deronda* (1876), and Charles Kingsley's (1819–1875) classic *The Water-Babies* (1863), among others.

In a very general way, the assumption that Darwinian thought could be extended to human history and the evolution of culture and civilization became known as Social Darwinism, a term that apparently originated in France about 1880 and was subsequently adopted in Italy, the United States, and Germany. Many different forms of Social Darwinism coexisted in nineteenth-century thought. Admirers of Darwin saw a political and economic message in evolutionary science: if the status quo is the result of eons of natural evolution, the existing state of society—with its hierarchical arrangement of race, class, and gender—must be inevitable. Whatever hardships might exist among the "lower orders" of the human race must, therefore, be accepted as the result of the impersonal workings of inexorable natural laws. Supporters of the status quo warned that any attempts to improve the conditions of the poor or mitigate harsh aspects of modern industrial society should be regarded as futile, misguided, and dangerous interference with natural processes and laws. On the other hand, Karl Marx (1818–1883) and Friedrich Engels (1820–1895) tried to use the very same evolutionary laws to justify the inevitability of communism. Ambiguities, contradictions, and the revised editions of Darwin's texts could be used to justify a variety of positions, but it was obvious that many who claimed to have an infallible interpretation of Darwinism had never directly confronted the pages of *Origin of Species* or *Descent of Man*.

The alleged relationship between the "Law of Evolution" and human society was explored in great detail by Herbert Spencer (1820–1903), engineer, inventor, writer, and social philosopher. Among Spencer's many books on philosophy, sociology, education, and science, *Principles of Biology* was particularly important as a link between the general reader and the scientist. Almost forgotten today, Spencer was one of the most widely read and influential philosophers of his era. Enthusiastic admirers of Spencer considered him a second Aristotle and an intellectual giant besides whom Darwin was a dwarf. The Spencerian Law of Evolution, which asserts that the entire universe is characterized by the inexorable development of complexity, richness, and excellence, is not found in *Origin of Species*. Indeed, Spencer's concept of evolution, especially social evolution, was Lamarckian rather than Darwinian. According to Spencer's law of universal history, gradual but inexorable progressive development encompassed all cosmic, evolutionary, biological, social, economic, and cultural processes. Progress was neither an accident nor something within human control; it was instead a "beneficent necessity," deeply embedded in the universe itself. Within Spencerian philosophy, society was analogous to an organism. Therefore, changes in the social organism must be measured on the scale of evolutionary progress. Followers of Spencer considered the great natural law of evolution to be as fundamental to the workings of human society as Newton's law of gravity was to the orderly movements of the heavenly bodies. The survival and advancement of the fittest was, therefore, an axiomatic corollary. Inevitably, without human interference, in the *very remote future* society would evolve into a better state. At that point the moral sense of humanity would be so highly developed that coercive government would no longer be needed.

While many Social Darwinists adopted a pessimistic view of social evolution as a Malthusian struggle, others were quite optimistic about the inevitability of progress. Despite the difficulties inherent in defining and delimiting the many doctrines that had come to be known as Social Darwinism, virtually all of those who applied evolutionary theory to human society agreed on one thing: nature selected the best and the fittest, and to them she gave her rewards. The best were those who were most fit, as demonstrated by their riches and their position in society. Spencer's followers identified the unfit as those who were weak, foolish, lazy, inefficient, reckless, and irresponsible. The workings of evolutionary law, they asserted, were immutable; the fit would be rewarded and the unfit would be destroyed. Such arguments led many skeptics to remark that the whole supposedly scientific package was nothing but biological Calvinism. The fit were urged to reproduce their own kind and do all that was possible to prevent the multiplication of the unfit. It is in this concern with the differential fertility of rich and poor that the analogy between biological and Social Darwinism most obviously breaks down. In biology, fitness and successful competition are measured in terms of survival and reproductive success, that is, leaving a larger number of offspring. In society, the measure of success was (and is) the accumulation of wealth, not the number of offspring. Indeed, the well-to-do feared that advanced nations were committing "race suicide" because the fit, so worn out in the struggle for wealth, were not doing their share in the struggle to produce more offspring. The unfit, in contrast, were allegedly doing nothing but reproducing their own unfit kind.

Opponents of pessimistic and Spencerian versions of Social Darwinism proposed alternative visions of human nature and the operation of evolutionary law. Still committed to a belief in Darwinism, thinkers such as John Dewey, Lester Frank Ward, Charlotte Perkins Gilman, and others challenged the view of society founded on struggle, competition, and the inevitability of human misery. The idea that cooperation and altruism were more powerful factors in evolutionary change than competition and the struggle for existence was advanced by Peter Kropotkin (1842–1921), author of *Mutual Aid* (1902). Kropotkin, a geographer and naturalist, was a Russian prince who became one of Europe's most prominent anarchist-nihilist revolutionaries. Many of Kropotkin's biological ideas were stimulated by his surveys of the flora and fauna of Manchuria and Siberia. In areas thought to be the very antithesis of the Garden of Eden, it seemed logical to expect the struggle for existence to be especially severe. Surprisingly, Kropotkin found much competition between species, but little evidence of intraspecific struggle. Thus, he proposed that in addition to the law of struggle, a primary factor in the evolution of many species must be a law of mutual aid and support. He argued that this law could be found, at least implicitly, in Darwin's *Descent of Man*. Consciousness of species solidarity produced an instinct that drove the social insects and led wolves to hunt in packs. Nature was, Kropotkin asserted, neither all pitiless struggle nor all peace and harmony. In Russia, biologists were generally evolutionists by the time the *Origin* appeared, but their concepts were based on French Lamarckian biology and Russian materialist philosophies. Darwinism was received warmly and led to discussions of how natural selection could complement Lamarckian mechanisms. Eventually, these disparate elements were drawn together in the form of Lysenkoism, a doctrine that had particularly destructive effects on Soviet biology, genetics, and agriculture.

Long after Darwin called attention to the different reception his work had received in France and Germany, scholars began to analyze the reception of Darwinism in great detail. While many interesting international differences have been found, no simple correlation seems to exist between the reception of Darwin's theory around the world and the special characteristics of specific countries. For example, in 1982, the centennial of Darwin's death, Italy held the largest number of commemorative symposia and produced the most centennial publications. Many Italian naturalists had been among Darwin's network of correspondents and advisors, and his work was still of general interest at the end of the twentieth century. Perhaps the major general conclusion of the scholarships on the international and comparative reception of Darwinism has been that in many areas where the biological sciences had not yet been firmly established, the discussion of Darwinism was essentially confined to arguments concerning social, political, and philosophical questions. It is ironic that, in virtually every part of the world, Darwin, who saw a world without goals, purpose, or progress, was generally misread and misinterpreted as having proved the inevitability of progress. In terms of non-Western philosophical systems, China and the Islamic world provide the most striking examples of reinterpretations of Darwinism. In the Islamic world, interest in Darwin was primarily social, political, and philosophical, rather than scientific. By the 1880s, Arab secularists were using Darwinism as a symbol of modernization.

A version of Darwinism more closely allied to the ideas of Herbert Spencer than those of Charles Darwin reached China by the end of the nineteenth century. By the 1880s, Christian missionaries were providing translations of the works of Charles Lyell, Charles Darwin, and other scientists, but so-called Darwinian concepts were more forcefully introduced to China by British diplomats and businessmen. As champions of colonialism, such individuals generally assumed that the British had discovered and demonstrated the truth of the natural and moral laws that governed individuals, nations, and races and invariably led to the triumph of the strong over the weak. After 1895, the year of China's defeat in the Sino-Japanese War, Spencer's slogan "the survival of the fittest" entered Chinese and Japanese writings as "the superior win, the inferior lose." Concerned with evolutionary theory in terms of the survival of China, rather than the origin of species, Chinese intellectuals saw the issue as a complex problem involving the evolution of institutions, ideas, and attitudes. Indeed, they concluded that the secret source of Western power and the rise of Japan was their mutual belief in modern science and the theory of evolutionary progress. Many adaptations of Darwinism evolved in China, including varieties that might be called Taoist Darwinism, Confucian Darwinism, Legalist Darwinism, and Buddhist Darwinism. Eventually, China absorbed, transformed, and was transformed by the intermixture of ideas, including those of Charles Darwin, Thomas Henry Huxley, Karl Marx, and Mao Zedong.

According to Japanese scholars, traditional Japanese culture was not congenial to Western science because the Japanese view of the relationship between the human world and the divine world was totally different from that of Western philosophers. Japanese philosophers envisioned a harmonious relationship between heaven and earth, rather than conflict. Traditionally, nature was something to be seen through the eyes of a poet, rather than as the passive object of scientific investigations. Thus, for cultural and philosophical reasons, many aspects of the transfer and adaptation of Western science to Japan were problematic. Nevertheless, Darwinian ideas were introduced to Japan by American zoologist Edward S. Morse (1838–1925), who arrived in Japan in 1877. Although

Morse was primarily interested in furthering his research on brachiopods while in Japan, he also presented lectures on evolutionary theory. Japanese translations of the *Origin* and the *Descent of Man* appeared in the 1880s and 1890s. The traditional Japanese vision of harmony in nature might have been uncongenial to a theory based on natural selection, but Darwinism was eagerly adopted by Japanese thinkers, who saw it as a scientific rationalization for Japan's intense efforts to become a modernized military and industrial power. Whereas European and American scientists and theologians became embroiled in disputes about the evolutionary relationship between humans and other animals, Japanese debates about the meaning of Darwinism primarily dealt with the national and international implications of natural selection and the struggle for survival. Late nineteenth-century Japanese commentators were likely to refer to Darwinism as an "eternal and unchangeable natural law" that justified militaristic nationalism directed by supposedly superior elites.

In German scientific circles Darwinian theory was most enthusiastically welcomed and publicized by Ernst Haeckel, who made Darwinism essentially synonymous with progress. Moreover, Haeckel felt that he was responsible for boldly applying Darwinian theory to aspects of biology and philosophy that Darwin himself had avoided. Generally, political conservatives assumed that Darwin's primary message was competition and the struggle for survival, while radicals and reformers found a message conducive to social evolution. When twentieth-century geneticists rejected the Lamarckian version of heredity and environmentalism, moderate versions of Social Darwinism lost their base of scientific support and strict new hereditarian versions were promoted by right wing radicals. Arguments about the nature of "superior" germplasm were, as might be expected, based on political factors rather than biological evidence. However, the relationship between Nazi ideology and Darwinism is not a simple one; an evolutionary framework might even be seen as contrary to Nazi racist assumptions about the "immutable superiority" of the German people.

In America, Darwinism found a sympathetic reception among scientists, intellectuals, businessmen, and politicians promoting the goals of competition, capitalism, and manifest destiny, even if evolutionary theory was not welcome in high school biology courses. Asa Gray (1810–1888), an eminent American botanist, was the first and foremost of Darwin's supporters in America. His approach to establishing a reconciliation between natural selection and natural theology has been called theistic Darwinism. During the 1860s and 1870s, Gray published numerous essays on natural selection and theology in leading scientific and literary journals; a collection of his essays was published in 1876 under the title *Darwiniana*.

As Gray's work indicates, despite the myth of perpetual warfare between science and religion, the way in which Darwin was interpreted by religious thinkers was not a simple matter. Darwin's theory resonated with many aspects of natural theology as well as economic science, which is not surprising given the role that Malthus allegedly played in the development of the idea of natural selection. Of course Darwin was well acquainted with the literature in natural theology through his earnest, if somewhat lackadaisical attempts to prepare himself for a career as a clergyman. It could be said that in Darwin's work nature serves as a substitute for God as a means of explaining design or the adaptation of organisms to their place in nature. By reasserting God's role as First Cause, orthodox Christian evolutionists, especially those with a Calvinist worldview, were able to accept a Darwinian view of the struggle of one against all in nature more readily than the so-called modernizing liberals who preferred a more hopeful Lamarckian approach to progressive, directed evolution. Agile theologians of various denominations could reclaim the Darwinian domain of a "theology without religion" by reversing the process and substituting God for nature. The tensions between science and religion were of particular interest to John William Draper (1811–1882), author of the classic *History of the Conflict Between Religion and Science* (1874), and Andrew Dickson White (1832–1918), the first president of Cornell University, who wrote the two-volume *History of the Warfare of Science with Theology in Christendom* (1896). This rather extreme view, which elevated controversy to warfare, was accepted by many other secular scholars and intellectuals in the early twentieth century.

During the 1870s and 1880s, many aspects of Darwinism were subject to challenges from scientists, as well as theologians. Physicists argued that the solar system was not old enough to provide the time frame implicit in Darwinian evolution. Moreover, attempts to understand the generation and inheritance of variation did not seem to support Darwinism. Scientists and theologians thought that it might be necessary to invoke the actions of God to create and direct variations. Advocates of theistic evolution believed that God formulated variations into a pattern of progressive development. Some of Darwin's critics condemned the "insufficiency" of natural selection and assumed that Lamarckian mechanisms accelerated the appearance of useful variations and evolutionary progress. Such alternatives to Darwin provided the comfort of purposeful variation and a reconciliation between evolutionary science and religion. Julian Huxley later referred to this period as the "eclipse of Darwinism." During the first two decades of the twentieth century, even Protestant fundamentalists writing for the series of booklets called *The Fundamentals* seemed to be leaning towards some accommodation with the "softer" versions of theistic evolution, as long as no doubts were raised about the "divine procedure in creating man." Textbooks written in the late nineteenth century generally included some discussion of evolution, but the approach was generally more Lamarckian, progressive, and theistic than Darwinian. By the 1920s, however, as biologists began the process of integrating evolutionary theory and Mendelian genetics, textbook writers began to present biology in terms of a modern evolutionary framework. The new focus on evolution, random variations, and natural selection revived the conflict between fundamentalists and evolutionists.

EVOLUTION AND SCIENCE EDUCATION

One of the enduring legacies of the first two decades of the twentieth century, a period often referred to as the progressive era in America, was an increased interest in education, literacy, science, and social reform. When examining the role of evolution in the high school biology curriculum, however, it is important to remember that before the 1920s very few students could aspire to more than a basic primary school education, especially in the rural South. In 1890, when the federal government began collecting census data on education, about 200,000 students were enrolled in high schools. By 1910 that number was closer to 2 million. Progressive era compulsory school laws encouraged, or forced, significant increases in the number of public high schools. For example, fewer than 10,000 students in Tennessee attended high schools in 1910. Fifteen years later more than 50,000 students in that state were enrolled in high schools.

By the 1920s, most scientists were sure that it was impossible to teach biology without references to evolution, but Christian fundamentalists saw evolution as a threat to religious belief. Fundamentalist texts, with provocative titles like *Hell and the High Schools: Christ or Evolution—Which?* and *God or Gorilla*, were widely disseminated, and William Jennings Bryan (1860–1925) was energetically campaigning against the teaching of the anti-Bible heresy known as evolution in state universities and public high schools. Although Bryan had suffered many defeats, including three unsuccessful campaigns for the presidency, he had established a solid reputation as a great orator and crusader for various causes, including women's suffrage, free silver, independence for the Philippines, Prohibition, and the income tax.

Even though many Americans endorsed Darwinian themes, such as the struggle for existence and manifest destiny, they were more than likely to condemn the idea that human beings and monkeys shared a common ancestry. Evidence of early hominids, or "man-like apes," discovered during the first quarter of the twentieth century further energized the fundamentalist crusade against teaching evolution in public schools. Whatever scientists might think about the nature of Darwin's work, a large majority of Americans knew that they did not want evolutionary theory discussed in their schools. The most famous symbol of the debate about the place of evolution in the biology curriculum was the infamous Scopes trial in Dayton, Tennessee. In 1925, John Thomas Scopes (1901–1970) was tried and convicted of the crime of teaching "the theory of the simian descent of man" in violation of state law.

Because of the opposition of clergymen, journalists, and educators, a bill prohibiting the teaching of "Darwinism, Atheism, Agnosticism, or the theory of

A portrait of the elderly Charles Darwin

Evolution as it pertains to man" was narrowly defeated in the Kentucky legislature in 1922, but there was little opposition to the fundamentalist campaign in Tennessee. Thus, in 1925 an antievolution bill written by State Representative John Washington Butler, farmer, part-time schoolteacher, and clerk of the Round Lick Association of Primitive Baptists, passed by a wide margin. The Butler Act made it a misdemeanor for a teacher in any school in the state that was supported by public funds "to teach any theory that denies the story of the Divine Creation of man as taught in the Bible, and to teach instead that man has descended from a lower order of animals." When Governor Austin Peay signed the bill, he said that in all probability the law would never be actively enforced. Opponents of the Butler Act warned that this attack on science would embarrass the state, but Bryan rejoiced that Tennessee had become the first state in the nation to adopt an antievolution law.

In response to the Butler Act, the American Civil Liberties Union (ACLU) placed an advertisement in the *Chattanooga News* offering to pay the expenses of anyone willing to test the constitutionality of the law. Some of the leading citizens of Dayton, apparently believing the old adage that any publicity is good publicity, decided that a trial would enhance the reputation of their little city. When the regular biology teacher refused to be the subject of a test case, Scopes was asked to volunteer. Although Scopes did not teach biology, he had conducted a review session while substituting for the biology teacher. Having read Darwin's *Origin of Species*, *Descent of Man*, and *Voyage of the Beagle*, Scopes considered evolution an essential part of the biology curriculum. Moreover, he knew that the textbook used in Dayton, George William Hunter's *Civic Biology*, contained a section called "Charles Darwin and Natural Selection." Hunter, who was chairman of the biology section at DeWitt Clinton High School in New York, believed that students needed to know about evolution, natural selection, random variations, and eugenics.

Journalists, like H. L. Mencken of the *Baltimore Sun*, were soon calling the Scopes Trial the "monkey trial," as well as "the trial of the century." Usually seen as a battle between evolutionists and creationists, the trial also reflected larger issues of constitutional law related to the conflict between majoritarian democracy and individual liberty. Dayton's "boosters" were thrilled when Bryan agreed to assist the prosecution on behalf of the World's Christian Fundamentalist Association. Zealous in his fight for the right of local majorities to pass laws that supported his crusade against teaching evolution, Bryan had little concern for minority rights and individual liberties. Insisting that the trial was a battle for popular control over public education, Bryan argued that the schools should not teach things that were offensive to the majority of taxpayers.

The defense team was joined by Clarence Darrow (1857–1938), America's most famous trial lawyer, who offered to assist without fees or expenses. The Scopes trial was conducted by Judge John T. Raulston in a special term of the Eighteenth Judicial Circuit. Experts on science, biology, evolution, and religion came to Dayton to testify against the Butler Act, but the prosecution successfully argued that expert testimony was irrelevant and inadmissible. The merits of evolution were not relevant to the trial, because the legislature and governor had already decided that issue. Denied the testimony of their expert witnesses, the defense lawyers moved to call Bryan to the stand as an expert witness on the Bible, thus creating the famous confrontation between Bryan and Darrow. Bryan agreed to debate Darrow in order to protect the word of God against the man he considered the foremost atheist in the United States. Darrow's interrogation of Bryan, which took place on the courthouse lawn, has been compared to the Punch and Judy puppet shows so often performed at outdoor festivals. Responding to Darrow's questions, Bryan disingenuously refused to admit that biblical passages might need to be interpreted. He insisted that his writings and speeches simply commented on lessons from the Bible, which he accepted literally, even if the sun did not revolve around the earth. Bryan not only rejected the theory that man evolved from lower animals, he objected to diagrams that portrayed man as a mammal. Darrow bombarded Bryan with questions about the age of the earth, Noah and the Flood, ancient civilizations, and comparative religion. When the prosecution asked Darrow to explain the purpose of his interrogation, Darrow bluntly replied that his goal was to prevent "bigots and ignoramuses from controlling the education of the United States." Given the widespread sympathy for Bryan and all that he stood for, Darrow's approach probably explains why he lost both the battle in Dayton and the war for the minds and hearts of a large part of the American public. Shortly after the trial, Bryan died of a stroke; his followers portrayed him as a martyr who had been killed by the stress of Darrow's hostile treatment. A Bryan Memorial University Association was founded in Dayton to raise money for William Jennings Bryan University. Classes began in September 1930 in the old Rhea County High School building where Scopes had taught.

The ACLU hoped to appeal the Scopes decision to the U.S. Supreme Court, but the Tennessee Supreme Court ruled that the Butler law only applied to public employees acting in their official capacity. Thus the law did not infringe on Scopes's individual liberty. As an employee he was obligated to perform in accordance with state statutes. The court also argued that the law did not require the teaching of a particular religious view; it prohibited the teaching of an evolutionary theory of human origins. Scopes's conviction was, however, overturned because Judge Raulston had determined the fine instead of the jury. The defense, therefore, had no conviction to appeal and the Butler Act remained intact and in effect was a constant warning and potential threat to teachers who might consider evolution an integral part of modern biology. When Mississippi passed an antievolution law in 1926, the ACLU could not find any teacher or taxpayer willing to participate in a challenge. Antievolution laws were quickly adopted in Arkansas, Louisiana, and Texas, and local school boards all over the South imposed similar restrictions.

The Scopes trial became the model for a play by Jerome Lawrence and Robert E. Lee. A movie version of *Inherit the Wind* premiered in Dayton on the thirty-fifth anniversary of the trial. Still seeking publicity, Dayton proclaimed a new Scopes trial day. Although Scopes was awarded the key to the city, he was convinced that a new trial would have had the same outcome. Indeed, teachers in Tennessee were still required to sign a pledge that they would not teach evolution. A plaque erected by the Tennessee Historical Society on the courthouse grounds in Dayton tells the whole story with admirable brevity: "Here, from July 10 to 21, John Thomas Scopes, a County High School teacher, was tried for expounding the theory of the simian descent of man, in violation of a lately passed state law. William Jennings Bryan assisted the prosecution: Clarence Darrow, Arthur Garfield Hayes [sic] and Dudley Field Malone the defense. Scopes was convicted."

While the infamous "Monkey Trial" created a national sensation, the outcome for science education was far from clear. Evolutionists viewed the publicity generated by the case as a triumph for modern science, but the teaching of evolution actually declined after the Scopes trial. The trial had a profound impact on textbook authors and publishers. When Tennessee dropped Hunter's *Civic Biology* from its list of approved texts, Hunter removed the section devoted to Darwinian evolution. Vague references to "change" replaced the discussion of evolution, and man was described as "the only creature that has moral and religious instincts." Most high school biology textbooks written after 1925 did not include the word evolution in the index or glossary. One exception was a book written by Alfred C. Kinsey (1894–1956), professor of zoology at Indiana University and founder of the Institute for Sex Research. His *Introduction to Biology* (1926) deliberately presented biology in an evolutionary framework. But for the most part, professional zoologists had little interest in high school biology texts, and publishers willingly censored potentially offensive materials.

Profound anxiety about the state of American scientific literacy was triggered by the launch of the Soviet satellite *Sputnik* in 1957. Driven by the need to compete with Russia, the U.S. federal government provided support through the National Science Foundation for the development of improved curriculum materials. The resulting Biological Sciences Curriculum Study (BSCS) series set a new standard for high school biology texts. Americans may prefer to remember the 1960s and 1970s as the glorious Space Age, but these decades also saw the resurgence of religious fundamentalism and the appearance of this movement's most zealous televangelists and textbook watchers. Attempts to improve science education galvanized the old antievolutionary forces and led fundamentalists to develop new tactics. The success of the antievolution movement was obvious in 1964 when Texas rejected the BSCS biology books and nervous

American textbook publishers rushed to camouflage or eliminate references to evolution.

When the U.S. Supreme Court invalidated all antievolution education laws in 1968, in the case *Epperson v. Arkansas*, on the grounds that such laws were an unconstitutional restriction of an educator's right to teach evolution and an unconstitutional establishment of religion, the antievolution movement responded by developing new tactics. The most successful strategy, based on the so-called Fairness Doctrine of the Federal Communications Commission, is known as the equal time or balanced treatment approach. Rather than call for a total ban on teaching evolution, fundamentalists demanded that if evolution was part of the biology curriculum, then "special creationism" must be taught as an alternative model.

The Creation Research Society (CRS), one of the organizations leading the campaign for balanced treatment, was founded in 1963 in order to "publish research evidence supporting the thesis that the material universe, including plants, animals, and man are the result of direct creative acts by a personal God." According to literature distributed by CRS, membership in the organization was granted only to Christian men with a postgraduate degree in science who openly professed a belief in the literal truth of the Bible. Various disputes led the organization to split into several factions, but the basic belief system of these groups remained essentially the same. One of the most active offshoots of the Creation Research Society was the Creation Science Research Center, formed in 1970 in San Diego, California, in order to campaign for the introduction of "scientific biblical creationism" in the public schools. In 1969 the Board of Education of California succumbed to fundamentalist demands that evolution and creationism be included in the state's public school biology curriculum as alternative theories.

Creationists convinced many legislators, politicians, parents, and students that their equal time doctrine was justified by American democratic principles of fairness and justice. Most scientists, however, find the idea that scientific theories should be evaluated by public debate and popular preferences contrary to the dictates of reason and the most fundamental principles of science. Scientists argue that to teach creationism rather than evolution is like teaching Aristotelian physics and Ptolemaic astronomy as if the Scientific Revolution and the Enlightenment had never happened. Although many scientists find it impossible to believe that the theory of evolution needs defending in the twenty-first century, public opinion polls generally reveal that the majority of Americans do not accept evolutionary theory, especially with respect to human origins. Creationists also demand that students be told that evolution is "only a theory." Scientists argue that evolution is a fact; the statement that natural selection is the mechanism of evolution is a theory. Moreover, as used by scientists, the term *theory* is not something to be denigrated or trivialized. Theory, as in atomic theory or evolutionary theory, refers to a set of explanation and not a mere hypothesis or guess.

Evolutionary theory, especially as it pertains to human origins and development, depends on evidence from geology, paleontology, physical anthropology, the human fossil record, archaeology, and so forth. Such records are admittedly imperfect and fragmentary and, when cautiously interpreted, give rise to useful, but always tentative theories. Creationism is based on a belief in the inerrancy of the Bible and takes its explanation of human origins from the Book of Genesis. Creationists believe creation occurred just once and that God created all possible species, which creationists call "kinds," for the duration of the world. Based on an essentially literal interpretation of biblical chronologies, creationists assume that the earth is only a few thousand years old and that human beings and all other life forms were directly created in the not too distant past, essentially in their present form. Fossils of extinct forms are explained as the remains of creatures that were destroyed by the Great Flood in the time of Noah.

Rejecting the methods of conventional, or "establishment" science, creationists and cult archaeologists have developed their own methods of ascribing specific dates to various events, artifacts, and sites. Creationists claim not only that their beliefs are justified by the word of God, as preserved in the King James English Bible, but that they can provide scientific evidence for this biblical version of creation. Convinced that there are no discrepancies between their interpretation of geology, paleontology, and biology and the Bible, leaders of the creation movement attempt to redefine science as "the search for truth." Given this archaic, essentially medieval definition of science, with its rejection of what creationists call naturalism, creationists then argue that the facts of science cannot contradict the Bible because biblical revelation is absolutely true and authoritative. Creationists have adopted other tactics, such as substituting terms like "intelligent design" for Creator and then arguing that because the term God is not used, the creation model is science rather than religion.

Modern scientists, when exhibiting appropriate humility, acknowledge that they do not traffic in ultimate truth, but in attempts to test hypotheses and construct theories. Leaders of the creationist movement, in contrast, see themselves as committed evangelists who must go out into the world and preach their version of truth. Unlike conventional modern scientists, creationists are not trying to test a hypothesis, but are involved in a grim battle between the forces of good and the forces of evil. Seeing the world in terms of absolutes, creationists argue that there can only be two possible ultimate worldviews: evolutionism or creationism. Even after the beginning of the new millennium, creationists still contend that the true author of the concept of evolution is Satan himself.

Creationists argue that scientists use evolution to promote a secular or atheistic view and that evolution should actually be called "evolutionary naturalism" and recognized as a form of religion.

During the 1970s and 1980s, several states, including Tennessee, Arkansas, and Louisiana, adopted "equal time" or "balanced treatment" laws that called for teaching alternative theories of origins if evolution was included in the biology curriculum. At a trial popularly labeled Scopes II, but officially known as *McLean v. Arkansas Board of Education*, Judge William R. Overton in the U.S. District Court of the Eastern District of Arkansas, Western Division, ruled on the "Balanced Treatment for Creation-Science and Evolution-Science Act" that had been signed into law on March 19, 1981, by Frank White, then governor of Arkansas. The purpose of the law was to enforce balanced treatment of "creation-science" and "evolution-science" in the public schools of Arkansas. According to the creation model selected by the Arkansas Board of Education, the universe and everything in it had been created suddenly, out of nothing, only a few thousand years ago. All living forms were deliberately designed and suddenly created. The geological structure of the modern earth, the distribution of existing species, and the fossil record were ascribed to a recent catastrophic, global flood.

Scientists, philosophers, and theologians felt compelled to organize and assert that evolution is a core theory in modern biology, whereas creation science is religion. In a suit filed on May 27, 1981, challenging the constitutionality of the Arkansas Balanced Treatment Law, the plaintiffs charged that the Act constituted an establishment of religion prohibited by the First Amendment to the Constitution, which was made applicable to the states by the Fourteenth Amendment. Second, the Act violated the right to academic freedom guaranteed to students and teachers by the Free Speech Clause of the First Amendment. Third, the plaintiffs argued that the Act was impermissibly vague and thus a violation of the Due Process Clause of the Fourteenth Amendment. The plaintiffs also charged that the Arkansas law threatened rather than enhanced religion by making the biblical literalism of a particular religious faction the basis of public education. Individuals and organizations listed as plaintiffs included the resident Arkansas Bishops of the United Methodist, Episcopal, Roman Catholic and African Methodist Episcopal Churches, the American Jewish Committee, parents of children attending Arkansas public schools, a high school biology teacher, the National Association of Biology Teachers, and the National Coalition for Public Education and Religious Liberty. Theologians from mainstream religious denominations tend to be among the most persuasive opponents of such laws because they regard Special Creation as religion, not science, and it is definitely not their religion. They say that Special Creation should be taught in comparative religion courses and not in biology. Catholic schools have accepted the teaching of evolution since the 1950s, and in 1996 Pope John Paul II declared that evolution did not conflict with Catholic doctrine. The defendants included the Arkansas Board of Education and its members, the Director of the Department of Education, and the State Textbooks and Instructional Materials Selection Committee.

Although the plaintiffs and the scientific community were not surprised that Judge Overton ruled against the Arkansas Balanced Treatment Act, the clear, decisive, strong ruling was unexpected. The judge stated without qualification or equivocation that so-called creation science "is simply not science." The two-model approach, he concluded, is "fallacious pedagogy." Students are supposed to conclude that any evidence alleged to call evolutionary theory into question constitutes evidence for "creation-science." In practice, when scientists disagree about the age of various fossils or the rate of mutation, creationists use the dispute as evidence for creationism; such logic is obviously specious. No reasonable person should accept disagreements about neutral mutations or punctuated evolution as evidence for a young earth and a worldwide flood. Judge Overton ruled against the Balanced Treatment Law on the following grounds: (1) Creation Science is a religion; (2) it is illegal to teach religion in the public schools; (3) Creation Science is not science.

Creationists complained that their case had not been well presented and that the judge was biased. Televangelists and leaders of the Moral Majority openly accused the Arkansas attorney general, who had defended the Balanced Treatment Law for the state, of being in collusion with the ACLU. A similar Louisiana Balanced Treatment law was brought to the U.S. Supreme Court in 1987. Although the court ruled against the Louisiana law, the decision was not as clear and forceful. Justice William J. Brennen, who wrote the majority report, said that Creation Science as presented in the case under consideration was religion; but the teaching of alternative scientific views on origins would be acceptable. Special Creationists, therefore, claimed that the decision made it possible to bring creationism into the biology classroom as a new science. After winning the Arkansas decision, evolutionists seemed to assume the war was over. In contrast, creationists continued to work tirelessly on the legal, popular, and political fronts, winning considerable support from politicians, journalists, parents, judges, and educators. Judicial rulings against creationism and equal time laws led to renewed fundamentalist efforts at the local and state levels and more pressure on school boards, curriculum committees, and textbook publishers.

Despite setbacks in the courts, most public opinion polls continued to show that Americans basically supported the position that the public schools should "teach both sides." Public opinion polls conducted in the 1980s found that over 80% of Americans supported inclusion of creationist theories in biology courses. Surveys of American college students indicated that many had been taught creationism along with evolution in high school and that very few actually understood the modern theory of evolution. Most students thought of evolution as a progressive Lamarckian advance from microbes to man. Significant numbers of students said that they rejected evolutionary theory because it contradicted their own convictions and most thought that the terms theory and guess were synonymous. More than half of all students polled said that dinosaurs and humans were contemporaries, and many believed that human beings were created about 10,000 years ago. Many believed that biblical creationism should be taught in public schools. Survey results among science teachers found that about 20% considered the Bible an authoritative source of information about the age of the earth and the origin of life. Many of the teachers thought that the earth is less than 20,000 years old.

Not surprisingly, studies of science education standards at the beginning of the twenty-first century revealed significant deficiencies in the treatment of evolution, molecular biology, geology, and cosmology. American students scored poorly on international tests in science and registered a decline on the National Assessment of Educational Progress, a science test known as the nation's report card. In some states, the guidelines for biology education deliberately avoided using the term "evolution," even if the basic concepts were part of the curriculum, in order to get the biology standards approved by the legislature. To accommodate the demands of creationists, the Kansas City school district adopted another approach in 1999. The Board of Education voted to allow the elimination of all references to the theory of evolution from science curriculums throughout the state, although the governor complained that this decision was tragic and embarrassing. After almost two years of debate and stinging international ridicule, the teaching of evolution was restored by the Kansas Board of Education despite protests from some board members. Kansas is one of several states, including Alabama, Arizona, Illinois, New Mexico, Texas, and Nebraska, where school boards have attempted to remove evolution from state science standards or, failing that, issue warnings that question the validity of evolution as part of modern science. For example, in 2001 the Alabama State Board of Education reviewed its policy of requiring a notice in science textbooks that identifies evolution as a controversial theory. The board voted to replace the statement that had been pasted into textbooks since 1995 with a new version of the disclaimer. Students are warned that evolution by natural selection is a controversial theory and urged to "wrestle with the unanswered questions and unresolved problems still faced by evolutionary theory." Fundamentalists have also called for removing references to the big bang theory of the universe's origin in earth science textbooks unless the biblical story of the creation is included.

According to a national survey conducted in 2000, an overwhelming majority of Americans still think that public schools should teach both evolution and creationism. About 30% of the respondents believe that creationism should be taught as a scientific theory, either with or without evolution in the curriculum. At the other end of the spectrum, 20% thought that evolution should be taught in science classes without any mention of creationism. Most respondents said that evolution should be taught as a scientific theory, while creationism should be discussed as a religious belief rather than a scientific theory. Young Americans 18–24 years old and Americans with relatively high education levels were more likely to support teaching evolution and less likely to favor teaching creationism. Although antievolutionary fervor has long been associated with the South, the survey found little variation in responses by geographic region, a finding that reflects the fact that creationist organizations have become more widely disseminated. Since the 1980s, antievolution activism has actually increased throughout the world. Young Earth Creationist movements, once regarded as a uniquely American phenomenon, are flourishing in other nations, including Australia, Korea, Russia, and Turkey.

The campaign for teaching Special Creationism as an acceptable alternative to Darwinian evolution raises many questions about the separation of church and state, academic freedom, and the ability of scientists to communicate with the public. The controversy over evolution proves more clearly than any other episode in the history of science that science is not limited to the laboratory, nor is it a catalog of facts; science is clearly an integral part of the history of ideas and culture. Educators, scientists, and philosophers are convinced that the debate about the teaching of evolution may take on many new forms in the twenty-first century, but it is unlikely that the issue will ever be totally settled.

SUGGESTED READINGS

Albritton, C. C. (1980). *The Abyss of Time: Changing Conceptions of the Earth's Antiquity After the Sixteenth Century*. San Francisco, CA: Freeman, Cooper.

Alexander, R. D. (1980). *Darwinism and Human Affairs*. Seattle, WA: University of Washington Press.

Alland, A., Jr. (1985). *Human Nature: Darwin's View*. New York: Columbia University Press.

Alter, S. G. (1999). *Darwin and the Linguistic Image: Language, Race and Natural Theology in the Nineteenth Century*. Baltimore, MD: Johns Hopkins University Press.

Appel, T. (1987). *The Cuvier-Geoffroy Debate: French Biology in the Decades Before Darwin*. New York: Oxford University Press.

Appleman, P., ed. (1976). *An Essay on the Principle of Population. Thomas Robert Malthus. A Norton Critical Edition*. New York: Norton.

Appleman, P., ed. (1979). *Darwin. A Norton Critical Edition*. New York: Norton.

Awbrey, F., and Thwaites, W., eds. (1984). *Evolutionists Confront Creationists*. San Francisco, CA: American Association for the Advancement of Science.

Bannister, R. C. (1979). *Social Darwinism: Science and Myth in Anglo-American Social Thought*. Philadelphia, PA: Temple University Press.

Barrett, P. H., and Gruber, H. E., eds. (1980). *Metaphysics, Materialism, and the Evolution of Mind: Early Writings of Charles Darwin.* Chicago, IL: University of Chicago Press.

Barthelemy-Madaule, M. (1984). *Lamarck the Mythical Precursor: A Study of the Relations Between Science and Ideology.* Cambridge, MA: MIT Press.

Beer, G. (2000). *Darwin's Plots: Evolutionary Narrative in Darwin, George Eliot, and Nineteenth-Century Fiction.* 2nd ed. Cambridge: Cambridge University Press.

Bell, G. (1982). *The Masterpiece of Nature. The Evolution and Genetics of Sexuality.* Berkeley, CA: University of California Press.

Boller, P. F. (1981). *American Thought in Transition: The Impact of Evolutionary Naturalism, 1865–1900.* Chicago, IL: Rand McNally.

Bonner, J. T. (1988). *The Evolution of Complexity by Means of Natural Selection.* Princeton, NJ: Princeton University Press.

Bowler, P. J. (1989). *Evolution. The History of an Idea.* Berkeley, CA: University of California Press.

Bowler, P. J. (1996). *Life's Splendid Drama. Evolutionary Biology and the Reconstruction of Life's Ancestry, 1860–1940.* Chicago, IL: University of Chicago Press.

Brackman, A. C. (1980). *A Delicate Arrangement. The Strange Case of Charles Darwin and Alfred Russel Wallace.* New York: Times Books.

Brent, P. (1981). *Charles Darwin. A Man of Enlarged Curiosity.* New York: Norton.

Brooks, J. L. (1984). *Just Before the Origin: Alfred Russel Wallace's Theory of Evolution.* New York: Columbia University Press.

Browne, J. (1995). *Charles Darwin. Voyaging.* New York: Alfred A. Knopf.

Burkhardt, R. W., Jr. (1977). *The Spirit of System. Lamarck and Evolutionary Biology.* Cambridge, MA: Harvard University Press.

Caudill, E. (2000). *The Scopes Trial: A Photographic History.* Knoxville, TN: The University of Tennessee Press.

Chambers, R. (1994). *Vestiges of the Natural History of Creation and Other Evolutionary Writings.* Edited and with an introduction by James A. Secord. Chicago, IL: University of Chicago Press.

Coleman, W. (1964). *Georges Cuvier, Zoologist.* Cambridge, MA: Harvard University Press.

Corsi, P. (1988). *The Age of Lamarck: Evolutionary Theories in France, 1790–1830.* Berkeley, CA: University of California Press.

Cronin, H. (1991). *The Ant and the Peacock: Altruism and Sexual Selection from Darwin to Today.* New York: Cambridge University Press.

Darwin, C. (1859). *The Origin of Species by Means of Natural Selection or The Preservation of Favoured Races in the Struggle for Life.* Edited and introduction by J. W. Burrow. Baltimore, MD: Penguin, 1968.

Darwin, C. (1881). *The Formation of Vegetable Mould, through the Action of Worms, with Observations on Their Habits.* Foreword by S. J. Gould. Chicago, IL: University of Chicago Press, 1985.

Darwin, C. (1958). *The Autobiography of Charles Darwin 1809–1882. With original omissions restored.* Edited with appendix and notes by his granddaughter Nora Barlow. New York: Harcourt, Brace and Company.

Darwin, C. (1962). *The Voyage of the Beagle*. Annotated and with an Introduction by Leonard Engel. Garden City, NY: Doubleday.

Darwin, C. (1965). *The Expression of the Emotions in Man and Animals*. With a preface by Konrad Lorenz. Chicago, IL: University of Chicago Press.

Darwin, C. (1981). *The Descent of Man and Selection in Relation to Sex*. With a new introduction by J. T. Bonner and R. M. May. Princeton, NJ: Princeton University Press.

Dawkins, R. (1986). *The Blind Watchmaker*. New York: Norton.

Desmond, A. (1984). *Archetypes and Ancestors: Palaeontology in Victorian London 1850–1875*. Chicago, IL: University of Chicago Press.

Diamond, J. (1991). *The Third Chimpanzee: The Evolution and Future of the Human Animal*. New York: HarperCollins.

DiGregorio, M. A. (1984). *Thomas Henry Huxley's Place in Natural Science*. New Haven, CT: Yale University Press.

Dundes, A., ed. (1988). *The Flood Myth*. Berkeley, CA: University of California Press.

Durant, J., ed. (1985). *Darwinism and Divinity: Essays on Evolution and Religious Belief*. Oxford: Basil Blackwell.

Eisley, L. (1958). *Darwin's Century: Evolution and the Men Who Discovered It*. Garden City, NY: Doubleday.

Ellegard, A. (1990). *Darwin and the General Reader: The Reception of Darwin's Theory of Evolution in the British Periodical Press, 1859–1872*. New Foreword by David Hull. Chicago: University of Chicago Press.

Farber, P. L. (2000). *Finding Order in Nature: The Naturalist Tradition from Linnaeus to E. O. Wilson*. Baltimore, MD: Johns Hopkins University Press.

Gillispie, C. C. (1951). *Genesis and Geology. A Study in the Relations of Scientific Thought, Natural Theology, and Social Opinion in Great Britain, 1790–1850*. New York: Harper & Row.

Ginger, R. (1958). *Six Days or Forever? Tennessee v. John Thomas Scopes*. Boston, MA: Beacon Press.

Glick, T. F., ed. (1988). *The Comparative Reception of Darwinism*. Chicago: University of Chicago Press.

Godfrey, L. R., ed. (1983). *Scientists Confront Creationism*. New York: Norton.

Gould, S. J. (1984). *Ontogeny and Phylogeny*. Cambridge, MA: Belknap/Harvard.

Gray, A. (1963). *Darwiniana. Essays and Reviews Pertaining to Darwinism*. Edited by A. Hunter Dupree. Cambridge, MA: The Belknap Press of Harvard University Press.

Greene, J. C. (1999). *Debating Darwin. Adventures of a Scholar*. Claremont, CA: Regina Books.

Gruber, H. E. (1974). *Darwin on Man. A Psychological Study of Scientific Creativity. Together with Darwin's Early and Unpublished Notebooks*. New York: E. P. Dutton.

Hanson, R. W., ed. (1986). *Science and Creation: Geological, Theological, and Educational Perspectives*. New York: Macmillan.

Haught, J. F., and Ayala, F. J., eds. (2000). *Science and Religion in Search of Cosmic Purpose*. Washington, DC: Georgetown University Press.

Hofstadter, R. (1955). *Social Darwinism in American Thought.* Boston, MA: Beacon Press.

Hull, D. L. (1983). *Darwin and His Critics: The Reception of Darwin's Theory of Evolution by the Scientific Community.* Chicago: University of Chicago Press.

Huxley, L., ed. (1900). *The Life and Letters of Thomas Henry Huxley.* 2 vols. New York: AMS Press, reprint 1979.

Huxley, T. H. (1959). *Man's Place in Nature.* Ann Arbor, MI: University of Michigan Press.

Irvine, W. (1955). *Apes, Angels and Victorians: The Story of Darwin, Huxley, and Evolution.* New York: McGraw-Hill Book Company.

Jones, G. (1980). *Social Darwinism and English Thought: The Interaction Between Biological and Social Theory.* Atlantic Highlands, NJ: Humanities Press.

Keynes, R. (2002). *Darwin, His Daughter & Human Evolution.* New York: Riverhead Books.

King-Hele, D. (1977). *Doctor of Revolution; The Life and Genius of Erasmus Darwin.* London: Faber and Faber.

Kitcher, P. (1983). *Abusing Science: The Case against Creationism.* Cambridge, MA: MIT Press.

Koerner, L. (1999). *Linnaeus: Nature and Nation.* Cambridge, MA: Harvard University Press.

Kropotkin, P. (1972). *Mutual Aid, A Factor in Evolution.* Ed. and with an intro. by P. Avrich. New York: New York University Press.

Lamarck, J. B. (1809). *Zoological Philosophy. An Exposition with Regard to the Natural History of Animals.* Translated by Hugh Elliot. Chicago, IL: University of Chicago Press, 1984.

Larson, E. J. (1997). *Summer for the Gods: The Scopes Trial and America's Continuing Debate over Science and Religion.* New York: Basic Books.

Leakey, R. E., and Lewin, R. (1993). *Origins Reconsidered. In Search of What Makes Us Human.* New York: Doubleday.

Lewin, R. (1997). *Bones of Contention. Controversies in the Search for Human Origins.* 2nd ed. New York: Simon and Schuster.

Lewis, C. (2000). *The Dating Game: One Man's Search for the Age of the Earth.* New York: Cambridge University Press.

Lindberg, D. C., and Numbers, R. L., eds. (1986). *God and Nature. Historical Essays on the Encounter Between Christianity and Science.* Berkeley, CA: University of California Press.

Lovejoy, A. O. (1936). *The Great Chain of Being: A Study in the History of an Idea.* New York: Harper, reprint 1960.

Lyell, C. (1863). *Geological Evidences of the Antiquity of Man: With Remarks on Theories of the Origin of Species by Variation.* New York: AMS Press, reprint 1973.

Lyell, C. (1830–33). *Principles of Geology: Being an Attempt to Explain the Former Changes of the Earth's Surface by Reference to Causes Now in Operation.* 3 volumes. A facsimile of the first edition, with an introduction by M. J. S. Rudwick. Chicago, IL: University of Chicago Press, 1990–1991.

Maienschein, J., and Ruse, M., eds. (1999). *Biology and the Foundation of Ethics.* New York: Cambridge University Press.

Margulis, L., and Sagan, D. (1986). *Microcosmos: Four Billion Years of Evolution from Our Microbial Ancestors*. New York: Summit Books.

Mason, S. F. (1991). *Chemical Evolution: Origins of the Elements, Molecules, and Living Systems*. New York: Oxford University Press.

Maupertuis, P. L. (1966). *The Earthly Venus*. Translated by S. B. Boas. New York: Johnson, reprint.

Mayr, E. (1991). *One Long Argument: Charles Darwin and the Genesis of Modern Evolutionary Thought*. Cambridge, MA: Harvard University Press.

McGowan, C. (2001). *The Dragon Seekers: How an Extraordinary Circle of Fossilists Discovered the Dinosaurs and Paved the Way for Darwin*. Cambridge, MA: Persus Pub.

McNeil, M. (1987). *Under the Banner of Science: Erasmus Darwin and His Age*. Manchester: Manchester University Press.

Millhauser, M. (1959). *Just Before Darwin: Robert Chambers and Vestiges*. Middletown, CT: Wesleyan University Press.

Moore, J. A. (2002). *From Genesis to Genetics: The Case of Evolution and Creationism*. Berkeley, CA: University of California Press.

Moore, J. R. (1979). *The Post-Darwinian Controversies*. New York: Cambridge University Press.

Morris, H. M., and Parker, G. E. (1987). *What Is Creation Science?* El Cajon, CA: Master Books.

National Academy of Sciences (1984). *Science and Creationism: A View from the National Academy of Sciences*. Washington, DC: National Academy Press.

Nelkin, D. (1982). *The Creation Controversy. Science or Scripture in the Schools*. New York: Norton.

Norell, M. A., Gaffney, E. S., and Dingus, L. (2000). *Discovering Dinosaurs: Evolution, Extinction, and the Lessons of Prehistory*. Berkeley, CA: University of California Press.

Numbers, R. L. (1993). *The Creationists: The Evolution of Scientific Creationism*. Berkeley, CA: University of California Press.

Numbers, R. L., ed. (1995). *Creationism in Twentieth-Century America: A Ten-Volume Anthology of Documents, 1903–1961*. New York: Garland Publishing.

Oparin, A. I. (1953). *The Origin of Life*. Translated by S. Morgulis. New York: Dover.

Outram, D. (1984). *Georges Cuvier: Vocation, Science, and Authority in Post-Revolutionary France*. Manchester: Manchester University Press.

Paradis, J., and Williams, G. C. (1989). *Evolution & Ethics. T. H. Huxley's Evolution and Ethics with New Essays on Its Victorian and Sociobiological Context*. Princeton, NJ: Princeton University Press.

Pennock, R. T. (1999). *Tower of Babel: The Evidence Against the New Creationism*. Cambridge, MA: MIT Press.

Pennock, R. T., ed. (2001). *Intelligent Design Creationism and Its Critics: Philosophical, Theological, and Scientific Perspectives*. Cambridge, MA: MIT Press.

Raven, C. E. (1986). *John Ray Naturalist: His Life and Works*. 2nd ed. New York: Cambridge University Press.

Rudwick, M. J. S (1985). *The Meaning of Fossils: Episodes in the History of Palaeontology*. 2nd ed. Chicago, IL: University of Chicago Press.

Ruse, M., ed. (1996). *But Is It Science?: The Philosophical Question in the Creation/ Evolution Controversy.* Amherst, NY: Prometheus Books.

Ruse, M. (1999). *The Darwinian Revolution: Science Red in Tooth and Claw.* 2nd ed. Chicago, IL: University of Chicago Press.

Ruse, M. (1999). *Mystery of Mysteries: Is Evolution a Social Construction?* Cambridge, MA: Harvard University Press.

Ruse, M. (2000). *The Evolution Wars: A Guide to the Debates.* Santa Barbara, CA: ABC-CLIO.

Ruse, M. (2001). *Can a Darwinian Be a Christian?: The Relationship between Science and Religion.* New York: Cambridge University Press.

Russell, N. (1986). *Like Engend'ring Like: Heredity and Animal Breeding in Early Modern England.* Cambridge University Press.

Russell, R. J., Stoeger, W.R., and Ayala, F. J., eds. (1998). *Evolutionary and Molecular Biology: Scientific Perspectives on Divine Action.* Berkeley, CA: Center for Theology and the Natural Sciences.

Scopes, J. T., and Presley, J. (1967). *Center of the Storm, Memoirs of John T. Scopes.* New York: Holt, Rinehart and Winston.

Shipman, P. (1994). *The Evolution of Racism: Human Differences and the Use and Abuse of Science.* New York: Simon & Schuster.

Stauffer, R. C., ed. (1975). *Charles Darwin's Natural Selection. Being the Second Part of His Species Book Written from 1856 to 1858.* New York: Cambridge University Press.

Steele, E. J. (1981). *Somatic Selection and Adaptive Evolution: On the Inheritance of Acquired Characters.* 2nd ed. Chicago, IL: University of Chicago Press.

Tanner, N. M. (1981). *On Becoming Human.* New York: Cambridge University Press.

Todes, D. P. (1989). *Darwin Without Malthus: The Struggle for Existence in Russian Evolutionary Thought.* New York: Oxford University Press.

Toumey, C. P. (1994). *God's Own Scientists: Creationists in a Secular World.* New Brunswick, NJ: Rutgers University Press.

Vucinich, A. (1989). *Darwin in Russian Thought.* Berkeley, CA: University California Press.

Wallace, A. R. (1891). *Darwinism.* London: MacMillan.

Wallace, A. R. (1905). *My Life.* New York: Dodd, Mead.

Webb, G. E. (1994). *The Evolution Controversy in America.* Lexington, KY: The University Press of Kentucky.

Weikart, R. (1998). *Socialist Darwinism: Evolution in German Socialist Thought from Marx to Bernstein.* Bethesda, MD: International Scholars Publications.

Wilder-Smith, A. E. (1975). *Man's Origin, Man's Destiny: A Critical Survey of the Principles of Evolution and Christianity.* Minneapolis, MN: Bethany Fellowship, 1975

Williams-Ellis, A. (1966). *Darwin's Moon: A Biography of Alfred Russel Wallace.* London: Blackie.

Wilson, E. O. (1975). *Sociobiology: The New Synthesis.* Cambridge, MA: Harvard University Press.

Wilson, E. O. (1992). *The Diversity of Life.* Cambridge, MA: Harvard University Press.

Winchester, S. (2001). *The Map That Changed The World: William Smith and the Birth of Modern Geology*. New York: HarperCollins.

Young, D. A. (1995). *The Biblical Flood: A Case Study of the Church's Response to Extrabiblical Evidence*. Grand Rapids, MI: Eilliam B. Eerdmans Publishing.

Young, R. M. (1985). *Darwin's Metaphor: Nature's Place in Victorian Culture*. New York: Cambridge University Press.

Ziadat, A. A. (1986). *Western Science in the Arab World: The Impact of Darwin, 1860–1930*. New York: St. Martin's Press.

9

GENETICS

Just as the term *biology* was coined at the beginning of the nineteenth century to replace *natural philosophy*, the term *genetics* was coined at the beginning of the twentieth century to separate new forms of scientific inquiry from previous studies of *generation, inheritance,* or *heredity.* Although genetics, embryology, and evolution were originally inseparable aspects of biology, during the twentieth century they became complex and separate disciplines. In the 1920s classical genetics, often referred to as Mendelism, was still a tentative and contested branch of natural history, barely separated from embryology, the study of development and growth. By the 1950s genetics had emerged as a powerful new unifying principle at the heart of the life sciences.

Genetics can be seen as a domain containing within its borders both the newest scientific developments and the oldest fragments of folklore. The Bible reflects many ancient ideas about plant and animal breeding. Jacob, for example, put to good use the belief that offspring are influenced by what their mothers experience during pregnancy. While tending sheep for his uncle, Jacob was allowed to keep all the striped and spotted lambs. Such variants were usually very rare, but Jacob increased his allotment by peeling wooden rods into appropriate designs and showing them to the ewes. In addition to increasing the number of striped lambs, he improved their quality by showing the design only to the best of the ewes so that his own herd grew in vigor as well as in number. Even in the twentieth century advocates of "prenatal culture" continued to believe that maternal impression could influence a baby's physical and mental traits.

The relationship between the act of sexual intercourse and reproduction was obviously known in very ancient times and was exploited for practical purposes. Thus, animals and men were castrated to make them more tractable or useful, and animals with valuable traits were selected for breeding purposes. But the diversity of means of reproduction and differences in development were a source of constant confusion. Fertilization could be internal or external, the young could appear first as eggs, worm-like larvae, miniature versions of their parents, or totally different in form from the adults. Stories of strange hybrids and monstrous births are common themes in ancient myths and folk tales. Creatures such as centaurs, a cross between horses and humans, were certainly only products of the imagination, but a Sumerian proverb comments on the distance between the mule and its parents, and Greek philosophers discussed the sterility of this hardy hybrid.

Despite a general belief in ancient times that inbreeding was beneficial, eminent naturalists from Aristotle to Conrad Gesner (1516–1565) gave credence to the possibility of matings more bizarre than that of a jackass and a mare. Allegedly reliable sources reported that different species often met and mated at the water holes in Libya. This strange proclivity to crossbreeding gave rise to the proverb "Something new is always coming from Libya." A cross between a camel and a leopard must have produced the giraffe, while a mating between a camel and a sparrow resulted in the ostrich. And was it not obvious that matings between humans and animals had produced such bizarre creatures as the Minotaur, the manatee, and the great apes? Early Western descriptions of the great apes depicted them as wild men or monsters, proportioned like humans, but gigantic in stature, hollow eyed, hairy, mute, strong, and impossible to capture. Indeed, it was not until 1847 that Thomas S. Savage (1804–1880), an American clergyman and naturalist, and Jeffries Wyman (1814–1874), physician and anatomist, published the first "scientific" description of the gorilla as a creature distinct from humans and other apes.

While many Greek myths contain examples of parthenogenesis or uniparental reproduction, such as the birth of Athena from the head of Zeus, Hippocrates argued that both parents contributed traits to their offspring. Aristotle rejected this theory and supported the idea that the male provided the *form* or blueprint for the embryo while the female simply provided *matter* and served as an incubator. If the heat of the womb was insufficient, the paternal blueprint could not be fully executed and the unfortunate embryo developed into a female. When it came to the breeding of kings and warriors, the Spartans were, according to the story of King Archidamos II (fifth century *B.C.*), concerned with the physique of both parents. Archidamos was reprimanded for marrying a short woman on the grounds that she would produce diminutive kinglets, rather than robust kings.

Nineteenth-century scientists were faced with the task of reconciling ancient ideas and observations with new scientific theories concerning cell structure and function, embryology, and evolution. Rudolf Wagner (1805–1864), the prolific editor of the multivolume *Handbook of Physiology* (1842–53), asserted that more than 300 different theories of procreation had already been published. None of them, he declared, had been satisfactory. Probably no scientist was ever more willing to pose hard questions about inheritance than Charles Darwin. In an attempt to accommodate heredity, variation, and evolution within a general framework, Darwin revived the Hippocratic theory of inheritance and renamed it the *provisional hypothesis of pangenesis*. Heredity, according to Darwin, reflected the direct transmission of qualities from parent to offspring, as influenced by external conditions. The prevailing view that inheritance involved the blending of characters from both parents created problems for Darwin's theory of natural selection, because matings between rare variants and ordinary members of the species would presumably produce offspring that were intermediate between the parental types. The favorable variation would, moreover, be even further diluted in succeeding generations. In defense of evolution, Darwin called attention to the astonishing array of odd and frivolous traits had been selected and maintained by animal breeders and gardeners. Perhaps, Darwin suggested, domestication represented an extreme example of environmental influences that affected the reproductive system and stimulated variation. In any case, Darwin argued, variation is essentially random.

According to a rather disingenuous account in his autobiography, Darwin began to arrange his notes for the *Variation of Animals and Plants under Domestication* just 2 months after the publication of the *Origin of Species*. By Darwin's own reckoning the two-volume *Variation*, which was finally published in 1868, cost him 4 years and 2 months of hard labor. Towards the end of the second volume, Darwin presented the theory that he ruefully referred to as his "well-abused hypothesis of Pangenesis." Well-abused was an apt description, because even though Darwin's theory was (and is) regarded as highly speculative and wholly erroneous, it was one of the most widely discussed theories of inheritance in the late nineteenth century. Whenever he began working out his views on pangenesis, he continually attempted to incorporate new ideas about cellular structure, physiology, and growth.

In preparation for attacks on his pet theory, Darwin began his exposition with a quotation from William Whewell (1794–1866), a British philosopher of science: "Hypotheses may often be of service to science, when they involve a certain portion of incompleteness, and even of error." Although Darwin probably began to formulate the provisional hypothesis in the early 1840s, he was always able to tell himself that his theory of natural selection did not require an explanation of exactly how variations arose or how they were transmitted. He did think, however, that the theory of pangenesis could explain major classes of generally accepted facts and observations concerning reproduction. Of course, some of the information about reproduction accepted without question as factual in the nineteenth century no longer enjoys that status. For example, Darwin assumed that the influence of the male on the female reproductive system was so profound that later offspring could be contaminated by the legacy of previous matings.

Darwin asserted that he had been "led, or rather forced" to believe that every organ, tissue, or cell of the whole body reproduced itself by giving off minute units that he called *gemmules*. After circulating through the body, the gemmules were collected in the reproductive organs and incorporated into the gametes. Thus, all ovules and pollen grains, fertilized seeds, sperm, eggs, and buds consisted of multitudes of gemmules given off by "each separate atom of the organism." At conception, gemmules from both parents combined to form the embryo. In other words, the particles, or gemmules, came from all parts of the parental body and grew into the cells of the embryo and, subsequently, the tissues and organs of the new individual.

Pangenesis did not seem to throw much light on hybridism, but it was not incompatible with the possibility that novel gemmules might be formed in hybrids. Such gemmules could be transmitted to the offspring along with latent ancestral gemmules. In conclusion, Darwin summarized the complex facts subsumed by his general hypothesis with the assertion that "the child, strictly speaking, does not grow into the man, but includes germs which slowly and successively become developed and form the man."

The provisional hypothesis of pangenesis gave Darwin a particulate theory of heredity, development, and growth. The characteristics of the new individual depended on whether the gemmules for particular traits came from the maternal or paternal line. But Darwin did not think of sexual reproduction as either the cause of variation or a means of providing diversity in a population. Usually, the parental traits blended, but in some cases one trait exerted *prepotency*, or dominance, over the other. Darwin accounted for the phenomenon known as *atavism* or reversion by postulating that some gemmules could be transmitted but not expressed. Latent ancestral gemmules might be transmitted for several generations before making their presence known. Reversion could, therefore, be attributed to dormant ancestral gemmules, which became active under certain conditions. Metaphorically speaking, each animal and plant was like a garden full of seeds, some of which germinated, some remained dormant, and some perished. With appropriate manipulations, pangenesis could be used to account for the prepotency of some traits, the latency of others, regeneration, and malformations. If the limb of a salamander were cut off, limb gemmules circulating in the blood could go to the site and direct the formation of a new limb. Malformations might occur when the wrong gemmules were expressed at a specific site.

Ambiguous and amorphous as it was, the hypothesis that Julian Huxley later called Darwin's great "edifice of speculation" could not be ignored. Skepticism about the reality of Darwin's gemmules prompted his cousin Francis Galton (1822–1911) to subject the provisional hypothesis to an experimental test. Unlike Darwin, Galton believed that the hereditary material was produced in the reproductive organs, with little if any contribution from other bodily tissues. If it were true that the gemmules were generated by body tissues and passed through the circulating fluids of the body to the reproductive organs, a blood transfusion should transfer gemmules from donor to recipient. Rabbits that had received blood transfusions from rabbits of a different color were used in Galton's experiments. Contrary to predictions drawn from Darwin's hypothesis, blood transfusions had no effect on the coat color of the offspring of the transfused rabbits. Although Darwin was disappointed by the transfusion experiments, he refused to believe that Galton had identified a fatal objection to his hypothesis. Certainly the gemmules were not necessarily associated with the circulatory system, he argued, because there must be some mechanism for the transmission of gemmules in organisms without blood, such as plants and lower animals. The possibility of testing the theory of pangenesis by the hybridization of plants occurred to Galton, but he abandoned this approach because of his lack of interest in the minutiae of gardening.

PLANT HYBRIDIZATION AND GENETICS

The controversial theory of evolution and Darwin's speculations about heredity emphasized the importance of the problem of variation, but the work most closely associated with the development of modern genetics was the ancient and wholly respectable study of plant hybridization.

Plant hybridizers can be divided into two traditions: practical horticulturists, who wanted to be able to establish new and commercially useful varieties, and philosophical naturalists, who were interested in basic questions such as: Was it possible to create new species by crossbreeding existing types? Carolus Linnaeus (1707–1778), who had a profound influence on plant and animal breeders, suggested that God might have originally created one prototype for each plant genus. Modern species might then have formed as a result of variation and hybridization.

Evidence for plant sexuality had been established in the 1690s by the German botanist Rudolf Jacob Camerarius (1665–1721), a physician and professor of botany and medicine at the University of Tübingen. Other scientists, including Theophrastus, Cesalpino, Malpighi, Nehemiah Grew, and John Ray, had speculated about sexuality in plants before Camerarius, but the issue remained unsettled. In a famous 1694 letter known as *De sexu plantarum* (On the Sex of Plants), Camerarius described the history of ideas about the parts of the flower and his own experiments on plant sexuality and reproduction. Having studied a variety of different plants, Camerarius carefully removed the male flowers from plants that produced separate male and female flowers. Flowers characterized by pistils did not produce seeds or fruits in the absence of flowers that contained stamens. When pollen was placed on the pistils of female flowers, they produced fruit. Therefore, he concluded that the anthers and their pollen grains (found on the flowers with stamens) were the male organs of the plant. The style and ovary found in the flowers with pistils were the female parts of the plant. Given an understanding of plant sexuality, Camerarius predicted, botanists could produce experimental hybrids by using the pollen of one species to fertilize the egg of another. This would result in progeny with a mixture of the characteristics of both parental species. Comparing reproduction in plants and animals, Camerarius argued that hermaphroditic plants fertilized themselves while hermaphroditic snails did not. Later research confirmed this. Describing the medicinal virtues of plants, Camerarius argued that plants with similar flower structures have similar healing properties, which suggested that they must contain similar substances. Modern chemotaxonomical research indicates that this approach was indeed valid.

Joseph Gottlieb Koelreuter (1733–1806) was one of the first botanists to develop a technique for artificial fertilization in plants, systematically test plant hybrids, and provide rigorous experimental evidence of plant sexuality. Born in Sulz, Germany, where his father was an apothecary, Koelreuter studied medicine at the University of Tübingen and graduated in 1755. While managing the natural history collections of the Imperial Academy of Sciences in St. Petersburg, Russia, Koelreuter began his research on the structure of flowers and pollen. After returning to Germany, Koelreuter was appointed professor of natural history and director of the gardens in Karlsruhe, where he continued his experiments on plant hybridization. Unfortunately, in 1786 he lost this position. Although Koelreuter often complained that he lacked financial support and professional recognition, his contemporaries respected his ingenious experimental approach. Despite limited resources, he carried out more than 500 different hybridization tests with 138 species and studied the shape, size, and color of pollen grains from more than 1000 species. To avoid errors caused by accidental self-pollination, Koelreuter pioneered techniques for artificial hybridization. After painstakingly removing the anthers and pollinating his experimental plants by hand, he carefully covered the flower to prevent contamination by extraneous pollen. These experiments and observations called into question prevailing ideas about hybrids. Following Linnaeus, botanists generally assumed that any plant with characteristics intermediate between two known species must be a hybrid. Therefore, they thought that breeding tests to determine purity of type were unnecessary.

In 1761 Koelreuter published an account of his pioneering studies of tobacco plant hybrids. For the most part, his hybrids were intermediate between both parents, but contrary to prevailing beliefs, reciprocal crosses usually produced identical kinds of offspring. According to Linnaeus, the leaves and the outer layers of the stem represented the paternal contribution, while the inner portions of the flower and the stem arose from the maternal line. Although some hybrids were sterile, other hybrids were more vigorous than the parental varieties. Because of his pious belief in the fixity of species, Koelreuter assumed that there must be barriers to hybridization in nature. Puzzled by the fact that his initial hybrids were usually uniformly intermediate between the parents, while the next generation was extremely diverse, Koelreuter rationalized this apparent anomaly as the result of experimental interference with nature. In addition to his botanical work, Koelreuter was intrigued by contemporary alchemical theories, and he tended to see the results of his experiments on fertilization in terms of the transmutation of plant species. Like Francis Bacon, Koelreuter hoped that systematic experimentation would lead to products of great economic importance. Perhaps, Koelreuter predicted, hybrid trees might produce timber in half the time previously required.

Even though Koelreuter was sure he had unequivocally settled the question of plant sexuality, critics questioned his conclusions on the grounds that he had mutilated plants by castrating the flowers. Contrary to God's plan for nature, his experiments forcibly mated species that should have been separate and distinct. Moreover, he confined his plants in pots rather than allowing them to grow in open fields. Surely these unnatural conditions must reduce fertility and produce degenerate monstrosities rather than true hybrids. New species, therefore, could not be produced by hybridization. Thurs, despite the work of Camerarius, Linnaeus, and Koelreuter, the debate about plant sexuality continued well into the nineteenth century, along with the belief that plant hybridization leads to the multiplication of species. To settle this controversy, several major scientific academies offered prizes for new research into the nature of plant hybridization. The winners of the most significant competitions were Carl Friedrich von Gaertner (1772–1850), Charles Naudin (1815–1899), and A. F. Wiegmann (1771–1853).

Gaertner, a physician who was the son of the eminent botanist Joseph Gaertner (1732–1791), confirmed and extended Koelreuter's work. In 1849 he published an extensive monograph on plant hybridization, reviewing the published findings of other researchers as well as reporting his own experimental results. Having performed thousands of experiments on hundreds of plant species, Gaertner rejected the notion that natural hybridization gave rise to new plant species. Convinced that form and essence were the same, he insisted that his results were consistent with the constancy of species. Pollination with foreign pollen, Gaertner concluded, altered the traits of the plant because of the form-building force inherent in the life-giving pollen. As Gregor Mendel politely noted, although Gaertner's *Experiments and Observations on Hybridization in the Plant Kingdom* (1849) was based on a formidable series of experiments with hundreds of species, it added little to the theoretical aspects of the debate about hybridization and heredity.

Many botanists observed the phenomena now recognized as dominance and segregation, but because they were accustomed to treating the plant as a whole, they had little incentive to pay attention to the distribution of nonessential traits among the progeny of hybrids. Nevertheless, Naudin suggested that ancestral traits might segregate during reproduction. Attempting to clarify taxonomic relationships in the potato and cucumber families, Naudin resorted to hybridization tests and became intrigued with the evolutionary significance of hybridization. Still, he continued to believe that hybrids were unnatural entities, abhorred by nature. When he discovered that the offspring of certain hybrids reverted to the parental types, he suggested that segregation of the two original species had occurred. That is, nature dissolved hybrid forms by separating the specific essences of the original species. From 1862 until 1882, Darwin and Naudin exchanged publications and letters, but Darwin did not think that Naudin's hypothesis concerning segregation in hybrids could explain reversion to distant ancestral traits. As all students of modern biology know, while Darwin was puzzling over Naudin's observations, Gregor Mendel was conducting a remarkably systematic experimental analysis of this phenomenon in an Augustinian monastery garden.

JOHANN GREGOR MENDEL

Johann Mendel (1922–1884), the son of poor peasants, achieved success beyond his family's wildest expectations as a member of the Augustinian order and the abbot of his monastery at Brno in Moravia. Within 50 years of his death, Gregor Mendel was virtually canonized as the patron saint of modern genetics. The basic generalizations of genetics became known as Mendel's laws, and the Mendelianum at the Moravian Museum in Brno has become a shrine to his memory. But in the 1960s, as scientists celebrated the Mendel Centennial, historians began an assault on the "Mendel myth" that raised questions about whether Mendel could be considered a Mendelian, much less the discoverer of the fundamental laws or theoretical framework of modern genetics. While such charges may be aimed at the wrong target, it is certainly true that the modern tendency to see Mendel's work as absolutely clear and compelling exacerbates the question of why it was ignored until about 1900 and obscures irreconcilable differences between natural science in the 1860s and the life sciences in the twentieth century. Perhaps all participants in the debate about Mendel's true goals and methods are fortunate that so few of his private papers have survived as a possible means of refuting mythmakers of all persuasions.

The story of Mendel's life has been told as both a mystery and a tragedy because of the way in which his work and ideas seemed to disappear. Johann Mendel was born in a village in an area that has been known as Austrian Silesia, Moravia, and Czechoslovakia. Because of Mendel's exemplary performance at the local primary school, the vicar urged his family to continue his education. Unfortunately, when Mendel's father was injured and could no longer manage the family farm, Mendel's financial situation became desperate. Stress, anxiety, and malnutrition led to a physical breakdown and forced him to abandon his studies until his sister Theresia generously gave him part of her dowry. With family finances still precarious, Mendel gratefully accepted the opportunity to join the Augustinian order in Brno, the capital of the province of Moravia. Entering the monastic community as Brother Gregor allowed Mendel to continue his education and escape the bitter struggle for existence. The Brno monastery was well known as a center of learning; many of the monks taught at the local *Gymnasium* or philosophical institute. Even before Mendel's time, there had been great interest in Moravia in the practical breeding of sheep, fruit trees, and so forth. Many of Brno's monks were active in natural history and agricultural societies, including Mendel's predecessor as abbot, F. C. Napp (1792–1867). Sheep breeders in the area were particularly interested in the transmission of desirable traits from parents to offspring. When the economic importance of sheep breeding in Moravia diminished because of the impact of the British wool industry, naturalists in the area turned their attention to plant hybridization.

Mendel did not claim to have had a special call to enter the religious life, but he conscientiously studied theology in preparation for his new role, while also taking courses in agricultural science at the philosophical institute. After completing his religious instruction, Mendel assumed his parish duties, but his attempts to meet the spiritual needs of parishioners suffering and dying in the local hospital left him depressed and ill. Seeking an appropriate alternative, Abbot Napp dispatched Mendel to a grammar school as a substitute teacher. The schoolmaster was impressed with Mendel's work and recommended that he take the university examination that would qualify him for a regular teaching appointment in the natural sciences. At his first attempt to qualify, Mendel passed the physics and meteorology sections but failed geology and zoology. To prepare for another attempt, Mendel was sent to the University of Vienna to attend lectures in physics, chemistry, zoology, botany, plant physiology, paleontology, and mathematics. One of Mendel's professors at the university, the plant physiologist Franz Unger (1800–1870), taught species transmutation and suggested hybridization as a source of new species. In addition to his formal courses, Mendel participated in the Zoological-Botanical Society of Vienna and published two short papers concerning insect damage to plants in the society's *Proceedings* of 1853 and 1854.

When Mendel returned to the monastery, he was appointed substitute teacher of physics and natural history at the Brno Technical School. He became a member of the natural science section of the local agricultural society and began a series of experiments on plant hybridization and methods of artificial pollination. In 1856, during his second attempt to pass the examination for his regular teaching license, Mendel apparently experienced a psychological break-down. Unable to cope with the stress of the written examinations, Mendel decided to forgo all further attempts at certification.

Increasingly fascinated by experimental puzzles, Mendel studied the pea weevil, grew many different strains of peas, established about 50 beehives, and attempted crosses between American, Egyptian, and European bees. Clearly, Mendel hoped to breed better bees, but instead he created a strain that was excessively belligerent. He did, however, produce a strain of large, tasty peas that were cultivated in the monastery's vegetable garden for many years. Although his superiors generally encouraged research, they apparently objected when Mendel undertook his initial breeding experiments on mice. In addition to his biological investigations, Mendel engaged in long-term studies of meteorology, storms, and sunspot activity. During the 1860s he published several papers on meteorology and exceptional storms. As Brno's correspondent for Austrian regional weather reports, Mendel collected daily records of temperature, humidity, rainfall, and barometric pressure. Using this primitive database, he attempted to prepare statistical analyses of common meteorological phenomena, sunspot activity, and unusual storms. With Mendel's support, a novel system of weather forecasts was established to assist Moravian farmers. While Mendel's work in meteorology might be dismissed as a fruitless enterprise, his serious treatment of complex weather patterns was not unlike his approach to plant hybridization in that it involved the careful collection and statistical treatment of large amounts of data.

The sacrifices that Mendel's family made for the sake of his education were amply rewarded in 1868 when Brother Gregor was elected abbot of the monastery. Unfortunately for science, the burdens of administrative work drastically curtailed his research. Despite his new responsibilities, Mendel remained involved with the Central Board of the Agricultural Society, the Brno Horticultural Society, the Society of Apiculturists, and other local science societies. From 1875 on he became embroiled in a dispute about taxes on monastery properties; as usual, conflict caused his health to deteriorate. His death was attributed to chronic inflammation of the kidneys and cardiac hypertrophy. The plaque that commemorates Mendel's grave in the central cemetery in Brno identifies him as the biologist who "discovered the laws of heredity in plants and animals." Historians argue that although scientists consider Mendel the father of modern genetics, the image of Mendel as "the first geneticist," is a complicated construction.

Failing to appreciate the importance of Mendel's scientific work, his successors at the monastery destroyed most of his private papers. According to his nephews, Mendel expected them to publish some of his papers posthumously, but when he died they received nothing from the monastery. Thus, we have very little direct information about Mendel's preliminary ideas, sources, and goals. What little evidence remains has been collected at the Brno Mendel Museum. In any case, Mendel never kept a diary and, given his position at the monastery, he had to be extremely cautious in expressing his philosophical views. He seems to have been very reserved in his relationships with his monastic colleagues, and his letters revealed little of a personal nature. Some of his books survived, and his annotations are rather suggestive. In his copy of Gaertner's treatise on plant hybridization, Mendel carefully marked all references to *Pisum* variants and the characteristics of pea hybrids. A copy of a German translation of the *Origin* with Mendel's own marginalia is preserved in the Mendelianum. While Mendel presumably accepted the fact of evolution and rejected the provisional hypothesis of pangenesis, his views on Darwin's attempt to prove that natural selection was the mechanism of evolution are uncertain. His nephew Ferdinand Schindler later recalled that Mendel had great esteem for English science and was glad to know that his nephew had learned the language of Darwin and Shakespeare.

Despite the existence of many accounts of Mendel and the origins of genetics, his inner life and sources of inspiration remain a mystery. Given the uncertainties about Mendel's ideas, some historians have argued that Mendel himself was not a Mendelian and that he was not even interested in heredity. Some scholars argue that Mendel's paper was incomprehensible in the 1860s because his contemporaries saw no conceptual distinction between heredity and development. Thus, Mendel's experiments contained the answer to a question that had not been articulated. Rather than wondering whether Mendel was himself a "Mendelian," perhaps it is better to simply acknowledge the difficulty of reading Mendel's work without imposing contemporary concepts on Gregor Mendel and the Mendelians of the early twentieth century. Other sources of information about the intellectual atmosphere of the area in which Mendel lived and worked, the university he attended, and the books and journals available clearly indicate that the naturalists of Brno did not work in isolation from the scientific community of their time. Nevertheless, no complete and satisfactory explanation has emerged to tell us why and how the milieu that nourished and stimulated Mendel's work also caused it to be misunderstood and neglected. In planning his experiments, Mendel was probably interested in the ongoing debate about whether hybridization was the source of new species. His famous paper was called "Experiments on Plant Hybrids," rather than "The Search for the Laws of Heredity," but claims that Mendel only addressed contemporary issues in botany seem to ignore his work with bees and mice and his long-term interest in meteorology.

In the summer of 1854, Mendel began working with 34 different strains of peas; 22 kinds of peas were selected for further experiments. All the traits chosen for study were distinct and discontinuous and exhibited clear patterns of dominance and recessiveness. These preliminary experiments were apparently done to test various strains for constancy in the transmission of selected traits. These experiments were followed by years of tedious and painstaking work. It is virtually impossible to believe that such a prodigious amount of effort would have been undertaken without a well-conceived hypothesis and a precise experimental program. From 1856 to 1863 Mendel grew and tested over 28,000 plants and analyzed seven pairs of traits. From the beginning, Mendel apparently realized that the putative hereditary particles, or "elements," occurred in pairs, with one member of each pair coming from one of the parents. The first generation of hybrids displayed the trait from only one of the parents, i.e., the dominant trait. By breeding these hybrids Mendel proved that the corresponding recessive trait, which had seemingly disappeared, had been dormant rather than diluted or destroyed. What is now called Mendel's first law, or the law of segregation, refers to Mendel's proof that recessive traits reappear in predictable patterns in subsequent generations.

In plants, the hereditary elements responsible for dominant or recessive traits are passed on to each generation by means of pollen grains and egg cells. For the sake of simplicity, the elements carried by the pollen grain and the egg cell may be represented by the symbols A and a. If the elements do not influence or contaminate each other, or blend during fertilization, the dominant trait (A) and the recessive trait (a) could meet at random and produce offspring of the following kinds: 1 AA, 2 Aa, and 1 aa. Mendel actually expressed the results of breeding tests using two different traits in the form $A + 2Aa + a$. Although there are three categories, because AA and Aa are indistinguishable, the ratio of offspring would appear to be three A to one a. This form of representation might not have occurred to many nineteenth-century botanists, but it would presumably be entirely natural to an experimentalist also teaching high school mathematics. Because Mendel routinely used the symbol A where a modern geneticist would use AA, some historians have concluded that Mendel did think about hereditary particles occurring in pairs. Others argue that Mendel used the symbols A and a in reference to plants that breed true for the dominant and recessive trait, respectively. In other words, if Mendel used the symbol A to stand for "that which creates the dominant trait," the use of AA for the homozygote would be redundant. True-breeding homozygotes were of little interest to Mendel. As his experimental program indicated, it was segregation among the heterozygotes that established the new laws of genetics.

In his simplest experiments, Mendel crossed strains of peas that differed in only one trait. This series of experiments demonstrated the famous 3:1 ratio, but Mendel also studied the offspring of parental types that varied in two or three traits. Such bi- and trifactorial crosses were apparently used to test predictions based on the hypothesis that independent nonblending factors were transmitted from generation to generation. When Mendel analyzed the offspring of such matings, he found complex, but now predictable patterns of recombination.

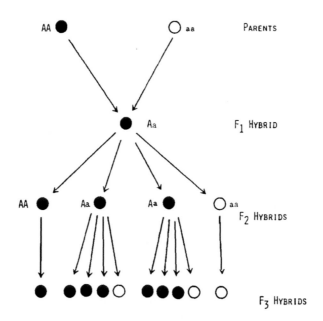

Mendel's hybridization experiments illustrating the 3:1 ratio. The diagram illustrates the results of crossing two strains of peas differing in a particular trait. The dominant trait appears in individuals with two A factors (genotype AA) or one factor A and one factor a (Aa). The recessive trait appears only in individuals with the genotype aa.

The law of independent assortment, also known as Mendel's second law, refers to this demonstration that the behavior of any pair of traits is independent of all other pairs of traits. Even after the rediscovery of Mendel's work, geneticists did not clearly explicate the second law until they were forced to cope with embarrassing deviations from independent assortment.

The results of Mendel's experiments on peas were discussed in a paper presented at the March 1865 meeting of the Brno Natural History Society and published in the 1866 issue of the Society's *Proceedings*, a journal that was distributed to over 100 scientific institutions in Europe and the United States. Mendel thought that his presentation at the Brno Natural History Society had encountered "divided opinion," but he was pleased to be asked to publish his lecture in the *Proceedings* of the society. Reflecting on the difficulty of reconciling his results with contemporary scientific knowledge, Mendel reflected on the brevity of exposition that was necessarily imposed on a public lecture. Perhaps the brevity of his presentation accounts for the subsequent confusion and misinterpretation. Some historians claim that Mendel was only interested in the role of hybrids in the generation of new species. Others argue that while Mendel was obviously interested in hybrids, he clearly told his audience that he had embarked on his experimental program in order to discover a "generally applicable law of the formation and development of hybrids." Moreover, he was confident that his discovery would provide important insights into "the evolutionary history of organic forms."

In order to reach experts on hybridization, including Karl von Nägeli (1817–1891), professor of botany at Munich, Mendel acquired 40 reprints of his article, "Research on Plant Hybrids." Some of these reprints were apparently still available as late as 1921, when a Japanese scientist making a pilgrimage to Brno acquired copies. But in the 1860s, neither the members of the Brno Natural History Society nor Nägeli showed much interest in Mendel's ideas. Firmly convinced that all hybrids produce variable offspring, Nägeli saw nothing of interest in Mendel's data on the mathematical relationships among discrete traits and he dismissed the analyses as "merely empirical, not rational." Obviously, Mendel and Nägeli differed in their understanding of the terms "rational" and "empirical." In his second letter to Nägeli (March 4, 1967), Mendel called the expression $A + 2Aa + a$ the "empirical simple series for two different traits." When this simple series was extended to any number of differences between the two parental plants, Mendel felt that he had "entered the rational domain." He felt that this claim was justified because his previous experiments proved that "the development of a pair of different traits proceeds independently of any other differences." However, he did attempt to verify the results obtained with *Pisum* by testing other plants.

Whatever Mendel thought he had discovered, he was certainly under the impression that it was important. In presenting his study to his colleagues, Mendel probably simplified his material as much as possible and used repetition and emphasis to convey his most important points. Analysis of the published form of his presentation suggests that he placed special emphasis on what he called the "law of combination of differing traits according to which hybrid development proceeds." Mendel also said that the main goal of his experiments was the establishment of a "generally applicable law of the formation and development of hybrids." Thus, some analysts believe that Mendel clearly understood the implications of his work.

Despite Nägeli's expressed reservations and the indifference of the scientific community, Mendel remained convinced of the fundamental value and universality of his work. His contemporaries and later critics, however, would say that Mendel's statistical approach to plant hybridization did not confirm, establish, or generate a general theory of inheritance, reproduction, development, and growth. Mendel had applied the methods of physics and mathematical thinking to a fundamental biological problem. His method generated, predicted, and tested specific numerical ratios, but the ratios were not of immediate interest to his

TABLE 1   A Summary of Mendel's Results

	Number of plants		Ratio of dominants/recessives
Trait studied	Dominants	Recessives	
Length of stem:			
tall/short	787	277	2.84:1
Position of flower:			
axial/terminal	651	207	3.14:1
Shape of pod:			
inflated/constricted	882	299	2.95:1
Color of pod:			
green/yellow	428	152	2.82:1
Shape of seed:			
round/wrinkled	5474	1850	2.96:1
Color of cotyledons:			
yellow/green	6022	2001	3.01:1
Color of seed-coat:			
gray/white	705	224	3.01:1

Results are tabulated from Mendel's results for the F_2 generation in his seven major series of experiments. For each trait studied, the characteristic listed first is the dominant one.

contemporaries. While Mendel's experiments clearly appear to disconfirm or falsify the theory of blending heredity, Mendel apparently interpreted his results cautiously and did not make explicit claims about the existence of material particles in the germ cells. The reproducible pattern of ratios of types found in the offspring of hybrids might be thought of as an "empirical regularity." Further testing and confirmation could raise the status of such a "regularity" to that of a new empirical law. Regular and predictable statistical patterns would not automatically reveal the nature of the mechanism generating such ratios, but they did support the concept that the inheritance of visible characters directly depended on the distribution of invisible factors.

Since Mendel was conducting further experiments with other species, Nägeli encouraged him to carry out more experiments on the hawkweeds (*Hieracium*). Although Nägeli is often cast in the role of the villain in the Mendel epic, his contemporaries knew him as a distinguished botanist. Nägeli had developed a theory of blending heredity, which involved the alleged fusion of the maternal and paternal *idioplasm* (hereditary material). If Nägeli were right, Mendel's allegedly pure types in the F_2 generation would have been mixtures. Further inbreeding, therefore, should reveal the presence of the hidden characteristics. Unwilling to consider the possibility that he might be wrong, Nägeli gave Mendel little attention or encouragement and no recognition in his own publications. The scientist who had earnestly promised in his 1853 survey of botanical research that "every new method brings us new results, every new result is a source of new methods" did not mention Mendel, or his methods and results, when he published his major treatise on inheritance and evolution in 1884. In 1905, Nägeli's student Carl Correns, who is best known as one of the rediscoverers of Mendel's laws, published the correspondence between Mendel and Nägeli.

Sent down the wrong garden path by Nägeli, Mendel spent 5 years trying to hybridize *Hieracium*. Despite his valiant attempts to reproduce, verify, and validate the orderly results he had obtained with peas, hawkweeds refused to conform to the laws of segregation and independent assortment. In a paper published in 1870, Mendel noted that he had chosen *Hieracium* because the genus contains a remarkable number of distinct forms. The results of his experiments with hawkweeds, however, did not conform to the rules deduced from his work with peas. By 1903 it became clear that Mendel had observed the paradigmatic case in peas. The aberrant results obtained with hawkweeds were due to peculiarities in the reproductive pattern of a genus in which parthenogenesis (apomixis) is not uncommon.

It is tempting to imagine that if Charles Darwin and Gregor Mendel had met and discussed their research interests and problems, a grand synthesis integrating genetics and evolution could have been achieved by the 1870s. Creating an imaginary scientific revolution, perhaps in a work of fiction or a "thought experiment," would, of necessity, require misinterpreting or misrepresenting complex issues embedded in nineteenth-century assumptions about generation, heredity, variation, and evolution. The modern synthesis in which Darwinism and Mendelism seem to be inseparable soulmates tends to obscure the nineteenth-century intellectual context that made it virtually impossible for Darwin and his contemporaries to see Mendelism as the solution to their problems. That is, evolutionary theory requires rethinking the problem of heredity. Instead of thinking about a process that produces offspring like the parents, modern geneticists had to think about a process that could transmit differences between parents and offspring and then stabilize some variations.

Interest in the qualities that parents transmit to their offspring is, of course, very ancient, but before the Mendelian revolution, naturalists did not see the study of heredity as a separate area of investigation. Heredity or inheritance was part of the broad domain referred to as generation, which included embryology, growth, development, and regeneration. Rather than ask how individual traits are transmitted, naturalists wanted to know how the new organism was constructed. On the other hand, although Darwin and Mendel proposed very different hypotheses about the nature of heredity, both were able to think about individual variations in terms of unit characters rather than taking a holistic view of the individual as a reflection of the "essence" of the species.

Apparently Mendel had a clear goal in mind at the outset of his tests of garden peas for purity of type and suitability as an experimental system. Mendel worked with simple, discrete traits in pure lines, and, rather than looking at only the interesting cases, he presumably counted all the progeny from his crosses. Indeed, the experimental design reported in Mendel's classic paper is so elegant that Sir Ronald A. Fisher (1890–1962), statistician and geneticist, suggested that the experiments were actually a confirmation or demonstration of a theory Mendel had previously formulated. Furthermore, Fisher claimed that the results that Mendel published were closer to the theoretical expectation than sampling theory would predict. Such results, Fisher insisted, could not be obtained without an "absolute miracle of chance." Considered one of the founders of modern statistics, Fisher introduced the concepts of variance and randomization as essential aspects of experimental design. Fisher's *Statistical Methods for Research Workers* (1925) and *The Genetical Theory of Natural Selection* (1930) became classic guides to planning and evaluating experiments. Rather than trying to discredit Mendel, Fisher publicized his statistical analysis in order to emphasize the power of abstract thought that led to Mendel's experimental strategy. Moreover, Fisher noted that as scientists of different generations read Mendel's paper, they would find what they expected to find and ignore that which did not confirm their own expectations. Thus, it is possible that Fisher himself had a particular agenda in mind when he set out to conduct his analysis of Mendel's data.

Some skeptics accepted Mendel's results as proof of the power of pure abstract thought rather than the Machiavellian manipulation of bad data, while full-blown cynics assumed that Mendel's experiments were entirely fictitious. More generous reviewers suggested that Mendel might have simplified his presentation, much as a teacher does in the classroom, so that his audience could understand his most fundamental concepts. Perhaps the good monk had been misunderstood by an assistant who deliberately "fudged" the results in order to simultaneously please Mendel and avoid some very tedious sorting and counting. Alternatively, Mendel might have gotten his problematic ratios by unconsciously misclassifying ambiguous cases or by counting particular traits until the results looked good. When the ratios looked bad, he presumably kept classifying samples until they improved. Knowing when experiments end could be considered an art form, or an important aspect of good experimental technique. In the 1980s, new statistical techniques suggested that Fisher's analysis had been faulty and that Mendel's data could be purged of the unwarranted aura of bias.

To the old charge that Mendel's ratios are too good to be true, skeptics added the argument that Mendel's independent assortment results are odd because some signs of linkage might be expected. That is, although the pea actually has seven pairs of chromosomes, the genes for the seven traits that Mendel selected do not lie on seven different chromosomes. In thinking about all these possibilities, perhaps we should remember where Mendel's garden was located and ask whether the prayers of this worthy monk were answered by ratios that came closer to theoretical expectations than secular statisticians would predict. Thus, the choice of traits, the arrangement of their genes, and the lack of complications due to linkage might well be part of the Mendel miracle.

MENDELIAN GENETICS: RECOGNITION OR REDISCOVERY?

Among the many attempts to explain the neglect of Mendel's work, the "obscure journal theory" might be considered the simplest. This theory is not, however, very compelling because the publications of the Brno Society were not sufficiently obscure. Copies of the *Proceedings* could be found in the libraries of many universities and scientific societies throughout Europe and America. Given the numbers of scientific papers published in the 1860s, it could be said that no paper published in any reputable journal in Mendel's era was as obscure as most papers published today. Accounts of several citings and sightings of Mendel's paper have survived. For example, in *Plant Hybrids* (1881), Wilhelm Focke (1834–1922), a German botanist, referred to Mendel's experiments on peas and the constant ratios found among his hybrid types. When George John Romanes (1848–1894), the Oxford biologist, asked Darwin for help with an article on hybridization, Darwin recommended Focke's treatise, which he considered unreadable because of its ponderous Germanic construction, but useful for its comprehensive bibliography. Another case, which ultimately led to a lifetime of devotion to Mendelian relics and documents, occurred in 1899 when a conscientious young man named Hugo Iltis began to prepare himself for graduate work in natural history by reading back issues of the *Brno Proceedings*. When Iltis discovered Mendel's paper, he excitedly showed it to his professor. Unfortunately, that learned man assured Iltis that Mendel's paper was of no importance, presenting as it did nothing but numbers and ratios. Later Iltis became a teacher at the Brno *Gymnasium*, the founder and custodian of the Mendel Museum, and the author of a classic biography of Gregor Mendel.

Such examples suggest that it is more proper to say that Mendel's paper was misunderstood rather than unknown. Naturalists who encountered Mendel's paper apparently saw it as a collection of "mere facts," unconnected with a theoretical structure that could relate its numbers and ratios to anything that naturalists would call a logical theory. Another factor operating against Mendel was that the traits he studied, while admirable in an experimental system, were not the kinds of qualities that interested most biologists at the time. Scientists, as well as plant and animal breeders, considered properties like size, vigor, strength, milk and beef production in cattle, wool in sheep, speed in race horses, or intelligence in humans the kind of traits worthy of study. Such traits, however, generally do not show simple Mendelian ratios because they are governed by many genes and influenced by environmental factors.

In a practical sense, therefore, the discipline now known as classical genetics ostensibly began not with the publication of Mendel's papers in the 1860s but at the beginning of the twentieth century with the rediscovery of the Mendelian laws of inheritance by Hugo de Vries (1848–1935), Carl Correns (1864–1935), and Erik von Tschermak (1871–1962). During the intervening years, developments in the study of cell division, fertilization, and the behavior of subcellular structures had established a new framework capable of accommodating Mendel's "ratios and numbers." Yet, given the constraints of the intellectual, social, and professional milieu of the early twentieth century, it was just as difficult for the "rediscoverers" to see Mendel's work in terms of its 1860s context as it is for us to interpret Mendel's ideas, and theirs, without the distorting lens of twenty-first-century concepts. The revival of interest in discontinuous or saltative evolution was probably a key factor in the approach the rediscoverers brought to the study of heredity. That is, if the sudden appearance of new characteristics led to the formation of new species, studies of the transmission of clearly defined characteristics, i.e., Mendelian breeding experiments, might furnish the key to a new science of heredity and evolution.

Some historians argue that the original story of triple independent discovery is a myth created by geneticists who have exaggerated the sudden enlightenment that allegedly occurred at the moment of rediscovery. Certainly de Vries, Correns, and Tschermak hoped to emphasize their creativity by claiming to have discovered the laws of inheritance *before* finding Mendel's paper. Having established the value of their own work, they could avoid a nasty priority battle with their own contemporaries by emphasizing Mendel's earlier work. Of this trio, de Vries had carried out the most extensive experimental program and was clearly the most disappointed at having credit for the discovery of the mathematical laws of inheritance ascribed to Mendel.

In the 1890s de Vries performed a series of hybridization experiments on poppies with distinctive black and white markings and observed the 3:1 ratio. Self-pollination of the F_2 generation showed that the white poppies bred true, while the black ones were of two kinds: some bred true, but others produced both black and white offspring. Like Mendel, de Vries had observed the 3:1 ratio in his F_2 hybrids, as well as the reappearance of the recessive trait, and had recognized that the F2 plants that displayed the dominant trait could be separated into two groups distinguished by studies of their offspring. De Vries discussed these experiments in his lectures at the University of Amsterdam as early as 1896, but he chose to postpone publication until he had collected additional evidence. The delay was to prove very costly to his reputation. By 1900 he had demonstrated his law of segregation in hybrids derived from some 15 different species.

At the turn of the century, the scientific world was interested in the concept of discrete hereditary factors and their independent segregation and recombination, but it was Gregor Mendel rather than Hugo de Vries who was honored as the founder of genetics. Correns and others, however, believed that de Vries exaggerated his claims of originality and distorted his role in the discovery of the Mendelian laws. How unfair it all was, de Vries complained, to have prepared his case so carefully and then find all the credit awarded to a man who had studied peas in the 1860s. In a postscript to one of his hybridization papers, de Vries acknowledged Mendel's work as "quite good for its time," but he considered his own work more significant in terms of its scope and its linkage to his own theory of intracellular pangenesis and August Weismann's (1834–1914) theory of the germplasm.

In *Intracellular Pangenesis* (1899), de Vries argued for a relationship between inherited characteristics and material particles or hereditary factors in the cell nucleus. Emphasizing the need to study traits as separate units, de Vries proposed a theory of inheritance based on factors he called *pangenes*. Although de Vries's pangenes were just as obscure and invisible as gemmules, unlike Darwin's gemmules, which supposedly circulated throughout the body, de Vries's hypothetical units remained intracellular. Relying on new developments in cell theory, de Vries proposed that pangenes were located in the nucleus of the cell, probably on the "chromatic threads." Presumably, pangenes multiplied within the cell nucleus during embryological development and passed into the cytoplasm where they became active. Microscopists had not yet observed such particles in the act of moving from the nucleus to the cytoplasm, but there was considerable evidence of the importance of the nucleus in cell function. For example, when protozoa were cut into pieces, only the part that contained the nucleus could regenerate into a new organism.

Unlike many of his contemporaries, de Vries focused his attention on a general theory of heredity and showed little interest in theories of embryological development. He was confident that a general theory of heredity would make it possible to explain the origin of species. In attempting to deal with ideas about evolution and the creation of new species, de Vries had been trying to transfer a characteristic from one species to another by hybridization. The kinds of traits that intrigued de Vries were those that exhibited large or discontinuous variations. In *Mutation Theory* (1901, 1903) de Vries expounded his theory that evolution occurs not through gradual change, as Darwin believed, but through large discrete steps called *saltations* or *mutations*. When *Mutation Theory* was translated into English, the section discussing Mendel's work was omitted.

Experiments with mutant and normal varieties of the evening primrose (*Oenothera*) seemed to support de Vries's idea that segregation did not occur in the case of "progressive mutations," which resulted in the formation of new species. Research on mutability, de Vries predicted, would not only solve the great puzzle of the origin of species, it would lead to practical methods of inducing a species to mutate in particularly desirable directions. At that point there would be no limit to the power scientists could hope to gain over nature. Later work with the evening primrose revealed that de Vries was actually seeing the effects of gross chromosomal rearrangements, rather than simple mutations. *Oenothera* has been called a bizarre and unique genus. The "progressive mutations" that de Vries thought of as leaps towards the formation of new species might more properly be called "monster anomalies." Unlucky in his timing, de Vries was also unfortunate in his choice of an atypical model system.

For the public record, Carl Correns and Erik von Tschermak were more generous towards Mendel than de Vries; indeed, they suggested the use of the terms *Mendelism* and *Mendel's laws*. But since de Vries had anticipated them, they did not have as much at stake in any potential priority war. Correns, who had studied with Nägeli, acknowledged that the task of discovering Mendel's laws in 1900 was much simpler than it had been in the 1860s. Growing hybrids of maize and peas for several generations and analyzing new developments in cytology apparently led Correns to think about the transmission of paired characters. Although he probably had evidence for the 3:1 ratio in his maize hybrids by the late 1890s, he later presented a rather dramatic account of how during a sleepless night the explanation of the 3:1 ratio came to him like a stroke of lightning. Critics, however, noted that he did not appreciate the significance of his statistical data until he read Mendel's paper. Despite his familiarity with Nägeli's studies on *Hieracium*, from which he might have learned of Mendel, Correns claimed that it was Focke's section on peas that belatedly led him to Mendel's paper. Shaken by the discovery that Mendel had anticipated him by 35 years, the unhappy Correns soon encountered evidence that Hugo de Vries was about to lay claim to the same discovery. While Correns was eager to praise Mendel, his attitude towards de Vries has been described as malicious. Correns persistently implied that de Vries had merely imitated Mendel and had tried to suppress references to his predecessor.

Surprisingly, although Correns is chiefly remembered as a "rediscoverer" of Mendel, he eventually came to the conclusion that the scope of Mendelism was not broad enough to provide a foundation for a general theory encompassing heredity, development, cytology, and evolution. His own experimental work on plants seemed to provide examples of non-Mendelian patterns of inheritance. Like many of his contemporaries, Correns was very interested in the possibility of the inheritance of acquired changes in plants and thought that the cytoplasm might play an important role in developmental processes.

Tschermak often complained that his contributions to genetics had been slighted. A botanist with a strong interest in practical aspects of plant breeding, Tschermak's training included work on commercial seed farms as well as advanced studies at the Universities at Ghent and Vienna. Tschermak's interest in the question of hybrid vigor led to a series of experiments on the growth-promoting effects of foreign pollen. On analyzing the results of these experiments, Tschermak discovered the 3:1 ratio for hybrids between peas with yellow and green cotyledons and for smooth and wrinkled seeds. He also noted the 1:1 ratio for backcrosses of peas with green cotyledons and hybrid pollen from the second seed generation. At this point, Tschermak claimed, he was led to Mendel's paper by the citation in Focke's book. When he read Mendel's paper, Tschermak was shocked to find that Mendel had already carried out similar, but more extensive experiments. Unlike Mendel's multigenerational studies, Tschermak had raised only two generations before he submitted his thesis for publication. In March 1900, Tschermak received another unwelcome surprise, a reprint of de Vries's paper "On the Law of Segregation of Hybrids." This bad news was followed by a copy of Correns's paper "Gregor Mendel's Law." Tschermak quickly prepared an abstract of his thesis and sent copies of his article on artificial hybridization to de Vries and Correns to establish himself as a participant in the rediscovery of Mendel's laws, but his claim has been regarded as marginal, and some critics argued that he did not really understand Mendel's laws.

The apparently simultaneous discovery of Mendel's laws by de Vries, Correns, and Tschermak suggests that by 1900 "rediscovery" had become inevitable. Indeed, others, such as William Bateson (1861–1926), were working along similar lines. Before assuming the role of champion of Mendelian genetics, Bateson had been an evolutionary morphologist. Many aspects of Bateson's changing attitudes towards evolution, variation, embryology, genetics, and materialism are ambiguous, but it is clear that his early efforts to deal with fundamental issues in evolution and embryology had produced more frustration than enlightenment. After some unproductive attempts to use the study of embryological stages as a way to understand phylogenetic pathways in evolution, Bateson turned to the problem of heredity and variation, especially the role of discontinuous variants in evolutionary change.

During the summers of 1883 and 1884, Bateson studied the classification and development of marine organisms under the direction of the American biologist William Keith Brooks (1848–1908). When Bateson returned to Cambridge as a Fellow of St. John's College, he took up the study of variation. In pursuit of relevant materials, Bateson traveled as far as Russia and Egypt. He was quite disappointed by his failure to find evidence of a causal relationship between changing environments and the appearance of new varieties. In 1908 he was appointed professor of biology at Cambridge University, but he soon left to become director of the John Innes Horticultural Institute. Research on the inheritance of variation remained his principal area of interest, but during the 1890s he essentially abandoned his fieldwork. Instead, he established an experimental breeding program that would allow him to analyze discontinuous variations, the relationship between parental traits and those found in the offspring, and traits found in the offspring of experimental hybrids. Although disease and bad weather disrupted many of Bateson's experiments, these studies led to the publication of *Materials for the Study of Variation* (1894), a remarkable book calling for systematic investigation of the inheritance of discontinuous variations. Like de Vries, Bateson found the Darwinian concept of selection acting on small, continuous variations quite unsatisfactory. Convinced that Darwin had made a totally "gratuitous assumption," Bateson argued that discontinuous variation was an important and fundamental property of living beings, independent of natural selection.

By the time Bateson read Mendel's paper, he was already thinking along the lines of traits as discrete units. A reprint sent by de Vries led Bateson to the 1865 Brno *Proceedings* and a new mission in life as the passionate advocate and popularizer of Mendelian genetics. It should be noted, however, that Bateson did not believe that Mendelism required a belief in the existence of a specific material substance. Long after Mendelism gave way to the modern chromosomal theory of genetics, Bateson continued to hope for the establishment of a general theory that would encompass both inheritance and embryological development. Despite some ambiguity about Bateson's personal beliefs concerning materialism, his struggle against the forces tending to separate embryology and genetics has been taken as evidence of his continued attachment to a more conservative and holistic philosophy of biology.

Many of the terms now used by geneticists were introduced by Bateson, including *genetics* (from the Greek for descent), *allelomorph* (which later became allele), *zygote*, *homozygote*, and *heterozygote*. After coining the term "genetics" in 1905, Bateson introduced the term to a broader audience in his presidential address on the new Mendelian science at the third International Conference on Hybridization and Plant Breeding. Impressed with Bateson's predictions, when the secretary of the Society edited the proceedings for publication, he used the term "genetics" in the title and included a brief account of Mendel's life and work. Proclaiming the potential of Mendelism for both theoretical and applied, practical programs, Bateson suggested that Mendel's ratios would provide the kind of mathematical order found in the stoichiometric rules of chemistry. Through research in genetics, Bateson predicted, biology would become a science of predictable and quantifiable processes, like physics and chemistry. Perhaps eventually, the laws of genetics would lead to the establishment of a "periodic table" of Mendelian factors. In America, Mendelian genetics was quickly incorporated into the agricultural sciences as a promising approach to breeding, selection, and hybridization. Preaching the gospel of Mendelism to agricultural scientists, hybridizers, and animal breeders, Bateson asserted that it was important for scientists to establish "*particular* knowledge of the evolution of *particular* forms" rather than continue to search for "*general* ideas about evolution."

In 1909 the Danish botanist Wilhelm L. Johannsen (1857–1927) introduced the term *gene* to replace older terms like factor, trait, and character. Challenging biometricians, Darwinians, and the followers of August Weismann, Johannsen introduced his own terminology of heredity in order to eliminate what he considered the misconceptions associated with terms like germplasm. To clarify various aspects of the new science of genetics, Johannsen coined the terms *phenotype* and *genotype*, which are now used to indicate the appearance of the individual and its actual genetic makeup, respectively.

Further experiments carried out by scientists in Europe and America extended Mendel's work and proved that Mendel's laws were applicable to animals as well as plants. Still, these studies did not resolve questions concerning the relationship between the inheritance of traits and evolutionary change. Indeed, orthodox Darwinists tended to view Bateson's approach as a direct challenge to the theory that evolution proceeded by the action of natural selection on continuous variations. Another assault on simplistic Mendelism was launched by scientists using a statistical approach to inheritance known as *biometrics*. Much ink was spilled in a wasteful war of words that raised doubts about the relative importance of continuous and discontinuous variations. Bateson counterattacked with a new book, *Mendel's Principles of Heredity* (1909). The text included a translation of Mendel's paper and Bateson's assertion that Mendel's laws would prove to be universally valid. While Bateson admitted that the biometricians might hope to contribute to the development of statistical theory, he warned readers that in the truly significant field of heredity, their efforts "resulted only in the concealment of that order which it was ostensibly undertaken to reveal." Deluding themselves and others, he warned, the biometricians were delaying recognition of the value of Mendelism.

Charles Darwin's cousin Francis Galton can be called the founder of *biometry*, the application of statistics to biological problems. Like Charles Darwin, Galton inherited enough money to abandon medical studies and the unpleasant prospect of working for a living. Discovering a talent for writing, he published some popular works about his travels in the Sudan, Syria, and Africa. Like Mendel, Galton devoted much of his energy to the study of weather conditions. Galton was instrumental in establishing the Meteorological Office and the National Physical Laboratory, but his primary interest was the inheritance of complex physical and mental characteristics in humans. This led to what he called anthropometric research, that is, quantitative and statistical studies of human populations, and the establishment of the Biometric Laboratory at University College, London. The use of fingerprints for identification was made possible by this approach.

Unlike Darwin, Galton was a talented mathematician who thought the problem of heredity could be solved through the statistical analysis of the distribution of traits in large populations. Like Francis Bacon, Galton believed in the virtues of collecting and analyzing data and hoped that the application of this methodology would eventually produce a rational society guided by a scientific priesthood. Galton seemed to think that almost any question would yield to statistical treatment. For example, as a test of the efficacy of prayer he analyzed mortality patterns among members of the royal family. Pious friends and relatives were distressed by his statistical demonstration that the prayers of entire kingdoms had no effect on royal longevity. Similarly, he compared the rates of disasters for ships with and without missionaries. He also tried to design a quantitative scale for the measurement of beauty and love. Not surprisingly, the Galtonian approach has been criticized as a rather narrow-minded commitment to measuring whatever heritable traits are measurable and mismeasuring those that are not.

Glorying in the evidence of progress in science and industry, Galton urged his countrymen to direct their future evolutionary path by applying the tools of science and mathematics to the study of human heredity and the origins of "natural ability." If human beings understood the laws of inheritance, Galton promised, the human race could be improved by getting rid of the "undesirables" and encouraging the multiplication of the "desirables." Statistical analyses of biographical encyclopedias supported his belief that distinguished individuals were most likely to be the offspring of distinguished families (not unlike the fruitful family tree that had produced Erasmus Darwin, Charles Darwin, and Francis Galton). Intellect, character, and talent, just as much as physical features, were, therefore, the products of heredity. In *Hereditary Genius* (1869) and other writings, Galton argued that genius was a conspicuous trait just as susceptible to measurement as height and eye color.

Statistical comparisons of eminent men in America and England were used to dismiss the possibility that the production of "men of genius" was related to extrinsic conditions like social advantage or disadvantage. Taking a hard position in the "nature versus nurture" controversy, Galton contended that intelligence was determined by heredity and that attempts at education and other remedial methods could have little or no effect. He argued that human races appeared to remain stable even after migrations and adaptation to life under altered conditions. Children must, therefore, reflect ancestral traits rather than their environment. His arguments were apparently quite persuasive, at least to men who shared his background, status, and social milieu, as indicated by Darwin's ready acceptance of Galton's belief that genius was inherited. Prior to assimilating Galton's work on hereditary genius, Darwin had assumed that all normal men were essentially the same in intellect but varied greatly in diligence, patience, and dedication to hard work.

Galton saw variation and heredity as different aspects of a mechanism for passing on a collection of characteristics from a series of ancestors and combining them through sexual reproduction, a process that maintained variation in the population. The mean of a population, therefore, reflected "stability of type" from generation to generation. Ancestral characters, according to Galton, were transmitted by particles that could be divided and passed on in ever-smaller but still effective units. Unable to explain just how this process might work, Galton focused on the statistical issues and attempted to deduce the laws of inheritance from studies of the distribution of deviations from the average in successive generations. Ancestral contributions diminished with each succeeding generation, but they were not lost. Just as certain poisons or drugs were very potent even in minute doses, tiny remnants of ancestral materials could make their presence known under appropriate circumstances.

In 1883, long after he had worked out the principles of the mathematical treatment of human heredity, Galton coined the word *eugenics* from a Greek root meaning "noble in heredity" or "good in birth." But in the wake of the many twentieth-century atrocities committed in the name of eugenics, from American sterilization laws to the murderous racist doctrines of the Nazi regime, the term later assumed an irrevocably evil connotation. The goal of Galtonian eugenics was to improve the human race by applying the principles of scientific breeding to human beings. Ultimately, this was supposed to lead to the triumph of the "more suitable races" or bloodlines over those disparaged as "less suitable." Despite their love of statistics, Galton and his followers seemed to forget that one half of the human race, when tested for any measurable trait, would fall below average. On a more intimate and immediate note, much to Galton's sorrow and regret, his own eugenically propitious marriage failed to produce any evidence of the heritability of genius. Ironically, probably as the result of his youthful indiscretions, Galton and his wife had no children. Admirers of Galton have expressed hope, perhaps facetiously, that with improvements in the techniques for recovering DNA from tissue samples, his unique genotype might be recovered and cloned by the nuclear substitution method that produced Dolly the sheep in 1997.

As early as 1902, the mathematician Udny Yule (1873–1949) had suggested that the same mechanism could explain both simple Mendelian patterns and complicated cases of blended inheritance if large numbers of factors were assumed to act together. Mendelians and biometricians, however, were not yet ready for any compromises. Gradually it became apparent that the warring parties could be united by the "multiple gene hypothesis." Sir Ronald A. Fisher expressed the union of the two schools of thought in 1918 in his paper "The Correlation Between Relatives on the Supposition of Mendelian Inheritance." His book *The Genetical Theory of Natural Selection* (1930) provides the classic introduction to population genetics, a discipline based on the statistical approach to patterns of inheritance. Slow, adaptive evolution, Fisher argued, could be explained in terms of the mathematics of shifting relative frequencies of genes in populations. Evolution by means of Darwinian natural selection could, therefore, be reconciled with Mendelian genetics. But because the statistical methods used by early population geneticists ignored the implications of complex interactions among different genes and between genes and the environment, this approach has been disparagingly referred to as "beanbag genetics."

CYTOLOGY: THE STRUCTURAL BASIS OF MENDELISM

As late nineteenth-century cytologists attempted to link cell theory to theories of evolution and heredity, they became increasingly convinced that the cell nucleus was very different in form and function from the cytoplasm. New methods of preparing and staining biological preparations made it possible to see the nucleus and the special staining bodies known as chromosomes during cell division. Such studies supported the concept that the nucleus played a key role in growth, development, and heredity. Based on these observations, August Weismann (1834–1914) enunciated the guiding principles that provided a new theoretical framework for studies of cytology and heredity. The first was his theory of the continuity of the germplasm and the second was his prediction of, and rationalization for, the reduction division of the chromosomes during the formation of the germ cells.

For his inaugural lecture at the University of Freiburg, Weismann discussed Darwinism, which he saw as comparable to the Copernican theory in its impact on human thought. Even more than Darwin, Weismann was committed to the concept that evolution must occur through gradual and small steps so that many coadaptations could occur. Changes that appeared to be large and abrupt saltations, he argued, were really the outcome of many previous small, subtle changes. In 1864, Weismann experienced the first episode of a painful eye disorder that eventually forced him to abandon experimental work. It was, therefore, as a theoretician that he made his greatest, and remarkably diverse, contributions to science.

Examining the major biological problems left in the wake of the Darwinian revolution, Weismann realized that variation and inheritance were the subjects most in need of logical analysis and reassessment. Darwin had provided a new mechanism for evolution; biologists must provide a mechanism of inheritance. Although Weismann was intensely interested in evolution, he realized that the stability of inheritance between generations was the most salient fact of heredity. Once that was understood, variation could be treated as a special case, or corollary. In other words, heredity must first be studied at the level of the cell and the individual, not in terms of evolving species or populations. When Weismann began his research, he accepted the prevailing theory of the inheritance of acquired characters, but as he struggled to find a source of genetic variation he gradually abandoned many assumptions about the relationship of "soft inheritance" to evolution. By the turn of the century he was convinced that understanding the germplasm would provide the key to evolution by natural selection.

A critical examination of existing facts and theories concerning heredity, cell structure, reproduction, plant and animal breeding, and his own observations led to the suggestion that the precursors of the germ cells might be distinguishable from other cells at a very early stage in development. Moreover, in the course of development, maturation, and adult life, the descendants of the germ cells might remain separate and distinct from the other cells of the body. According to Weismann's theory of the continuity of the germplasm, there is a continuous line of descent from the germ cells of one generation to another. In other words, every organism is made up of two distinct constituents: the *germplasm* and the *somatoplasm*. While the germplasm was continuously handed down to successive generations, the somatoplasm died away with the passing of each individual like the leaves of a tree in autumn. Sexual reproduction must, therefore, serve as a source of genetic variability. If the germ cells were not derived from the body of the parent, but from the ancestral germplasm, the concept of the inheritance of acquired characteristics could not play a role in heredity or evolution.

The germplasm was not presented as a Platonic idea to be imposed on mere matter, but as a specific component of the germ cell whose chemical and physical properties, especially its molecular structure, allowed the fertilized germ cell to become a new individual. The chemical nature of this putative cellular component was obviously unknown when Weismann published *The Germ-Plasm; a Theory of Heredity* (1893), but his theory provided a framework for studies of many aspects of cell behavior, particularly the division of the nucleus and the chromosomes. Whereas *mitosis*, the division of the somatoplasmic cells, required keeping the chromosome number constant, Weismann predicted that a special reduction division must occur during the maturation of the germ cells in order to divide the number of chromosomes by half. Fertilization would restore the normal chromosome complement when the egg and sperm nuclei fused. By 1892 Weismann could feel secure in the conviction that his hypothesis had been vindicated by microscopic observations of the complicated steps involved in cell division. Indeed, Weismann insisted, his theory of heredity was not "theoretically superficial and cytologically impractical" like Darwin's, but was a logically based biochemical model. Ultimately, Weismann thought that the laws of inheritance would be explicable in terms of molecular movements, as indicated by his laudatory references to the doctrine of Hermann von Helmholtz (1821–1894), physicist, physiologist, and physician, that "all laws must be reduced in the last analysis to laws of motion."

By the turn of the century, some scientists had begun to suspect that there was a relationship between the classical factors tabulated in breeding experiments and the behavior of the chromosomes in mitosis and meiosis. Chromosomes were known to occur in homologous pairs that conjugated and then separated during the formation of germ cells. The process of chromosome separation during meiosis is properly called disjunction. (Although this is the process that leads to the segregation of factors or genes into different germ cells, disjunction of chromosomes and segregation of factors should not be confused.) The chromosomes of the germ cells were thought to fuse during fertilization in order to reconstruct the whole essence of the species. Chromosome pairing was thought to provide the rejuvenating effect that many nineteenth-century observers attributed to sex, but the maternal and paternal chromosome set were thought to separate as a group. By 1910, some biologists were quite sure that the behavior of the chromosomes during gamete formation and the process of fertilization indicated that the paired factors could be on the paired chromosomes contributed by egg and sperm. This cytological approach was supported by the work of Walter S. Sutton (1877–1916), Theodor Boveri (1862–1915), Nettie M. Stevens (1861–1912), and Edmund B. Wilson (1856–1939).

In 1889, Boveri described experiments that seemed to prove that the nucleus controlled development. Boveri fertilized enucleated egg fragments of *Sphaerechinus granularis* with the sperm of another sea urchin, *Echinus microtuberculatus*. Because of differences in the early development in these species, Boveri could tell by observing the skeletons of his experimental larvae whether the nucleus (from the sperm) or the cytoplasm (from the enucleated egg fragment) controlled development. According to Boveri, the larvae resembled only the male parent. Thomas Hunt Morgan translated this paper into English for *The American Naturalist* and attempted to test Boveri's conclusions. Apparently unable to reproduce some of Boveri's experiments, Morgan remained convinced that the cytoplasm controlled embryonic development.

In another series of experiments, Boveri confirmed previous reports that eggs fertilized by two spermatozoa did not develop into normal larvae. He used this system to explore the individuality of the chromosomes and their relationship to development. If the four blastomeres of these "dispermic eggs" were separated by placing them in calcium-free seawater, each cell seemed to have abnormal numbers of chromosomes. Noting that each embryo developed somewhat differently, Boveri investigated what he called the "physiological properties" of the chromosomes in germ cell formation and embryological development. These experiments were briefly described in "Multipolar Mitosis as a Means of Analysis of the Cell Nucleus" (1902) and more fully elaborated over the next 5 years. Having analyzed hundreds of specimens, Boveri reiterated his conclusion that the nucleus controls inheritance. Because cells apparently needed a full complement of chromosomes for normal development, Boveri concluded that the individual chromosomes must possess different qualities. It was impossible for him to identify the role of individual chromosomes, but it was clear that they were not functionally equivalent. The question of the individuality of the chromosomes was complicated by the possibility that between cell divisions, when the well-defined chromosomes were transformed into an apparently formless tangled mass of chromatin threads, their individuality might be lost. The reconstituted chromosomes that appeared at the next cell division might, therefore, be quite different individuals.

Inspired by observations of the behavior of the sex chromosomes and Boveri's experiments on double-fertilized eggs, William Sutton turned to the analysis of the chromosomes of grasshoppers for signs of constant and significant morphological differences among the chromosomes. Sutton had been a student of Clarence E. McClung (1870–1946), a pioneer in the study of male gametogenesis and the sex chromosomes. In 1902 McClung suggested that a special chromosome, known as the "accessory chromosome," was responsible for determining maleness or femaleness. While assisting McClung in his studies of the relationship between sex determination and the accessory chromosome in the lubber grasshopper, Sutton discovered that the accessory chromosome was present in only half of the developing spermatocytes. Recognizing the significance of the 1:1 ratio, McClung suggested that this chromosome was the determinant of sex. After leaving McClung's laboratory in Kansas, Sutton joined Wilson's group at Columbia University.

While still a student, Sutton realized that the distribution of the chromosomes in meiosis might provide the cytological basis of the segregation of Mendelian factors. According to Sutton's hypothesis, the regular correspondence between the apparent size and shape of the "mother series" of chromosomes and the "daughter series" suggested the existence of morphologically distinct individual chromosomes. In "The Chromosomes in Heredity" (1903), Sutton predicted that the behavior of pairs of chromosomes, especially their separation during the reduction division in gamete formation, constituted the "physical basis of the Mendelian law of heredity." Furthermore, Sutton suggested, the random assortment of different pairs of chromosomes could account for the independent segregation of pairs of genes. If, as suggested by Boveri and Sutton, each chromosome carried a different portion of the genetic and developmental program, researchers might be able to demonstrate nuclear control by determining how a specific chromosome determined sex in a wide variety of species.

Between 1900 and 1910, cytologists and embryologists puzzled over the role of the cytoplasm, the nucleus, the chromosomes, and the factors that established sexual differentiation. General explanations of sex determination included hypotheses that implicated external factors (such as nutrition, or temperature), those that postulated internal causes, and others that involved a combination of environmental and internal factors. After the rediscovery of Mendel's laws, several biologists attempted to explain sex as a Mendelian character. For example, Correns suggested that sex could be thought of as a unit character, and that the male was a hybrid in which maleness was dominant to femaleness. Matings between a heterozygote and a homozygous recessive individual would produce the expected 1:1 ratio. This theory could not, however, explain the appearance of male aphids after generations of females produced by parthenogenesis, or why unfertilized eggs produced male bees and fertilized eggs produced female bees.

Cytological studies of peculiar "unpaired" chromosomes suggested that the presence or absence of these "accessory," or "X" chromosomes might determine sex in various insects. The term "X" was first applied to a chromosome in 1890 when Hermann Henking (1858–1942) used it to designate a particular "chromatin element" identified during spermatogenesis. Studies of the sex chromosomes provided a model system that made it possible to establish the relationship between specific chromosomes and specific characters. Moreover, studies of the sex chromosomes were considered important aspects of the controversy over whether the nucleus or the cytoplasm controlled heredity and development. The idea that sex might be determined by such nuclear structures was a critical departure from the dominant view of the 1890s, which held that sex was determined by environmental factors. Eventually, the chromosomal theory replaced the old theory that environmental factors determined sexual development.

Nettie Stevens (1861–1912) is of special interest as an example of a very rare species, the early twentieth-century American woman scientist. A native of Vermont and a graduate of Westfield State Normal School in Massachusetts, Stevens was able to earn both a B.A. (1899) and an M.A. (1900) from Stanford University. Her doctoral research was performed at Bryn Mawr. Although Bryn Mawr was a small women's college, during this time period two of America's leading biologists, Edmund B. Wilson and Thomas Hunt Morgan, were associated with its faculty. Wilson went to Columbia in 1891, but he remained in close contact with Morgan, who taught at Bryn Mawr until 1904. After earning her Ph.D. in 1903, Stevens won a fellowship to study abroad and work with Boveri at the University of Würzburg. An award from the Carnegie Institution of Washington made it possible for her to conduct research on sex chromosomes at Bryn Mawr, free of teaching responsibilities, from 1904 to 1905. During this period she began publishing her reports on sex determination in insects.

Stevens saw her work as an investigation of the "histological side of the problems in heredity connected with Mendel's Law." Studies of the common mealworm demonstrated that the somatic cells of the females contained 20 large chromosomes, while those of the male contained 19 large chromosomes and one small one. Sex determination might, therefore, be caused by some unique property of an unusual pair of chromosomes. The male insects produced two kinds of spermatozoa: eggs fertilized by those containing 10 large chromosomes gave rise to females, while those fertilized by spermatozoa with the small chromosome developed into males. From these observations, Stevens concluded that there must be some intrinsic difference affecting sex in the chromatin contributed by the two kinds of spermatozoa. (Conventionally, the large sex chromosome is referred to as X and the corresponding small chromosome is Y; females are then XX and males are XY.) Within a few years Stevens had discovered the mode of sex determination in over 50 species of insects and one primitive chordate. Among the insect species, about 85% had "unequal partners" (X and Y), and the others had the single accessory chromosome (XO). Based on these data, Stevens concluded that sex was determined not by an accessory chromosome, but by the characteristics of a specific pair of chromosomes. This chromosomal method of sex determination, Stevens wrote, was consistent with Mendel's laws of heredity.

When Nettie Stevens's work is not totally ignored in modern textbooks, it is usually presented as if she had worked with Wilson or had confirmed his previous discovery. (Stevens did not even get an entry in the *Dictionary of Scientific Biography*, but she does appear in *Notable American Women*.) Shortly after seeing Stevens's paper, Wilson published his own study of sex determination in several insect species. In some species, the male has one less chromosome than the female, but in most bugs the males had a large and a small sex chromosome. In addition to confirming McClung's hypothesis about the relationship between sex and the accessory chromosomes, Wilson corrected previous suggestions that the large accessory chromosome was the male determinant. The small Y chromosome, which was always found in the male, was absent in some species. Wilson attempted to relate sex determination to physiological differences arising from different chromosome constitutions. He believed that the chromosomes determined the sex of the individual and that the smaller chromosome of the male represented a degenerate female chromosome.

The work of Boveri, Sutton, Stevens, and others suggested that the chromosomes were individuals in the morphological and the functional sense. Just how these individual entities differed and what unique hereditary qualities they carried could not be ascertained by cytological studies alone. But both Boveri and Sutton commented on the way in which the facts from cytology and Mendelian breeding studies complemented each other in supporting the hypothesis that the heredity Mendelian factors were actually constituents of the chromosomes. Sutton also predicted that each chromosome must carry many Mendelian factors because the number of characters that had already presented themselves for genetic analysis was much greater than the number of chromosomes in any cell nucleus. This meant that characters on the same chromosome should be inherited together, rather than independently. De Vries went further and predicted that during the formation of germ cells, some pangenes could be exchanged between pairs of homologous chromosomes, which he called "nuclear threads." It is unclear whether he thought that actual pieces of the chromosomes moved, or whether the chromosomes were simply vehicles, like buses or trains, that carried pangenes and remained intact when their passengers departed. At the time there was little evidence of linkage, but as more and more traits were closely studied by breeding tests, it became obvious that independent assortment was not a universal phenomenon. By 1905 there was good evidence that certain genes were almost always transmitted together.

Although the Boveri-Sutton hypothesis seemed to explain many aspects of cytology and heredity, it was not universally accepted. Wilson, however, was an enthusiastic proponent. Having previously come to the conclusion that the nucleus controlled development and heredity, Wilson accepted the Sutton-Boveri hypothesis as evidence for the relationship between Mendelian genetics and the chromosome. As the highly regarded author of *The Cell in Development and Heredity* and the mentor of many prominent scientists, Wilson was instrumental in converting several skeptical scientists into advocates of the chromosome theory. Praising Boveri for linking cytology, embryology, and genetics, Wilson dedicated his 1896 edition of *The Cell in Development and Inheritance* to him. Moreover, assuming that the cell was the primary unit of life, development, and heredity, Wilson reasoned that the development of all organisms, unicellular and multicellular, was essentially the same. It would, therefore, be possible to apply lessons learned from experiments on simple protozoans to the development of embryos. Thus, Wilson argued that the discovery that when certain protozoans were divided, only the pieces that contained the nucleus could regenerate should also be taken as evidence that the nucleus must control inheritance.

THOMAS HUNT MORGAN AND THE *DROSOPHILA* GROUP

The experimental program that was most successful in exploiting correlations between breeding data and cytological observations was established by Thomas Hunt Morgan (1866–1945) in the famous "fly room" at Columbia University. Morgan and his associates—Alfred H. Sturtevant, Calvin B. Bridges, Hermann J. Muller, Curt Stern, and others—proved that genes are located on the chromosomes in a specific linear sequence. Sturtevant remembered the days in the fly room as a time of camaraderie and cooperation, but historians have focused on signs of tension, interpersonal friction, and differences in scientific, political, and philosophical issues.

Morgan was a zoologist with broad interests, including cytology, descriptive and experimental embryology, and evolutionary theory, but he is primarily remembered for his contributions to the chromosome theory of heredity, the work for which he won the Nobel Prize in 1933. Ironically, while the original support for the chromosomal theory of inheritance came largely from the work of Boveri and Wilson, Morgan initially resisted this concept. Historians have argued, however, that since each of these scientists began his career as an embryologist, the evolution of the theory of the gene should be analyzed from the viewpoint of embryology. Focusing on embryology rather than hybridization studies emphasizes the historical controversy about the relative importance of the nucleus and the cytoplasm in the control of development and heredity. Many embryologists thought that the chromosome theory revived preformationism by assuming that adult characteristics were somehow preformed in the nucleus. On the other hand, even though the cytoplasm of the cell might appear homogeneous under the most powerful microscopes available, it might contain a complex, but invisible physiological matrix. In contrast to Wilson, who believed that the chromosomes were the critical determinants of hereditary and development, Morgan assumed that the cytoplasm determined heredity and developmental phenomena. Morgan entered genetics as an embryologist, and later in his life he returned to those problems of marine embryology that he had abandoned for his *Drosophila* studies. The *Drosophila* experiments allowed him to separate the problem of heredity (transmission) from that of development, a position that Wilson adopted during the 1890s. Presumably, Morgan would have been pleased to see the reunion of genetics and embryology celebrated at the end of the twentieth century with the 1995 award of the Nobel Prize to three developmental biologists for work that led to the discovery of a family of genes involved in determining the architecture of the body during embryological development in *Drosophila*.

Between 1903 and 1905, Morgan published several articles about sex determination in which he argued that, contrary to the conclusions of Stevens and Wilson, sex is determined by internal factors rather than chromosomes. His skepticism about the universality of Mendelian genetics was exacerbated by Lucien Cuénot's (1866–1951) work on coat color in the house mouse. Cuénot's early research in comparative anatomy had encompassed studies of a broad range of living and extinct species, but during the 1890s he became interested in the mechanism of sex determination. When Mendel's laws were rediscovered, Cuénot planned a series of ingenious crossbreeding experiments on mice that proved that Mendel's laws were applicable to animals as well as plants. In 1905, Cuénot discovered that when a yellow-haired variant of the house mouse known as Agouti was bred with normal mice, the offspring included black and yellow mice. Matings between these yellow mice always produced litters of black and yellow mice. These matings never produced the expected Mendelian ratios, because all the yellow mice were hybrids. It seemed to be impossible to establish a pure-breeding strain of Agouti mice. To the dismay of Mendelians, Cuénot's finding of a non-Mendelian 2:1 ratio seemed to represent what philosophers of science call a "monster anomaly." Morgan argued that the yellow trait never dissociated from black and that this disproved the Sutton-Boveri-Mendelian model. Later studies, however, proved that the Agouti gene is lethal in homozygous embryos. In addition to providing exceptions to Mendelian ratios, these findings called into question the previously unexamined assumption that embryos of all possible genotypes were equally viable and would appear in the final count of phenotypes.

Thomas Hunt Morgan was born in Lexington, Virginia. He attended the State College of Kentucky and was awarded his B.S. in zoology in 1886. As a graduate student at Johns Hopkins University, Morgan studied biology, anatomy, physiology, morphology, and embryology. Both Wilson and Morgan received doctorates from the Johns Hopkins University in the laboratory of William Keith Brooks, and both conducted research at the Naples Marine Biology Station. In 1885 Wilson became the first professor of biology at Bryn Mawr. When he left to go to Columbia University, Morgan took his place. In 1904 Morgan followed Wilson to Columbia, but he remained involved in the work of former colleagues at Bryn Mawr, including Nettie Stevens and the physiologist Jacques Loeb (1858–1924). During his frequent trips to Europe, Morgan established friendships with eminent embryologists, such as Hans Driesch. Morgan's text *The Development of the Frog's Egg; An Introduction to Experimental Embryology* (1897) was the first major book on the subject to be published in English. Morgan left Columbia in 1928 to establish the Division of Biological Sciences at the California Institute of Technology, where he remained until his death in 1945.

Until a visit to Hugo de Vries's garden aroused an interest in mutation theory, Morgan's primary research area was experimental embryology. Like Bateson and de Vries, Morgan was skeptical of the Darwinian doctrine of natural selection and the primacy of continuous variation as the raw material of evolution. This interest in mutations led to breeding experiments in various animals, including mice, rats, pigeons, and lice. Finally, in the fruit fly, *Drosophila melanogaster*, Morgan discovered the ideal system for elucidating the complex relationships among genes, traits, chromosomes, and the statistical data produced by breeding experiments. Indeed, Morgan exploited *Drosophila* so effectively it was later said that God had created this Lilliputian creature especially for him. Actually, other scientists, including W. E. Castle at Harvard, W. J. Moenkhaus at Indiana University, and Nettie Stevens at Bryn Mawr, were breeding *Drosophila* in the laboratory some years before Morgan adopted it as his model system. In the 1980s, after being eclipsed by molds, bacteria, and bacteriophage, *Drosophila* played a leading role as experimental animal of choice in studies of molecular genetics, developmental biology, neurobiology, and other basic problems in metazoan biology.

Drosophila melanogaster, the species of fruit fly most commonly used by geneticists, has a life cycle of about 2 weeks, breeds prolifically, mates enthusiastically and indiscriminately, and is easily maintained in the laboratory. It is only about 1/8 of an inch in size, small enough so that thousands can be kept in a few bottles on inexpensive banana mash, but large enough to be easily studied with a low-powered microscope. A single pair of parents will quickly produce hundreds of offspring. More important, the fruit fly has only four pairs of chromosomes per nucleus and has scores of easily recognizable, even strikingly bizarre, heritable traits. Although *Drosophila* seemed a promising model of saltative evolution, Morgan discovered that even the most preternatural mutations did not establish a new species as de Vries had suggested.

In 1910 Morgan sent a paper to the journal *American Naturalist* in which he argued that the chromosomes could not possibly carry Mendelian factors, because, if they did, characters on the same chromosome would have to "Mendelize" together. Despite Morgan's objections, the chromosomal theory of heredity had strong partisans in his laboratory group. Almost all of Morgan's early *Drosophila* researchers had been undergraduates at Columbia University, where Wilson's influence was pervasive. Experiments on fruit flies, especially the remarkable white-eye mutant, played a large role in converting Morgan into a supporter of the chromosome theory and Mendelian genetics. Members of the "fly group" were also actively involved in convincing Morgan to accept the chromosome theory. Indeed, Muller later insisted that Morgan did so essentially "against his will." By the summer of 1911, however, Morgan's group had shown that factors affecting eye color, body color, wing shape, and sex all segregated together with the X chromosome. Although Morgan still thought that cytoplasmic factors might play some role in heredity and development, he was convinced that the best way to explain his *Drosophila* data was to posit the physical connection of these traits on the chromosome.

Experiments with the fruit fly provided a powerful new way of demonstrating that apparent deviations from the law of independent assortment were due to linkage, that is, two factors being carried on the same chromosome. Linkage was first demonstrated for the sex-linked traits of white eyes and rudimentary wings. Studies of double mutants revealed another important phenomenon called recombination, which could be explained in terms of "crossing over," that is, the reciprocal exchange of genes between homologous paternal and maternal chromosomes. A cytological basis for reciprocal exchanges of bits of the chromosome had been provided by Frans-Alfons Janssens's (1863–1924) studies of "chiasma," that is, places where homologous chromosomes seemed to twist around each other during the early phase of meiosis. In 1909, Janssens suggested the chiasmatype hypothesis, that is, chiasma might represent sites of chromosome breakage and healing. If the physical exchange of segments of chromosomal material actually occurred at such sites, the study of recombinants would represent another useful bridge between cytology and Mendelian genetics.

Morgan's group confirmed the chiasmatype hypothesis and used statistical studies of recombination, a reflection of the frequency of crossing over, to produce maps of the linear order of the genetic factors on the chromosomes. Using chromosomes that were heterozygous for cytological and genetic markers, cytogeneticists provided further evidence that the correlation between the occurrence of genetic recombination and the frequency of cytologically observable chiasmata was very strong in Drosophila as well as maize.

In fruit flies the white eye, yellow body, and rudimentary wing mutations were found almost exclusively in males because the X chromosome carries these traits. Because females have two X chromosomes (XX) and males (XY) have only one X chromosomes, reciprocal crosses involving these mutants did not produce identical results. X-linked mutations also provided a powerful test of the idea that the behavior of chromosomes establishes the physical basis of Mendelism. If genes are carried by the chromosomes in a linear array, then the degree of linkage established by mating tests should serve as a measure of the distance between genes on a chromosome. Based on the proposition that there must be a quantifiable relationship between the strength of linkage between genes and their linear sequence on the chromosome, Alfred Sturtevant (1891–1970) constructed the first chromosome map. "The Linear Arrangement of Six Sex-Linked Factors in Drosophila, as Shown by Their Mode of Association" was published in the *Journal of Experimental Zoology* in 1913. By 1915 Morgan and his associates had described four groups of linked factors that corresponded to the four pairs of *Drosophila* chromosomes. In place of the "bean bag" image previously associated with Mendelian genetics, the chromosome theory of inheritance conjured up a picture of genes as beads on a string. Based on statistical studies of inheritance in *Drosophila*, Morgan assigned five principles to the gene: segregation, independent assortment, crossing over, linear order, and linkage groups.

In 1926 Morgan published *The Theory of the Gene* as a summation of the developments in genetics since the rediscovery of Mendel's laws. He also used the book as a means of defending his theory, refuting criticism, and marshaling evidence for the physical reality of the gene. Morgan and his associates saw the theory of the gene as a powerful and sophisticated generalization that explained the nature of the link between generations. Mendel's laws and the apparent exceptions to Mendel's laws could now be attributed to linkage groups, crossing over, multiple alleles, and so forth. Through his extensive writings and those of his disciples, Morgan exerted a profound influence on the development of genetics and cytology. Nevertheless, other theories, especially those based on the belief that cytoplasmic, or extranuclear, inheritance was at least as significant as the genetic material sequestered in the cell nucleus, continued to compete with Mendelian genetics. Even though the chromosome theory, or, as Morgan preferred to call it, the theory of the gene, was not immediately accepted by all geneticists, after visiting Morgan's laboratory in 1922 William Bateson professed his readiness to renounce his doubts about the chromosome theory of inheritance and pay respectful homage to the stars that had arisen in the West.

Critics of Mendelism often pointed to the kinds of mutants studied by the fly group as evidence that mutations were basically pathological and could not, therefore, play an important role in evolution. In rebutting this argument, Morgan and his colleagues admitted that the kinds of traits studied in the laboratory were generally deleterious, but not all mutations were as striking as rudimentary wings or white eyes. Presumably, many mutations produced subtle changes that were undetected, whether they were advantageous or deleterious. Some geneticists thought that the study of mutations and their effect on physiological genetics might restore the traditional relationship between genetics and embryology. Most notably, Richard Goldschmidt (1878–1958) introduced the idea that geneticists might discover mutations that affected the rates of major embryonic processes and thus give rise to "hopeful monsters."

When de Vries visited the United States in 1904, he was asked to give the inaugural lecture at the Station for Experimental Evolution at Cold Spring Harbor, New York. Emphasizing his mutation theory, de Vries suggested that the recently discovered roentgen and curie rays produced by radium might be used to induce mutations in plants and animals. Shortly afterwards, Morgan made some attempts to induce "Devriesian mutations" in various animals, including *Drosophila*, by subjecting them to radium, acids, alkalis, salts, sugars, and proteins. Failing to find macromutations or new species among the immediate offspring of the treated animals, Morgan decided that the results were not worth publishing. Mutants were obviously valuable in genetic analysis, but the natural rate of mutation was too low for practical quantitative studies of mutation as a process. To overcome this obstacle, Hermann Joseph Muller (1890–1967) focused his energies on finding methods that significantly increased the rate of mutation in *Drosophila*. The mutations investigated by Muller certainly did not support de Vries's belief that new species could be created in one step, but they did confirm his predictions about the unlimited horizons that would result from the experimental study of mutation.

While still a high school student in New York City, Muller became interested in the sciences, particularly evolution. Majoring in genetics at Columbia University, he came under the influence of E. B. Wilson. In 1910 Muller earned his B.A. and enrolled in Cornell Medical School, but he soon decided to join Morgan's *Drosophila* group at Columbia University. Known for his ingenuity in experimental design, Muller clarified obscure aspects of chromosome behavior and genetic mapping. Muller earned his Ph.D. in 1915 for research on crossing over, but he thought his contributions were not sufficiently appreciated. From 1921 to 1932 Muller, the quintessential New Yorker, was a member of the faculty of the University of Texas. When Muller began his systematic study of mutations, the term *mutation* was applied to a confusing collection of distinct phenomena, which, from the genetic point of view, were totally unrelated. Some so-called mutations were special cases of Mendelian recombination, some were due to abnormalities in chromosome distribution, and others were caused by changes in individual genes or hereditary units. Muller argued that, in the interests of scientific clarity, the term should be limited in usage and redefined as an alteration within an individual gene.

Beginning a full-scale attack on the problem of true mutations, Muller tested various agents in an attempt to increase the frequency of mutations. When Morgan embarked on this program, *Drosophila* geneticists had painstakingly identified about 100 spontaneous mutations. By 1927 he had successfully demonstrated that ionizing radiation could be used to induce hundreds of mutations in fruit flies. Most of these induced mutations were stable over many generations and behaved like typical Mendelian factors when subjected to breeding tests. Muller's presentation of his research at the Fifth International Congress of Genetics in Berlin in 1927 was a landmark event in the establishment of radiation genetics. Although the importance of his discovery was immediately understood, Muller did not receive his Nobel Prize until 1946. Brooding over his feelings of isolation in Texas and his failure to get elected to the National Academy of Sciences (a notoriously arbitrary and capricious procedure), Muller accused his former mentor of sabotaging his professional advancement. His state of mind in 1932 can be judged by the suicide note found in his pocket when he was found alive, but dazed, after taking an overdose of sleeping pills. Perhaps this sign of emotional distress, as well as problems caused by his outspoken defense of radical ideas and socialist causes, expedited Muller's departure from Texas.

While working in Berlin in 1932, Muller and Nikolai Timoféeff-Ressovsky (1900–1981) attempted to use the induction of mutations as a means of gaining insight into the identity of the gene. If the gene is a real physical entity of a particular size and shape, bombarding it with radiation should provide some information about its nature. These rather frustrating preliminary experiments led to a collaborative exposition of the "hit" or "target" theory of mutation by Timofeeff-Ressovsky and Max Delbrück in 1935. Recognizing the threat posed by the Nazi regime, Muller left Germany and accepted Nikolai Ivanovitch Vavilov's (1887–1943) invitation to work at the Institute of Genetics, which was part of the USSR Soviet Academy of Sciences. At the time, Vavilov had achieved an international reputation as an outstanding botanist, plant geographer, and geneticist. After graduating from the Agricultural Institute, Vavilov worked at the Department of Agriculture, the Bureau for Applied Botany, and the Bureau of Mycology and Phytopathology of the Agricultural Scientific Committee. In 1913–1914, Vavilov traveled to Europe where he studied with William Bateson and others. In 1917 Vavilov was appointed deputy head of the Bureau for Applied Botany, which became a major research institute under his guidance. He also served as professor and head of the Department of Applied Botany at Saratov University. Later Vavilov and his associates moved to Petrograd (St. Petersburg). In 1924, the Department was transformed into the Institute of Applied Botany and New Crops. The primary mission of the Institute was collecting plants throughout the world in order to study their properties and potential uses. Vavilov directed the Institute of Experimental Agronomy and was named president when it became the V. I. Lenin All-Union Academy of Agriculture in 1930. From 1930 to 1940, he was director of the Institute of Genetics. Vavilov organized and participated in over 100 collecting missions. He collected plants from Iran, the United States, Central and South America, the Mediterranean, Ethiopia, and Afghanistan. From 1931 to 1940, Vavilov was president of the Russian Geographic Society.

Muller and Vavilov hoped to encourage the study of genetics and fight the growth of Lysenkoism, a pseudoscientific dogma that denied the very existence of the gene and the whole theoretical framework of "idealistic-decadent" Mendel-Morganist genetics. The neo-Lamarckian doctrine promoted by Russian horticulturist Trofin Denisovich Lysenko (1898–1976) was endorsed by Joseph Stalin (1879–1953) and the Central Committee of the Russian Communist Party as true Marxist biology. Lysenkoism seemed to resonate with the writings of Engels and the tenets of dialectical materialism. Western geneticists denounced Lysenko as a charlatan and an unscrupulous careerist whose theories had catastrophic effects on Soviet agriculture, education, and science. Nevertheless, Lysenko enjoyed the full support of the Soviet government, party, courts, police, and the press. Although he was born a peasant in Czarist Russia, Lysenko graduated from the Uman School of Horticulture, earned a doctorate from the Kiev Agriculture Institute, and became director of the Odessa All-Union Selection and Genetics Institute and president of the Lenin Academy of Agricultural Sciences. Lysenko was awarded the Order of Lenin and served as director of the Soviet Academy of Science's Institute of Genetics until 1964. Opposition to Lysenko began to grow when Nikita Khrushchev (1894–1971) became premier.

Combining aspects of Lamarckian biology, the teachings of Russian horticulturist I. V. Michurin (1855–1935), peasant beliefs, and Stalinist doctrines, Lysenko created his own theory of agrobiology. Lysenko claimed that cultivated plants could be transformed into different species by means of "revolutionary leaps." From 1929 to 1964, with the support of Soviet officials, Lysenko dominated collectivized peasants and scientists. Despite the catastrophic failures associated with his agricultural techniques, Lysenko claimed that his methods of adapting crops to the brutal Russian climate would result in great increases in productivity. While Lysenko was in power, it was impossible to teach or practice genetics throughout the Soviet domain. Denying the existence of genes and chromosomes, Lysenko and his followers denounced and persecuted the "agents of international fascism" who believed in such "bourgeois constructs." Mendelian geneticists were persecuted and imprisoned for supporting a "bourgeois biology." In 1941, as a result of his opposition to Lysenko, Vavilov was arrested and sent to a prison camp, where he died 2 years later.

The influence of Lysenkoism was felt throughout the Communist world. Indeed, when Lysenkoism was imposed on Chinese geneticists in 1949, Mendel-Morganist theory was banished from schools, laboratories, and the press until the Qingdao Genetics Conference in 1956. Lysenkoism was officially renounced by the Czechoslovak government in 1966. With the fall of Lysenkoism, Vavilov was officially rehabilitated. Many books and articles paid tribute to his memory. His name was commemorated by the N. I. Vavilov All-Russian Scientific Research Institute of Plant Industry and attached to the Russian Society of Geneticists and Breeders, the Institute of General Genetics of the Academy of Sciences, the Institute of Plant Industry, and the Saratov Agricultural Institute.

At a meeting of the Lenin All-Union Academy of Agricultural Sciences in 1936, Muller compared the choice between Mendelian-Morganist genetics and Lysenkoism to a choice between medicine and shamanism, or astronomy and astrology. Of course, Muller would be very disappointed to know that even after the close of the twentieth century Americans were still likely to favor astrology over astronomy and creationism over evolution. Thoroughly disillusioned by the losing battle against Lysenkoism, after 2 years in Leningrad and 4 years in Moscow, Muller left the Soviet Union. Eventually, he found a position at the University of Edinburgh, where he studied the chromosomal basis of embryonic death from radiation damage. The outbreak of World War II forced him to return to the United States. After the war, Muller studied radiation sickness, worked out estimates of spontaneous and induced mutation rates in humans, and campaigned for radiation protection. Muller exploited the publicity generated by his 1946 Nobel Prize to campaign against the medical, industrial, and military abuse of radiation. He remained a vocal critic of Lysenko and resigned from the Soviet Academy of Science as a protest. In 1945 Muller became professor of zoology at Indiana University, where he remained until his death in 1967. The first volume of the *Annual Review of Genetics* was published that year and dedicated to Muller.

As a student, Muller developed an intense interest in evolution and human genetics, especially the factors that might improve, preserve, or degrade the human gene pool. Most mutations, he noted, were stable, deleterious, and recessive. Knowledge of mutation, therefore, had profound implications for eugenics and human reproduction. The eugenics movement in the United States was, however, a strange mélange of conservative and radical elements. On many occasions Muller publicly condemned the "mainline creed" of the eugenics movement as a perversion based on pseudoscience and racial prejudice that served the interests of fascists and reactionaries.

Galtonian eugenics held that since civilized values did not permit a return to nature where the unfit were simply eliminated by starvation, disease, and so forth, in order to maintain and elevate the human race, society must limit the reproduction of the unfit and encourage the fit to procreate. Such ideas required replacing the gospel of *laissez-faire* with that of social management. If the problems of the "unfit" were due to hereditary factors, then expensive attempts to improve their lot through remediation and education were inevitably futile. The hereditarian position supports the proposition that the only biologically sound solution for social inequalities is to keep the unfit from producing ever more unfit progeny.

In the early twentieth century interest in eugenics and intelligence testing coincided with the establishment of Mendelian genetics. Coincidence and sequence are not proof of causality, but some historians claim that the rising tide of hereditarian attitudes in Western culture was, at least in part, responsible for the establishment of hereditarian theories in biology. The new science of genetics was certainly compatible with the concept that human characteristics were biologically determined, but the relationship between genotype and human behaviors was problematic. As the continuing controversy over evolutionary theories indicates, human beings are quite capable of seeing themselves as separate and distinct entities, above and outside the "animal" world where genes and traits could be tested.

Muller's commitment to eugenics was very strong, but his ideas were more subtle and complex than those associated with the mainstream of eugenics, which were largely based on race and class prejudice. For Muller, the goal of the true science of eugenics was the admittedly distant one of using scientific understanding to consciously guide human biological evolution rather than the immediate goal of purging the world of the "unfit." The "Geneticists' Manifesto," signed by Muller and about 20 other American and British scientists, called for changing attitudes towards sex and procreation so that improved social and scientific attitudes would lead to a better genetic endowment for future generations. Muller was one of the founders of the American Society of Human Genetics.

In lectures, essays, and books, Muller warned his audience that, in the absence of natural selection, undesirable genes would inevitably accumulate in the human gene pool until the germplasm became "riddled through with defects." His concern for the quality of the human gene pool was forcefully expressed in his 1949 presidential address to the American Society of Human Genetics on "Our Load of Mutations." Modern medicine and technology were such dysgenic forces, he warned, that our genetic load had already reduced the evolutionary fitness of the human race. The forces of natural selection that had previously purged the human gene pool of harmful genes could no longer operate in the advanced nations. It was, therefore, necessary for geneticists to provide direction and guidance in order to establish voluntary, but socially sanctioned eugenic reproductive controls. As a social duty, people with bad genes should refrain from reproducing. Those blessed with a good genetic endowment should be encouraged to participate in positive eugenic programs, including "germinal choice." That is, suitable women should be artificially inseminated with the sperm donated by great men. In order to be sure that the men involved were truly worthy, their semen could be collected, frozen, and then used to create better babies after history had made its judgment. Open to the idea that nurture as well as nature affected human development, Muller attempted to determine the relative importance of environment and heredity, primarily by studying twins. Nevertheless, he predicted that there would be a "complete and permanent collapse of the evolutionary process" unless human beings, or nature, restored the operation of natural selection.

With the success of mutation studies, Muller saw the question of the basic mechanism of evolution translated into the problem of the nature, frequency, and mechanism of mutation. His successful work on mutations allowed him to serve as a gadfly to the scientific community and the eugenics movement. Because he saw eugenics as a special branch of evolutionary science, he was convinced that mutations must be of fundamental concern to eugenics. Certainly Muller raised many serious questions about human evolution and the burden of mutations, and in doing so he challenged other scientists to question the wisdom and propriety of attempting to apply selective or eugenic measures to human beings.

CYTOPLASMIC INHERITANCE

Most accounts of the history of genetics claim that Mendelism was generally accepted by 1915, but some historians argue that cytoplasmic inheritance remained a strong alternative theory until the establishment of molecular genetics in the 1950s and 1960s, when the chromosome theory was reborn as the nucleic acid theory of heredity. Even though most American geneticists generally accepted Morgan's argument that, for genetic purposes, the cytoplasm could be ignored, many biologists, especially in Germany and France, continued to support cytoplasmic theories of inheritance. Embryologists, in particular, argued that Mendelian genetics failed to explain the most fundamental characteristics of the organism, its development, and major differences between higher taxonomic categories. Perhaps, they argued, the cytoplasm determined the essential heredity of the species, whereas the chromosomes determined rather trivial differences between members of the same species. For example, in *The Organism as a Whole* (1916), Jacques Loeb (1859–1924) emphasized the profound gap between a mosaic of discrete, randomly associating Mendelian factors and the harmonious development of an organism from the fertilized egg. Perhaps the organization of the egg cytoplasm provided the "mold" that structured the development of the embryo. Advocates of cytoplasmic inheritance claimed that, although their techniques were extremely difficult, their approach to inheritance and development was more holistic, sensitive, and integrative. Whatever the strength of the challenge that cytoplasmic inheritance may have presented to

Mendelian genetics, the doctrines of this field became increasingly controversial in the 1940s because of their association with Lysenkoism and Stalinism.

Advocates of cytoplasmic inheritance challenged the "nuclear monopoly" and looked for evidence that the most important hereditary materials or principles were located outside the cell nucleus. Some characteristics of organisms do not show Mendelian segregation and, among higher organisms, are inherited through the maternal line only. The variegated leaves of some plants, for example, are related to the presence or absence of chlorophyll, the green pigment found in the cytoplasmic organelles called chloroplasts. Cytoplasmic organelles, such as the chloroplast and the mitochondrion, were discovered in the early decades of the twentieth century. Because pollen grains and sperm cells do not contain significant amounts of cytoplasm, only the egg cell contributes chloroplasts and mitochondria to the next generation. During the 1960s, mainstream geneticists discovered that certain self-replicating subcellular organelles in plant and animal cells contain their own DNA, ribosomes, transfer RNAs, and the enzymes that allow them to synthesize some of their own unique proteins. These discoveries changed the nature of the debate, and seemed to offer a way to subsume the existence of nonnuclear heredity within mainstream genetic theory. Historians of cytoplasmic inheritance, however, argue that evidence of subcellular organization and extranuclear inheritance provides an increasingly sophisticated challenge to "nucleocentric genetics."

SUGGESTED READINGS

Adams, M. B., ed. (1990). *The Wellborn Science: Eugenics in Germany, France, Brazil, and Russia.* New York: Oxford University Press.

Allen, G. (1978). *Thomas Hunt Morgan. The Man and His Science.* Princeton, NJ: Princeton University Press.

Baltzer, F. (1967). *Theodor Boveri: Life and Work of a Great Biologist, 1862–1915.* Translated by D. Rudnick. Berkeley, CA: University of California Press.

Bateson, W. (1902). *Mendel's Principles of Heredity—A Defense.* Cambridge, England: Cambridge University Press.

Bateson, W. (1928). *William Bateson, F.R.S., His Essays and Addresses, with a Memoir by Beatrice Bateson.* Cambridge, England: Cambridge University Press.

Bennett, J. H., ed. (1965). *Experiments in Plant Hybridization. Gregor Mendel: Mendel's original paper in English translation with commentary and assessment by the late Sir Ronald A. Fisher together with a reprint of W. Bateson's "Biographical Notice of Mendel."* Edinburgh: Oliver and Boyd.

Berg, R. L. (1990). *Acquired Traits. Memoirs of a Geneticist from the Soviet Union.* Translated from the 1983 Russian edition by David Lowe. New York: Penguin.

Blacher, L. I. (1982). *The Problem of the Inheritance of Acquired Characters. A History of a priori and Empirical Methods Used to Find a Solution.* Translated from Russian. English translation edited by F. B. Churchill. Published for the Smithsonian Institution Libraries, and the National Science Foundation. New Delhi, India: Amerind Publishing Co.

Bowler, P. J. (1989). *The Mendelian Revolution: The Emergence of Hereditarian Concepts in Modern Science and Society*. Baltimore, MD: Johns Hopkins University Press.

Brink, A. P., and Styles, E. D., eds. (1967). *Heritage from Mendel: Proceedings of the Mendel Centennial Symposium*. Madison, WI: University of Wisconsin Press.

Brush, S. (1989). Nettie M. Stevens and the discovery of sex determination by chromosomes. *Isis* 69:163–172.

Burnham, J. C., ed. (1971). *Science in America. Historical Selections*. New York: Holt, Rinehard and Winston, Inc.

Carlson, E. A. (1966). *The Gene: A Critical History*. Philadelphia, PA: Saunders.

Carlson, E. A. (1982). *Genes, Radiation and Society: The Life and Work of H. J. Muller*. Ithaca, NY: Cornell University Press.

Carlson, E. A. (2001). *The Unfit: A History of a Bad Idea*. Cold Spring Harbor, NY: Cold Spring Harbor Laboratory Press.

Corcos, A. F., and Monaghan, F. V. (1993). *Gregor Mendel's Experiments on Plant Hybrids. A Guided Study*. New Brunswick, NJ: Rutgers University Press.

Darden, L. (1991). *Theory Change in Science. Strategies from Mendelian Genetics*. New York: Oxford University Press.

Darwin, C. (1868). *The Variation of Plants and Animals Under Domestication*. 2 volumes. New York: Orange Judd and Co.

Davis, B. D., ed. (1991). *The Genetic Revolution. Scientific Prospects and Public Perceptions*. Baltimore, MD: Johns Hopkins University Press.

Dillon, L. S. (1987). *The Gene: Its Structure, Function, and Evolution*. New York: Plenum.

Dobzhansky, T. (1970). *Genetics of the Evolutionary Process*. New York: Columbia University Press.

Dunn, L. C., ed. (1951). *Genetics in the Twentieth Century*. New York: MacMillan.

Fisher, R. A. (1930). Has Mendel's work been rediscovered? *Annals of Science* 1:115–137.

Garber, E. D., ed. (1985). *Genetic Perspectives in Biology and Medicine*. Chicago, IL: University of Chicago Press.

Graham, L. R. (1981). *Between Science and Values*. New York: Columbia University Press.

Grun, P. (1976). *Cytoplasmic Genetics and Evolution*. New York: Columbia University Press.

Hartl, D. L., and Orel, V. (1992). What did Gregor Mendel think he discovered? *Genetics* 131:245–253.

Harwood, J. (1993). *Styles of Scientific Thought. The German Genetics Community, 1900–1933*. Chicago, IL: University of Chicago Press.

Huxley, J. S. (1949). *Soviet Genetics and World Science: Lysenko and the Meaning of Heredity*. London: Chatto and Windus.

Iltis, H. (1932). *Life of Mendel*. Translated by E. and C. Paul. New York: W. W. Norton & Company, Inc. (Reprinted, New York: Hafner & Co., 1966.)

Jacob, F. (1973). *The Logic of Life: A History of Heredity*. Translated by B. E. Spillman. New York: Pantheon.

Joravsky, D. (1970). *The Lysenko Affair*. Cambridge, MA: Harvard University Press.

Kevles, D. J. (1986). *In the Name of Eugenics. Genetics and the Uses of Human Heredity*. Berkeley, CA: University California Press.

Kohler, R. E. (1994). *Lords of the Fly. Drosophila Genetics and the Experimental Life*. Chicago, IL: University of Chicago Press.

Lewis, E. B., ed. (1961). *Selected Papers of A. H. Sturtevant: Genetics and Evolution*. San Francisco, CA: W. H. Freeman.

Lubrano, L. L., and Solomon, S. G., eds. (1980). *The Social Context of Soviet Sciences*. Boulder, CO: Westview Press.

Ludmerer, K. (1972). *Genetics and American Society: A Historical Appraisal*. Baltimore, MD: Johns Hopkins University Press.

Lynn, R. (2001). *Eugenics, A Reassessment*. New York: Praeger Publishers.

Margulis, L. (1981). *Symbiosis in Cell Evolution. Life and Its Environment on the Early Earth*. San Francisco: W. H. Freeman.

Mayr, E. (1982). *The Growth of Biological Thought*. Cambridge, MA: Harvard University Press.

Mayr, E., and Provine, W., eds. (1980). *The Evolutionary Synthesis*. Cambridge, MA: Harvard University Press.

McKusick, V. A. (1998). *Mendelian Inheritance in Man. Catalogs of Autosomal Dominant, Autosomal Recessive, and X-Linked Phenotypes*. 12th ed. Baltimore, MD: Johns Hopkins University Press.

Medvedev, Z. A. (1971). *The Rise and Fall of T. D. Lysenko*. Translated by I. M. Lerner. New York: Anchor Books.

Morgan, T. H. (1926). *The Theory of the Gene*. New Haven, CT: Yale University Press.

Morgan, T. H., Sturtevant, A. H., Muller, H. J., and Bridges, C. B. (1915). *The Mechanism of Mendelian Heredity*. New York: Henry Holt and Company.

Nardone, R. M., ed. (1968). *Mendel Centenary: Genetics, Development and Evolution*. Washington, DC: The Catholic University of America Press.

Olby, R. (1985). *Origins of Mendelism*. 2nd ed. Chicago, IL: University of Chicago Press.

Orel, V. (1996). *Gregor Mendel: The First Geneticist*. Translated by Stephen Finn. New York: Oxford University Press.

Peters, J. A., ed. (1959). *Classic Papers in Genetics*. Englewood Cliffs, NJ: Prentice-Hall.

Provine, W. B. (1971). *The Origins of Theoretical Population Genetics*. Chicago, IL: University of Chicago Press.

Reilly, P. R. (1991). *The Surgical Solution. A History of Involuntary Sterilization in the United States*. Baltimore, MD: Johns Hopkins University Press.

Russell, N. (1986). *Like Engend'ring Like: Heredity and Animal Breeding in Early Modern England*. Cambridge: Cambridge University Press.

Sachs, J. (1890). *History of Botany (1530–1860)*. Translated by Henry E. F. Garnsey. Oxford: Clarendon Press.

Sager, R. (1972). *Cytoplasmic Genes and Organelles*. New York: Academic Press.

Sapp, J. (1987). *Beyond the Gene. Cytoplasmic Inheritance and the Struggle for Authority in Genetics*. New York: Oxford University Press.

Schneider, L., ed. (1986). *Lysenkoism in China: Proceedings of the 1956 Quingdao Genetics Conference*. New York: M. E. Sharpe, Inc.

Shine, I., and Wrobel, S. (1976). *Thomas Hunt Morgan. Pioneer of Genetics*. Lexington, KY: The University Press of Kentucky.

Soyfer, V. N. (1994). *Lysenko and the Tragedy of Soviet Science*. Translated by Leo Gruliow and Rebecca Gruliow. New Brunswick, NJ: Rutgers University Press.

Stern, C., and Sherwood, E., eds. (1966). *The Origin of Genetics: A Mendel Source Book*. San Francisco, CA: W. H. Freeman.

Stubbe, H. (1972). *History of Genetics, From Prehistoric Times to the Rediscovery of Mendel's Laws*. 2nd ed. Translated by T. R. W. Waters. Cambridge, MA: MIT Press.

Sturtevant, A. H. (1965). *A History of Genetics*. New York: Harper & Row.

Sumner, A. T. (1990). *Chromosome Banding*. Boston, MA: Hyman.

Todes, D. P. (1989). *Darwin Without Malthus: The Struggle for Existence in Russian Evolutionary Thought*. Oxford: Oxford University Press.

Vries, H. de (1901–1903). *The Mutation Theory*. 2 volumes. Leipzig: von Veit.

Wallace, B. (1991). *Fifty Years of Genetic Load. An Odyssey*. Ithaca, NY: Cornell University Press.

Weismann, A. (1893). *The Germ-Plasm; A Theory of Heredity*. Translated by W. N. Parker and H. Ronnefeldt. New York: Charles Scribner's Sons.

Wilson, E. B. (1928). *The Cell in Development and Inheritance*. 3rd ed. New York: Macmillan. (Reprint, New York: Garland Pub., 1987.)

Zirkle, C. (1959). *Evolution, Marxian Biology, and the Social Scene*. Philadelphia, PA: University of Pennsylvania Press.

10

MOLECULAR BIOLOGY

Modern genetics can be seen as the result of the integration of three lines of investigation: statistical analysis of patterns of inheritance; microscopic studies of the intriguing behavior of subcellular entities; and the biochemical researches that elucidated the nature of various cellular components. Classical genetics could address the question of how the gene was transmitted; however, it could not answer the question of how the gene works. Until the mechanism of gene action could be analyzed, geneticists essentially used the gene as a symbol for analyzing Mendelian phenomena. Many geneticists regarded the gene as an abstract concept that was useful in organizing patterns detected by breeding experiments, rather than a material entity. Of course, in the early twentieth century many chemists felt the same way about the atom. But well before the nature of the gene had been clarified, Mendelian-Morganist genetics triumphed over alternative concepts such as soft heredity, cytoplasmic inheritance, and the inheritance of acquired characteristics. A blurring of focus concerning the actual nature of the gene allowed geneticists to think of chromosomal genes and Mendelian factors as if they were known to be the same entity before molecular biology transcended classical genetics and directly attacked the problem of what genes are and how they work.

The reception accorded to Mendelism, like Darwinism, varied in different countries in response to subtle social, intellectual, and professional priorities and constraints. Some historians posit a fierce battle between those who thought in terms of the new nuclear genetics and those who maintained traditional, broader concerns about growth and development. This dichotomy does not seem to account for the long-term popularity and respect for Edmund B. Wilson's book *The Cell in Development and Inheritance*, which first appeared in 1896. Subsequent editions provide valuable guides to prevailing views of the cell, before and after the rediscovery of Mendel's laws, as well as the state of cytology and genetics. At first the gene represented a necessary, but still shadowy and hypothetical entity. That is, the new vocabulary of genetics helped clarify the distinction between the characters seen in breeding tests and the units transmitted through the germ cells. Early twentieth-century geneticists attempted to resolve questions about the difference between the nature of the "unit-factor" and that of the "unit-character." After Thomas Hunt Morgan established the chromosome theory, the language of biology shifted towards a new dialect of genes, alleles, genotypes, and phenotypes. Although Morgan's version of the chromosome theory answered important questions about the physical basis of inheritance, it could not answer major questions about the chemical nature of the gene and the relationship between the genetic units and the structure of the chromosomes.

Nevertheless, at the turn of the century cytologists were able to phrase questions about cell function in terms of the role of the nucleus, chromatin, and the chromosomes. Explorations of the behavior of the chromosomes during mitosis (somatic cell division) and in meiosis (the formation of germ cells) proved to be especially fruitful. Microscopists knew that the sperm and the egg contained equal numbers of chromosomes, which was half the number of chromosomes found in body cells. During mitosis, the chromosomes appeared to divide longitudinally, while a special reduction division that occurred in meiosis halved the number of chromosomes distributed to the gametes. Microscopic studies had shown that fertilization involved the combination of male and female germ cells. The behavior of the chromosomes during the cell cycle and gamete formation was certainly intriguing, but despite the work of Boveri, Sutton, Stevens, and others, very little could be said about the nature of individual chromosomes.

WHAT IS THE GENE?

The work of T. H. Morgan, H. J. Muller, and others on the genetics of *Drosophila* established the program for increasingly detailed studies of the mechanism of inheritance. More powerful approaches to the production and analysis of mutations, in particular, led to a quest for understanding the actual physical nature of the gene. A virtual obsession with mutations did not, however, obscure a fundamental feature of the gene, that is, its stability. Genes normally produce tens of thousands of accurate copies. Errors or mutations are rare, but once they occur these new forms of the gene are stable in turn. After a period in which concern for genetics overshadowed interest in evolution, the study of mutations was recognized as an integral part of the Darwinian puzzle; the mutational event

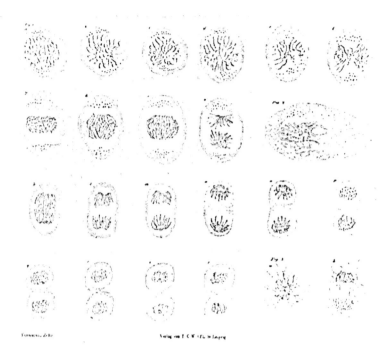

Dividing salamander larvae cells as depicted in Walter Flemming's *Cellsubstance, Kern und Zellthielung,* Leipzig, 1882

was the source of variation. Natural selection acted ultimately to choose among the infrequent, but critical mutations that affect the fitness of the individuals carrying them.

After Muller's discovery of the mutagenic effect of x-rays, other kinds of radiation, such as ultraviolet light, were found to have similar effects. Physicists interested in the mechanism of heredity surveyed these observations and framed new hypotheses about the nature of the gene; such speculations led to collaborations between geneticists and physicists. For example, in a paper entitled "On the nature of gene mutation and gene structure" (1935), the geneticist Nicolai Timoféeff-Ressovsky and physicists Karl G. Zimmer and Max Delbrück drew attention to the applicability of hit and target theories of radiobiology to the mutational process. Timoféeff-Ressovsky and Muller had previously collaborated on attempts to understand the nature of the gene by inducing mutations. Much of the inspiration for Delbrück's formulation of a quantum-mechanical approach to the putative gene molecule was drawn from the essay "On Light and Life" (1933) composed by Niels Bohr (1885–1962), a founder of quantum theory. Bohr was interested in the concept of paradoxical scientific models, such as the gene, which must be both stable and mutable. The Austrian physicist Erwin Schrödinger (1887–1962) later made Delbrück's speculations about the gene molecule and mutation better known through his book *What Is Life?* (1944), a work that has been called the *"Uncle Tom's Cabin"* of molecular biology."

An understanding of the hereditary substance, Schrödinger admitted, would not soon come from physics, but from advances in biochemistry, acting in conjunction with physiology and genetics. This in itself was a remarkable admission for one of the founders of wave mechanics, but Schrödinger also confessed that he could only see one general conclusion coming from Delbrück's molecular model and that presenting this idea was his primary motive for writing *What Is Life?* This was the tantalizing possibility that while living matter might not actually evade the known laws of physics, an exploration of Delbrück's model of the hereditary substance might well reveal other, as yet unknown laws of physics.

Although Schrödinger later became the semi-legendary hero in the foundation myths of what has been dubbed the informational school of molecular biology, many biochemists found Schrödinger's attempts to fuse chemistry and biology superficial, naive, and misleading. Citing Schrödinger's uncritical use of analogies between the growth of crystals and cells, chemists noted that Schleiden and Schwann had been ridiculed for expressing similar ideas almost 100 years before. While several eminent molecular biologists later claimed that their move from physics to biology was influenced by Schrödinger's book, some biochemists were repulsed by the "temerity" of a physicist who wrote about the cell and the gene without an appreciation of the relevant chemical knowledge. Echoing Shakespeare, scientists who respected Schrödinger's contributions to physics attempted to warn readers of *What Is Life?* that there might well be "more things in chromosomes than are dreamt of even in wave mechanics."

Studies of cytology in the 1930s also began to focus on the question of the nature of the hereditary material. Various plants and animals provided promising materials for such studies, but just as *Drosophila*'s mating habits proved advantageous for breeding experiments, its giant salivary gland chromosomes seemed specially designed for cytologists. Working in Morgan's famous "fly room" at Columbia University, Calvin Bridges (1889–1938) demonstrated a relationship between specific bands arrayed along the chromosomes and the linear sequence of genes on linkage maps. In addition to finding correlations between certain peculiar traits and duplications of particular chromosomal sections, Bridges noted that the effect of certain genes appeared to be related to their position on the chromosome. The recognition of chromosome banding and the discovery of means of resolving bands along mammalian chromosomes were key events in the development of modern cytogenetics. Additional insights into chromosome structure and organization were provided by the development of ultraviolet microspectrophotometry and special staining methods. Ingenious use of such techniques brought cytochemists tantalizingly close to understanding the chemical nature of the chromosomes.

In retrospect it is possible to trace a direct path to the DNA double helix that seemed to solve the mystery of the gene. During the first half of the twentieth century, however, the complexity of the chromosomes and the incommensurable methodologies and assumptions that characterized chemistry and genetics created an apparently inescapable labyrinth of paradoxes and misconceptions. Indeed, some geneticists argued that it was impossible to imagine that *any* particle or molecule could constitute the genetic material. As William Bateson complained in his review of *The Mechanism of Mendelian Heredity* (1915) by Morgan and his associates, it was inconceivable that particles of chromatin or any other particular substance, no matter how complex, could possess the qualities appropriate to the gene. Even the most dedicated materialist, Bateson argued, could not believe that particles of chromatin, which were apparently homogeneous and indistinguishable from each other by any known test, could "by their material nature" carry all the properties of life. Given the state of biochemical knowledge and methodology at the time, this rather pessimistic statement was not totally unreasonable. If fundamental differences between particles of chromatin did exist, they were not susceptible to chemical investigation.

Nevertheless, some biologists were sure that the gene must be a chemical entity with a specific, defined molecular arrangement. Both August Weismann and Hugo de Vries, for example, thought of the gene as a special chemical entity. The gene might be a minute particle, or a single large molecule with the property of self-duplication; the action of hereditary factors might be comparable to that of cellular ferments (i.e., catalytic proteins or enzymes). Genes might be proteins, or genes might make proteins, but the idea that the genetic material could be nucleic acid was, until the 1950s, generally considered quite unlikely. During the early decades of the twentieth century the nucleic acids, which are now known to be the material basis of heredity, were thought to be simple, repetitive, and rather uninteresting chemicals.

In many respects, the story of Johann Friedrich Miescher (1844–1895), the discoverer of the nucleic acids, is not unlike that of Gregor Mendel. But Mendel was rediscovered, his work vindicated, his genius eulogized, and his name was attached to the fundamental laws of genetics, whereas Miescher remains a forgotten man. Indeed, Miescher's discovery of nucleic acid is often confused with the elucidation of the double-helical structure of DNA by James D. Watson (1928–) and Francis Crick (1916–). Perhaps the lack of respect for chemistry that figures so prominently in many accounts of the discovery of the double helix and the neglect of experimental methodology, advances in instrumentation, and what practitioners of biochemistry honor as "craftsmanship," all contributed to the neglect of Miescher's work. That is, just as premodern physicians who used their heads rather than their hands were regarded as superior to surgeons, scientists who revolutionize theory and challenge disciplinary boundaries receive more attention from historians than those who introduce new methods, materials, and instruments. Nevertheless, Alfred Hershey, one of the founders of molecular biology, found the development of new methods so satisfying that he told colleagues: "Ideas come and go, but a method lasts!"

After receiving his M.D. degree (1868), Miescher began his remarkable studies of the physiological chemistry of pus cells, which were obtained by washing out used bandages from the surgical clinic. Given the high rate of postsurgical infection found in European hospitals at the time, a plethora of pus cells was always available. Attempting to purify pus cell nuclei, Miescher subjected his preparations to an acid extract of pig gastric mucosa (a crude preparation of the protein-digesting enzyme called pepsin). After this treatment, Miescher obtained an organic acid that had a remarkably high phosphorus content. The solubility properties of this substance, and its resistance to pepsin, suggested that it was a previously unknown cell constituent. Although this novel material was not well characterized, it was given the name *nuclein*. Later Miescher demonstrated that nuclein could also be isolated from salmon sperm.

The difficulties involved in preparing nuclein should not be underestimated. Nuclein was unstable and great care was needed to isolate it. Although Miescher recognized the problems inherent in his studies, especially the issue of the purity of his nuclein preparation, he was exasperated by what he considered excessively cruel attacks on his work. Indeed, he complained that he had strug-

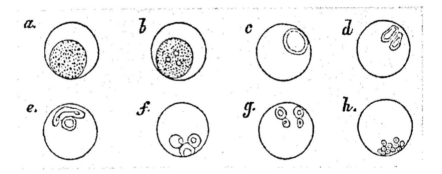

An engraving of pus cells showing the various stages of disintegration of the nucleus (Edmond Montgomery, *On the Formation of So-Called Cells in Animal Bodies,* London, 1867)

gled with biological substances so complex, under conditions that were so difficult, that "the real chemists shun them." Critics charged that Miescher's so-
called nuclein was a crude, chemically undefined mixture, or merely albuminous
material contaminated with phosphate salts. Stung by harsh criticism, cautious,
and reluctant to publish preliminary data, Miescher turned to safer studies of
the physiology of the Rhine salmon. About 5 years before his death, however,
Miescher resumed his chemical work on nuclein. These studies were published
posthumously.

Just what physiological function Miescher ascribed to nuclein is not entirely clear. However, he did suggest that if some substance in the sperm were
the specific cause of fertilization, it would have to be nuclein. Walter Flemming
(1843–1905), who suspected that nuclein was an important component of the
cell nucleus, coined the term *chromatin* because the relationship between nuclein and the chromatic threads had not been clearly established. Miescher, however, considered his own chemical work superior to the studies of histologists
whom he contemptuously described as the "guild of dyers." Privately, Miescher
speculated on the possible role of nuclein in the transmission of heritable traits.
He thought its atoms could form isomers, or alternative spatial arrangements,
which could account for variations. Still, it was not easy to see how enough
variation could exist in such molecules to explain hereditary phenomena. Proteins were better known and understood and, because of their amazing variability, seemed the more logical vehicle to serve as the physical basis of heredity.

Despite uncertainty about the chemistry of nuclein, by the end of the nineteenth century suspicion was growing that it was identical to chromatin and that
it might have some intimate relationship to the material basis of heredity. Oscar
Hertwig (1849–1922) suggested in 1885 that nuclein was probably responsible
for fertilization and the transmission of hereditary characteristics. He argued
that fertilization was a physical-chemical and morphological process and ridiculed those who claimed that fertilization was a kind of fermentation process in
which the sperm merely acted as a catalyst. Ten years later, Edmund B. Wilson
pointed out that the chromosome complements contributed by the two sexes
were equivalent to each other. Breeding tests indicated that the two sexes play
an equal role in heredity, even though different species vary in other aspects of
reproduction and development. Because the physical basis of heredity must reside in something that the sexes contribute equally in all types of reproduction,
it must reside in the chromatin. Since chromatin seemed to be essentially the
same as Miescher's nuclein, this substance must be the genetic material.

Studying nuclein from thymus and yeast, Albrecht Kossel (1853–1927)
proved that there were two kinds of nucleic acids. He also identified the purines
and pyrimidines that are the nitrogenous components of nuclein (adenine,
guanine, thymine, cytosine, and uracil). Extending his work to a variety of plant
and animal substances, Kossel demonstrated that both forms of nucleic acid are

fundamental constituents of cell nuclei. Investigating the proteins associated with nuclein, Kossel characterized histone and identified several amino acids. Considered a pioneer of physiological chemistry, Kossel was awarded the 1910 Nobel Prize in Medicine or Physiology for isolating and identifying the chemical components of the cell nucleus. The two types of nuclein prepared by Kossel, which were long known as thymus nucleic acid and yeast nucleic acid, are now designated deoxyribonucleic acid (DNA) and ribonucleic acid (RNA). Both nucleic acids contain the bases adenine, cytosine, and guanine, but in DNA the fourth base is thymine and in RNA it is uracil. DNA and RNA also differ in their sugar component, deoxyribose and ribose, respectively. RNA was initially thought of as the exclusive property of yeast cells, but better chemical techniques demonstrated its presence in all cells, including bacteria. Some scientists thought that RNA might serve as an energy reservoir, capable of supplying phosphorus atoms for cellular metabolism. Waves of confusion rather than enlightenment seemed to follow in the wake of these discoveries, but even as early as the 1830s the redoubtable Justus von Liebig had been driven to complain that the rapid pace of scientific progress could make one crazy.

Paradoxically, further chemical studies of the nucleic acids and microscopic studies of chromatin during the period from about 1910 to 1930 militated against the concept that nuclein or chromatin could serve as the hereditary material. Chromatin seemed to behave very peculiarly during the cell cycle; at times it apparently vanished. This did not seem consistent with the characteristics proper to the genetic material, particularly the need for stability from generation to generation. Finally, work carried out by Phoebus Aaron Levene (1869–1940) seemed to rob nuclein of complexity, another fundamental requisite of any chemical claiming to be the genetic material. After receiving his M.D. degree from St. Petersburg Imperial Medical Academy in 1891, Levene left Russia and emigrated to New York. While practicing medicine in New York City, he studied chemistry at Columbia University. Five years later, while recuperating from the effects of overwork and tuberculosis, he decided to abandon his medical practice. Totally dedicated to chemical research, Levene joined the newly formed Rockefeller Institute for Medical Research. During his association with that institution he investigated almost every important class of biological compounds, from sugars and lipids to proteins and nucleic acids. His prodigious labors resulted in more than 700 papers in the *Journal of Biological Chemistry*.

Levene's analyses of the nucleic acids suggested that the four bases were present in equimolar amounts in nucleic acids obtained from many different sources. These data led to the *tetranucleotide interpretation*, an explanatory device that began as a working hypothesis, but soon became the primary paradigm of nucleic acid chemistry. A minor revolution in biochemical concepts, and a great deal of hard chemical work, was needed to dismantle the multiple layers of explanation subsumed by the *tetranucleotide theory*. In essence, Levene's chemical analyses indicated that there were equal amounts of all four bases in DNA from all sources. On a more sophisticated level of organization, the tetranucleotide interpretation claimed that all DNA polynucleotides were combinations of a fixed and unchangeable sequence of units that were themselves a combination of four nucleotides. Ultimately, this led to the conclusion that DNA was a highly repetitious polymer, analogous to glycogen, and, therefore, incapable of generating the diversity that was essential in the genetic material.

Although several studies conducted in the 1940s suggested that DNA might be the hereditary material, this possibility was generally dismissed because nucleic acids were thought to be rather boring tetranucleotide polymers. When scientists thought about the chemical complexity required by the genetic material, proteins seemed to be the only logical candidates, at least until the work of Erwin Chargaff (1905–) challenged prevailing ideas about DNA. But as Chargaff said, it was only in the post–Watson-Crick era that the nucleic acids rose from minor curiosities to the center of biological thinking. Chargaff's work can be seen as an example of the linkage between changing ideas and changing methodologies that facilitates the emergence of new experimental possibilities, if not immediate revolutions in theory. In addition, Chargaff's initial assumption that nucleic acids might be as complicated, highly polymerized, and chemically diverse as proteins reflects the fruitful intersection of very different research programs.

According to Chargaff, Oswald T. Avery's 1944 paper on the transformation of pneumococcal types led him to redirect his research to DNA. Since reading Schrödinger's *What Is Life?*, Chargaff had been thinking about the chemical nature of the gene. Thus, Avery's paper convinced him that DNA might be the "hereditary code-script." If different DNAs did indeed have different biological activities, as indicated by Avery's report on the transforming principle, they must differ chemically, and the tetranucleotide theory must be wrong. If DNA was as complex as protein, its biological activity might depend on specific structural patterns, or sequences, but at the time there were no chemical techniques capable of revealing the architecture of the nucleic acids. Nevertheless, Chargaff later argued that even in the 1940s it was possible to see the beginnings of the new biochemical language of genetics.

Using the methods that first became available after World War II—paper chromatography, ultraviolet spectrophotometry, and ion-exchange chromatography—Chargaff proved unequivocally that the four bases found in DNA were not necessarily present in equal amounts. Precise comparative studies of nucleic acids from different sources, such as calf thymus, spleen, liver, human sperm, yeast, and tubercle bacilli, disproved the assumption that all DNAs were identical to calf thymus DNA. However, the base composition of DNA from different organs of the same species was constant and characteristic of the species. Although at the time no theory could account for his observations, Chargaff noted

Erwin Chargaff in 1953

that the molar ratio of total purines to total pyrimidines and the molar ratio of adenine to thymine and guanine to cytosine from different sources were always about one. Given the state of DNA chemistry in 1949, such "regularities" could only be described as "striking," even if they were "perhaps meaningless," not unlike Mendel's "numbers and ratios." Further evidence of these regularities suggested that DNA from different sources must share some fundamental structural principles. Indeed, Chargaff actually proposed four rules on DNA composition. These have been called the first parity rule, the cluster rule, the second parity rule, and the GC rule. His base-pairing rule, or first parity rule, is well known through its role in the discovery of the DNA double helix. Close inspection of the genomes of different species at the beginning of the twenty-first century renewed interest in Chargaff's other suggestions about regularities in the composition of nucleic acids. That is, nonrandom sequences of bases might reflect evolutionary selection pressures related to control of transcription or to posttranscriptional events.

From Chargaff's work it was clear that, like the proteins, nucleic acids were complex, interesting, and distinct entities. Further research proved that DNA was a macromolecule, that is, a long-chain structure of surprisingly high molecular weight, as revealed by measurements of viscosity and the rates of diffusion and sedimentation of carefully prepared solutions of DNA. The sizes of DNA "molecules" seemed to increase as methods of preparation improved and as techniques for the analysis of large, linear polymers became more sophisticated. The recognition that shear forces could easily break large DNA molecules was especially important.

The nucleic acids were of particular interest to scientists involved in research on plant viruses. In 1935 Wendell Meredith Stanley (1904–1971) applied the methods that had recently been used to crystallize proteins to purify tobacco mosaic virus (TMV), the pathogen that was largely responsible for the establishment of virology. TMV had been intensively studied by plant pathologists; it was also a favored model system in the development of the techniques essential to modern biochemistry and molecular biology, including centrifugation, crystallography, x-ray diffraction, immunology, and electron microscopy. Stanley found that even after repeated recrystallizations, TMV retained its physical, chemical, and biological properties. One year later, the presence of nucleic acid in highly purified TMV was established, but until the 1950s it was unclear whether TMV protein, nucleic acid, or nucleoprotein was the infectious component. Most of the plant viruses examined by the 1940s contained RNA rather than DNA. Stanley's success in crystallizing the virus made TMV the "right organism for the job," if the job involved physical and biochemical aspects of genetics, but made it more difficult to think about viruses as organisms, that is, living systems. Biochemists were able to come to terms with this paradox by thinking about or implicitly redefining "viral life" in terms of the possession of a genetic system in which mutation and recombination of characters could occur.

During the 1940s most chemists, physicists, and geneticists thought that the genetic material must be a protein, but another significant line of research on a phenomenon known as *transformation* suggested that nucleic acids played a fundamental role in inheritance. The first well-known series of experiments to challenge the assumption that genes must be proteins was carried out in the laboratory of Oswald T. Avery (1877–1955). In some respects, Avery's work was a refinement of an approach previously developed by Fred Griffith (1877–1941), a well-respected British bacteriologist. Various strains of *Diplococcus pneumoniae* were known to cause pneumonia, a leading cause of death. By the 1920s, several strains of pneumococci had been identified and differentiated by serological (immunological) methods. Avery and others had shown that the serological tests reflected differences in the polysaccharide capsule that envelops the bacterium. Encapsulated forms produced smooth, shiny, rounded colonies

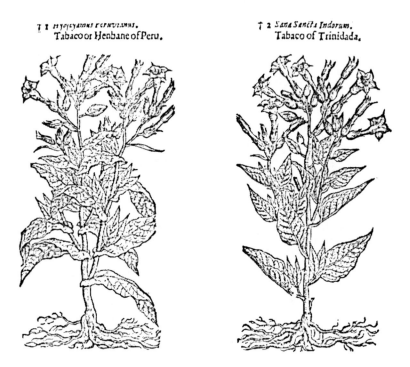

Tobacco plants as depicted in John Gerard's *The Herball* (London, 1633)

(S) when grown in vitro. Less virulent strains produced rough colonies of bacteria (R) that lacked the slimy polysaccharide capsule.

An intriguing, but puzzling set of observations of pneumococci was reported by Griffith in 1928; these experiments suggested that various strains could undergo some kind of transmutation of type. From the blood of mice injected with living R pneumococci together with a heat-killed S type, Griffith isolated living virulent bacteria. This indicated that the living nonvirulent R pneumococci had acquired something from the dead S type that transformed them into deadly organisms with the S-type polysaccharide capsule. In retrospect, it can be said that Griffith had observed genetic transformation, but he probably did not realize that the phenomenon he had discovered involved the transfer of hereditary material. Unfortunately, Griffith died during World War II when a German bomb blew up the laboratory where he was working. Several scientists confirmed Griffith's observations and proved that pneumococcal transformations could be effected in the test tube as well as in the animal body. The agent responsible was found to pass through a filter that could hold back bacteria. Intrigued by the possibility of finding some means of modifying the virulence of pathogenic bacteria, bacteriologists followed these experiments with great interest. A paper published by Oswald Avery, Colin MacLeod (1909–1972), and Maclyn McCarty (1911–) in 1944, however, demonstrated that the phenomenon demanded the attention of geneticists and biochemists.

Shortly after graduating from Columbia University College of Physicians and Surgeons in New York City and establishing a private medical practice, Avery accepted a position at the Rockefeller Institute for Medical Research. Most of Avery's research career was devoted to the study of pneumococcus. In the 1920s, Avery and Michael Heidelberger (1888–1991) proved that the substances on the surface of pneumococci that determined their virulence were polysaccharides. In addition to providing insights into the nature of different strains of pneumococci, this work indicated that the immune system could respond to polysaccharides as well as proteins. Avery was initially skeptical of Griffin's observations, but eager to determine whether a specific chemical was responsible for transforming pneumococci from one type to another. By 1933, his associate James Lionel Alloway (1900–1954) had prepared crude aqueous solutions of the transforming principle. Purifying the transforming principle, however, proved to be very difficult. About 10 years later, Colin MacLeod and Maclyn McCarty, working in Avery's laboratory, isolated the active "transforming factor" from S-type pneumococci and identified it as a "highly polymerized and viscous form of sodium desoxyribonucleate." Tests for the presence of protein proved negative. Furthermore, the transforming principle was inactivated by crude preparations of an enzyme known to attack DNA, but it was unaffected by enzymes that broke down RNA and proteins. Although Avery noted that the transforming principle had been compared to genes and viruses, he was more cautious about publishing speculations concerning the genetic implications of these findings than his younger colleagues would have liked.

In private letters Avery acknowledged that his studies of the transforming principle had profound implications for the biochemistry of DNA, genetics, enzyme chemistry, cell metabolism, and so forth, but he was reluctant to publish speculations that might later prove embarrassing. The possibility that the biological activity of his preparations could be due to minute amounts of some other substance associated with DNA could not be rigorously excluded. Nevertheless, Avery concluded that the available evidence strongly suggested that DNA was the transforming principle. If further tests confirmed this hypothesis, scientists would have to think about DNA not merely as a structural component of the chromosomes, but as "functionally active in determining the biochemical activities and specific characteristics of pneumococcal cells." The chemistry of DNA would also, Avery noted, require further study in order to explain the biological specificity that must be inherent in the genetic material. That is, Avery and his associates realized that it was not possible to reconcile the biological activity of their DNA preparations with prevailing ideas about DNA chemistry. Because the possibility existed that minute amounts of contaminants in nucleic acid preparations might be the actual genetic material, a great deal of effort was devoted to purification of transforming factors. This enigma would not be solved until the acceptance of the double-helical model of DNA proposed by Francis Crick and James D. Watson in 1953. But between the publication of Avery's paper and that of Watson and Crick, many other observations were made that provided clues about the molecular basis of inheritance.

Work with bacterial viruses was especially significant in providing the basis for a direct assault on what Watson and Crick rather melodramatically referred to as the "secret of life." The path that led to the Watson-Crick collaboration was largely determined by Watson's membership in the "American Phage Group." Members of the group, which had essentially coalesced by 1943, saw themselves as a special inner circle of scientists plugged into a privileged network of instant information that did not reach outsiders until many months later in the form of obsolete papers published in learned journals. The leaders of the Phage Group, Max Delbrück (1906–1981), Salvador Luria (1913–), and Alfred Hershey (1908–1997), who shared the Nobel Prize in 1969 for "discoveries concerning the replication mechanism and genetic structure of viruses," have been called the "Three Bishops" of the "Phage Church." Because of Delbrück's profound influence on the development of molecular biology, he was sometimes referred to as the Pope of the Phage Church, and he reserved the right to denounce ideas and results reported from outside the group as heresy.

Although Chargaff defined molecular biology as "the practice of biochemistry without a license," Max Delbrück's genealogy included the eminent chemist Justus von Liebig. Delbrück was born in Berlin and earned his Ph.D. in theoretical quantum mechanics at Göttingen in 1930. His studies and research brought him into contact with eminent physicists including Max Born, Wolfgang Pauli, Niels Bohr, Otto Hahn, and Lise Meitner. In 1933, with the Russian geneticist Nikolai Timoféeff-Ressovsky and the physicist Karl G. Zimmer, Delbrück attempted to explore questions about gene structure and mutation in terms of quantum physics. Like Bohr, Delbrück was fascinated by the paradox of the gene as an entity that expressed both stability and mutability. In 1937 Delbrück left Germany to work at the California Institute of Technology, where he was introduced to the advantages of using bacteriophage as an experimental model. During the 1940s, Delbrück, Salvador Luria, and Alfred Hershey demonstrated that phages could produce mutants, or stable variants, and, therefore, bacteriophages provided a valuable model system for studying genetic recombination. Moreover, Hershey proved that the genetic determinants of phages are linearly linked to each other in a manner similar to the genes on the chromosomes of higher organisms. Phage researchers have called bacteriophage genetics "Mendelism at the molecular level," because studying the multiplication of phage particles was essentially equivalent to studying naked genes at work.

One of Delbrück's key contributions to the evolution of phage research was the organization of summer phage courses and symposia on quantitative genetics at the Cold Spring Harbor Laboratory for Quantitative Biology. Cold Spring Harbor and Caltech thus became the Mecca and Medina of the Phage Church. According to Delbrück's admirers, the phage course led to the emergence of the American Phage Group and established Delbrück's place as its intellectual leader. Scientists who attended the phage course later recalled becoming permanently "infected." Indeed, according to Watson, "most of the basic facts about the gene and how it functions were learned through studies of bacteriophages."

Bacterial viruses provided the perfect model system for a new approach to genetics, especially appealing to scientists who had originally been trained as physicists and could think about heredity and reproduction in terms of the replication and multiplication of virus particles. Different kinds of phages have provided different kinds of insights into the nature of recombination and the nature of the genetic material. Unlike classical breeding experiments with peas, mice, or fruit flies, experiments with bacteriophage were simple, rapid, and precise. Most of the early studies of phage genetics were carried out by a small number of Americans working within the shared ideology and paradigms of Delbrück's Phage Group.

Other scientists, such as the French microbiologist André Lwoff (1902–1994), adopted quite different approaches to phage studies, viewing them as part of a research program examining the biochemistry and physiology of microorganisms. Indeed, the sardonic term "American Phage Church" has been attributed to Lwoff. After earning his M.D. (1927) and his doctorate of science (1932)

Max Dulbrück, Nobel Laureate, 1969 (California Institute of Technology)

from the Université de Paris, Lwoff spent most of his research career at the Pasteur Institute. Honored by his colleagues as an artist and "master craftsman," Lwoff was especially interested in the role of growth factors in the life cycle of microorganisms, the concept of biochemical evolution, and the phenomenon known as *lysogeny*. Although Lwoff was known for his diverse interests, it was his work on lysogeny that was honored by the 1965 Nobel Prize. Lysogenic bacteria appeared to contain viruses that could be induced to multiply and "lyse" (destroy) their bacterial hosts. Through his studies of lysogenic strains of bacteria, Lwoff proved that the virus was hidden in the genetic material of the host in a form he called the "prophage." The prophage and its host could coexist peacefully for many generations, but under appropriate circumstances the latent virus could be induced to produce large numbers of bacteriophages, which eventually lysed the infected bacterium. By exposing lysogenic bacteria to ultraviolet light, Lwoff could reliably induce activation of the prophage. Lwoff's work on the replication of bacterial viruses and their association with the bacterial genome provided the foundation for experiments on genetic control mechanisms conducted by François Jacob and Jacques Monod. For Lwoff, the concept of a virus that often remained quietly integrated into the host's DNA led to the study on oncogenes, cancer-causing genes that might have originated as viruses. In contrast, until 1949, when Élie Wollman (1917–), working with Delbrück at Caltech, confirmed the reality of lysogeny, Delbrück denounced this concept as heresy.

Microorganisms had been the objects of biochemical studies by the 1930s, but they did not seem to have the kind of sex life that Mendelian geneticists used in breeding tests. In 1942 Julian Huxley asserted that bacteria and viruses were totally asexual and did not have the genetic system found in multicellular organisms. It was thought that Mendelian principles did not apply to the world of microorganisms. During the 1940s, however, the experiments of Joshua Lederberg (1925–) and Edward L. Tatum (1909–1975) challenged the idea that bacteria were peculiar organisms with "no genes, nuclei, or sex." While a medical student at Columbia University, Lederberg became aware of Avery's work on the transforming principle and the possibility that DNA might be the genetic material. Lederberg decided to work with Tatum, who was investigating the genetics of *Escherichia coli* and establishing a new program in microbiology at Yale University. From mixed cultures of two different strains of *Escherichia coli* (K-12) with known mutations, Lederberg and Tatum isolated strains of bacteria that seemed to prove that bacterial cells did indeed participate in sexual mating behavior leading to genetic recombination. Because of the possibility that transformation, such as that observed by Avery, had occurred, Lederberg and Tatum showed that contact between bacteria was necessary for the exchange of genetic information. The choice of strain K-12 proved to be very fortunate, because subsequent investigations revealed that most strains of bacteria were sexually incompatible.

At the 1946 Cold Spring Harbor Symposium on Microbial Genetics, Lederberg initiated a new phase in genetic research by convincing biologists that microbial genes were subject to recombination and replication processes very much like the genes of plants and animals. By studying the frequency of recombinations of genetic determinants in bacterial matings, Lederberg began the task of constructing a genetic map of *Escherichia coli*. In 1953, Lederberg and Norton Zinder (1928–) demonstrated that bacteriophages were capable of transferring bacterial genes from one bacterium to another; this phenomenon was called *transduction*. These discoveries demonstrated that in many ways bacteria and bacteriophages were more favorable systems for an attack on the chemical nature of the gene than higher organisms.

To explain certain irregularities in bacterial genetics, William Hayes (1913–1994) suggested that bacterial mating involved an asymmetrical process involving a gene donor and a gene acceptor, rather than the fusion of equivalent male and female genetic complements found in sexual reproduction. When Hayes presented this concept in 1952, many biologists, including Lederberg and Delbrück, were highly skeptical. But after further research by Hayes and the experimental demonstrations by Wollman and Jacob, even Lederberg admitted that the concept of unidirectional gene donation clarified many aspects of bacterial genetics. In explaining how he had arrived at his innovative concept, while others were still struggling to explain bacterial genetics in terms of cell fusion, Hayes said that he "had the great advantage of knowing virtually no genetics while Lederberg knew too much." Indeed, Lederberg later admitted that he had found many complicated ways to rationalize the genetic data in terms of the exchange of nuclei, even though he had to resort to complications that were "a little bit like Ptolemy's epicycles."

Within 2 years Hayes and Lederberg independently discovered that the ability to act as a donor in a bacterial mating is controlled by a factor that seemed to behave as an infectious particle, independent of the chromosome. Although bacterial cells do not have a nucleus, they contain structures analogous to the chromosomes of higher cells. When the fertility factor (F) became associated with the bacterial chromosome, it promoted the transfer of chromosomal genes to the acceptor cell. Several other entities were discovered that also could exist in association with the bacterial chromosome or independently. In an article written for *Physiological Reviews* in 1952, Lederberg proposed using the term *plasmid* for the extrachromosomal hereditary determinants found in bacteria. Such factors have also been called episomes and plasmagenes. Further studies proved that bacterial matings involve the transfer of DNA from the donor cell to the recipient and that plasmids are DNA. With the development of recombinant DNA procedures in the 1970s, plasmids became important tools as carriers of foreign genetic material for use in genetic engineering.

In 1958 Lederberg was awarded the Nobel Prize in Medicine or Physiology for his discoveries of sexual reproduction and genetic recombination in bacteria. He shared the prize with his former mentor Edward Tatum and George Beadle. Active in public service, Lederberg used his scientific expertise to call attention to social and public policy issues, such as preventive medicine, epidemiology, emerging diseases, genetic testing, space exploration, and nuclear and biological weapons. Working with NASA on biological issues related to space travel, Lederberg campaigned for the application of the doctrine of quarantine to space missions in order to prevent the interchange of material between the earth and other planets. As an advisor to the Arms Control and Disarmament Agency, Lederberg was involved in the negotiations that led to a biological weapons disarmament treaty in the 1970s. During the 1990s he was a major force in raising awareness of the dangers of biological weapons and bioterrorism. Events at the turn of the twenty-first century confirmed his warnings about the threat of chemical and biological weapons.

Unlike Chargaff and Lederberg, Delbrück and Hershey were among the many skeptics who resisted the idea that Avery's pneumococcal transforming factor could be DNA. But DNA was taken very seriously when Hershey's own experiments with bacteriophage implicated DNA as the genetic material. Between 1951 and 1952 Alfred Hershey and his associate Martha Chase (1927–) carried out the famous "Waring blender" experiment that members of the phage group accepted as proof that DNA was the genetic material. Apparently Hershey began these experiments with considerable skepticism about the idea that nucleic acids could serve as the genetic material. According to Hershey DNA was a "monotonous tetranucleotide scaffold," and Avery's transforming principle must have contained a highly specific protein structure that served as the genetic material. Thomas F. Anderson (1911–) later remembered how he and Hershey discussed the "wildly comical possibility" that phage DNA could enter the bacterial cell and act like the transforming principle. Many scientists were surprised when Hershey's most famous experiment indicated that the joke was "not only ridiculous but true."

By the 1950s, phage workers were beginning to think that bacteriophages might act like tiny hypodermic needles full of genetic material, which could be injected into their bacterial victims, leaving empty viral coats outside. To test this possibility, Hershey and Chase used radioactive sulfur to label phage proteins and radioactive phosphate to label the DNA. After allowing bacteriophages to attack bacterial cells, the infected cultures were spun in a blender and centrifuged in order to separate intact bacteria from smaller particles. On finding that most of the phage DNA remained with the bacterial cells while the labeled protein was released into the medium, Hershey and Chase concluded that the viral protein served as a protective coat that facilitated adsorption to the bacteria so that phage DNA could be injected into the host cell. Because their technique

Alfred D. Hershey, Nobel Laureate, 1969 (Director, Genetics Research Unit, Carnegie Institute of Washington)

for separating labeled phage protein coats from infected bacterial cells introduced some uncertainty, Hershey and Chase could not claim that their experiment provided unequivocal proof that DNA was the genetic material. Somewhat tentatively, they concluded that further chemical work was needed to determine the identity of the genetic material. Other members of the Phage Group, including James D. Watson, quickly accepted the concept that DNA was the genetic material.

When the U.S. State Department refused Salvador Luria permission to attend the meeting of the Society for General Microbiology held at Oxford in April 1952, Watson went instead of his major professor. In presenting an account of the Hershey-Chase experiment, Watson defended the position that DNA was the genetic material, probably much more vigorously than Hershey himself would have. Nevertheless, although the evidence that DNA might be the genetic material was becoming more persuasive, because its structure was still obscure and misunderstood it was impossible to establish a logical relationship between the chemical nature of the gene and the mechanism by which the gene acted as the vehicle of inheritance. This dilemma was resolved when James D. Watson (1928–) and Francis Crick (1916–) described a model of DNA structure that immediately suggested explanations for its biological activity. The success of the Watson-Crick model amply confirms one of the aphorisms attributed to Francis Crick: "If you can't study function, study structure."

WATSON-CRICK AND THE DNA DOUBLE HELIX

Rejecting any show of undue modesty, molecular biologists have hailed the elucidation of the three-dimensional structure of DNA as one of the greatest achievements of twentieth-century biology, comparable to the legacy of Darwin and Mendel. Just one year after the Hershey-Chase experiment, Watson and Crick provided the long-sought relationship between the logic of the genetic material and the structure of DNA. Their elegant double-helical model for DNA was based on the biochemical and x-ray crystallographic studies that other scientists had performed. Several x-ray crystallographers had previously attempted to determine the three-dimensional structure of DNA. Outstanding for his pioneering work, and for his own estimate of his place in the history of science, William T. Astbury (1898–1961) referred to himself as the "alpha and omega, the beginning and end of the whole thing." Early, crude x-ray diffraction patterns of DNA suggested an arrangement of long polynucleotide chains with the rather flat nucleotides arranged perpendicularly to the long axis of the molecule. In 1947, about the time that Erwin Chargaff was disproving the tetranucleotide hypothesis, Astbury concluded that the degree of organization indicated by his x-ray findings supported Levene's theory of the regularity of DNA composition. Although his speculations about the regularity of the sequence of bases in DNA were in error, Astbury demonstrated that DNA had a regular crystalline structure that might eventually be clarified by the techniques of x-ray crystallography.

In 1951, James D. Watson, an American postdoctoral fellow, met Francis Crick, 12 years older but still a graduate student at Cambridge. Two years later both Watson and Crick were world famous. Both attributed their success to their special relationship: their ability to complement, criticize, and stimulate each other. Nonetheless, despite their intellectual compatibility, the contrasts between these two scientists were striking.

A former Quiz Kid, Watson entered the University of Chicago when only 15 years of age for an experimental early admissions program that allowed him to take special courses in the biological sciences. Although enthusiastic about ornithology, Watson did not seem especially interested in anything else. Teachers remembered Watson as a student who appeared to be completely indifferent to everything going on in the classroom; nevertheless, he could always rise to the top of the class. Formal training in genetics was negligible at the undergraduate level at the time, but Watson claimed that after reading Erwin Schrödinger's *What Is Life?*, finding the secret of the gene became his mission and obsession. Having been rejected by Harvard and Caltech, Watson decided to do his graduate work at Indiana University, which was a major center of genetic research in the 1940s. Distinguished members of the faculty included H. J. Muller, the *Drosophila* geneticist, Tracy Sonneborn (1905–1981), a pioneer in paramecium genetics, and Salvador Luria, who was working with phage and bacteria. While Muller seemed the obvious choice as a thesis advisor, Watson decided that *Drosophila*'s better days were over. The more progressive geneticists were now working with microorganisms, and sure that phage were preferable to paramecia, Watson decided to work for Luria, a friend of Max Delbrück, the hero of Schrödinger's *What Is Life?*. Thus, while still a graduate student, Watson became an acolyte of the phage church.

Based largely on a paper Watson wrote for a course taught by Sonneborn, historians have suggested that the controversial German biologist Franz Moewus (1908–1959) might have had a greater influence on Watson than Schrödinger. Given the fact that Moewus was accused of having fabricated much of his data, Schrödinger would provide a more respectable inspirational figure. In the 1940s many biologists considered Moewus one of the outstanding scientists of the twentieth century. When other researchers were unable to confirm his results, Moewus blamed their faulty experimental techniques. By the mid-1950s, however, even supporters of Moewus were coming to the conclusion that Moewus was a great fraud rather than a great scientist. In his 1948 essay on "The Genetics of *Chlamydamonas* with Special Regard to Sexuality," Watson suggested that while Moewus's ideas were certainly important, it was hard to believe that he could have performed all of the experiments he described. The genetic analyses reported by Moewus, Watson concluded, seemed to represent "wishful thinking" rather than facts obtained by experimentation. While Watson was at the Marine Biological Laboratory at Woods Hole in the summer of 1954, Moewus was participating in the botany course. Questions about whether Moewus had faked his data were still a matter of debate, but by the end of the summer Watson, along with Ruth Sager, Boris and Harriet Ephrussi, and even Tracy Sonneborn, concluded that Moewus "had irretrievably blown his last opportunity to prove his innocence."

While Watson's remarkable memoir *The Double Helix* (1968) disingenuously suggests that he left Indiana University totally innocent of chemistry, physics, and the scientific literature, his transcripts and doctoral dissertation on the effects of x-rays on phage replication provide some evidence to the contrary. Having received his doctorate at 22, Watson was sent to Europe in hopes that further training in the Old World would produce a veneer of maturity and sophistication. After an unsatisfactory year in Copenhagen, Watson transferred to Max Perutz's group at the Cavendish Laboratory to work on the molecular struc-

James D. Watson, Nobel Laureate, 1962

ture of nucleic acids extracted from plant viruses. At Cambridge, Watson found a kindred soul in Francis Crick, who shared his enthusiasm for speculating about the secret of the gene. In his second memoir, *Genes, Girls, and Gamow: After the Double Helix* (2001), Watson was still defending the importance of being willing to speculate. "If I have a good idea, I tend to believe it's true. An idea is better than no idea." Reflecting on his role in creating a revolution in biological thought, Watson concluded that "what I've said has generally been totally right."

Francis Crick grew up in Northampton, England, where his father and uncle ran a shoe factory that had been founded by their father. As a child Crick became interested in science and began to worry that everything of interest would have been discovered by the time he grew up. Rejecting his family's strict Congregationalist beliefs, Crick became a chronic skeptic and agnostic. Physics and chemistry, rather than vitalistic philosophies, would, he decided, provide the solution to problems thought to be utterly mysterious. After receiving a degree in physics from University College, London, in 1937, Crick began research on a rather dull problem concerning the viscosity of water under pressure. His research project was interrupted by World War II. As a member of the Admiralty Research Laboratory at Teddington, Crick carried out work on mine detection and development. When a mine blew up the apparatus he had constructed for his graduate work, Crick happily abandoned this research project. After the war, Crick hoped to do basic research in particle physics or physics applied to biology.

Like several other physicists who made the transition to biology, Crick attributed his conviction that physicists could solve fundamental biological problems to Schrödinger's *What Is Life?*. Topics that called into question assumptions about the borderline between living and nonliving, such as the nature of the gene and the workings of the brain, were of most interest to Crick. In 1949, Crick joined Max Perutz (1914–) and John Kendrew (1917–1997) at the Cavendish Laboratory in order to study the structure of proteins and other complex biological molecules by x-ray techniques. Although colleagues acknowledged Crick's intelligence and intensity in argumentation, he seemed to find it impossible to settle down to one topic long enough to construct a doctoral thesis. Indeed, the 1953 landmark paper describing the DNA double helix was published while Crick was still struggling to convert his "ragbag" of research topics into a presentable doctoral thesis.

In personality, Crick and Watson may have been poles apart, but when they met each immediately felt he had found a collaborator who shared a special approach to biology. Watson was described as a loner, quiet, and introverted; in contrast, Crick was famous for his loud voice and laugh. Watson described Crick as the brightest person he had ever worked with, an extroverted and ingenious theoretician who never had a "modest moment." Their enthusiastic collaboration illustrates Watson's talent for establishing useful scientific relationships, while his rather calculated and devious relationship with Maurice Wilkins (1916–), an x-ray crystallographer working at King's College, London, provides another example of a peculiar but ultimately productive association. The Watson-Crick collaboration was a fortunate one, for Crick doubted that either he or Watson could have discovered the structure of DNA alone. In contrast, Wilkins and Rosalind Franklin (1920–1958), another x-ray crystallographer at King's College, were unable to collaborate effectively; indeed, Wilkins and Franklin were barely on speaking terms. According to Crick, the structure of DNA was there waiting to be discovered, and if he and Watson had not done so, the puzzle might have been solved by Rosalind Franklin, Maurice Wilkins, Linus Pauling, or by further refinements of biochemistry. Instead, working together, Watson and Crick arrived at the structure of DNA by a combination of guessing, model building, and the unacknowledged exploitation of Rosalind Franklin's x-ray crystallographic data.

Many scientists were left to wonder why Watson and Crick had been so successful in discovering a model for DNA when they began what has been called the race for the double helix so far behind their competitors Rosalind Franklin and Maurice Wilkins, in London, and the ingenious chemist Linus Pauling, in California. Significantly, most of the explanations have focused on personality issues and interactions rather than specific scientific knowledge and technical opportunities. For 2 years Watson and Crick stumbled along the wrong path, making all possible incorrect choices for the structure of DNA. When they finally put together Chargaff's data on the base ratios and Franklin's study of the B-form of DNA with information about the proper forms of the bases, they were able to build a rational, elegant, and biologically interesting model of DNA in about 3 weeks. Despite the plethora of books and articles about the discovery of the structure of DNA, the most entertaining and controversial account of this remarkable burst of activity is still Watson's book *The Double Helix*. While Watson's account of the personalities involved in the story spared no one, his portrait of Rosalind Franklin was particularly harsh and, according to others who worked with her, totally fallacious. After undergraduate studies of physical chemistry at Cambridge University, Franklin performed structural studies of coal and coke in the laboratories of the British Coal Utilization Research Association. She returned to Cambridge and earned her Ph.D. in 1945. Because of her expertise in x-ray crystallography, Franklin was asked to join the research group at King's College, London in 1951. Unfortunately, rather than collaborating, Franklin and Wilkins worked separately and competitively. In 1953 Franklin moved to the laboratory of J. D. Bernal at Birkbeck College to conduct further studies of DNA and the structure of tobacco mosaic virus and the polio virus. Franklin died of cancer in 1958, only 38 years of age. She did not know that Watson, Crick, and Wilkins would share the 1962 Nobel Prize for the discovery

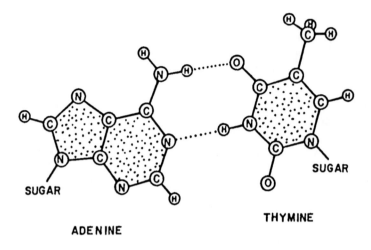

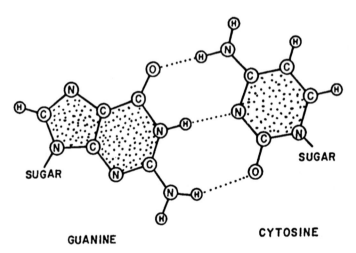

Diagram of the complementary base pairs found in the DNA double helix. Adenine pairs with thymine and guanine pairs with cytosine.

of the DNA double helix, nor did she realize how her unpublished data had been used by Watson and Crick, because Watson's autobiography was not published until 1968, 10 years after her death.

In their first *Nature* paper (April 25, 1953), Watson and Crick described their "radically different structure for the salt of deoxyribose nucleic acid." Incorporating the usual chemical assumptions and evidence that the structure was a two-stranded double helix, the novel feature of the structure was "the manner in which the two chains are held together by the purine and pyrimidine bases." The planes of the bases in the proposed double helix were perpendicular to the fiber axis of the polynucleotide chain and the bases on the two strands were joined by hydrogen bonds. The dimensions of the double helix indicated that a purine on one chain always paired with a pyrimidine on the other chain; specifically, adenine always paired with thymine and guanine with cytosine. Although any sequence of bases was possible on one chain, the rules of base pairing automatically determined the sequence of bases on the other chain. Thus, the Watson-Crick double helix immediately explained Chargaff's data concerning the molar ratios of purines to pyrimidines and how DNA exhibits order and stability, as well as variety and mutability.

Despite the irresistible elegance of the double-helical DNA model, many scientists felt that the Watson-Crick paper had a rather hollow tone. The x-ray data available in the literature at the time were, as the authors admitted, insufficient for a rigorous test of the hypothetical structure. More exact data, however, appeared in the two communications that followed the Watson-Crick proposal. One paper by Maurice Wilkins and his coworkers discussed the available x-ray data for calf thymus DNA, nucleoprotein preparations from sperm heads, and bacteriophage. A paper by Rosalind Franklin and R. G. Gosling presented the most refined x-ray diffraction patterns of DNA then available, and Franklin demonstrated that the double helix was consistent with the x-ray patterns. Although Watson and Crick only alluded to the question of how the genetic material might function at the molecular level, the double helix immediately suggested means by which DNA molecules could be used to make copies exactly like the original. In a sentence described as one of the most coy statements in the scientific literature, Watson and Crick proclaimed: "It has not escaped our notice that the specific pairing we have postulated immediately suggests a possible copying mechanism for the genetic material." Crick, who insisted on some discussion of the genetic implications of the model, had settled for this "compromise" because of his co-author's lingering uncertainty. In a series of papers published shortly after their proposal of the double helix, Watson and Crick elaborated on the genetic implications of the complementary structure of DNA. It took about 5 years for experimental tests to prove the essential validity of the DNA replication scheme proposed in 1953. In 1957 Matthew Meselson (1930–) and Franklin W. Stahl (1929–) carried out a series of ingenious experiments that traced the fate of parental strands of DNA and proved that DNA replication was semiconservative, that is, each old DNA strand became associated with a newly synthesized strand of DNA.

With the promulgation of the Watson-Crick model, researchers were soon engaged in transforming the chromosome theory into the nucleic acid theory.

The gene was no longer an abstract factor, but an information molecule. In his memoir *What Mad Pursuit* (1988), Crick assessed what he called the "classical period" of molecular biology, from the discovery of the DNA double helix in 1953 to about 1966 when the genetic code was elucidated. The 1966 annual Cold Spring Harbor meeting was devoted to the genetic code. For Crick it marked the end of classical molecular biology. He had begun biological research in 1947 and had seen the questions about the gene that had most intrigued him essentially answered by 1966.

Satisfied that the chemical nature of the gene had been discovered, scientists were able to link knowledge of what the gene was to studies of how the gene worked. Watson and Crick played an important role in formulating the general principles explaining how information stored in DNA is replicated and passed on to daughter molecules and how information was transferred from DNA to the metabolic activities of the cell. Even before the structure of DNA had been elucidated and confirmed, Watson wrote himself a message: "DNA → RNA → protein." The arrows in this slogan did not refer to actual chemical transformations, but represented the transfer of genetic information from the base sequences of DNA to the amino acid sequences in proteins. Watson's cryptic slogan about the uni-directional flow of information from DNA to RNA to protein has been called the Central Dogma of molecular biology. Already certain that DNA was the template for RNA, Watson predicted that RNA chains might serve as templates for protein synthesis. Determining how nucleic acids could determine the structure of protein molecules was a problem that seemed to elude the approach that led to the Watson-Crick DNA model. Eventually, scientists would discover that information could flow from RNA to DNA by means of an enzyme called reverse transcriptase; this insight has been critical to understanding a group of viruses called retroviruses, one of which is the human immunodeficiency virus (HIV), the virus associated with the acquired immunodeficiency syndrome (AIDS).

After the discovery of the DNA double helix, Watson returned to the United States for a 2-year stay at the California Institute of Technology. In 1956 Watson accepted a position as professor of biology at Harvard University. He became the director of the Cold Spring Harbor Biological Laboratories in 1968 and its president in 1994. Watson's best known books include his memoir *The Double Helix* (1968), *The Molecular Biology of the Gene* (1965), the first widely used textbook on molecular biology, and *The Molecular Biology of the Cell* (with John Tooze and David Kurtz, 1983).

Shortly after the publication of the DNA model in *Nature*, George Gamow (1904–1968), a physicist known for his contributions to cosmology, nuclear physics, and his attempts to popularize physics through a series of books about the adventures of "Mr. Tomkins," wrote to Watson with the suggestion that DNA might serve as a direct template for polypeptide synthesis. Working out variations of an overlapping code, Gamow proposed that DNA had 20 kinds of "cavities" that could serve as direct templates for the 20 amino acids. After ruling out Gamow's hypotheses, Crick composed a manuscript entitled "On Degenerate Templates and the Adaptor Hypothesis." Given the unlikelihood that DNA and RNA could serve as direct templates for the amino acids, Crick proposed the existence of 20 adaptors and 20 special enzymes that could join an amino acid to its adaptor. The combination would then fit itself to the mRNA and the amino acids would be lined up to be linked as proteins. Gamow did not solve the problem of the genetic code, but he and Watson worked together to form the RNA Tie Club. Each member of the club corresponded to one of the amino acids.

While Watson and his associates founded the RNA Tie Club and speculated about how the genetic code might be solved by logic, model building, and guesswork, biochemists showed that with truly herculean labors the puzzle could be solved experimentally. During the 1950s work by Severo Ochoa (1905–1993), Arthur Kornberg (1918–), and others led to the discovery of enzymes that could be used to synthesize nucleic acids in vitro, thus creating new ways to investigate DNA and RNA. Work in enzymology and high-energy phosphates eventually led Ochoa to the enzymes involved in the biosynthesis of the nucleic acids. After earning his M.D. from the University of Madrid in Spain, Ochoa's interest in biochemistry took him to laboratories in Glasgow, Heidelberg, London, and Washington University, St. Louis, where he worked with Carl and Gerty Cori, before accepting a position at the New York University School of Medicine. While working with the Coris in 1947, Kornberg decided to abandon nutritional studies and became a lifelong "enzyme hunter." In his autobiography, Kornberg recalled that biochemistry had seemed a very dreary field when he was a medical student, but once he embarked on studies of enzymology with Ochoa and the Coris, he became fascinated by biological macromolecules and complex metabolic pathways. During the 1950s, Kornberg determined the pathways of nucleotide biosynthesis and began his research on nucleic acid biosynthesis and replication. In 1955 Ochoa and Marianne Grunberg-Manago isolated polynucleotide phosphorylase, a bacterial enzyme with which they could synthesize RNA in vitro. One year later Kornberg isolated the enzyme DNA polymerase from the bacterium *Escherichia coli*. When provided precursors and an appropriate primer, the enzyme catalyzed the formation of DNA molecules. Although DNA polymerase was a particularly exciting discovery, Kornberg said that he had "never known a dull enzyme." Ten years later he synthesized biologically active viral DNA. In 1959 Kornberg shared the Nobel Prize with Ochoa for their "discovery of the mechanisms in the biological synthesis of ribonucleic acid and deoxyribonucleic acid." Further studies of RNA and DNA biosynthesis made it possible to create synthetic polynucleotides that could be used to determine the genetic code.

THE GENETIC CODE

Given the assumption that genes determined the amino acid sequence of proteins, the double-helix structure suggested that instructions for building proteins must be encoded in the base sequence of DNA. In a paper presented to the Society for Experimental Biology in 1957, Crick referred to this as the "sequence hypothesis," that is, the sequence of nucleic acids determined the sequence of amino acids in proteins. Because DNA was found in the nucleus, whereas protein synthesis seemed to occur in the cytoplasm, some means of transmitting a copy of a gene from the nucleus to the cytoplasm was necessary. It was fairly easy to imagine that the hypothetical copy of the gene existed in the form of an RNA molecule made from DNA according to the rules of base pairing. The process of transferring information from a DNA template to an RNA copy is known as *transcription*; the information is still present in the language of nucleic acids. The really difficult conceptual problem was imagining how the information in nucleic acid could dictate the sequence of a protein. Transferring information from the language group of nucleic acids to that of proteins is known as *translation*; the difficulty factor in this step can be compared to translating ancient Egyptian hieroglyphics into modern English. In other words, how can about 20 amino acids, and a seemingly infinite number of proteins, be described by nucleic acid molecules containing only four different bases?

Although Gamow and Crick were unsuccessful in deciphering the genetic code, logic did seem to offer ways of resolving the differences between the language of the nucleic acids, which was communicated by four nucleotide bases, and that of proteins, which was expressed in terms of about 20 amino acids. Presumably this could be done in a manner analogous to the way that the Morse code uses dots and dashes to represent all the letters of the alphabet, an alphabet that can be used to generate an infinite number of meaningful (and meaningless) texts. In the case of the genetic code, taking the bases three at a time provides $4 \times 4 \times 4 = 64$ triplet combinations, which became known as codons. Speculations about the possibility that special "cavities" along the DNA molecule could serve as direct templates for the 20 amino acids led Crick to the adaptor hypothesis. Given the unlikelihood that strands of nucleic acid could serve as direct templates for the assembly of amino acids into polypeptide chains, Crick proposed the existence of 20 adaptors and 20 enzymes that could join an amino acid to its adaptor. The combination would then fit itself to a specific messenger RNA (mRNA) so that the amino acids could be linked together as proteins. By 1957 biochemists had confirmed the existence of adapter molecules, which were called transfer RNAs (tRNA); each amino acid had at least one specific tRNA. In 1968 Robert William Holley (1922–1993), Har Gobind Khorana (1922–), and Marshall Nirenberg (1927–) shared the Nobel Prize for Medicine or Physiology "for their interpretation of the genetic code and its function in protein synthesis." As a result of their independent, but interrelated researches, they made it possible to determine all the three-letter words, or codons, that make up the genetic code. When the codons were first worked out and studied in several different species, scientists exuberantly called it the *universal genetic code*. As exceptions and complications eventually emerged, the more modest term *standard code* came into use.

After majoring in chemistry at the University of Illinois, Holley began his graduate work in organic chemistry at Cornell University, where he was a member of the team that first synthesized penicillin. Although Holley was originally interested in the organic chemistry of natural products, his research program broadened to include work on amino acids, peptides, nucleic acids, and the biosynthesis of proteins. In 1955, while at Caltech, Holley studied protein synthesis and attempted to detect the acceptor of activated amino acids. This work led to the discovery of the special type of nucleic acid called transfer RNA. Although all the tRNAs seemed to be relatively small molecules, the laborious work of isolating and sequencing alanine transfer RNA took about 10 years. In order to determine the sequence of nucleotides in alanine tRNA, Holley developed procedures conceptually similar to those introduced by Fred Sanger for sequencing the amino acids in proteins. In 1965 Holley published his classic paper containing the total nucleotide sequence of yeast alanine transfer RNA. This work represented the first determination of the complete structure of a biologically active nucleic acid and provided key insights into the mechanism by which the information in the genetic code is used to determine the amino acid sequence of proteins. By 1968 a dozen tRNAs had been sequenced and the relationship between the configuration of transfer RNAs and their role in protein was beginning to emerge.

Marshall Nirenberg, at the National Institutes of Health, initiated the direct experimental approach that led to deciphering the genetic code. Working with Johann Heinrich Matthaei (1929-), Nirenberg began an ingenious series of experiments on protein synthesis in a cell-free bacterial system. At the Fifth International Congress of Biochemistry, held in Moscow in 1961, Nirenberg announced that by using a synthetic RNA containing only uracil (polyuridylic acid), he had obtained a polypeptide containing only the amino acid phenylalanine. In other words, UUU was the codon for the amino acid phenylalanine.

Invariably described as shy, quiet, and unassuming, Nirenberg was virtually unknown, and his initial talk, entitled "The dependence of cell-free protein synthesis in *E. coli* upon naturally occurring and synthetic template RNA," was poorly attended. Fortunately, scientists who heard his presentation and understood its implications arranged for Nirenberg to speak at the final session of the full Congress. According to Watson, Nirenberg's "bombshell" stunned and convinced almost all members of the audience that this technique would soon lead to cracking the genetic code. Although many scientists attempted to solve the code by using synthetic RNAs, the sequence of bases within many of the triplets remained elusive. At the Sixth International Congress of Biology, held in New York in 1964, Nirenberg described further refinements of the tRNA binding method that could be used to decipher the sequence of bases within each triplet. Within a few years the task of determining all 64 codons was essentially complete; the code did indeed consist of three-letter, nonoverlapping words, including many that were redundant, and some that served as punctuation marks for beginning and terminating polypeptide chains.

Har Gobind Khorana was born in Raipur, a village in the Punjab region of India, which is now part of West Pakistan. In 1948 he earned a Ph.D. from the University of Liverpool, England. Originally trained as an organic chemist, Khorana became particularly interested in proteins and nucleic acids. Khorana developed methods for the synthesis of both RNA and DNA polynucleotide chains in which the base sequences were precisely known. By synthesizing each of the 64 possible triplets, Khorana confirmed and extended Nirenberg's original assignment of the codons, proving that the code did indeed consist of three-letter, nonoverlapping words, read in a specific linear manner. These detailed studies also established that the code was redundant, that is, some amino acids corresponded to more than one triplet. Some codons appeared to serve as "punctuation marks" to begin and end the synthesis of particular polypeptide chains. In 1970 Khorana announced that his team had accomplished the complete synthesis of the first wholly artificial gene, the DNA sequence that coded for alanine transfer RNA. In the popular press Khorana's accomplishment was widely described as one of the greatest breakthroughs of molecular biology and the possible prelude to therapeutic or dangerous manipulations of human genes. During the 1980s, Khorana turned to research on the chemistry and molecular biology of rhodopsin, the light-transducing pigment of the retina, and bacteriorhodopsin, a form of rhodopsin found in bacteria.

The *Journal of Molecular Biology* (*JMB*) was founded by John Kendrew in 1959 when the subject was so new that, according to Sydney Brenner (1927–), few people were willing to call themselves molecular biologists in public. By the 1970s molecular biology had become so pervasive that almost all biological researchers might be called molecular biologists. Many of the most fundamental discoveries of molecular biology appeared in *JMB* between 1959 and 1975, including the discovery of messenger RNA; the mechanism of protein biosynthesis; the structure of the genetic code; DNA replication; the methodology that established nucleic acid sequencing, cloning, transfection, restriction enzymes, DNA mapping, site-directed mutagenesis, and intricate cellular control systems. Technical aspects of the transformation of molecular biology are best grasped by comparing the heroic protocols described in the "classic" papers in *JMB* with the advertisements for commercially available kits and instruments in later issues of the journal.

For Francis Crick, the Cold Spring Harbor annual meeting in 1966, which was devoted to the genetic code, marked the end of the classical era of molecular biology. With molecular biology ostensibly firmly established, Crick decided it was time to move on and revolutionize other backward fields, such as embryology and developmental biology, neurobiology, and biogenesis. In any case, his attempts to demonstrate the relationship between a mutation in a gene and its altered protein had literally produced little but frustration and tears. Working with the British biochemist Vernon M. Ingram (1924–), Crick planned to look for alterations in lysozyme, an enzyme found in egg white and human tears. Unfortunately, the project was a failure and no mutants were found. The concept was, however, a valuable one and Ingram was later involved in studies of mutations in hemoglobin, the oxygen-carrying protein found in red blood cells.

Initially attracted to the study of development and differentiation, Crick found the obstacles to progress in this field unacceptably intractable. Eventually other problems, such as the nature of mind and the origin of life, claimed Crick's attention. With the biochemist Leslie Orgel (1927–), Crick published a book called *Life Itself: Its Origin and Nature* (1981) based on the panspermia theory that had been introduced by the physical chemist Svante Arrhenius (1859– 1927). According to Crick's version of "directed panspermia," life on earth might have arisen, not by random biochemical reactions in the primordial soup, as postulated by Russian chemist A. I. Oparin (1894–1980), but from bacteria or "seeds" deliberately sent to earth via a rocket launched by a higher civilization that had evolved elsewhere in the universe. This willingness to speculate on untestable ideas has been lauded as evidence of Crick's eclectic curiosity, creativity, and boldness in pursuit of theories, even theories that appear to be closer to science fiction than to orthodox science. After spending a sabbatical year at the Salk Institute for Biological Studies in La Jolla, California, Crick was offered an endowed chair and persuaded to stay. He decided to turn to research on the brain and soon discovered that the state of the neurosciences in the 1980s was rather like that of genetics and embryology in the 1920s and 1930s: the major questions were unanswered and probably unanswerable without new techniques and new ideas. Many years later Crick admitted that he had not anticipated the remarkable developments that would occur in molecular biology in response to the discovery of recombinant DNA, rapid DNA sequencing, monoclonal antibodies, and ribozymes.

While the Central Dogma originally placed RNA in a rather passive and subservient role between DNA and protein, in the 1980s Thomas R. Cech (1947–) and Sidney Altman (1939–) independently demonstrated that RNA could function as a biocatalyst as well as an information carrier. The discovery of ribozymes, RNA molecules with catalytic activity, prompted some RNA enthusiasts to speculate that RNA molecules, under the appropriate circumstances, might perform virtually all the functions previously ascribed to DNA and proteins. In 1989, Cech and Altman shared the Nobel Prize in Chemistry for the discovery that RNA could act as a catalyst as well as an information carrier.

HOW GENES WORK

When the double helix was announced it served as a guide for much research, but the question of where and how protein synthesis occurred remained obscure. Protein synthesis seemed to require a messenger from nucleus to cytoplasm, as summarized in Watson's slogan: "DNA makes RNA makes protein." In the 1950s it was known that most of the cytoplasmic RNA was found in small particles called ribosomes, which consisted of RNA and proteins. Some scientists thought that the RNA in the ribosomes was the messenger and that each ribosome was involved in synthesizing one protein. Experimental evidence, however, indicated that ribosomal RNA was very stable and that it seemed to come in two fixed lengths rather than the assortment that should characterize the messengers that would code for very diverse proteins. Later studies revealed the complex interconnections between ribosomal RNAs, proteins, and translation factors, but ribosomes apparently functioned as protein synthesizing factories.

One of the most fruitful approaches to the relationship between gene activity and the hypothetical messenger resulted from collaboration between François Jacob (1920–), and Jacques Monod (1910–1976) at the Pasteur Institute. At the time, Monod was studying bacterial growth factors and their inducible enzymes, such as beta-galactosidase. Jacob and Élie Wollman, who were investigating the mechanism of the transmission of genetic information during bacterial mating, devised an ingenious experiment to determine the sequence of genetic transfer between bacterial cells. Joshua Lederberg, who had been skeptical of the Hayes hypothesis concerning bacterial matings, found the "interrupting mating" experiments conducted by Jacob and Wollman absolutely convincing. Jacob and Wollman realized that if bacterial conjugation involved the sequential transfer of genetic material from a "male" donor cell to a "female" recipient, then the time taken for genes to appear in the recipient cells could be used to create a genetic map. After allowing appropriate strains to mate for measured intervals, bacterial cultures were put into a blender. Jacob had purchased the blender while he was in the United States for a meeting at Cold Spring Harbor Laboratory. When his wife refused to have such an appliance in her kitchen, he brought it to the laboratory. Vigorous agitation separated bacteria engaged in conjugation and stopped the transfer of DNA. Crick referred to this violent disruption of bacterial conjugation as "molecular coitus interruptus." Indeed, Jacob and Wollman's "blender experiment" was rather reminiscent of Spallanzani's experiments on frogs also involved in the act of mating. Transfer of the first gene took about 7 minutes; the most distant gene was transferred after about 100 minutes.

In a landmark experiment carried out by Jacob, Monod, and Arthur Pardee (1921–), this gene transfer system was used to determine what happened when a gene that coded for an inducible enzyme was transferred to recipient bacteria that lacked this gene. If new ribosomes, specific for the gene, had to be created, as suggested by Crick and others, protein synthesis could not occur until the ribosomes had accumulated. Reflecting on discussions of work being carried out independently by Monod and André Lwoff, Jacob realized that his colleagues were working on different phases of the same genetic mechanism. Jacob's initial insight led to a new theoretical framework known as the "operon theory," as well as support for the messenger concept. The classic demonstration of the operon theory, know as the "PaJaMo experiment," was named for Pardee, Jacob, and Monod, who published their results in the *Journal of Molecular Biology* (1959). Jacob and Monod's operon theory provided an explanation for the induction of enzymes and the genetic control of the synthesis of enzyme molecules. The operon, which appeared to be the fundamental unit of bacterial gene expression and regulation, consisted of structural genes, regulator genes, and control elements. According to the operon theory of gene regulation, genes were endowed with the dual functions of encoding and regulating proteins. If this concept applied to both unicellular and multicellular organisms, it essentially eliminated the intractable paradox of how identical genomes could exist in biochemically differentiated cells. Modification of cells during embryological development or adaptation to new environments could occur despite identical nuclear genomes; the induction of specific proteins would establish fundamental differences in the cytoplasm of differentiated cells. Jacob, Monod, and Lwoff shared the Nobel Prize in 1965 for their "discoveries concerning the genetic regulation of enzyme and virus synthesis." In addition to explaining adaptive mechanisms in terms of gene regulation, Monod proposed a very fruitful general theory of the regulation of enzyme activity known as allosteric control.

The collaboration between Jacob and Monod was very different from the better known relationship between Watson and Crick; it was more complex, subtle, and productive. Drawn to theories that were simple and elegant, Monod believed that "a beautiful model or theory may not be right; but an ugly one must be wrong." In contrast, Jacob reflected that he "did not find the world so strict and rational." Some historians have argued that the perspective that French scientists, such as Jacob and Monod, brought to studies of the mechanism of gene expression was quite different from that of their American and British counterparts. That is, French biologists did not abandon complex questions of growth, nutrition, and development when they adopted modern Mendelian-Morganist genetics. Indeed, some historians argue that neo-Lamarckism and Lysenkoist biology were "compatible" with the broad concerns of French biologists who presumably found classical genetics too restrictive. Jacob, however, had no sympathy for Lysenkoism and denounced its progenitor as a "charlatan" and an "unscrupulous careerist" who had imposed an "idiotic theory on biology and a catastrophic practice on agriculture."

Nevertheless, the work of French biologists was particularly important in resolving questions about the relationship between classic Mendelian genetics and theories of cytoplasmic inheritance, that is, theories concerning hereditary materials or principles located outside the cell nucleus. Oversimplification had become a danger for Mendelian-Morganist genetics as well as for alternative theories of cytoplasmic inheritance. Distinctions between nucleus and cytoplasm are essentially irrelevant in the case of many microorganisms, but geneticists also needed to remember that the eukaryotic cell is much more complicated than a nucleus suspended in homogeneous cytoplasmic enzyme soup. From several different research traditions, evidence of cytoplasmic inheritance accumulated slowly until the 1960s, when the techniques of biochemistry, biophysics, Mendelian genetics, and molecular genetics were brought to bear on the problem.

Since the rediscovery of Mendel's principles, research in genetics has generally been guided by the belief that the fundamental units of inheritance are located on chromosomes in the cell nucleus and that, therefore, the cytoplasm could be ignored genetically. Embryologists, however, objected that nuclear genes, identical in every cell, could not explain how cells differentiated from each other in the course of development. They argued that differences between cells in the developing embryo must have a basis in the cytoplasm. Some biologists suggested that nuclear genes determined relatively minor differences between individual within a species, while cytoplasmic determinants were responsible for more fundamental characteristics and the harmonious development of the organism.

Both parents contribute equally to the inheritance of nuclear genes, but hereditary factors in the cytoplasm (extranuclear genes) are more likely to be transmitted through the maternal line because the egg is rich in cytoplasmic organelles. Therefore, extranuclear genes would not follow Mendel's statistical laws of segregation and recombination. It is interesting that Carl Correns, primarily remembered as one of the rediscoverers of Mendel, carried out many experiments on plants that suggested non-Mendelian inheritance. As early as 1909 geneticists were reporting examples of non-Mendelian inheritance in higher plants, such as green-and-white variegated patterns. These patterns seemed to be related to the behavior of the chloroplasts, organelles that contain the photosynthesizing apparatus of green plants. Because of their rather large size, scientists have been able to study the behavior of chloroplasts in dividing cells with the light microscope since the 1880s.

During the 1930s, some scientists were already investigating the possibility of using various unicellular organisms as model systems for genetic research. Such studies led Victor Jollos (1887–1941) to claim that environmentally induced modifications, such as resistance to heat and arsenic, could be transmitted for many generations after the removal of the inducing agent. According to Jollos, the "acquired modifications" were transmitted by the cytoplasm, rather than the nucleus. Herbert Spencer Jennings (1868–1947), an unorthodox advocate of Lamarckian ideas, pioneered the study of various microscopic organisms, including *Paramecium*, a unicellular organism that could reproduce asexually or engage in sexual conjugations. The sexual reproductive cycle of these unicellular organisms was, however, poorly understood. After studying with Jennings, Tracy M. Sonneborn (1905–1981) worked out the mating types of *Paramecium* so that he could analyze the interactions of genes, cytoplasm, and environment. Sonneborn's studies of the kappa factor, or "killer" trait, seemed to provide evidence of cytoplasmic inheritance that challenged classical genetic theory. Cytoplasmic determinants, often referred to as "plasmagenes," however, became associated with the neo-Lamarckian concept of the "inheritance of acquired traits" and Lysenkoist genetics.

Geneticists soon discovered that many cases of apparent cytoplasmic heredity could be explained in terms of the distribution of well-known cytoplasmic organelles and mechanisms that were compatible with Mendelian genetics. Genes located in mitochondria and chloroplasts became the most extensively studied examples of extranuclear inheritance. The "petite" mutants in yeast and the "poky" mutants of the bread mold *Neurospora*, for example, were traced to mitochrondrial genes. Some researchers believe that mitochondria, chloroplasts, and perhaps other cytoplasmic organelles are basically endosymbionts, that is, descendants of free-living organisms that have become permanent constituents of plant and animal cells. Sonneborn's killer particles, for example, were identified as bacterial symbionts.

In the 1950s, Ruth Sager (1918–1997) and Boris Ephrussi (1901–1979) were particularly important in clarifying cytoplasmic inheritance associated with chloroplasts and mitochondria, respectively. Sager's research combined genetic analysis of the recombination and linkage of cytoplasmic genes with biophysical studies of organelle DNA and biochemical studies of chloroplast mutations. Using cytoplasmic mutants as an analytical tool, rather than an ideological issue, Sager helped clarify the interdependent network of nuclear and organelle gene products and signals. Advances in experimental methods made in the 1960s and 1970s allowed scientists to detect cytoplasmic genes and to demonstrate that mitochondria and chloroplasts contain their own DNA and RNA. Mitochondrial DNAs, however, appear to differ from nuclear DNAs in various fundamental ways. Recognition of extranuclear genetic systems raised very interesting questions about the possible evolutionary advantage of maintaining such genetic complexity within the cell. In 1981 scientists reported the complete nucleotide sequence of human mitochondrial DNA, making it possible to add the analysis of mitochondrial DNA to the study of ancient bones and tools as a means of studying our prehistoric ancestors and constructing family trees. Studies of mitochondrial DNA led to reports that scientists had found Adam and Eve, or at least a female common ancestor of modern human beings who had lived some 200,000 years ago. This ancestral mother of the human race was dubbed "Mitochondrial Eve."

The discovery of the nature of the genetic material in the 1950s served as a stimulus and a framework guiding investigations of genetic mechanisms at the molecular level. Since the beginning of the twentieth century, however, the study of inherited metabolic disorders has made it possible to gain a glimpse of the way the gene works, or fails to work. That is, certain inherited diseases are due to defective enzymes, which represent the altered gene products of mutant genes. George W. Beadle (1903–1989) and Edward L. Tatum (1909–1975) systematically exploited this correlation in their studies of the genetic control of biochemical reactions in *Neurospora*, the red bread mold. As Beadle pointed out in his reviews of the field, however, Archibald Edward Garrod (1857–1936) had previously called attention to the link between Mendelian patterns of inheritance and certain "inborn errors of metabolism" found in the most refractory but most interesting of all biological systems—*Homo sapiens*.

It might, therefore, be said that the study of biochemical genetics began in 1902 with Garrod's report on alcaptonuria. Individuals with alcaptonuria cannot completely metabolize the amino acid tyrosine and, therefore, excrete an intermediate of tyrosine metabolism known as homogentisic acid. The condition is more frightening to new parents than it is distressing to afflicted individuals, because the urine of babies with alcaptonuria darkens when exposed to air and causes peculiar stains in their diapers. From tests of alcaptonuric individuals, Garrod concluded that the disorder was not the manifestation of an acquired disease, but the result of some alteration in a metabolic process. Other human "chemical abnormalities," such as albinism, cystinuria, and pentosuria, seemed to be similar kinds of disorders attributable to "inborn errors of metabolism." Analyzing the family trees of afflicted individuals suggested that such disorders were inherited as Mendelian recessives. Indeed, as would be likely for recessive genes, a large proportion of the affected individuals were the children of first cousins.

Eventually, Garrod traced the defect in alcaptonuria to the penultimate step in the metabolic breakdown of specific amino acids (tyrosine and phenylalanine). Individuals with congenital alcaptonuria, Garrod concluded, lacked a special enzyme that was normally involved in splitting the benzene ring of tyrosine and phenylalanine. Garrod proved that his patients could metabolize precursors of homogentisic acid, but that the metabolic pathway was interrupted at that point. Homogentisic acid, therefore, accumulated and was excreted in the urine. Genes were related to enzymes as indicated by the fact that individuals who had inherited two recessive genes lacked the enzyme needed for specific metabolic reactions. The parents of the affected individuals carried the recessive gene, but did not display the condition because they also had one normal gene for the enzyme in question.

Since the publication of Garrod's observations, many other human disorders have been shown to be inborn errors of metabolism, or genetic diseases. But unlike Garrod, who was primarily a clinician, modern geneticists have generally turned to smaller and smaller creatures for their investigations. Human beings, with their extended childhood, peculiar mating customs, and small family size, hardly seem to represent likely experimental animals. Nevertheless, George Beadle and Edward Tatum, who shared the 1958 Nobel Prize in Medicine or Physiology for establishing the "one gene, one enzyme theory," later said that their research would have been less circuitous if they had known of Garrod's studies before they began their research on *Neurospora*. Although Garrod may have been a relatively obscure figure for biochemists, William Bateson had included references to Garrod's work in *Mendel's Principles of Heredity* (1909) and *Problems of Genetics* (1913). According to Beadle, Garrod was the true father of biochemical genetics. While Beadle called Garrod the father of biochemical genetics for his work on inborn errors of metabolism, Mendelian inheritance, and defective enzymes, some historians have argued that this is just another "precursor myth" fabricated by scientists. There is, however, no doubt that Garrod was interested in applying Mendelian laws to the pattern of transmission of congenital defects, and his careful attention to the chemical literature is obvious in his detailed examination of the biochemical basis of the clinical disorders found in his patients.

Born on a small farm in Wahoo, Nebraska, Beadle earned his B.S. and M.S. from the College of Agriculture of the University of Nebraska. At Cornell University, Beadle studied maize genetics with Rollins A. Emerson (1873–1947), head of the department of plant breeding and an early American advocate of Mendelian genetics. During postdoctoral studies in T. H. Morgan's laboratory, Beadle met the French embryologist Boris Ephrussi, who had come to Caltech to learn *Drosophila* genetics. Before accepting a faculty appointment at Harvard University, Beadle decided to spend a year in Ephrussi's laboratory investigating biochemical genetics. Beadle and Ephrussi attempted to analyze the biochemical basis of heredity through studies of the genetic control of eye color pigments in fruit flies. They demonstrated that a larval eye taken from the mutant-type vermilion fly developed into an adult structure with the wild-type eye color when it was implanted in the abdomen of a wild-type larva. Tests with various mutants proved that the steps involved in the sequential synthesis of eye pigment were controlled by several distinct genes. Some of the biochemical intermediates in the biosynthetic pathway to pigment were chemically identified. In 1937 Beadle became professor of biology at Stanford University. When Morgan died in 1945, Beadle was awarded the position of chairman of the Division of Biology at Caltech.

After earning his Ph.D. in biochemistry at the University of Wisconsin and performing postdoctoral research at the University of Utrecht in the Netherlands, Tatum moved to Stanford to work in Beadle's laboratory as a research associate. Frustrated with *Drosophila* work, Beadle and Tatum turned to the bread mold *Neurospora crassa* as a means of unraveling the biochemistry of the relationship between genes and gene products. Beadle and Tatum decided to reverse the procedures generally used to identify specific genes with particular biochemical reactions. Instead of taking a mutant as their starting point and then searching for the chemical reaction it controlled, they decided to begin with known chemical reactions and then look for the genes controlling them. They, therefore, needed an organism with well-known biochemical pathways. *Neurospora* was ideal; it had a fairly short life cycle, and techniques for genetic analysis had already been worked out. Furthermore, the wild-type mold could be grown on a simple synthetic medium containing only a carbon source, certain inorganic salts, and the vitamin biotin. After using x-rays to generate mutations, Beadle and Tatum selected mutants that had lost the ability to synthesize certain organic substances. Biochemical and genetic analyses proved that the mutant strains were genetically different from the parental type.

During the 1940s, the work done by Beadle and Tatum was received with considerable skepticism. As often happens with ideas that are originally regarded as heretical, eventually there were attacks on their originality and priority. In particular, Archibald Garrod and Franz Moewus were cited for having previously proposed that genes controlled the action of enzymes and for using microorganisms to analyze biochemical genetics. Moewus had published a paper on the biochemical genetics of sexuality in *Chlamydomonas*, a unicellular green algae, in 1938, three years before Beadle and Tatum's first *Neurospora* paper. However, statisticians attacked the data published by Moewus, and other biologists were unable to reproduce his experimental results. In spite of the allegations of fraud, some biologists credited Moewus with creating interest in studies of the genetics of microorganisms.

In 1945 when Beadle enunciated the one gene, one enzyme hypothesis, the chemical identity of the gene was still unknown. After 1953 the question "What is the gene?" was generally framed in terms of explaining how information encoded in DNA could be used to direct the synthesis of proteins. Viruses, which could now be seen as nucleic acids, or genes, wrapped in protein coats, seemed to be the perfect models for exploring the flow of genetic information. Nevertheless, despite the prominence of bacteriophage studies in the evolution of molecular genetics during this period, it was the analysis of another human disease that led to the first significant demonstration of the relationship between a heritable gene mutation and a specific molecular alteration in a gene product. By analyzing the abnormal hemoglobin produced by patients with sickle cell anemia (SCA), Linus Pauling (1901–1994) and Vernon M. Ingram established the existence of a specific molecular disease.

According to a survey of scientists conducted in the 1970s, Linus Pauling ranked with Newton and Einstein among the 20 most important scientists of all time. Willing to tackle social, political, and medical questions, as well as science, Pauling could also be ranked as one of the most controversial scientists of all time. Linus Pauling was the only scientist to win two unshared Nobel Prizes. In 1954 he won the Chemistry Prize "for his research into the nature of the chemical bond and its application to the elucidation of the structure of complex substances." The biochemist John Edsall (1902–) called Pauling's formulation of the alpha helix "one of the greatest triumphs of creative thinking in all of the field of protein chemistry." In 1962 Pauling was awarded the Nobel Peace Prize for warning of the dangers of radioactive fallout in nuclear weapons testing and warfare. His activities in part led to the banning of atmospheric testing of nuclear weapons by the United States and the Soviet Union.

Pauling grew up in a small town and began his higher education at Oregon Agricultural College. After the Agricultural College had became Oregon State University, Pauling joined the faculty and accepted a heavy teaching load. A prolific author, he published hundreds of papers and many books on a wide range of subjects, including the structure of crystals, the nature of the chemical bond, quantum mechanics, proteins and antibodies, biology and medicine, war and peace, the dangers of atmospheric testing of nuclear weapons, science and world affairs. His antiwar activities led to intensive and hostile questioning by a congressional subcommittee in 1960. Later in life he became involved in the holistic or alternative health movement and a leader in the battle for orthomolecular medicine and megadose vitamin C therapy and prophylaxis for colds, cancer, and schizophrenia. His controversial ideas about vitamin C and cancer led to the unprecedented action taken by the Editorial Board of the *Proceedings of the National Academy of Sciences* (*PNAS*) in rejecting a paper co-authored by Pauling and Ewam Cameron, a Scottish surgeon and cancer specialist. This was probably the first time that *PNAS* refused to publish a paper by a member of the Academy. Pauling's advocacy of orthomolecular medicine resulted in a loss of conventional sources of funding for his research and his decision to leave Stanford University. In 1973 he established the Institute for Orthomolecular Medicine (now the Linus Pauling Institute of Science and Medicine). The American Orthomolecular Association was founded in 1975, with Pauling as its Honorary President. The results of clinical trials of the effect of vitamin C on the common cold and cancer remain controversial, with claims of bias on both sides of the issue.

Biographers and historians of science have been especially interested in Pauling's thought processes and their origins, as well as his awesome body of work, which recognized no boundaries between physics, chemistry, and biology. Generally, when people asked Pauling about his remarkable achievements, he said that he believed in having lots of ideas and throwing away the bad ones. According to Crick, the key to understanding Pauling's approach to diverse problems was to realize that he was constantly thinking at the molecular level. Because of this orientation, Pauling was the first to realize that sickle cell anemia is a molecular disease and the first to suggest that a molecular clock regulates gene mutation.

Sickle cell anemia is a severe, chronic, and crippling disorder. Under the microscope, the red blood cells of individuals with SCA assume a peculiar crescent or sickle shape. This ancient genetic disorder was not rare in Africa and the southern Mediterranean, but the first fully documented report of SCA appeared in 1910 when Dr. James B. Herrick (1861–1954), a prominent American cardiologist, published his case study of Walter Clement Noel, a 20-year-old West Indian student studying dentistry at the Chicago College of Dental Surgery. Noel died at the age of 32, 9 years after returning to Grenada. Although the general idea that "sicklemia" was an inherited disease was accepted by the 1930s, the relationship between *sickle cell anemia*, the disease that afflicts individuals with two copies of a gene for the abnormal hemoglobin known as hemoglobin S, and *sickle cell trait* was not fully clarified until the 1950s. The red blood cells of individuals who have inherited only one gene for hemoglobin S display sickle cell trait. Although these cell usually appear to be normal, they can undergo the reversible change in shape known as sickling in response to changes in oxygen levels. Individuals with sickle cell trait or sickle cell anemia appear to be relatively resistant to malarial parasites, but children weakened by SCA often succumbed to other diseases. Genes for abnormal hemoglobins that appear to offer some protection against malaria are found in virtually every region affected by the disease, including Greece, India, East and West Africa, and the Middle East.

When T. H. Morgan came to Caltech in 1929, Pauling became actively interested in a wide variety of biological problems representing what he called "molecular architecture and the processes of life." During the 1930s Pauling was involved in studies of the relationship between the structure and function of hemoglobin, the oxygen-carrying protein present in red blood cells, which has been described as a "molecular lung." Thinking about the physical properties of hemoglobin led to a long-term interest in the structure of proteins and the nature of antibody-antigen reactions. Pauling became interested in sickle cell anemia when he heard a lecture about the red blood cells of patients with the illness. In 1949 Pauling and his associates published a paper entitled "Sickle-Cell Anemia, a Molecular Disease," which demonstrated that the sickling phenomenon was caused by an alteration in the physical properties of hemoglobin.

Despite the obvious importance of Pauling's venture into the new world of "molecular medicine," the precise nature of the alteration in the hemoglobin molecule remained unclear until 1957, when Vernon Ingram demonstrated the specific chemical difference between normal and sickle cell hemoglobin. Basing his approach on the amino acid sequencing techniques developed by Frederick

Sanger (1918–) in the 1940s, Ingram was able to determine and compare the amino acid sequences of normal and sickle cell hemoglobin. In 1955, Sanger published the complete amino acid sequence of insulin, the first protein to be characterized in this way. Sanger was awarded the Nobel Prize in Chemistry in 1958 for his work on the structure of proteins. During the 1960s, Sanger became interested in the nucleic acids. His success in developing new approaches to sequencing RNA and DNA was recognized by his second Nobel Prize in Chemistry (1980), which was shared with Walter Gilbert and Paul Berg.

Breaking down the hemoglobin molecule into peptide fragments, Ingram isolated one altered peptide and found that in hemoglobin S the amino acid valine had been substituted for the normally occurring glutamic acid. Given the fact that the two forms of hemoglobin differed with respect to only one amino acid, Ingram postulated that a corresponding change, a true point mutation, had occurred in the gene coding for hemoglobin. Since geneticists had already shown that sickle cell anemia was transmitted as a simple Mendelian recessive factor, the Mendelian inheritance of a significant human disease, genetic mutation, and the precise alteration in a gene product had been linked together for the first time. The development of transgenic mouse models in 2001 may lead to gene therapy for sickle cell anemia.

The importance of malaria throughout human history appears to be reflected in the evolution of protective gene mutations found among human populations in Africa, Asia, and the Mediterranean. At the beginning of the twenty-first century, malaria still infects about 500 million people per year and kills over 2 million. Resistance to malaria is also associated with a mutated form of a gene that codes for glucose-6-phosphate dehydrogenase (G6PD). Another genetic disease that correlates with endemic malaria is known as beta-thalassemia. Before the introduction of modern therapeutic interventions, affected children usually died before reaching the age of 5. Very high incidence rates of beta-thalassemia carriers are found among both the Greek and Turkish populations of Cyprus; before the 1970s, one of every 158 infants was born with the condition. Beginning in the early 1970s, Cypriot pediatricians began programs for population screening. Combined with prenatal diagnosis and abortion, these programs resulted in almost complete prevention of new births of children with thalassemia. The program helped relieve the heavy public burden caused by the high incidence of thalassemia, but raised difficult ethical issues.

Despite the prevalence and wide distribution of genetic diseases associated with an evolutionary response to malaria, sickle cell anemia is probably the best known and the one that has consistently been treated as a "racial disease," as well as a molecular disease. Thus, the history of sickle cell anemia reveals many aspects of the politics of race, as well as the history of disease. In the United States, for example, the association between SCA and race became so strong that a diagnosis of SCA in a "white" person generally led to a search for "black blood" in the genealogy. When statistics seemed to prove that SCA was more prevalent in North America than Africa, some observers claimed that it proved the "dysgenic effects of race-mixing." In order to promote research and reverse decades of myths and misinformation, the 1972 National Sickle Cell Anemia Prevention Act established a fund for medical research and genetic counseling.

REVERSE TRANSCRIPTASE, RECOMBINANT DNA, AND THE NEW CENTRAL DOGMA

As they eventually went their separate ways, James Watson, Francis Crick, and other members of the Phage Group devoted much of their attention to determining the mechanism by which information encoded in DNA was incorporated into the amino acid sequence of proteins. In this search they were guided by the Central Dogma encoded in Watson's cryptic note: "DNA → RNA → protein."

In 1958 Crick published an article entitled "On Protein Synthesis," which addressed special aspects of theoretical biology. Primarily, Crick used this occasion to discuss the central dogma and analyze the validity of what he called the "sequence hypothesis"—the sequence of nucleic acids determines the sequence of amino acids in proteins. The paper provides an excellent example of Crick's conviction that theories are valuable guides, even if they are later found to be incorrect. Proteins, Crick remarked, are the macromolecules that do most of the jobs needed for life. Indeed, proteins can do almost anything; proteins serve as enzymes and have vital structural functions. Yet it is DNA that is most fascinating because it serves as the material basis of inheritance and is ultimately responsible for the synthesis of proteins. In looking at the base sequence of DNA and the amino acid sequence of proteins, Crick made many predictions about how DNA could direct the synthesis of proteins, including his famous adaptor hypothesis.

Analyzing the logic of the colinearity between the sequence of bases in nucleic acids and the amino acids in proteins, Crick elaborated on the implications of the central dogma. According to Crick, the central dogma stated that information is transferred from nucleic acid to nucleic acid, or nucleic acid to protein, but not from protein to nucleic acid. In this context, information refers to the precise determination of sequence. When information passes into protein it cannot get out again. However, within 10 years it became clear that the central dogma, at least in its original form, had failed to predict the possibility that information could be transferred from RNA to DNA.

In 1975 the Nobel Prize in Medicine or Physiology was awarded to Howard Temin (1934–1994), Renato Dulbecco (1914–), and David Baltimore (1938–) for "discoveries concerning the interaction between tumor viruses and the genetic material of the cell, and the possible relationship between viruses and human cancers." As a graduate student at the California Institute of Technology, Temin planned to study experimental embryology, but was drawn to Dulbecco's work on animal virology. In 1958 Temin and Harry Rubin, a postdoctoral fellow, developed the first reproducible in vitro assay for a tumor virus. Continuing his work on Rous sarcoma virus (RSV) at the University of Wisconsin, Temin noted that the inhibition of DNA synthesis and DNA-dependent RNA synthesis blocked RSV infection. This observation was surprising because the RSV genome is single-stranded RNA, and it was assumed that the RNA served as the genetic material. To explain his observations, Temin suggested that a DNA intermediate must be involved in RSV infection. Temin's initial explanatory framework was called the provirus hypothesis.

According to the provirus hypothesis, after an RNA virus enters a host cell, a DNA provirus is synthesized. The DNA provirus contains the genetic information of the RNA viral genome and progeny viral RNA is synthesized from the DNA provirus. This hypothesis could also account for the integration of RSV genetic material into the host chromosomes. The use of various inhibitors in these experiments suggested possible artifacts, and Temin's hypothesis was not readily accepted. The major impediment, however, seems to have been the way in which the provirus hypothesis conflicted with the central dogma of molecular genetics. New ways of thinking about the problem appeared in 1970 when Temin and Satoshi Mizutani proved that virions of RSV contain an enzyme that transcribes single-stranded viral RNA into DNA. An anonymous reviewer for *Nature* gave the enzyme the name "reverse transcriptase." Soon after Temin and Mizutani suggested the existence of reverse transcriptase, David Baltimore demonstrated the same enzyme in Rauscher murine leukemia virus. The new enzyme provided an obvious mechanism for the formation of a DNA intermediate during RNA tumor virus infection and transformation. Less than a month after Baltimore and Temin announced their discovery of reverse transcriptase, Solomon Spiegelman (1914–1983), who had been one of the most outspoken critics of Temin's provirus hypothesis, confirmed the existence of reverse transcriptase in eight additional RNA cancer viruses.

Baltimore was introduced to biological research methods as a high school student enrolled in a summer program at Jackson Memorial Laboratory in Bar Harbor, Maine, where Howard Temin had already established a reputation as a promising young scientist. As a graduate student in biophysics at MIT, he became interested in animal viruses and transferred to Rockefeller University, where he earned his Ph.D. in biology in 1964. After postdoctoral experience at MIT and Albert Einstein College of Medicine, Baltimore moved to the Salk Institute for Biological Studies, at La Jolla, California, to work with Renato Dulbecco on animal viruses. Baltimore joined the faculty of MIT in 1968 and became president of Rockefeller University in 1990. In addition to demonstrating the action of reverse transcriptase in RNA tumor viruses, Baltimore investigated RNA virus replication, poliovirus, leukemia virus, vaccines, genetic con-

Howard Temin, Nobel Laureate, 1975

trol mechanisms, genetic engineering, and immunology. Reflecting on the pleasure found in research, Baltimore said that "a virologist is among the luckiest of biologists because he can see into his chosen pet down to the details of all its molecules." In his Noble Prize lecture, Baltimore attributed the award to the work he and his colleagues had done in "bringing molecular biology to bear on the awful and awesome problems of cancer." The Nobel Prize made it possible for Baltimore to reach a broader audience in his campaigns for responsible guidelines for research in genetic engineering, establishing national scientific goals, and international recognition of the dangers of biological warfare. A paper published in the journal *Cell* in 1986 brought Baltimore into a highly notorious dispute when Thereza Imanishi-Kari, one of the co-authors, was accused of scientific misconduct by Margot O'Toole, a postdoctoral student who questioned some of the experimental results. Devoting much of his time and energy to the defense of his co-worker and his vision of scientific integrity, Baltimore felt compelled to resign from the presidency of Rockefeller University. After a lengthy and bitterly contested investigation of the charges, the National Institutes of Health found no evidence of "fraud, misconduct, manipulation of data, or serious conceptual errors" in the controversial paper. Imanishi-Kari was officially exonerated of all charges in 1996. One year later, Baltimore became president of Caltech. Many of Baltimore's friends and supporters believe that Baltimore had been the victim of a deliberate and vindictive witch hunt by Representative John Dingell, Chairman of the House Committee on Energy and Commerce and its Subcommittee on Oversight and Investigations and others, including Walter Steward and Ned Feder, Staff Scientists at NIH, who considered themselves specialists in discovering scientific fraud and misconduct.

The RNA tumor viruses and reverse transcriptase quickly became basic tools for studying the molecular biology of mammalian cells. Reverse transcriptase was used to prepare DNA copies of messenger RNA, which could be used as a specific probe to detect viral genetic information in normal and malignant cells. Crick has argued that reverse transcriptase does not really contradict the central dogma in as much as the transcription of information from one form of nucleic acid to another proceeds through essentially the same base-pairing process. If the discovery of reverse transcriptase, ribozymes, transposable genetic elements, and other novel aspects of "postclassical" molecular biology do not overturn the most cautiously elaborated formal versions of the central dogma, perhaps they have dealt a major blow to a more informal version of the dogma, which essentially held that "what is true for a phage is true for an elephant."

Indeed, the molecular biology of the gene in higher organisms, along with many aspects of cell growth, differentiation, and regulation, proved to be much more complicated than that of the microorganisms that served as the first successful model systems for molecular biologists. Recognition of this fundamental fact made it possible for scientists to pursue entirely new avenues of research

David Baltimore, Nobel Laureate, 1975

and helped to explain previously misunderstood and obscure studies of the genetics of plants and animals, such as Barbara McClintock's (1902–1992) pioneering experiments on the relationship between structural aspects of plant chromosomes and the effects of various genes. As a high school student in Brooklyn, New York, McClintock became interested in science. She earned her Ph.D. in 1927 for studies of maize cytogenetics. McClintock carried out pioneering studies of the relationship between the structure of plant chromosomes and their genetic effects and described previously unknown chromosomal rearrangements, chromosome breakage, and proof of chromosomal crossing over and recombination. Finding tenured academic appointments closed to women, McClintock became a staff member of the Cold Spring Harbor Laboratory in 1941 and remained there for the rest of her life.

In the 1940s, McClintock began the work that led to the discovery of genetic elements that could move within the genome and control the expression of other genes. By means of intricate studies of complex chromosomal rearrangements and the consequences of chromosome breakage and repair, McClintock discovered the existence of transposable genetic elements. Although McClintock concluded that these transposable factors acted as regulators of neighboring genes, "transposons" were generally ignored or humorously referred to as "jumping genes." Transposons were treated as a minor curiosity or aberration peculiar to maize genetics until similar factors were discovered in bacteria and fruit flies in the 1960s and 1970s, respectively. When McClintock was finally awarded the Noble Prize in 1983 in recognition for the work that led to the concept of the dynamic genome, she became the first geneticist since H. J. Muller to win such an unshared award. Scientists who had long respected her work regarded this as a fitting tribute to a woman scientist who had worked alone for so many years, almost as much a curiosity as the unique problems she had chosen as her own.

In the 1970s, scientists learned how to cut and splice DNA to form recombinant DNA molecules and place these unprecedented forms of DNA into host organisms such as *E. coli*. DNA from totally unrelated species could be chemically linked into a single molecule and then cloned, that is, induced to reproduce within bacteria. At a small but prestigious conference on nucleic acids held in June 1973, Stanley Cohen (1922–) of Stanford University and Herbert W. Boyer (1936–) of the University of California announced that they had used recombinant DNA to create a plasmid that was inserted into bacterial cells where the recombinant hereditary material could multiply. News of this novel technique spread rapidly through the scientific grapevine, and within months scientists were publicly calling for widespread and deliberate discussions of the possible risks involved in creating unprecedented forms of genetically engineered DNAs and recombinant organisms. Foreseeing the commercial potential of genetic engineering, Stanford University and the University of California,

San Francisco, applied for a patent on the recombinant DNA technique in 1974. Despite the fact that the papers that established the basis of the patent were co-authored with four other scientists, only Cohen and Boyer were cited as inventors. The U.S. Patent Office granted a patent for the Cohen-Boyer recombinant DNA process, the first major patient for the new era of biotechnology, on December 2, 1980. Boyer, with the help of venture capitalists, had already established Genentech, Inc., in 1976, for the commercial exploitation of recombinant DNA techniques.

As the debate about the relative benefits and biohazards associated with "gene-splicing" became increasing volatile, Paul Berg (1926–), who had created the first recombinant DNA molecules in 1972, emerged as one of the leaders of the movement to have scientists evaluate and control the potential hazards for laboratory workers and the public of the recombinant DNA techniques. In part, this movement was an attempt to prevent governmental agencies from controlling or prohibiting the use of genetic engineering. After earning his Ph.D. in biochemistry at Western Reserve University (now Case Western Reserve University) in 1952, Berg investigated various aspects of protein synthesis before becoming interested in tumor viruses. During the 1950s, Renato Dulbecco (1914–) had developed techniques that made it possible to study the molecular biology of animal viruses in a fairly simple and quantitative manner. Based on these studies, Dulbecco later discovered that the DNA of certain tumor viruses could become integrated into the DNA of a host cell as a provirus. Berg, therefore, thought that he might be able to use animal tumor viruses to probe the mechanism of gene expression in higher organisms in a manner analogous to the way in which bacterial viruses had been used to work out the mechanism of gene expression in bacterial systems. To pursue this question, Berg adopted the monkey tumor virus SV40 as a model system. Using restriction enzymes (enzymes that cut DNA at specific sites) and DNA ligase (an enzyme that links two strands of DNA), Berg created the first recombinant DNA molecules in 1972. After isolating a gene from a cancer-causing monkey virus, Berg linked this oncogene to the genetic material of one of the bacterial viruses that can infect *Eschericia coli*. When he thought about the potential danger of inserting a cancer gene into a bacterium that is commonly found in the human gut, Berg decided to stop his work and urge his colleagues to consider the implications of this technique. In 1980 Berg was awarded the Nobel Prize in Chemistry "for his fundamental studies of the biochemistry of nucleic acids with particular regard to recombinant DNA."

Immediately after the Gordon Conference at which Cohen and Boyer had presented their revolutionary technique, the co-chairs of the conference, Maxine Singer and Dieter Soll, sent letters to the National Academy of Science and the Institute of Medicine calling for the establishment of a committee to investigate the potential risks of recombinant DNA research. In order to publicize the issue more widely, Singer and Soll also published the letter in *Science* (September 21, 1973). The National Academy of Sciences created the Committee on Recombinant DNA, with Berg as its chairman. In July 1974, the Committee on Recombinant DNA published an open letter in *Science*, signed by Berg and 10 other scientists, calling for an unprecedented voluntary moratorium on recombinant DNA research pending evaluation of the risks and the establishment of research guidelines. In 1975 more than 100 molecular biologists met at a landmark conference at the Asilomar Conference Center in California to discuss potential hazards of recombinant DNA technologies and propose guidelines to minimize risks.

Despite the confusion and fears expressed in the 1970s, there was no doubt that the new techniques of molecular biology would make it possible to create genetic information that could be used to treat genetic diseases and manipulate the genetic materials of plants, animals, and microorganisms. Public debate about the potential uses and abuses of recombinant DNA, cloning, genetic engineering, and gene therapy had been raised even before more powerful, rapid, and simplified techniques became available. Although some concerns remained even after two decades of experience with recombinant DNA, by the end of the twentieth century the anticipated benefits of molecular medicine, gene therapy, and biotechnology generally outweighed the fear and uncertainty raised by the new techniques of molecular biology.

Eventually, when scientists and historians look back at the debates that marked the debut of recombinant DNA technology, they may wonder why such preliminary findings should have generated so much controversy. It could be argued that recombinant DNA need not have been singled out from related areas as uniquely dangerous. Some observers argued that the fear of Frankenstein-like biological monsters escaping from biology laboratories was merely a product of media attention during an era of generalized and exaggerated anxiety, exacerbated by widespread misunderstanding of the nature of science. After all, despite the universal fear of everything possibly related to cancer, there was little interest in the potential hazards of working with tumor viruses and pathogens before the prospect of genetic engineering was addressed. Many laboratories lacked the minimal safety equipment for working with pathogenic microbes, but this problem was not unique to the new molecular biology. Scientists, however, expressed concern about the potential dangers of combining work on molecular biology, cancer, and tumor viruses as it became clear that the exciting new field was attracting researchers with little or no background in microbiology and pathology who were ready to work in or establish laboratories lacking the safety equipment and conventions traditionally used when dealing with known and potential pathogens.

Just as remarkable as the initial controversy has been the virtual disappearance of the near-hysteria that peaked during the 1970s and the subsequent euphoria that developed in the course of the next decade with the prospect of a cornucopia of molecular biology–based commercial developments. The focus of media and public attention rapidly shifted from theoretical risks to the expectation of miraculous "breakthroughs" such as cures for cancer and genetic diseases, designer drugs, fruits, and vegetables, monoclonal antibodies, spider silk, bioinsecticides, and unlimited quantities of human gene products such as insulin, growth hormone, clotting factors, interferons, interleukins, and so forth.

Through the techniques of molecular biology, scientists could study the "morbid anatomy" of the chromosome and locate defective genes, much as Giovanni Morgagni's (1682–1771) postmortems located specific pathological lesions at a specific site. Indeed, by the 1980s scientists had established the existence of some 3000 human genetic diseases, which suggested that developments in genetics would make it possible to manage many diseases at the level of the gene. In theory, the ability to produce genes and gene products in the laboratory should allow "replacement therapy" for the hundreds of genetic diseases for which the defective gene and gene product are known. Diseases that seemed likely candidates for intervention at the level of the gene included thalassemia, Gaucher's disease, Tay-Sachs disease, Fabry's disease, hemophilia, adenosine deaminase deficiency (ADA), familial hypercholesterolemia and cancers of the breast, ovary, brain, and skin, but attempts to perform gene therapy in the 1990s were inconclusive. Bioethicists warned that the devastating consequences of genetic diseases made patients and their families very vulnerable to appeals to volunteer for experiments that might well be premature, ineffective, and dangerous. Some critics argued that human gene therapy was totally unacceptable because of unresolved social, ethical, moral, and evolutionary questions.

At the end of the twentieth century, gene therapy remained an experimental procedure associated with significant risks, as indicated by the death of 17-year-old Jesse Gelsinger, the first person to die as a result of experimental gene therapy. Although Gelsinger had a severe genetic disorder (ornithine transcarbamylase deficiency), at the time of the experiment he was in fairly good health and was controlling his disorder with drugs and diet. Gelsinger died on September 17, 1999, 4 days after the administration of a weakened cold virus, used to transmit the corrective gene. Shortly after Gelsinger's death, an NIH advisory panel proposed more stringent rules for gene therapy research. Critics argued that gene therapy should only be used as a last resort for patients with severe disorders that cannot be controlled by conventional therapies.

Despite hypothetical obstacles and uncertainties, the potential of genetic engineering became real enough by the 1980s to interest biotechnology companies and regulatory agencies. Scientists learned to see ideas and research results as potentially patentable products, while entrepreneurs rushed to create biotechnology firms, attract venture capitalists, and trade shares on the stock exchange. Although the possibility of human gene therapy attracted major media attention, genetic engineering techniques had immediate implications in industry, pharmacology, and agriculture. Genetic manipulation in plants, for example, might well have a more immediate and widespread impact on human health and well-being throughout the world than a direct attack on the genetic diseases of human beings. Indeed, many medical demographers believe that general improvements in patterns of morbidity and mortality owe more to nutrition than to medical interventions. Changes in the composition of major food crops could prevent much of the malnutrition that plagues impoverished children in developing countries. Genetic engineering applied to animals, as well as plants, could increase food production and provide more nutritious foods with better quality proteins, lower levels of saturated fat and cholesterol, and higher vitamin contents. And since plants have always been nature's apothecary shops, genetic manipulation of plants might provide a new generation of pharmacologically active and commercially valuable products. Plants might even serve as "factories" for the production of the genes and proteins of other species, including mammals. Genetically engineered plants could contain genes from different animals, bacteria, or insects in order to create varieties that are resistant to pests, parasites, certain herbicides, or extreme temperatures. Such plants could produce better yields under adverse conditions, cause less damage to the environment, and provide better nutrition. Nevertheless, by the end of the twentieth century, anxiety about genetically engineered foods was increasingly prevalent in the United States and Europe. Critics of biotechnology argued that not enough was known about the potential impact of genetically modified products on human health and the environment. Because of widespread public opposition, in 1998 Britain and the European Union banned new genetically engineered crops. Polls taken in the United States found that a majority of those polled did not want to buy genetically modified foods. The U.S. Food and Drug Administration, however, adopted the position that genetically engineered foods were not "substantially" different from conventional foods and did not require regulation or labeling.

Genetic engineering has also been used to alter the genome of various species of animals, such as mice and primates, in order to provide better models for medical research. By inserting genes associated with human diseases, such as Alzheimer's disease, diabetes, sickle cell anemia, and breast cancer into animal models, researchers hope to find better ways to deal with disease. Cloning techniques might also be used to save endangered animals, or even to create copies of extinct animals. Cloned cells or cell nuclei from rare animals, or preserved tissues of extinct species, would be inserted into enucleated eggs of a common species, such as a cow. Cloning would then be combined with interspecies gestation and birth.

THE HUMAN GENOME PROJECT

Refinements in the techniques of molecular biology made it possible to plan and execute the Human Genome Project (HGP), an international effort dedicated to mapping and sequencing all of the estimated 3 billion bases that make up the genes and interconnecting segments on the 23 pairs of chromosomes that carry the human genetic legacy. The Human Genome Project was officially launched in 1990, although the possibility of carrying out such a project had been raised in the 1970s when the first primitive maps of human chromosomes were established.

At a workshop on the human genome project held in Los Alamos in March 1986, Nobel Laureate (1980) Walter Gilbert (1932–), a pioneer in DNA-sequencing techniques and entrepreneurial biotechnology, declared that sequencing the entire human genome was the ultimate response to the primary commandment: "Know thyself." Scientists were urged to join the great quest to seize and sequence the "Holy Grail" of human genetics. Many geneticists assumed that only about 5% of the base pairs in human DNA actually coded for genes; the remaining noncoding regions were called *introns*, or *junk DNA*. By estimating the cost of sequencing at one dollar per base pair, Gilbert predicted that a crash program could provide the completed sequence for $3 billion. Some scientists warned against mortgaging the future of biology to a mindless routine of "Big Science" sequencing, but entrepreneurs and biotechnology firms eagerly anticipated exploiting the huge market for DNA diagnostic kits and novel drugs that even a partial sequencing of the genome was likely to generate.

Although the prospect of locating and sequencing the estimated 100,000 human genes seemed overwhelming, new techniques for rapid and automated sequencing had emerged in the 1980s and further improvements were inevitable. One of the most eagerly adopted new tools was the polymerase chain reaction (PCR) invented by Kary Mullis (1944–) in 1983. By allowing DNA replication to take place under controlled conditions, PCR rapidly and selectively amplifies minute samples of DNA. In 1973, Mullis earned his Ph.D. in biochemistry from the University of California, Berkeley. After teaching at the University of Kansas Medical School in Kansas City, he accepted a position as a research scientist with Cetus Corporation, a biotechnology company in Emeryville, California, that synthesized chemicals used in genetic cloning. While driving from San Francisco to Mendocino, California, Mullis suddenly pictured a way to find and replicate specific sequences of DNA. Remembering that moment, he said that he immediately realized that the method would "spread into every biology lab in the world." The universal adoption of this simple but revolutionary technique as an aid research and development in virtually every phase of molecular biology proved that Mullis was right. Cetus, rather than Mullis, derived enormous profits from the commercial version of PCR. In 1986, Mullis moved to Xytronyx, in San Diego, as director of research in molecular biology, but 2 years later he left to become a private research consultant. The 1993 Nobel Prize in Chemistry was awarded to Kary Mullis for the discovery of PCR and to Michael Smith (1932–) for the invention of site-directed mutagenesis. Enjoying the celebrity that followed the award, Mullis became known as much for his interest in surfing, women, and hallucinogenic drugs as for his original mind and outspoken opinions on a wide range of issues.

Many of the other technological innovations that made it possible to sequence the human genome were developed by Leroy E. Hood (1938–), an advocate of an approach to research that has been called "discovery science." Instead of testing hypotheses with experiments, discovery science involves generating big databases of information, such as the Humane Genome Project and molecular immunology. Born in Missoula, Montana, Hood earned his M.D. from Johns Hopkins (1964) and his Ph.D. from Caltech (1968). After 2 years at the National Institutes of Health, Hood became a professor of biology at Caltech. With a $12 million bequest from Bill Gates of Microsoft, in 1992 Hood established a department of molecular biotechnology at the University of Washington Medical School in Seattle. Hood was involved in establishing several biotechnology companies, including Amgen, Applied Biosystems, Darwin Molecular, and a nonprofit research center called the Institute for Systems Biology. After leaving academia, Hood complained that universities were unfit for the new age of biology.

Hood was best known for the development of instrumentation for sequencing and synthesizing proteins and genes, and his career demonstrated the wisdom of a mentor who told him: "If you really want to change a discipline, invent some new technology that will let you go beyond what people have seen before." Using this approach made it possible for Hood to pursue research interests in molecular immunology and autoimmunity, molecular evolution, cancer biology, and the analysis of the HIV genome. The Human Genome Project would have been totally impractical without high-speed automatic DNA sequencers. Using the automatic sequencer developed by Hood in 1985, scientists could collect more data overnight than they could previously generate in a week or more. Further improvements of the device produced more data more rapidly and more accurately.

In 1988 James Watson agreed to serve as director of the National Center for Human Genome Research established by the National Institutes of Health. Watson supported funding of research dealing with the ethical, legal, and social implications (ELSI) of the genome project. In the same year, an international council known as the Human Genome Organization (HUGO) was established to coordinate research and encourage the exchange of information like a virtual United Nations for work on the human genome. One year later, the sequence of chromosome 22 was completed, along with about a third of the base pairs in human DNA. Chromosome 22 is a small chromosome, but mutations traced to this chromosome have been associated with various heart defects, cancers, immune system disorders, schizophrenia, and mental retardation. Moreover, by 2000 the genomes of several microbes and eukaryotic organisms, including yeast (*Saccharomyces cerevisiae*), the nematode worm (*Coenorhabditis elegans*), the fruit fly (*Drosophila melanogaster*), the thale cress (*Arabidopsis thaliana*), the mouse (*Mus musculus*), and rice (*Oryza sativa*), were established. Thus encouraged, researchers expected to establish a complete map of the location, function, and exact sequence of each human gene by 2005.

In 1998 the consortium of scientists involved in the HGP was challenged by a for-profit corporation called Celera Genomics, founded and run by J. Craig Venter (1946–), a former NIH researcher. Venter predicted that Celera would sequence the human genome by 2001, 4 years ahead of the HGP's schedule. Like Leroy Hood, with whom he had published a paper entitled "A New Strategy for Genome Sequencing" in *Nature* in 1996, Venter was interested in DNA sequencing, supercomputers, bioinformatics, and entrepreneurial biotechnology. Many geneticists, including Watson, were extremely skeptical of Venter's "whole genome shotgun" approach, which involved using high-speed computers to reassemble the genome of an organism without using maps. However, Celera demonstrated the power of the shotgun method by sequencing the genome of the fruit fly and the mouse. By mid-1999, Celera's rapid progress forced the public HGP into expediting its efforts to complete and openly publish the sequence of the human genome. Following the "Bermuda Principles" established in 1996 by representatives of international genome researchers, which called for openly publishing all data on an accessible database, the HGP provided Celera with a significant advantage as the two groups raced to complete the first draft of the human genome. According to Celera officials, the public genome project was essentially administered as a "cottage industry," in contrast to Celera, which was centrally managed in a way that validated the concept of "economies of effort and money."

In March 2000, U.S. President Bill Clinton and Prime Minister Tony Blair of Great Britain made a joint declaration that all genome information should be free to the public. However, efforts to establish some degree of cooperation between the public and private projects were generally resisted by Venter. Although the rival teams agreed to "coordinate" the announcement of the first complete draft of the human genome, Celera never agreed to collaborative efforts.

Despite tension caused by the intense competition, on June 26, 2000, President Clinton was able to bring together leaders of the public genome project and Celera for a press conference at the White House announcing that both groups had completed working drafts of the complete human genome and that both maps would be released simultaneously. In February 2001 the public se-

James D. Watson at a National Institutes of Health Conference on DNA

quence data was published in the British journal *Nature* and the Celera sequence was published in the American journal *Science*. Arranging this fairly amicable separate but simultaneous publication of the final data was complicated, because Celera sells its genomic data to subscribers. DNA data collected by members of the HGP consortium were submitted to GenBank, a public repository, freely accessible through the Internet. After intense negotiations, the editors of *Science* agreed to grant an exception to the GenBank rule in order to publish Celera's paper, despite objections from leaders of the public consortium. Shortly after completion of the first draft of the genome, Celera acknowledged that its original business plan for selling access to the genetic data to drug companies was no longer very profitable or attractive to investors. In January 2002, Venter announced that he was stepping down from the presidency of Celera Genomics. In a move that reflected changes in the marketplace for biotechnology companies, Celera planned to shift its focus to drug development.

When the Human Genome Project began, scientists generally thought the gene map would include 100,000 to about 150,000 genes. Thus it was surprising that both the public and private projects indicated that the number of human genes was only about 30,000 to 40,000 and only slightly different from that of humbler creatures. Indeed, the number of human genes seemed to be only about twice as many and the number of genes needed to describe the fruit fly. Moreover, the map of the human genome suggested that only about 1% of the chromosomal material consists of genes. Although the remaining material has been called "junk DNA," many scientists believe it must have some functional significance or role in the structural integrity or evolution of chromosomes. Until further studies confirm or revise the first drafts of the genome, the possibility remains that the original gene maps might have seriously underestimated the number of genes. Even though Celera and the National Human Genome Project reported similar estimates for the number of genes, which seemed to lend validity to the surprisingly low numbers, they were apparently counting two different sets of genes. By August 2001 estimates as high as 60,000 genes were proposed, but the numbers were still controversial. The relatively small number of genes seems to support the theory that the genome is a "parts list" rather than a blueprint. Assembly of the parts might generate level upon level of complexity, such as the remarkable number of connections between neurons in the brain and nervous system and the diversity of antibodies produced the immune system.

With the completion of the Human Genome Project, scientists immediately used the partial maps to locate, isolate, and clone specific disease genes. The information can be used to improve diagnostic methods, help prevent the disease, design specific agents to treat patients, and, in some cases, it might lead to gene therapy that could correct defective genes. Genetic data on hereditary forms of cancer, for example, has allowed at-risk individuals to undergo preemptive surgical removal of organs such as the stomach, breast, ovary, uterus, colon, thyroid gland, and so forth. Another product of the HGP is the development of forensic genomics. Originally thought of as a way to establish databases for identifying criminals, following the September 11, 2001, attack on the World Trade Center and the Pentagon, Celera Genomic Group and Myriad Genetics Inc., two of the major American genomics companies, offered their expertise in DNA sequencing and typing to assist in the identification of victims of the attack. Forensic DNA analysis can help identify remains even after significant tissue decomposition has occurred by sequencing mitochondrial DNA from hair, teeth, and bones.

In addition to supporting efforts to sequence the human genome, the National Center for Human Genome Research promoted research dealing with the ethical, legal, and social implications of the genome project. Interest in the issues studied by the ELSI Working Group led to the passage of the Genetic Privacy Act in 1994 to regulate the collection, analysis, storage, and use of DNA samples and the genetic information derived from them. In 1995, the Equal Employment Opportunity Commission published guidelines that extended the protections specified by the Americans with Disabilities Act to cover discrimination based on genetic information related to illness, disease, or other conditions. Proof that protection against discrimination based on genetic information was necessary was demonstrated in a landmark 2001 case in which the U.S. Equal Employment Opportunity Commission went to court to stop a company from testing its employees for genetic defects. In this unprecedented legal battle over medical privacy in the workplace, the EEOC argued that basing employment decisions on the results of genetic tests violated the Americans with Disabilities Act. Concerns about the potential abuse of genetic data led many states to ban the use of genetic screening for making employment-related decisions. Because genetic information could lead to new forms of discrimination, many scientists and ethicists have supported the Universal Declaration of the Human Genome and Human Rights, which states: "No one shall be subjected to discrimination based on genetic characteristics that is intended to infringe or has the effect of infringing human rights, fundamental freedoms and human dignity."

GENOMICS AND PROTEOMICS

The Human Genome Project stimulated the rapid development of new disciplines, as well as a new vocabulary. Throughout the history of science, the establishment of new journals has generally marked the creation of a new discipline. Thus, the development of genomics, the science of the identification and characterization of genes and their arrangement in chromosomes, was reflected in the publication in 2000 of the first volume of the series *Annual Review of Genomics and Human Genetics*. With the completion of the first major phase of the Human Genome Project, scientists could directly confront the task of analyzing the tens of thousands of human genes and their relationship to the hundreds of thousands of human proteins. As a tribute to the success of the Genome Project, scientists suggested organizing a complete inventory of human proteins, which would be known as the Human Proteome Project.

Proteome is a term coined by Australian scientists Marc Wilkins and Keith Williams in 1995 to mean the "set of PROTEins encoded by the genOME," the total set of proteins expressed in a given organelle, cell, tissue, or organism at a given time with respect to properties such as expression levels, posttranslational modifications, and interactions with other molecules. *Proteomics* refers to the science and the process of analyzing and cataloging all the proteins encoded by a genome. Some scientists suggested that proteomics is about everything in biology that was not part of the Human Genome Project. Others refer to proteomics as "protein-based genomics." Although the word proteome is new, the idea of cataloging the proteins expressed in an organism can be traced back to the 1970s, with the development of two-dimensional gels. By the mid-1990s, improvements in this technique made it possible to separate and analyze thousands of proteins on a single gel, making it possible to analyze the proteomes or partial proteomes of various organisms and tissues. Ultimately, the science of proteomics would provide a complete description of cells, tissues, or whole organisms in terms of proteins, not as static entities, but in terms of patterns that change in response to developmental and environmental signals. Because proteins are involved in disease states, complete descriptions of proteins will stimulate rational drug design as well as the discovery of new disease markers and therapeutic targets.

One of the first major conferences devoted to the possibility and advisability of establishing a major proteome project was held in April 2001 in McLean, Virginia. Calling attention to the immense challenge that such a project would entail, organizers chose to title the conference "Human Proteome Project: Genes Were Easy." Speakers included scientists, officials, and executives from academia, government, and business. The conference was also the occasion for the first meeting of the Human Proteome Organization (HUPO), which was described as "an international effort to bring commercial and academic groups together to study the output of all human genes." Skeptics warned that a single Human Proteome Project could not match the success of the HGP, because the proteome is necessarily a more nebulous entity than the genome, which contains a fixed number of base pairs. It is clear, however, that the debates about access to the data generated by structural biologists analyzing the human proteome will be just as intense as those that marked the rivalry between the public genome project and Celera.

Given the publication of the first draft of the human gene map in 2001, perhaps broader claims that the Human Genome Project would provide the complete script from which scientists could read and decode the "nature" of man (and perhaps even the nature of woman) can be put into historic perspective by remembering that Vesalius once claimed that his anatomical studies constituted a true reading of the book of man. Just as the *Fabric of the Human Body* did not mark the end of the search for understanding the nature of life, only the most malignant sort of hubris could make anyone believe that the publication of the complete human genome marks the end of the quest rather than the beginning of a new dissertation. Reading the complete book of the human genome will certainly involve a more complete grammar of biology and a new avalanche of knowledge, but it will not in itself establish a new way of understanding or a new wellspring of wisdom. To avoid future disappointment, disillusionment, and a great deal of serious mischief, if not the great apocalypse some have predicted, it might be useful to establish a new version of the central dogma, which will simply remind us that "Knowledge is not equivalent to Wisdom."

SUGGESTED READINGS

Allen, G. E. (1975). *Life Science in the Twentieth Century.* New York: Wiley.

Barrett, J. T., ed. (1986). *Contemporary Classics in the Life Sciences.* Philadelphia, PA: ISI Press.

Beadle, G. W. (1963). *Genetics and Modern Biology.* Philadelphia, PA: American Philosophical Society.

Bohr, N. (1933). On Light and Life. *Nature* 131:421–423, 457–459.

Brenner, S., compiler (1989). *Molecular Biology: A Selection of Papers. Seminal Papers Reprinted from the Journal of Molecular Biology.* London: Academic Press.

Brock, T. D. (1990). *The Emergence of Bacterial Genetics.* New York: Cold Spring Harbor Laboratory Press.

Cairns, J., Stent, G. S., and Watson, J. D., eds. (1992). *Phage and the Origins of Molecular Biology,* 2nd ed. New York: Cold Spring Harbor Laboratory of Quantitative Biology.

Carmen, I. H. (1985). *Cloning and the Constitution. An Inquiry into Governmental Policymaking and Genetic Experimentation.* Madison, WI: University of Wisconsin Press.

Cavalieri, L. F. (1981). *The Double-Edged Helix: Science in the Real World.* New York: Columbia University Press.

Chargaff, E. (1963). *Essays on Nucleic Acids.* New York: Elsevier.

Chargaff, E. (1978). *Heraclitean Fire: Sketches from a Life Before Nature.* New York: Rockefeller University Press.

Charles, D. (2001). *Lords of the Harvest: Biotech, Big Money, and the Future of Food.* Cambridge, MA: Perseus Publishing.

Creager, A. N. H. (2001). *The Life of a Virus: Tobacco Mosaic Virus as an Experimental Model, 1930–1965.* Chicago, IL: University of Chicago Press.

Creager, A. N. H., Lunbeck, E., and Schiebinger, L. (2002). *Feminism in Twentieth-*

Century Science, Technology, and Medicine. Chicago, IL: University of Chicago Press.

Crick, F. (1988). *What Mad Pursuit. A Personal View of Scientific Discovery.* New York: Basic Books.

Crotty, S. (2001). *Ahead of the Curve: David Baltimore's Life in Science.* Chicago, IL: University of Chicago Press.

Dubos, R. J. (1976). *The Professor, the Institute, and DNA. A Biography of Oswald T. Avery.* New York: The Rockefeller University Press.

Dulbecco, R. (1987). *The Design of Life.* New Haven, CT: Yale University Press.

Echols, H. (2001). *Operators and Promoters: The Story of Molecular Biology and Its Creators.* Edited by Carol A. Gross. Berkeley, CA: University of California Press.

Fedoroff, N., and Botstein, D., eds. (1992). *The Dynamic Genome: Barbara McClintock's Ideas in the Century of Genetics.* Cold Spring Harbor, NY: Cold Spring Harbor Laboratory Press.

Fischer, E. P., and Lipson, C. (1988). *Thinking About Science. Max Delbrück and the Origins of Molecular Biology.* New York: W. W. Norton.

Florkin, M. (1972). *A History of Biochemistry.* New York: Elsevier.

Fredrickson, D. S. (2001). *The Recombinant DNA Controversy: A Memoir: Science, Politics, and the Public Interest 1974–1981.* Washington, DC: ASM Press.

Fruton, J. S. (1972). *Molecules and Life: Historical Essays on the Interplay of Chemistry and Biology.* New York: Wiley-Interscience.

Fruton, J. S. (1992). *A Skeptical Biochemist.* Cambridge, MA: Harvard University Press.

Garrod, A. E. (1909). *Inborn Errors of Metabolism.* London: Oxford University Press. Reprinted 1963.

Gehring, W. J. (1998). *Master Control Genes in Development and Evolution: The Homeobox Story.* New Haven, CT: Yale University Press.

Gibbons, J. H. (1988). *Mapping Our Genes.* Washington, DC: Office of Technology Assessment.

Glover, D. M., and Hames, B. D., eds. (1989). *Genes and Embryos.* New York: IRL Press.

Goertzel, T., and Goertzel, B. (1995). *Linus Pauling: A Life in Science and Politics.* Harper Collins.

Gore, A. (1985). *Biotechnology, Implications for Public Policy.* Edited by Sandra Panem. Washington, DC: Brookings Institution.

Hager, T. (1998). *Linus Pauling and the Chemistry of Life.* New York: Oxford University Press.

Hill, W. E., ed. (1990). *The Ribosome. Structure, Function, and Evolution.* Washington, DC: American Society for Microbiology.

Huemer, R. P., ed. (1986). *The Roots of Molecular Medicine: A Tribute to Linus Pauling.* New York: W. H. Freeman.

Hughes, S. S. (2002). Making dollars out of DNA. The first major patent in biotechnology and the commercialization of molecular biology, 1974–1980. *Isis* 92:541–575.

Jacob, F. (1976). *The Logic of Life. A History of Heredity.* New York: Vintage.

Jacob, F. (1988). *The Statue Within. An Autobiography.* New York: Basic Books.

Judson, H. F. (1979). *The Eighth Day of Creation: The Makers of the Revolution in Biology*. New York: Simon and Schuster.

Kay, L. E. (2000). *Who Wrote the Book of Life? A History of the Genetic Code*. Stanford, CA: Stanford University Press.

Keller, E. F. (1983). *A Feeling for the Organism: The Life and Work of Barbara McClintock*. San Francisco, CA: W. H. Freeman.

Keller, E. F. (2002). *The Century of the Gene*. Cambridge, MA: Harvard University Press.

Kevles, D. J. (1998). *The Baltimore Case: A Trial of Politics, Science, and Character*. New York: W.W. Norton.

Kevles, D. J., and Hood, L., eds. (1992). *The Code of Codes. Scientific and Social Issues in the Human Genome Project*. Cambridge, MA: Harvard University Press.

Kohler, R. E. (1982). *From Medical Chemistry to Biochemistry*. Cambridge: Cambridge University Press.

Kornberg, A. (1989). *For the Love of Enzymes. The Odyssey of a Biochemist*. Cambridge, MA: Harvard University Press.

Kornberg, A. (1995). *The Golden Helix: Inside Biotech Ventures*. Sausalito, CA: University Science Books.

Kornberg, A., and Baker, T. A. (1992). *DNA Replication*, 2nd ed. New York: W. H. Freeman.

Kornberg, A., Horecker, B. L, Cornudella, L., and Oro, J., eds. (1976). *Reflections on Biochemistry in Honour of Severo Ochoa*. Oxford: Pergamon Press.

Krimsky, S. (1984). *Genetic Alchemy. The Social History of the Recombinant DNA Controversy*. Cambridge, MA: MIT Press.

Krishnamurthy, R. S., ed. (1996). *The Pauling Symposium: A Discourse on the Art of Biography*. Corvalis, OR: Oregon State University Libraries.

Kuczewski, M. G., and Polansky, R., eds. (2000). *Bioethics: Ancient Themes in Contemporary Issues*. Cambridge, MA: MIT Press.

Lederberg, J., compiler (1990). *The Excitement and Fascination of Science: Reflections by Eminent Scientists*. Vol. III. Palo Alto, CA: Annual Reviews.

Lederberg, J., ed. (1999). *Biological Weapons: Limiting the Threat*. Foreword by William S. Cohen. Cambridge, MA: MIT Press.

Leicester, H. M. (1974). *Development of Biochemical Concepts from Ancient to Modern Times*. Cambridge, MA: Harvard University Press.

Lerner, K. L., and Lerner, B. W., eds. (2002). *World of Genetics*. 2 vol. New York: Gale Group.

Levene, P. A., and Bass, L. W. (1931). *Nucleic Acids*. New York: Chemical Catalogue.

Luria, S. E. (1974). *Life: The Unfinished Experiment*. New York: Charles Scribner's Sons.

Luria, S. E. (1984). *A Slot Machine, A Broken Test Tube. An Autobiography*. New York: Harper & Row.

Lwoff, A., and Ullmann, A., eds. (1988). *Origins of Molecular Biology: A Tribute to Jacques Monod*. New York: Academic Press.

Mayr, E. (1982). *The Growth of Biological Thought*. Cambridge, MA: Harvard University Press.

McCarty, M. (1985). *The Transforming Principle. Discovering That Genes Are Made of DNA*. New York: Norton.

McClintock, B. (1987). *The Discovery and Characterization of Transposable Elements: The Collected Papers (1938–1984) of Barbara McClintock*. New York: Garland Publishing.

Monod, J. (1971). *Chance and Necessity. An Essay on the Natural Philosophy of Modern Biology*. New York: Knopf.

Monod, J. and Borke, E., eds. (1971). *Of Microbes and Life*. New York: Columbia University Press.

Moore, W. (1989). *Schrödinger: Life and Thought*. New York: Cambridge University Press.

Morange, M. (1998). *A History of Molecular Biology*. Trans. By Matthew Cobb. Cambridge, MA: Harvard University Press.

Mullis, K. B. (1998). *Dancing Naked in the Mind Field*. New York: Pantheon Books.

Mullis, K. B., Ferré, F., and Gibbs, R. A., eds. (1994). *The Polymerase Chain Reaction*. Foreword by James D. Watson. Boston, MA: Birkhäuser.

Murphy, T. F., and Lappé, M. A., eds. (1994). *Justice and the Human Genome Project*. Berkeley, CA: University of California Press.

National Academy of Sciences. (1977). *Research with Recombinant DNA: An Academy Forum. Washington, DC*: National Academy of Sciences.

Nelkin, D., and Lindee, M. S. (1995). *DNA Mystique: The Gene as a Cultural Icon*. New York: W. H. Freeman.

Nelkin, D., and Tancredi, L. (1989). *Dangerous Diagnostics: The Social Power of Biological Information*. New York: Basic Books.

Nichols, E. K. (1988). *Human Gene Therapy*. Cambridge, MA: Harvard University Press.

Olby, R. (1974). *The Path to the Double Helix*. Seattle, WA: University of Washington Press.

Perutz, M. (1992). *Is Science Necessary? Essays on Science and Scientists*. New York: Oxford University Press.

Plant, D. W., Reimers, N. J., and Zinder, N. D., eds. (1982). *Patenting of Life Forms*. New York: Cold Spring Harbor Laboratory.

Portugal, F. H., and Cohen, J. S. (1977). *A Century of DNA*. Cambridge, MA: MIT Press.

Rabinow, P. (1996). *Making PCR. A Story of Biotechnology*. Chicago, IL: University of Chicago Press.

Richards, J., ed. (1978). *Recombinant DNA: Science, Ethics, and Politics*. New York: Academic Press.

Rogers, M. (1972). *Biohazard*. New York: Knopf.

Ruse, M., and Sheppard, A., eds. (2001). *Cloning: Responsible Science or Technomadness?* Amherst, NY: Prometheus Books.

Sager, R. (1972). *Cytoplasmic Genes and Organelles*. New York: Academic Press.

Sapp, J. (1987). *Beyond the Gene. Cytoplasmic Inheritance and the Struggle for Authority in Genetics*. New York: Oxford University Press.

Sapp, J. (1990). *Where the Truth Lies. Franz Moewus and the Origins of Molecular Biology*. New York: Cambridge University Press.

Sayre, A. (1975). *Rosalind Franklin and DNA*. New York: Norton.

Schrödinger, E. (1992). *What Is Life? The Physical Aspects of the Living Cell. With Mind and Matter and Autobiographical Sketches*. New York: Cambridge University Press.

Scriver, C. R. (1989). *Garrod's Inborn Factors in Disease*. Facsimile of the 1931 edition with commentary. New York: Oxford University Press.

Semenza, G., and Jaenicke, R., eds. (1990). *A History of Biochemistry. Selected Topics in the History of Biochemistry, Personal Recollections, III*. New York: Elsevier Science Publishers.

Serafini, P. (1989). *Linus Pauling: A Man and His Science*. New York: Paragon House.

Sloan, P., ed. (2000). *Controlling Our Destinies: Historical, Philosophical, Ethical, and Theological Perspectives on the Human Genome Project*. South Bend, IN: University of Notre Dame Press.

Söderqvist, T., ed. (1997). *The Historiography of Contemporary Science and Technology*. Harwood Academic Publishers.

Srinivasan, P. R., Fruton, J. S., and Edsall, J. T., eds. (1979). *The Origins of Modern Biochemistry: A Retrospect on Proteins*. New York: New York Academy of Sciences.

Stahl, F. W., ed. (2000). *We Can Sleep Later: Alfred D. Hershey and the Origins of Molecular Biology*. Cold Spring Harbor, NY: Cold Spring Harbor Laboratory Press.

Stent, G. S., ed. (1980). *The Double Helix. A Personal Account of the Discovery of the Structure of DNA*. New York: Norton.

Stent, G. S. (1999). *Nazis, Women and Molecular Biology: Memoirs of a Lucky Self-Hater*. Kensington, CA: Briones.

Summers, W. C. (1999). *Félix d'Herelle and the Origins of Molecular Biology*. New Haven, CT: Yale University Press.

Sumner, A. T. (1990). *Chromosome Banding*. Boston, MA: Hyman.

Tapper, M. (1999). *In the Blood: Sickle Cell Anemia and the Politics of Race*. Philadelphia, PA: University of Pennsylvania Press.

Teich, M., with Needham, D. M. (1992). *A Documentary History of Biochemistry, 1770–1940*. Rutherford, NJ: Farleigh Dickinson University Press.

Tiley, N. A. (1983). *Discovering DNA: Meditations on Genetics and a History of the Science*. New York: Van Nostrand Reinhold.

Watson, E. L. (1991). *Houses for Science. A Pictorial History of Cold Spring Harbor Laboratory. With Landmarks in Twentieth Century Genetics, A Series of Essays by James D. Watson*. Plainview, NY: Cold Spring Harbor Laboratory Press.

Watson, J. D. (2000). *A Passion for DNA: Genes, Genomes, and Society*. Cold Spring Harbor, NY: Cold Spring Harbor Laboratory Press.

Watson, J. D. (2002). *Genes, Girls, and Gamow: After the Double Helix*. New York: Alfred A. Knopf.

Watson, J. D., and Tooze, J. (1981). *The DNA Story: A Documentary History of Gene Cloning*. San Francisco, CA: Freeman.

Watson, J. D., Gilman, M., Witkowski, J. and Zoller, M. (1992). *Recombinant DNA*, 2nd ed. New York: Scientific American Books.

Wills, C. (1991). *Exons, Introns, and Talking Genes. The Science Behind the Human Genome Project*. New York: Basic Books.

Wright, S. (1994). *Molecular Politics. Developing American and British Regulatory Policy for Genetic Engineering, 1972–1982*. Chicago, IL: University of Chicago Press.

Yoxen, E. (1984). *The Gene Business: Who Should Control Biotechnology?* New York: Harper & Row.

Zilinskas, R. A., and Zimmerman, B. K., eds. (1986). *The Gene-Splicing Wars: Reflections on the Recombinant DNA Controversy*. New York: Macmillan, for AAAS.

Zimmerman, B. K. (1984). *Biofuture: Confronting the Genetic Era*. Foreword by Francis Crick. New York: Plenum Press.

Index